Advances in Neurobiology

Volume 10

Advances in Neurobiology covers basic research in neurobiology and neurochemistry. It provides in-depth, book-length treatment of some of the most important topics in neuroscience including molecular and pharmacological aspects. The main audiences of the series are basic science researchers and graduate students as well as clinicians including neuroscientists (neurobiologists and neurochemists) and neurologists.

More information about this series at http://www.springer.com/series/8787

Marta C. Antonelli
Editor

Perinatal Programming of Neurodevelopment

Springer

Editor
Marta C. Antonelli
Instituto de Biologia Celular y
Neurociencias
Facultad de Medicina
Universidad de Buenos Aires
Buenos Aires
Argentina

ISSN 2190-5215 ISSN 2190-5223 (electronic)
ISBN 978-1-4939-1371-8 ISBN 978-1-4939-1372-5 (eBook)
DOI 10.1007/978-1-4939-1372-5
Springer New York Heidelberg Dordrecht London

Library of Congress Control Number: 2014944808

Printed on acid-free paper

Springer is part of Springer Science+Business Media (www.springer.com)

Preface: Biological Embedding: Long-Term Effects of Early-Life Experiences and Approaches to Prevention Remediation

Introduction

Until recent years, what happens to an individual early in life was largely ignored because it was falsely believed that the brain and body were shaped by experiences when the child becomes able to respond rationally to the social environment. We now know in increasing detail, in both animal models and studies on our own species, that prenatal stress can have adverse effects that are manifested in prematurity or low birth weight at term, as well as in behavioral characteristics that are manifested throughout the life course. Furthermore, we now know that postnatal parental care and abuse and neglect in humans, and nest disruption and separation of infants from their mothers in animal models, play a powerful role in later mental and physical health. This volume addresses many aspects of adverse pre- and early postnatal influences on subsequent physical and mental health and this introductory overview will discuss the role of both animal and human studies and the translation and cross-talk between them in achieving a better understanding of underlying processes and mechanisms so that interventions can be developed to present or, when necessary, treat disorders that may arise. In all of this, the brain is the central organ of stress and adaptation and the "lived experiences" of an individual are an important contributor to physical and mental health outcomes and the brain represents an important target for prevention and amelioration of early-life adversity (McEwen and Getz 2013).

Towards an Understanding of Mechanisms and Consequences of Biological Embedding

Animal models have contributed enormously to our understanding of how, through the brain, the development of body and the brain are affected. This began with the "neonatal handling" studies of Levine and Denenberg (Levine et al. 1967) and led to the recent, elegant work of Meaney, Syzf and colleagues (Meaney and Szyf 2005). Epigenetic, transgenerational effects transmitted by maternal care are central to these findings (Hackman et al. 2010). Besides the amount of maternal care, the consistency over time of that care and the exposure to novelty are also very important, not only in rodents (Akers et al. 2008; Tang et al. 2006), but also in monkey models (Parker et al. 2006). Prenatal stress impairs hippocampal development in rats, as does stress in adolescence (Isgor et al. 2004). Abusive maternal care in rodents and the surprising attachment shown by infant rats to their abusive mothers appears to involve an immature amygdala (Moriceau and Sullivan 2006), activation of which by glucocorticoids causes an aversive conditioning response to emerge. Maternal anxiety in the variable foraging demand (VFD) model in rhesus monkeys leads to chronic anxiety in the offspring, as well as signs of metabolic syndrome (Coplan et al. 2001; Kaufman et al. 2005). Nest disruption of mothers nursing mouse pups impairs development of the hippocampus and other brain systems (Rice et al. 2008).

In studies of adverse childhood experiences (ACE) in human populations, there are reports of increased inflammatory tone, not only in children, but also in young adults related to early-life abuse, that includes chronic harsh language, as well as physical and sexual abuse (Danese et al. 2009; Miller and Chen 2010). Chaos in the home is associated with development of poor self-regulatory behaviors, as well as obesity (Evans et al. 2005). It should be noted that the ACE study was carried out in a middle class population (Anda et al. 2010), indicating that poverty is not the only source of early-life stressors.

Nevertheless, low socioeconomic status (SES) does increase the likelihood of stressors in the home and neighborhood, including racial isolation, chaos, noise and ugliness as, well as toxic chemical agents, such as lead and air pollution (Chang et al. 2009; McEwen and Tucker 2011; Theall et al. 2013). Without a determination of exact causes, it has been reported that low SES children are found to be more likely to be deficient in language skills, as well as self-regulatory behaviors and also in certain types of memory that are likely to be reflections of impaired development of parasylvian gyrus language centers, prefrontal cortical systems and temporal lobe memory systems (Farah et al. 2006; Hart 1995). Low SES is reported to correlate with smaller hippocampal volumes (Hanson et al. 2011). Lower subjective SES, an important index of objective SES, is associated with reduction in prefrontal cortical gray matter (Gianaros et al. 2007) and with increased inflammatory tone in serum along with altered white matter in the brain that is also associated with increased adiposity (Gianaros et al. 2012; Verstynen et al. 2013).

Moreover, having grown up in a lower SES environment is accompanied by greater amygdala reactivity to angry and sad faces (Gianaros et al. 2008b), which may be a predisposing factor for early cardiovascular disease (Gianaros et al. 2008a) that is known to be more prevalent at lower SES levels (Adler et al. 1993). Finally, depression is often associated with low SES, and children of depressed mothers, followed longitudinally, have shown increased amygdala volume while hippocampal volume was not affected (Lupien et al. 2011).

Yet, on the positive side, there are the "reactive alleles" also referred to as "biological sensitivity to context" that lead to beneficial outcomes and even better outcomes in nurturing environments compared to less reactive alleles, even though those same alleles can enhance adverse outcomes in a stressful early-life environment (Boyce and Ellis 2005; Caspi et al. 2003; Obradovic et al. 2010; Suomi 2006). Regarding adverse outcomes and good and bad "environments," the active process of adaptation to stressors ("allostasis" (McEwen and Stellar 1993; Sterling and Eyer 1988)) is adjusted via epigenetic influences to optimize the individuals adaptation to, and resulting fitness for, a particular environment, whether more or less threatening or nurturing as described in the Adaptive Calibration model (Del Giudice et al. 2011).

It is important to note that the conceptual models of allostasis and allostatic load are orthogonal to the model of Adaptive Calibration (Del Giudice et al. 2011) and provide complementary ways of understanding individual developmental trajectories and their adaptive value as well as their consequences. One lesson from these two models is that there are "trade-offs" in terms of physical and mental health that, on the one hand, may increase the likelihood of passing on one's genes by improving coping with adversity and enhancing mental health and overall reproductive success, but, on the other hand, may impair later health, e.g., by eating of "comfort foods" (see for example (Jackson et al. 2010)).

Understanding and attempting to modify such individual health outcomes is an important component of "personalized medicine" and must be considered along with pharmacogenomics in the development of therapies (Davidson and McEwen 2012; McEwen and Getz 2013). In this connection, it should be noted that resilience means not only the ability to resist stress-induced change, but also the ability to show experience-related recovery and adaptation or compensation, for example, when an individual from a safe environment is placed into a dangerous one or vice versa. It is the plasticity of the brain and body that are keys to the amelioration of early-life adversity.

Interventions to Ameliorate Early-Life Adversity

What can be done to remediate the effects of chronic stress, as well the biological embedding associated with early-life adversity? Interventions may involve pharmaceutical, as well as behavioral, or "top-down," interventions (i.e., interventions that involve integrated CNS activity, as opposed to pharmacological agents) that include cognitive-behavioral therapy, physical activity and programs that promote social support and integration and meaning and purpose in life (Carlson et al. 2009; Fried et al. 2004; Ganzel et al. 2010; McEwen and Gianaros 2011). More targeted interventions for emotional and cognitive dysfunction may arise from fundamental studies of such developmental processes as the reversal of amblyopia and other conditions by "releasing the brakes" that retard structural and functional plasticity (Bavelier et al. 2010). It should be noted that many of these interventions that are intended to promote plasticity and slow decline with age, such as physical activity and positive social interactions that give meaning and purpose, are also useful for promoting "positive health" and "eudamonia" (Ryff and Singer 1998; Singer et al. 2005) independently of any notable disorder and within the range of normal behavior and physiology. It should also be noted that, while complete reversal of early-life adversity may not be possible, compensatory changes in neural architecture and molecular and neurochemical processes in key brain regions such as amygdala and prefrontal cortex can be envisioned (Caldji et al. 1998). Thus it is important to explore the strategies, possibilities and limits of adult brain plasticity, as will be discussed below.

As noted above, "top down" therapy is one strategy and one example is regular physical activity, which has actions that improve prefrontal and parietal cortex blood flow and enhance executive function (Colcombe et al. 2004). Moreover, regular physical activity, consisting of walking an hour a day, 5 out of 7 days a week, increases hippocampal volume in previously sedentary adults (Erickson et al. 2011). This finding complements work showing that physically fit individuals have larger hippocampal volumes than sedentary adults of the same age-range (Erickson et al. 2009). It is also well known that regular physical activity is an effective antidepressant and protects again cardiovascular disease, diabetes and dementia (Babyak et al. 2000, Snyder et al. 2010). Moreover, intensive learning has also been shown to increase volume of the human hippocampus (Draganski et al. 2006).

Other "top down" activities include social integration and support, and finding meaning and purpose in life, and these are known to be protective against allostatic load (Seeman et al. 2002) and dementia (Boyle et al. 2010). Programs such as the Experience Corps that promote these, along with increased physical activity, have been shown to slow the decline of physical and mental health and to improve prefrontal cortical blood flow in a similar manner to regular physical activity (Carlson et al. 2009; Fried et al. 2004).

Depression and anxiety disorders are examples of a loss of resilience, in the sense that changes in brain circuitry and function, caused by the stressors that precipitate the disorder, become "locked" in to a particular state and thus need external intervention. Indeed, prolonged depression is associated with shrinkage of the hippocampus (Sheline 1996; Sheline 2003) and prefrontal cortex (Drevets et al. 1997). While there appears to be no neuronal loss, there is evidence for glial cell loss and smaller neuronal cell nuclei (Rajkowska 2000; Stockmeier et al. 2004), which is consistent with a shrinking of the dendritic tree described above after chronic stress. Indeed, a few studies indicate that pharmacological treatment may reverse the decreased hippocampal volume in unipolar (Vythilingam et al. 2004) and bipolar (Moore et al. 2000) depression, but the possible influence of concurrent cognitive-behavioral therapy in these studies is unclear.

Depression is more prevalent in individuals who have had adverse early-life experiences (Anda et al. 2010). BDNF may be a key feature of the depressive state and elevation of BDNF by diverse treatments ranging from antidepressant drugs to regular physical activity and may be a key feature of treatment (Duman and Monteggia 2006). Yet, there are other potential applications, such as the recently reported ability of fluoxetine to enhance recovery from stroke (Chollet et al. 2011). However, a key aspect of this new view (Castren and Rantamaki 2010) is that the drug is opening a "window of opportunity" that may be capitalized by a positive behavioral intervention, e.g., behavioral therapy in the case of depression or the intensive physiotherapy to promote neuroplasticity to counteract the effects of a stroke.

This is consistent with animal model work that shows that ocular dominance imbalance from early monocular deprivation can be reversed by patterned light exposure in adulthood that can be facilitated by fluoxetine, on the one hand (Vetencourt et al. 2008) and food restriction, on the other hand (Sanacora et al. 2012), in which reducing inhibitory neuronal activity appears to play a key role (Dhabhar et al. 2012). Investigations of underlying mechanisms for the re-establishment of a new window of plasticity are focusing on the balance between excitatory and inhibitory transmission and removing molecules that put the "brakes" on such plasticity (Tanaka et al. 2001).

In this connection it is important to reiterate that successful behavioral therapy, which is tailored to individual needs, can produce volumetric changes in both prefrontal cortex in the case of chronic fatigue (de Lange et al. 2008), and in amygdala, in the case of chronic anxiety (Holzel et al. 2010). This reinforces two important messages: (i) that plasticity-facilitating treatments should be given within the framework of a positive behavioral or physical therapy intervention; and (ii) that negative experiences during the window may even make matters worse (Castren and Rantamaki 2010). In that connection, it should be noted that BDNF also has the ability to promote pathophysiology, as in seizures (Heinrich et al. 2011; Kokaia et al. 1995; Scharfman 1997).

Conclusions

Pre- and postnatal experiences have a profound and lasting effect upon physical and mental health acting via the brain and the biological embedding of positive and negative experiences. The chapters in this volume document many aspects of this in both animal models and humans, and this introductory chapter has outlined treatment strategies and their potential efficacy and limitations for ameliorating the effects of early-life adversity. However, the best solution is to prevent the adversity from happening in the first place. The Nurse-Family Partnership (http://www.nursefamilypartnership.org/) is a primary example of a program designed to educate expectant parents on optimal ways of interacting with their infants and children to promote healthy development, and the National Scientific Council on the Developing Child (http://developingchild.harvard.edu/index.php/activities/council/) provides a rich website on this topic and is working actively to bring such preventative programs into practice.

Laboratory of Neuroendocrinology, Bruce S. McEwen, Ph.D.
The Rockefeller University, NY, NY 10065

References

Adler NE, Boyce TW, Chesney MA, Folkman S, Syme L (1993) Socioeconomic inequalities in health. JAMA 269:3140–3145

Akers KG, Yang Z, DelVecchio DP, Reeb BC, Romeo RD et al (2008) Social competitiveness and plasticity of neuroendocrine function in old age: influence of neonatal novelty exposure and maternal care reliability. PLoS ONE 3(7):e2840

Anda RF, Butchart A, Felitti VJ, Brown DW (2010) Building a framework for global surveillance of the public health implications of adverse childhood experiences. Am J Prev Med 39:93–98

Babyak M, Blumenthal JA, Herman S, Khatri P, Doraiswamy M et al (2000) Exercise treatment for major depression: maintenance of therapeutic benefit at 10 months. Psychosom Med 62:633–638

Bavelier D, Levi DM, Li RW, Dan Y, Hensch TK (2010) Removing brakes on adult brain plasticity: from molecular to behavioral interventions. J Neurosci 30:14964–14971

Boyce WT, Ellis BJ (2005) Biological sensitivity to context: I. An evolutionary-developmental theory of the origins and functions of stress reactivity. Dev Psychopathol 17:271–301

Boyle PA, Buchman AS, Barnes LL, Bennett DA (2010) Effect of a purpose in life on risk of incident Alzheimer disease and mild cognitive impairment in community-dwelling older persons. Arch Gen Psychiatry 67:304–310

Caldji C, Tannenbaum B, Sharma S, Francis D, Plotsky PM, Meaney MJ (1998) Maternal care during infancy regulates the development of neural systems mediating the expression of fearfulness in the rat. Proc Natl Acad Sci U S A 95:5335–5340

Carlson MC, Erickson KI, Kramer AF, Voss MW, Bolea N et al (2009) Evidence for neurocognitive plasticity in at-risk older adults: the experience corps program. J Gerontol A Biol Sci Med Sci 64:1275–1282

Caspi A, Sugden K, Moffitt TE, Taylor A, Craig IW et al (2003) Influence of life stress on depression: moderation by a polymorphism in the 5-HTT gene. Science 301:386–389

Castren E, Rantamaki T (2010) The role of BDNF and its receptors in depression and antidepressant drug action: reactivation of developmental plasticity. Dev Neurobiol 70:289–297

Chang VW, Hillier AE, Mehta NK (2009) Neighborhood racial isolation, disorder and obesity. Social forces: a scientific medium of social study and interpretation 87:2063–2092

Chollet F, Tardy J, Albucher JF, Thalamas C, Berard E et al (2011) Fluoxetine for motor recovery after acute ischaemic stroke (FLAME): a randomised placebo-controlled trial. Lancet Neurol 10:123–130

Colcombe SJ, Kramer AF, Erickson KI, Scalf P, McAuley E et al (2004) Cardiovascular fitness, cortical plasticity, and aging. Proc Natl Acad Sci U S A 101:3316–3321

Coplan JD, Smith ELP, Altemus M, Scharf BA, Owens MJ et al (2001) Variable foraging demand rearing: sustained elevations in cisternal cerebrospinal fluid corticotropin-releasing factor concentrations in adult primates. Biol Psychiatry 50:200–204

Danese A, Moffitt TE, Harrington H, Milne BJ, Polanczyk G et al (2009) Adverse childhood experiences and adult risk factors for age-related disease: depression, inflammation, and clustering of metabolic risk markers. Arch Pediatr Adolesc Med 163:1135–1143

Davidson RJ, McEwen BS (2012) Social influences on neuroplasticity: stress and interventions to promote well-being. Nat Neurosci 15:689–695

de Lange FP, Koers A, Kalkman JS, Bleijenberg G, Hagoort P et al (2008) Increase in prefrontal cortical volume following cognitive behavioural therapy in patients with chronic fatigue syndrome. Brain 131:2172–2180

Del Giudice M, Ellis BJ, Shirtcliff EA (2011) The adaptive calibration model of stress responsivity. Neurosci Biobehav Rev 35:1562–1592

Dhabhar FS, Malarkey WB, Neri E, McEwen BS (2012) Stress-induced redistribution of immune cells—From barracks to boulevards to battlefields: a tale of three hormones—Curt Richter Award Winner. Psychoneuroendocrinology

Draganski B, Gaser C, Kempermann G, Kuhn HG, Winkler J et al (2006) Temporal and spatial dynamics of brain structure changes during extensive learning. J Neurosci 26:6314–6317

Drevets WC, Price JL, Simpson Jr JR, Todd RD, Reich T et al (1997) Subgenual prefrontal cortex abnormalities in mood disorders. Nature 386:824–827

Duman RS, Monteggia LM (2006) A neurotrophic model for stress-related mood disorders. Biol Psychiatry 59:1116–1127

Erickson KI, Prakash RS, Voss MW, Chaddock L, Hu L et al (2009) Aerobic fitness is associated with hippocampal volume in elderly humans. Hippocampus 19:1030–1039

Erickson KI, Voss MW, Prakash RS, Basak C, Szabo A et al (2011) Exercise training increases size of hippocampus and improves memory. Proc Natl Acad Sci U S A 108:3017–3022

Evans GW, Gonnella C, Marcynyszyn LA, Gentile L, Salpekar N (2005) The role of chaos in poverty and children's socioemotional adjustment. Psychol Sci 16:560–565

Farah MJ, Shera DM, Savage JH, Betancourt L, Giannetta JM et al (2006) Childhood poverty: specific associations with neurocognitive development. Brain Res 1110:166–174

Fried LP, Carlson MC, Freedman M, Frick KD, Glass TA et al (2004) A social model for health promotion for an aging population: initial evidence on the experience corps model. J Urban Health: Bull NY Acad Med 81:64–78

Ganzel BL, Morris PA, Wethington E (2010) Allostasis and the human brain: integrating models of stress from the social and life sciences. Psychol Rev 117:134–174

Gianaros PJ, Hariri AR, Sheu LK, Muldoon MF, Sutton-Tyrrell K, Manuck SB (2008a) Preclinical atherosclerosis covaries with individual differences in reactivity and functional connectivity of the amygdala. Biol Psychiatry

Gianaros PJ, Horenstein JA, Cohen S, Matthews KA, Brown SM et al (2007) Perigenual anterior cingulate morphology covaries with perceived social standing. Soc Cogn Affect Neurosci 2:161–173

Gianaros PJ, Horenstein JA, Hariri AR, Sheu LK, Manuck SB et al (2008b) Potential neural embedding of parental social standing. Soc Cogn Affect Neurosci 3:91–96

Gianaros PJ, Marsland AL, Sheu LK, Erickson KI, Verstynen TD (2012) Inflammatory pathways link socioeconomic inequalities to white matter architecture. Cereb Cortex

Hackman DA, Farah MJ, Meaney MJ (2010) Socioeconomic status and the brain: mechanistic insights from human and animal research. Nat Rev Neurosci 11:651–659

Hanson JL, Chandra A, Wolfe BL, Pollak SD (2011) Association between income and the hippocampus. PLoS ONE 6:e18712

Hart B, Risley TR (1995) Meaningful differences in the everyday experience of young american children. Brookes Publishing Company, Baltimore, 304 pp

Heinrich C, Lahteinen S, Suzuki F, Anne-Marie L, Huber S et al (2011) Increase in BDNF-mediated TrkB signaling promotes epileptogenesis in a mouse model of mesial temporal lobe epilepsy. Neurobiol Dis 42:35–47

Holzel BK, Carmody J, Evans KC, Hoge EA, Dusek JA et al (2010) Stress reduction correlates with structural changes in the amygdala. Soc Cogn Affect Neurosci 5:11–17

Isgor C, Kabbaj M, Akil H, Watson SJ (2004) Delayed effects of chronic variable stress during peripubertal-juvenile period on hippocampal morphology and on cognitive and stress axis functions in rats. Hippocampus 14:636–648

Jackson JS, Knight KM, Rafferty JA (2010) Race and unhealthy behaviors: chronic stress, the HPA axis, and physical and mental health disparities over the life course. Am J Public Health 100:933–939

Kaufman D, Smith ELP, Gohil BC, Banerji MA, Coplan JD et al (2005) Early appearance of the metabolic syndrome in socially reared bonnet macaques. J. Clin Endocrin Metab 90:404–408

Kokaia M, Ernfors P, Kokaia Z, Elmer E, Jaenisch R, Lindvall O (1995) Suppressed epileptogenesis in BDNF mutant mice. Exp Neurol 133:215–224

Levine S, Haltmeyer G, Kara G, Denenberg V (1967) Physiological and behavioral effects of infantile stimulation. Physiol Behav 2:55–59

Lupien SJ, Parent S, Evans AC, Tremblay RE, Zelazo PD et al (2011) Larger amygdala but no change in hippocampal volume in 10-year-old children exposed to maternal depressive symptomatology since birth. Proc Natl Acad Sci U S A 108:14324–14329

McEwen BS, Getz L (2013) Lifetime experiences, the brain and personalized medicine: an integrative perspective. Metabolism 62(Suppl 1):S20–S26

McEwen BS, Gianaros PJ (2011) Stress- and allostasis-induced brain plasticity. Annu Rev Med 62:431–445

McEwen BS, Stellar E (1993) Stress and the individual. Mechanisms leading to disease. Arch Intern Med 153:2093–2101

McEwen BS, Tucker P (2011) Critical biological pathways for chronic psychosocial stress and research opportunities to advance the consideration of stress in chemical risk assessment. Am J Public Health 101(Suppl 1):S131–S139

Meaney MJ, Szyf M (2005) Environmental programming of stress responses through DNA methylation: life at the interface between a dynamic environment and a fixed genome. Dialogues Clin Neurosci 7:103–123

Miller GE, Chen E (2010) Harsh family climate in early life presages the emergence of a proinflammatory phenotype in adolescence. Psychol Sci 21:848–856

Moore GJ, Bebehuk JM, Wilds IB, Chen G, Manji HK (2000) Lithium-induced increase in human brain grey matter. Lancet 356:1241–1242

Moriceau S, Sullivan R (2006) Maternal presence serves as a switch between learning fear and attraction in infancy. Nature Neurosci 8:1004–1006

Obradovic J, Bush NR, Stamperdahl J, Adler NE, Boyce WT (2010) Biological sensitivity to context: the interactive effects of stress reactivity and family adversity on socioemotional behavior and school readiness. Child Dev 81:270–289

Parker KJ, Buckmaster CL, Sundlass K, Schatzberg AF, Lyons DM (2006) Maternal mediation, stress inoculation, and the development of neuroendocrine stress resistance in primates. Proc Natl Acad Sci U S A 103:3000–3005

Rajkowska G (2000) Postmortem studies in mood disorders indicate altered numbers of neurons and glial cells. Biol Psychiatry 48:766–777

Rice CJ, Sandman CA, Lenjavi MR, Baram TZ (2008) A novel mouse model for acute and long-lasting consequences of early life stress. Endocrinology 149:4892–4900

Ryff CD, Singer B (1998) The contours of positive human health. Psychological Inquiry 9:1–28

Sanacora G, Treccani G, Popoli M (2012) Towards a glutamate hypothesis of depression: an emerging frontier of neuropsychopharmacology for mood disorders. Neuropharmacology 62:63–77

Scharfman HE (1997) Hyperexcitability in combined entorhinal/hippocampal slices of adult rat after exposure to brain-derived neurotrophic factor. J Neurophysiol 78:1082–1095

Seeman TE, Singer BH, Ryff CD, Dienberg G, Levy-Storms L (2002) Social relationships, gender, and allostatic load across two age cohorts. Psychosom Med 64:395–406

Sheline YI (1996) Hippocampal atrophy in major depression: a result of depression-induced neurotoxicity? Mol Psychiatry 1:298–299

Sheline YI (2003) Neuroimaging studies of mood disorder effects on the brain. Biol Psychiatry 54:338–352

Singer B, Friedman E, Seeman T, Fava GA, Ryff CD (2005) Protective environments and health status: cross-talk between human and animal studies. Neurobiol. Aging 26S:S113–S118

Snyder MA, Smejkalova T, Forlano PM, Woolley CS (2010) Multiple ERbeta antisera label in ERbeta knockout and null mouse tissues. J Neurosci Methods 188:226–234

Sterling P, Eyer J (1988) Allostasis: a new paradigm to explain arousal pathology In Handbook of life stress, cognition and health (Fisher S, Reason J (eds)). Wiley, New York, pp 629–649

Stockmeier CA, Mahajan GJ, Konick LC, Overholser JC, Jurjus GJ et al (2004) Cellular changes in the postmortem hippocampus in major depression. Biol Psychiatry 56:640–650

Suomi SJ (2006) Risk, resilience, and gene x environment interactions in rhesus monkeys. Ann N Y Acad Sci 1094:52–62

Tanaka J, Miyakubo H, Okumura T, Sakamaki K, Hayashi Y (2001) Estrogen decreases the responsiveness of subfornical organ neurons projecting to the hypothalamic paraventricular nucleus to angiotensin II in female rats. Neurosci Lett 307:155–158

Tang AC, Akers KG, Reeb BC, Romeo RD, McEwen BS (2006) Programming social, cognitive, and neuroendocrine development by early exposure to novelty. Proc Natl Acad Sci U S A 103:15716–15721

Theall KP, Brett ZH, Shirtcliff EA, Dunn EC, Drury SS (2013) Neighborhood disorder and telomeres: connecting children's exposure to community level stress and cellular response. Soc Sci Med 85:50–58

Verstynen TD, Weinstein A, Erickson KI, Sheu LK, Marsland AL, Gianaros PJ (2013) Competing physiological pathways link individual differences in weight and abdominal adiposity to white matter microstructure. Neuroimage 79:129–137

Vetencourt JFM, Sale A, Viegi A, Baroncelli L, De Pasquale R et al (2008) The antidepressant fluoxetine restores plasticity in the adult visual cortex. Science 320:385–388

Vythilingam M, Vermetten E, Anderson GM, Luckenbaugh D, Anderson ER et al (2004) Hippocampal volume, memory, and cortisol status in major depressive disorder: effects of treatment. Biol Psychiatry 56:101–112

Contents

Contributors

Marta C. Antonelli Instituto de Biología Celular y Neurociencias, Facultad de Medicina, Universidad de Buenos Aires, Buenos Aires, Argentina

Jared Bagley Department of Psychological and Brain Sciences, University of California, Santa Barbara, CA, USA

Chantal Beijers Department of Psychiatry, University of Groningen,Centre Groningen, Groningen, The Netherlands

Claudi L. H. Bockting Department of Clinical and Developmental Psychology, University of Groningen, Groningen, The Netherlands

Greta Bodzai Department of Anatomy, PTE-MTA Lendulet PACAP Research Team, University of Pecs, Pecs, Hungary

Huibert Burger Department of General Practice, University of Groningen, University Medical Centre Groningen, Groningen, The Netherlands

Department of Psychiatry, University of Groningen,Centre Groningen, Groningen, The Netherlands

D. Bustamante Department of Psychopharmacology, Central Institute of Mental Health J5, Mannheim, Germany

Jaonnalee C. Campbell Department of Psychological and Brain Sciences, University of California, Santa Barbara, CA, USA

Center for Neuropharmacology and Neuroscience, Albany Medical College, Albany, NY, USA

Michol Consolazione International Associated Laboratory—Prenatal Stress and Neurodegenerative Diseases, Neural Plasticity Team-UMR CNRS/USTL n 8576 Structural and Functional Glycobiology Unit, University of Lille 1, Lille, France

Carla Dalmaz Department of Biochemistry, Institute of Basic Health Sciences, Universidade Federal do Rio Grande do Sul, Porte Alegre, RS, Brazil

Paula A. Desplats Department of Neurosciences, University of California San Diego, La Jolla, CA, USA

Néstor Alejandro Dinerstein Division Neonatology, Hospital Materno Infantil R. Sardá, Ciudad Autónoma de Buenos Aires, Argentina

Francesca Fagioli Department of Psychiatry, Sapienza University of Rome, Roma, Italy

AziendaSanitaria Locale, RM.E. Unità Operativa Complessa Adolescent, Rome, Italy

Dorottya Farkas Department of Anatomy, PTE-MTA Lendulet PACAP Research Team, University of Pecs, Pecs, Hungary

Jozsef Farkas Department of Anatomy, PTE-MTA Lendulet PACAP Research Team, University of Pecs, Pecs, Hungary

Alonso Fernández-Guasti Department of Pharmacobiology, Center for Research and Advanced Studies (CINVESTAV), México, D.F., Mexico

Annabel Ferreira Seccita Fisiologg y Nutricigg, Facultad de Ciencias, Universidad de la Repversid, Montevideo, Uruguay

Balazs Gaszner Department of Anatomy, PTE-MTA Lendulet PACAP Research Team, University of Pecs, Pecs, Hungary

Eleonora Gatta International Associated Laboratory—Prenatal Stress and Neurodegenerative Diseases, Pozzilli, Italy

P. Gebicke-Haerter Millenium Institute BNI-Chile, Programme of Molecular & Clinical Pharmacology, ICBM, BNI, Medical Faculty, University of Chile, Santiago 7, Chile

Ana M. Genaro Department of Pharmacology, School of Medicine, University of Buenos Aires, Buenos Aires, Argentina

Center of Pharmacological and Botanical Studies (CEFYBO-CONICET-UBA), Buenos Aires, Argentina

Vivette Glover Institute of Reproductive and Developmental Biology, Imperial College London, Hammersmith Campus, London, UK

María Aurelia González Mental Health Section, Hospital Materno Infantil R. Sardá, Ciudad Autónoma de Buenos Aires, Argentina

M. Herrera-Marschitz Millenium Institute BNI-Chile, Programme of Molecular & Clinical Pharmacology, ICBM, BNI, Medical Faculty, University of Chile, Santiago 7, Chile

Gabor Horvath Department of Anatomy, PTE-MTA Lendulet PACAP Research Team, University of Pecs, Pecs, Hungary

Peter de Jonge Department of Psychiatry, University of Groningen, Centre Groningen, Groningen, The Netherlands

Suzanne King Department of Psychiatry, McGill University, Montreal, QC, Canada

Douglas Hospital Research Centre, Verdun, QC, Canada

Tod E. Kippin Department of Psychological and Brain Sciences, University of California, Santa Barbara, CA, USA

Neuroscience Research Institute, University of California, Santa Barbara, CA, USA

The Institute for Collaborative Biotechnology, University of California, Santa Barbara, CA, USA

Peter Kiss Department of Anatomy, PTE-MTA Lendulet PACAP Research Team, University of Pecs, Pecs, Hungary

Blanka Kiss-Illes Department of Anatomy, PTE-MTA Lendulet PACAP Research Team, University of Pecs, Pecs, Hungary

Chris P. Knight Department of Psychological and Brain Sciences, University of California, Santa Barbara, CA, USA

Rachel Krolow Department of Biochemistry, Institute of Basic Health Sciences, Universidade Federal do Rio Grande do Sul, Porte Alegre, RS, Brazil

Timea Kvarik Department of Anatomy, PTE-MTA Lendulet PACAP Research Team, University of Pecs, Pecs, Hungary

David P. Laplante Douglas Hospital Research Centre, Verdun, QC, Canada

A. Miguel Larguía Division Neonatology, Hospital Materno Infantil R. Sardá, Ciudad Autónoma de Buenos Aires, Argentina

L. Leyton Millenium Institute BNI-Chile, Programme of Molecular & Clinical Pharmacology, ICBM, BNI, Medical Faculty, University of Chile, Santiago 7, Chile

Eva M. Loomans Department of Psychology, Tilburg University, Tilburg, The Netherlands

Department of Epidemiology and Health Promotion, Public Health Service, Amsterdam, The Netherlands

Aldo. B. Lucion Department of Biochemistry, Institute of Basic Health Sciences, Universidade Federal do Rio Grande do Sul, Porte Alegre, RS, Brazil

Stefania Maccari International Associated Laboratory– Prenatal Stress and Neurodegenerative Diseases, Neural Plasticity Team-UMR CNRS/USTL n 8576 Structural and Functional Glycobiology Unit, University of Lille 1, Lille, France

International Associated Laboratory—Prenatal Stress and Neurodegenerative Diseases, Pozzilli, Italy

Jérôme Mairesse International Associated Laboratory—Prenatal Stress and Neurodegenerative Diseases, Neural Plasticity Team-UMR CNRS/USTL n 8576 Structural and Functional Glycobiology Unit, University of Lille 1, Lille, France

International Associated Laboratory—Prenatal Stress and Neurodegenerative Diseases, Pozzilli, Italy

Jleep Mairesse International Associated Laboratory—Prenatal Stress and Neurodegenerative Diseases, Neural Plasticity Team-UMR CNRS/USTL n 8576 Structural and Functional Glycobiology Unit, University of Lille 1, Lille, France

Barbara Mammel Department of Anatomy, PTE-MTA Lendulet PACAP Research Team, University of Pecs, Pecs, Hungary

Sridharan Manavalan Department of Anatomy, PTE-MTA Lendulet PACAP Research Team, University of Pecs, Pecs, Hungary

Gisele Gus Manfro Departamentos de Psiquiatria, Faculdade de MedicinaHospital de Clínicas de Porto Alegre, Universidade Federal do Rio Grande do Sul, Porto Alegre, RS, Brazil

Jordan Marrocco IRCCS Centro Neurolesi "Bonino-Pulejo", Messina, Italy

Attila Matkovits Department of Anatomy, PTE-MTA Lendulet PACAP Research Team, University of Pecs, Pecs, Hungary

Damian G. Maur Department of Pharmacology, School of Medicine, University of Buenos Aires, Buenos Aires, Argentina

Center of Pharmacological and Botanical Studies (CEFYBO-CONICET-UBA), Buenos Aires, Argentina

Patrick O. McGowan Department of Biological Sciences, Centre for Environmental Epigenetics and Development, University of Toronto, Toronto, ON, Canada

Judith Meijer Department of Epidemiology, University of Groningen,Centre Groningen, Groningen, The Netherlands

Angel I. Melo Centro de Investigación en Reproducción Animal, CINVESTAV-Laboratorio Tlaxcala, Universidad Autónoma de Tlaxcala, Tlaxcala, Mexico

Maarten Mennes Donders Institute for Brain, Cognition and Behavior, Department of Cognitive Neuroscience, Radboud University, Nijmegen, The Netherlands

Françoise Molenat Association for the Education and the Research on the Child and its Environment (AFREE), Montpellier Cedex 05, France

P. Morales Department of Psychopharmacology, Central Institute of Mental Health J5, Mannheim, Germany

Sara Morley-Fletcher International Associated Laboratory—Prenatal Stress and Neurodegenerative Diseases, Neural Plasticity Team-UMR CNRS/USTL n 8576 Structural and Functional Glycobiology Unit, University of Lille 1, Lille, France

T. Neira-Peña Millenium Institute BNI-Chile, Programme of Molecular & Clinical Pharmacology, ICBM, BNI, Medical Faculty, University of Chile, Santiago 7, Chile

Ferdinando Nicoletti International Associated Laboratory—Prenatal Stress and Neurodegenerative Diseases, IRCCS Neuromed, Pozzilli, Italy

International Associated Laboratory—Prenatal Stress and Neurodegenerative Diseases, Department of Pharmacology and Human Physiology, Sapienza University of Rome, Rome, Italy

Cristie Noschang Department of Biochemistry, Institute of Basic Health Sciences, Universidade Federal do Rio Grande do Sul, Porte Alegre, RS, Brazil

Sandra Olvera-Hernández Department of Pharmacobiology, Center for Research and Advanced Studies (CINVESTAV), México, D.F., Mexico

Johan Ormel Department of Psychiatry, University of Groningen,Centre Groningen, Groningen, The Netherlands

María Eugenia Pallarés Instituto de Biología Celular y Neurociencias, Facultad de Medicina, Universidad de Buenos Aires, Buenos Aires, Argentina

Cecilia G. Pascuan Department of Pharmacology, School of Medicine, University of Buenos Aires, Buenos Aires, Argentina

Center of Pharmacological and Botanical Studies (CEFYBO-CONICET-UBA), Buenos Aires, Argentina

Mariana Pereira Center for Molecular and Behavioral Neuroscience, Rutgers, The State University of New Jersey, Piscataway Township, USA

Kyle Ploense Department of Psychological and Brain Sciences, University of California, Santa Barbara, CA, USA

Charlis Raineki Department of Biochemistry, Institute of Basic Health Sciences, Universidade Federal do Rio Grande do Sul, Porte Alegre, RS, Brazil

Dora Reglődi Department of Anatomy, PTE-MTA Lendulet PACAP Research Team, University of Pecs, Pecs, Hungary

Marie-Line Reynaert International Associated Laboratory—Prenatal Stress and Neurodegenerative Diseases, Neural Plasticity Team-UMR CNRS/USTL n 8576 Structural and Functional Glycobiology Unit, University of Lille 1, Lille, France

International Associated Laboratory—Prenatal Stress and Neurodegenerative Diseases, Pozzilli, Italy

E. Rojas-Mancilla Department of Psychopharmacology, Central Institute of Mental Health J5, Mannheim, Germany

Patrícia Pelufo Silveira Departamentos de Pediatria, Faculdade de MedicinaHospital de Clínicas de Porto Alegre, Universidade Federal do Rio Grande do Sul, Porto Alegre, RS, Brazil

Constanza Soto Conti Division Neonatology, Hospital Materno Infantil R. Sardá, Ciudad Autónoma de Buenos Aires, Argentina

A. Dennis Stant Department of Epidemiology, University of Groningen,Centre Groningen, Groningen, The Netherlands

Ronald P. Stolk Department of Epidemiology, University of Groningen,Centre Groningen, Groningen, The Netherlands

Andrea Tamas Department of Anatomy, PTE-MTA Lendulet PACAP Research Team, University of Pecs, Pecs, Hungary

Gyongyver Vadasz Department of Anatomy, PTE-MTA Lendulet PACAP Research Team, University of Pecs, Pecs, Hungary

Gilles Van Camp International Associated Laboratory—Prenatal Stress and Neurodegenerative Diseases, Neural Plasticity Team-UMR CNRS/USTL n 8576 Structural and Functional Glycobiology Unit, University of Lille 1, Lille, France

International Associated Laboratory—Prenatal Stress and Neurodegenerative Diseases, Pozzilli, Italy

Bea R. H. Van den Bergh Department of Psychology, Tilburg University, Tilburg, The Netherlands

Department of Psychology, KU Leuven, Leuven, Belgium

Mariëlle G. van Pampus Department of Obstetrics and Gynecology, Onze Lieve Vrouwe Gasthuis, Amsterdam, The Netherlands

Tjitte Verbeek Department of Epidemiology, University of Groningen,Centre Groningen, Groningen, The Netherlands

Marta Weinstock Department of Pharmacology, Institute of Drug Research, Hebrew University Medical Centre, Jerusalem, Israel

Maria A. Zorrilla Zubilete Department of Pharmacology, School of Medicine, University of Buenos Aires, Buenos Aires, Argentina

Center of Pharmacological and Botanical Studies (CEFYBO-CONICET-UBA), Buenos Aires, Argentina

Part I
Perinatal Programming: Studies in Laboratory Animals

Chapter 1
Changes Induced by Prenatal Stress in Behavior and Brain Morphology: Can They Be Prevented or Reversed?

Marta Weinstock

Abstract This chapter presents a critical analysis of the behavioral alterations reported in the offspring of women exposed to stress and/or depression during pregnancy and the neurochemical and structural changes underlying them. Among the alterations attributed to prenatal stress in humans and experimental rats of both sexes is impaired regulation of the hypothalamic–pituitary–adrenal (HPA) axis, anxiety and exaggerated fear of novelty, and decreased social interaction. Learning and attention deficits are more prevalent in boys and male rats. Fear of novelty and anxiety are associated with enlargement of the amygdala and its corticotropin-releasing factor content, and decreased socialization, with lower oxytocin activity in the amygdala. Learning deficits are associated with a decrease in neurogenesis, dendritic complexity, and spine number in the dorsal hippocampus. Fostering prenatally stressed (PS) pups onto control mothers prevents the dysregulation of the HPA axis and heightened anxiety, indicating a role for postnatal factors in their etiology. By contrast, learning impairment and decreased socialization are not affected by this fostering procedure and are therefore prenatally mediated.

In spite of their widespread use in depressed pregnant women, selective serotonin reuptake inhibitor (SSRI) antidepressants do not normalize the behavior of their children. When administered during gestation to stressed rats, SSRIs do not reduce anxiety or learning deficits in their offspring. Moreover, when given to unstressed mothers, SSRIs induce anxiety in the offspring. The detrimental effect of SSRIs may result from inhibition of the serotonin transporter exposing the brain to excess amounts of 5-hydroxytryptamine (5-HT) at a critical time during fetal development.

Abbreviations

ADHD	Attention deficit hyperactivity disorder
BrdU	5-bromo-2′-deoxyuridine
CeA	Central nucleus of the amygdala
COR	Corticosterone

M. Weinstock (✉)
Department of Pharmacology, Institute of Drug Research, Hebrew University Medical Centre, Ein Kerem, 91120 Jerusalem, Israel
e-mail: martar@ekmd.huji.ac.il

M. C. Antonelli (ed.), *Perinatal Programming of Neurodevelopment,*
Advances in Neurobiology 10, DOI 10.1007/978-1-4939-1372-5_1,

CRF	Corticotropin-releasing factor
CRF-BP	Corticotropin-releasing factor binding protein
CRFR1	Corticotropin-releasing factor receptor 1
CRFR2	Corticotropin-releasing factor receptor 2
DG	Dentate gyrus
DCX	Doublecortin
EE	Environmental enrichment
EPM	Elevated plus maze
HPA	Hypothalamic pituitary adrenal
11β-HSD-2	11β-hydroxy steroid dehydrogenase-2
MWM	Morris water maze
PS	Prenatally stressed
SERT	Serotonin transporter
SSRI	Selective serotonin reuptake inhibitor

1.1 Introduction

Observations from retrospective studies starting in the 1960s suggested that prolonged uncontrollable stress during pregnancy may cause alterations in the development and behavior of the offspring which can be detected in infancy, childhood, and adulthood. These alterations include a reduction in birth weight, delay in early developmental milestones, withdrawn or disruptive behavior, attention and learning deficits, anxiety, depression, and schizophrenia (see reviews by Koenig et al. 2002; Kofman 2002; Weinstock 1997, 2001, 2008). In the past decade, prospective studies were initiated in women who had been exposed to natural (Laplante et al. 2008) or man-made disasters (Imamura et al. 1999), marital discord (Lereya and Wolke 2012), and adverse social or work-related conditions (Khashan et al. 2008). Exposure to such adverse risk factors can also increase the incidence of depression in pregnant women (Giardinelli et al. 2012; Husain et al. 2012; Miszkurka et al. 2012; Qu et al. 2012). Both depression and gestational stress can each adversely affect child development and behavior. Therefore, other studies focused on offspring from birth through to adolescence of women with anxiety and depression during and after pregnancy (Bergman et al. 2007; Davis and Sandman 2012; Van den Bergh and Marcoen 2004; Van den Bergh et al.2008). However, these prospective studies underscored the difficulty in defining maternal stress and allowing for differences in the reaction of women to the same objective stress. While some reported an association between maternal distress and behavioral changes in children at different ages, none could differentiate unequivocally between prenatal, genetic, and postnatal factors in mediating the behavioral outcome.

A clearer assessment of the contribution of pre- and postnatal factors to the behavioral outcome that is less influenced by genetic factors can be achieved by studies in experimental animals. The majority has been performed in rats in which more comprehensive behavioral, morphological, and histological information is available than in other species. Several were able to replicate the increased anxiety, depressive-like behavior (Alonso et al. 1991; Morley-Fletcher et al. 2004; Poltyrev et al. 2005), learning (Yaka et al. 2007; Yang et al. 2006) and attention deficits (Wilson et al. 2012), reduced social interaction (Lee et al. 2007), and some of the characteristic neuronal changes of schizophrenia (Koenig et al. 2005). Like in humans (Van den Bergh et al. 2008), gestational stress in rats impaired the regulation of the response to stress of the hypothalamic–pituitary–adrenal (HPA) axis in the offspring (Barbazanges et al. 1996; Weinstock et al. 1992).

By fostering prenatally stressed (PS) pups onto control mothers, it was also possible to differentiate behavioral alterations arising from gestational stress per se from those ascribed to inadequate mother–infant interactions (Barros et al. 2006; Yang et al. 2006). Other procedures like housing the stressed mothers (Li et al. 2012) or their offspring in an enriched environment were able to reduce the effects of gestational stress on several aspects of the offspring behavior (Lui et al. 2011; Yang et al. 2007). This chapter discusses more recent research that has examined the effect of gestational stress on neurochemical, structural, gene, and proteomic changes in different brain regions of the offspring of both sexes. It also describes procedures that have been used to prevent or reverse the behavioral and structural changes induced by prenatal stress.

1.2 Gestational Stress and Activity of the HPA Axis in the Mother and Her Offspring

Subjects with anxiety and depression have hypercortisolemia and impairment of negative feedback by cortisol on the HPA. This has been attributed to the increased action of corticotropin-releasing factor (CRF; Keck 2006; Reul and Holsboer 2002). Hypercortisolemia also occurs after chronic stress. It has been postulated that prenatal stress produces alterations in brain structure and behavior through the action of "stress" hormones, CRF, glucocorticoids, and catecholamines arising in the maternal adrenal gland and placenta (reviewed in Jansson and Powell 2007; Sandman et al. 2011; Weinstock 2005). During a normal pregnancy, very little cortisol (in humans) and corticosterone (COR; in rodents) reaches the fetal brain because they are converted to inactive metabolites by the placental enzyme 11β-hydroxy steroid dehydrogenase-2 (11β-HSD-2). In addition, about 90 % of circulating corticosteroids are sequestered by a corticosteroid-binding globulin (CBG), thereby limiting their access to the fetus. However, chronic gestational stress reduces the level of CBG in rats (Takahashi et al. 1998) and downregulates the activity of 11β-HSD-2 in humans (O'Donnell et al. 2012) and rats (Jensen Pena et al. 2012). This is accomplished by DNA methylation at specific sites within the 11β-HSD-2 gene promoter, thereby increasing the concentration of free steroids that can reach the developing fetal brain. Gestational stress also releases adrenaline and noradrenaline into the circulation which can reduce placental blood flow causing hypoxia and ischemia that could adversely influence fetal brain development (Delcour et al. 2012; Fan et al. 2009). Higher levels of these catecholamines have been found in the fetal circulation in response to maternal stress (Ohkawa et al. 1991) and can reach the brain because of the absence of a blood brain barrier.

Several clinical studies have attempted to relate elevations in cortisol to the presence of chronic stress, anxiety, and/or depression (assessed by questionnaires) during pregnancy. No relation was found between the magnitude of the increase in maternal cortisol in plasma (Baibazarova et al. 2012) or saliva between gestational weeks 15–37 and the level of stress, anxiety, depression, or pregnancy-specific anxiety at any of the times that cortisol was measured (Davis and Sandman 2010). However, a significant relation between salivary cortisol and maternal mood was found in subjects with comorbidity of anxiety and depression but not in those with only one of these conditions (Evans et al. 2008). It is not clear why most studies failed to relate maternal anxiety and/or depression at a specific time during pregnancy to elevation of plasma cortisol. This may depend on the method of sample collection or its timing during the day, which may differ in subjects with alterations in their circadian rhythms due to depression. It is probable that the ongoing chronic emotional state of anxious, depressed women does not lend itself to the detection of a clearly defined increase in plasma cortisol, unlike that in response to stress.

Others have tried to relate the time of occurrence of stress, anxiety, and/or depression during gestation to the behavioral outcome in the offspring. Here too, there is little consensus among the earlier studies. For example, low birth weight, increased infant anxiety, and fear of novelty were associated with stress at 28–30 (Wadhwa et al. 1993), 15–17, 27–28, and 37–38 weeks (Huizink et al. 2003) and at 18 and 32 weeks of gestation (O'Connor et al. 2002). More recently, high maternal anxiety and elevated cortisol early in pregnancy were shown to be associated with a deleterious effect on infant cognitive development, while those occurring towards the end of pregnancy were associated with improved cognitive development (Davis and Sandman 2010). Likewise, maternal anxiety at 12–22 weeks of pregnancy was a significant predictor of symptoms of attention deficit hyperactivity disorder (ADHD), aggressive and delinquent behavior, and anxiety in 8–9-year-old children (Van den Bergh and Marcoen 2004). Alterations in the reactivity of the HPA axis were found in adolescent boys and girls, but depressive symptoms, only in girls (Van den Bergh et al. 2008). More recently, this group has shown that prenatal maternal-state anxiety measured around the 16th week of gestation resulted in hyperactivity/inattention, emotional symptoms, problems with peer relationship, and social interaction, which were more prevalent in boys than in girls aged 5 years (Loomans et al. 2011). Pregnancy-specific anxiety and a higher level of maternal cortisol measured at 20, 25, and 30 weeks were associated with increased anxiety in preadolescent children of both sexes (Davis and Sandman 2012). The fetal cortical and limbic systems develop during the first 10 weeks of pregnancy (Bayer et al. 1993). It is therefore most probable that any changes in their programming by elevated cortisol, and the resulting effects on behavior, occur during that period. Cortisol levels may remain elevated as long as maternal anxiety and depression continue. If they increase only at a later stage of fetal brain development, the outcome may be different, and improvement in cognition can occur, as indicated in the study by Davis and Sandman (2010).

1.2.1 Experimental Animals

In the rat, the HPA axis, cortex, and limbic systems develop from day 13 of gestation (Bayer et al. 1993); therefore, in most studies, stress was administered during the 3rd (last) week of pregnancy. When the rats were stressed randomly, thrice weekly by noise and flashing lights (Weinstock et al. 1988), or on alternate days throughout gestation (Takahashi et al. 1998), COR levels increased in the maternal and fetal blood after each stress. However, when the rats were subjected to noise and flashing lights once daily at the same time during the last week of gestation, COR no longer increased in the mother or fetuses by the 3rd day (Weinstock et al. 1988). Very few studies have assessed whether or not the rats adapted to the form of stress that was used. Varied short-acting stressors during the last week of gestation (Salomon et al. 2011), or psychosocial stress on days 16–20 (Brunton and Russell 2010), continued to increase plasma COR until the last day of stress. Adaptation to the stress after 2 or 3 days could partially explain the inconsistency in the behavioral data in the offspring when different stress paradigms were used.

In order to obtain direct evidence that maternal adrenal hormones mediate the alterations induced by gestational stress in the offspring, pregnant rats were adrenalectomized prior to the initiation of stress and given saline and maintenance levels of COR. This prevented the dysregulation of the response of the HPA axis to stress (Barbazanges et al. 1996) and the heightened anxiety and learning deficits in the offspring (Zagron and Weinstock 2006). Administration of COR to the pregnant rats to mimic the increase induced by stress reinstated the altered response of the HPA axis to stress (Barbazanges et al. 1996) and the increased anxiety but did not restore the learning deficits in the offspring (Salomon et al. 2011). Thus, while glucocorticoids mediate the anxiety and impaired regulation of the HPA axis induced by prenatal stress, other adrenal hormones appear to be responsible for the genesis of learning deficits.

The slower return of COR to baseline levels in response to stress in PS rats (McCormick et al. 1995; Weinstock et al. 1992; Weinstock et al. 1998) results from reduced efficiency of the feedback mechanism because of downregulation of hippocampal glucocorticoid (Weinstock et al. 1992) and mineralocorticoid receptors (Barbazanges et al. 1996; Tamura et al. 2011). Increased activation of the HPA axis in PS rats is also associated with a higher expression of CRF mRNA in the paraventricular nucleus (PVN) of the hypothalamus in females and a reduction in its corticotropin-releasing factor binding protein (CRF-BP) in males which is believed to limit the action of CRF (Zohar and Weinstock 2011).

1.3 Anxiety and Depressive-Like Behavior in Rats

The heightened anxiety seen in infants of stressed, anxious, and depressed mothers is paralleled in 10-day-old PS infant rats and detected by the increase in ultrasonic vocalizations in response to isolation from the mother (Laloux et al. 2012). PS males are also more anxious than control rats, as indicated by the longer latency to emerge from a dark cage into a brightly lit box (Ward et al. 2000), or by avoidance of the center area in the field (Abe et al. 2007). In 1988, following the description by Pellow and File (1986) of the use of the elevated plus maze (EPM) for detecting anxiolytic drugs, we used the test to demonstrate heightened anxiety in the offspring of both sexes of mothers subjected to unpredictable noise throughout gestation (Fride and Weinstock 1988). This finding was replicated in both sexes after variable forms of stress during the last week of gestation by Richardson et al. (2006) and Zohar and Weinstock (2011), but only in female offspring in a study by Schulz et al. (2011). When maternal stress consisted of thrice daily restraint, anxiety was detected in the EPM in which it was tested only in males (Baker et al. 2008; Estanislau and Morato 2005; Li et al. 2012; Vallee et al. 1997), or was found selectively in males, but not in females (Zuena et al. 2008) or in neither sex (Richardson et al. 2006; Rimondini et al. 2003). Maternal psychosocial stress also produced conflicting results in male offspring. These were found to be either less anxious (Gotz and Stefanski 2007) or more anxious than controls (Brunton and Russell 2010).

The disparate effect of prenatal stress demonstrated in these studies may arise from the amount by which plasma COR increased in response to the stress, and if this remained elevated during the period of the development of the limbic system. They could also result from the environmental conditions in which anxiety was assessed in the offspring, as demonstrated in the following experiment. Offspring of control mothers and those subjected to varied stress from day 14 of gestation were tested in the EPM under bright light, under dim light, or were housed from weaning under a reversed light cycle and tested under red light during the active phase of their cycle (Fig. 1.1). No difference was detected in the behavior of PS and controls of either sex in the EPM under bright light since the controls spent relatively little time in the open arms of the maze. However, under dim light, rats of both sexes ventured more into the open arms and a significant anxiogenic effect was detected only in females. In the third group, male and female controls spent even more time in the open arms, enabling clearer detection of anxiety in PS rats. A similar difference between PS and controls was also reported by others who assessed behavior in rats housed under reversed light (Brunton and Russell 2010; Zohar and Weinstock 2011).

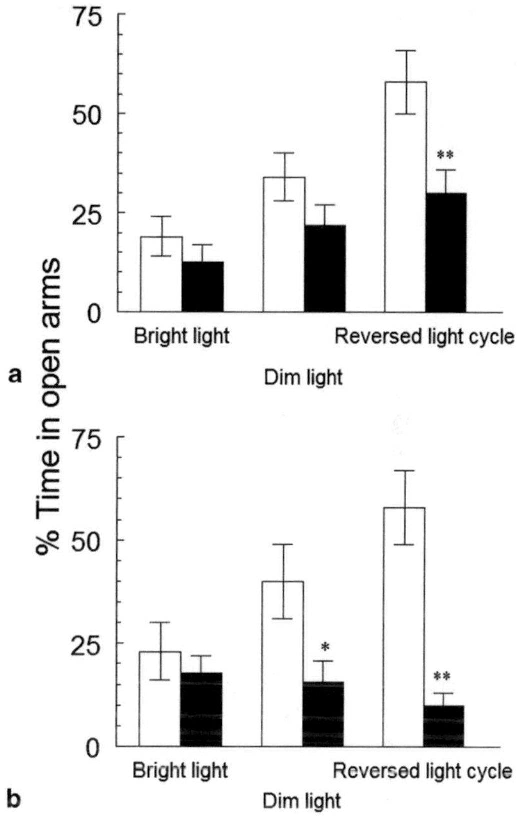

Fig. 1.1 Effect of different environmental conditions on behavior in the elevated plus maze. **a** Males. **b** Females. *White columns:* controls; *black columns:* prenatally stressed. Significantly different from controls, $*p<0.05$; $**p<0.01$

1.4 Neurochemical and Structural Changes Associated with Increased Anxiety Induced by Prenatal Stress

The amygdala plays a primary role in the formation and storage of memories associated with emotional events that are imprinted in the lateral nuclei. Anxious behavior is elicited through connections between the lateral nuclei, the central nucleus of amygdala (CeA), and the bed nuclei of the stria terminalis. The volume of the amygdala is increased in children with generalized anxiety disorders (De Bellis et al. 2000) or in those born to women with anxiety and depression during pregnancy (Buss et al. 2012), as shown by structural magnetic resonance imaging. In a study performed only in male offspring, prenatal stress caused an increase in the volume of the lateral amygdaloid nucleus which contained more neurons than that of controls (Salm et al. 2004). The CeA is a major extra hypothalamic site of CRF expression (Merchenthaler et al. 1982). Injection of CRF into the amygdala was shown to induce anxiogenic behavior (Gray and Bingaman 1996), while administration of a CRF receptor antagonist into the brain selectively reduced the heightened anxiety of PS males in the cage emergence test (Ward et al. 2000). In keeping

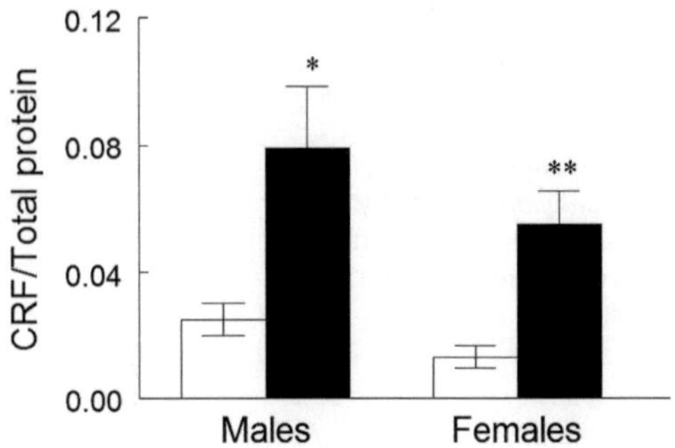

Fig. 1.2 Relative amount of corticotropin-releasing factor (*CRF*) protein in the amygdala of control and prenatally stressed rats *White columns:* controls; *black columns:* prenatally stressed. Significantly different from controls; $*p<0.05$; $**p<0.01$

with their anxiogenic behavior, we showed that PS males and females had higher levels of CRF protein in amygdala extracts (Fig. 1.2). In comparison with controls, PS males also had a reduction in the levels of CRF-BP mRNA and of CRF 2 receptors (CRFR2) mRNA (Zohar and Weinstock 2011), the activation of which reduces anxiety. The increase in CRF protein in the amygdala in PS rats could have resulted from excess COR activity via glucocorticoid receptors which are expressed in the CeA during development (Honkaniemi et al. 1992). Maternal psychological stress that only increased anxiety in the male offspring also reduced the expression of CRFR2 and increased that of CRF 1 receptors (CRFR1), which mediates anxiety, in the medial nucleus of the amygdala of males (Brunton et al. 2011).

1.5 Alterations in Cognitive Function and Spatial Learning Induced by Prenatal Stress in Humans and Rats

1.5.1 Humans

Cortical neurogenesis in humans occurs between gestational weeks 6 and 16 (Sidman and Rakic 1973) and neuronal migration and synaptogenesis leading to distinguishable cortical layers between weeks 24 and 26 (Meyer et al. 2000).

It has been shown that prenatal stress affects cognitive abilities in humans from infancy to adulthood. Exposure of mothers to the severe stress of a freezing ice storm in Canada resulted in some degree of cognitive retardation in their 2-year-old children, while exposure to moderate levels of stress enhanced development of their cognitive ability (DiPietro et al. 2006). At 5.5 years of age, severe stress exposure resulted in lower full-scale IQs and language abilities (Laplante et al. 2008). Likewise, others found that high maternal antenatal anxiety, but not low or moderate levels, was associated with poorer performance of 17-year-old adolescent offspring when the cognitive load of the task was increased (Mennes et al. 2006). Young adults whose mothers had experienced major adverse life events during pregnancy used a more rigid strategy to solve a navigational task that depended on the caudate nucleus, in contrast to the flexible, hippocampus-based one used by unstressed individuals (Schwabe et al. 2012). These studies did not provide any information about possible sex differences in the effect of prenatal stress on cognitive performance. Nevertheless, data accumulated so far show that adverse life events may have permanent effects on cognitive function and on the way in which spatial problems are solved. The outcome of maternal stress appears to depend on its intensity and the time during pregnancy of its occurrence as confirmed by studies in experimental animals described below.

1.5.2 Rats

In the rat, neurogenesis begins in various cortical regions on day 14 and continues until birth on day 21; in hippocampal fields CA 1–3, it starts on day 15, and in the granule cells of the dentate gyrus (DG) on day 19, continuing until postnatal day 19 (Rice and Barone 2000). There are clear sex differences in normal brain morphology in rats, particularly in the hippocampus, both during development (Munoz-Cueto et al. 1990) and in adulthood (Andrade et al. 2000; Madeira and Lieberman 1995; see below). As in humans (Newhouse et al. 2007; Sneider et al. 2011), sex differences underlie the performance of spatial learning in rodents and are related to different strategies used by each sex (Sandstrom et al. 1998). Several brain regions may participate in the execution of these strategies, including the medial prefrontal cortex (de Bruin et al. 2001) and the hippocampal formation. The latter is particularly involved in spatial navigation guided by distal cues to which females tend to respond better than males (Blokland et al. 2006).

Most reports of the effect of prenatal stress on spatial memory are based on experiments in which pregnant rats were restrained once or thrice daily for periods ranging from 30 to 120 min during the last week of gestation when the cortical and hippocampal neurons develop. Spatial memory of their adult offspring was assessed in the Morris water maze (MWM) test. Unlike the effect of prenatal stress on anxiety, there was much more agreement between studies, with most of them reporting learning deficits in males (Hosseini-Sharifabad and Hadinedoushan 2007; Lemaire et al. 2000; Li et al. 2012; Lui et al. 2011; Salomon et al. 2011; Szuran et al. 2000; Zagron and Weinstock 2006), and only one, selectively in females (Wu et al. 2007). In another study in which control females performed the task less well than males, prenatal stress actually improved the performance of females (Zuena et al. 2008). The relative immunity of adult PS females from impairment of spatial learning may result from the presence of estradiol which can increase neurogenesis, spine density, and their spatial performance (Gould et al. 1990; Phan et al. 2012).

When the effect of varied forms of prenatal stress was examined in prepubertal or juvenile rats, spatial learning was selectively impaired in females (Li et al. 2008; Weinstock 2011). However, the search strategy employed by both young PS male and female rats was shown to be less efficient than that of their respective controls (Wu et al. 2007). In adulthood, the same varied stress only caused spatial memory deficits in males (Weinstock 2011), possibly due to the positive influence of estradiol in females (Markham et al. 2010; Yaka et al. 2007). In association with learning impairments in PS males, there was a decrease in hippocampal long-term potentiation (LTP) (Yaka et al. 2007; Yang et al. 2006) and in the expression of the NR2B subunit of the glutamate-type *N*-methyl-D-aspartate (NMDA) receptor (Lui et al. 2011; Yaka et al. 2007) and in the GluR1 subunit of the α-amino-3-hydroxy-5-methyl-4-isoxazolepropionic acid (AMPA) receptor (Yaka et al. 2007).

Attempts have been made to relate alterations induced by prenatal stress in spatial learning to those in hippocampal morphology. In accordance with the selective effect on learning, we found a reduction in dendritic length, complexity, and number of spines in the DG in juvenile PS females. In the hippocampal CA1 and CA3 regions, the length and complexity of apical dendrites were decreased in both sexes (Bock et al. 2011). There do not appear to be any studies on the effect of prenatal stress on dendritic morphology in adult females. However, in adult males, a reduction was found in the complexity of dendrites and their length in the hippocampal CA3 region after once daily maternal restraint of 1 h (Hosseini-Sharifabad and Hadinedoushan 2007) and in the CA1, CA3, and DG after maternal restraint of 2 h (Fujioka et al. 2006). By contrast, as in humans (DiPietro et al. 2006), mild maternal restraint stress, which consisted in rats of only 30 min, improved cognitive function (Fujioka et al. 2001), and increased dendritic length in the CA1 and DG and dendritic complexity in the CA3 region in PS males (Fujioka et al. 2006).

Neurogenesis in the subgranular zone (SGZ) of the DG of the hippocampus continues throughout life and plays an important role in cognition (Abrous et al. 2005). When measured by the incorporation of cells labeled with 5-bromo-2′-deoxyuridine (BrdU), neurogenesis was decreased in the DG of male PS rats that showed learning deficits (Lemaire et al. 2000). When the same restraint stress used by this group failed to induce changes in spatial learning, BrdU was not decreased significantly in males in the dorsal hippocampus (that is involved in spatial learning) but only in the ventral part in females associated with increased anxiety (Zuena et al. 2008). We used doublecortin (DCX), a reliable marker of newly generated neurons (Rao and Shetty 2004), and found that it was selectively reduced in the SGZ of the DG in adult males (Fig. 1.3). On the other hand, GAP 43, a protein that positively influences axonal guidance and synaptic plasticity, was increased in PS females compared to that in controls (Fig. 1.3). Prenatal stress also diminished neurogenesis in the DG of monkeys but no information was given about their spatial learning ability (Coe et al. 2003).

Prenatal stress has been shown to reduce the size of the anogenital distance in prepubertal males (Holson et al. 1995; Pereira et al. 2006; Salomon et al. 2011), a sign of a relative lack of testosterone. If the amounts of testosterone synthesized by their hippocampal neurons are also reduced in PS males, it would adversely affect their synaptic plasticity (Ooishi et al. 2012) and could explain their greater susceptibility to learning and memory deficits. In summary, the majority of studies support a selective reduction in spatial learning by prenatal stress in adult males but not females. This is associated with a reduction of dendritic length, complexity, and

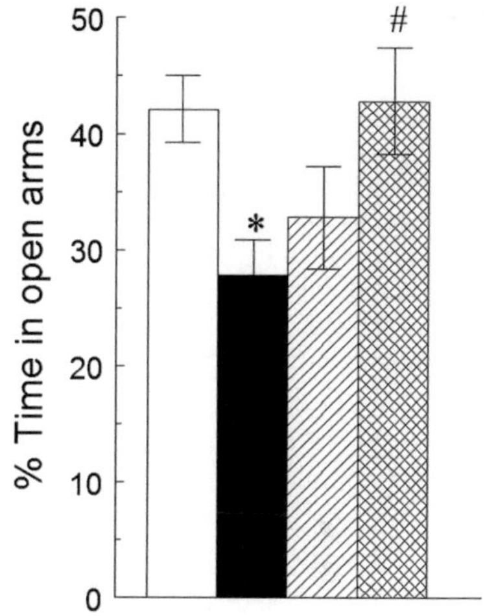

Fig. 1.3 Behavior in the elevated plus maze of stressed and control mothers with and without citalopram treatment. *White column:* control mothers; *black column:* stressed mothers; *hatched column:* controls with citalopram (10 mg/kg/day); *cross-hatched column:* stressed with citalopram (10 mg/kg/day). Significantly different from controls; $^{*}p<0.05$; significantly different from stressed $^{\#}p<0.05$

spines in the DG, and of neurogenesis in the SGZ which may occur through a relative lack of testosterone at a critical time during development.

1.6 Contribution of Pre- and Postnatal Factors to Human Offspring Behavioral Pathology

A clear association was found between the presence of anxiety disorder during pregnancy and depressive symptoms after birth, which together affected the outcome, increasing the incidence of separation anxiety disorder and ADHD in the offspring (Martini et al. 2010). It was also shown that maternal anxiety before birth was also associated with anxiety and depression during and after pregnancy and resulted in anxiety in the children, making it impossible to differentiate genetic personality traits from pregnancy-related stress in this outcome (Martini et al. 2010). In an attempt to assess the relative contributions of prenatal and postnatal maternal anxiety, a full range of child psychopathology and functioning was assessed in over 3000 mother–child pairs. Maternal depression was found to have a more significant impact on different types of child maladjustment than maternal anxiety in either the prenatal and postnatal periods. Internalizing difficulties in the child were linked to postnatal depression, while externalizing difficulties and impaired verbal IQ were associated with adverse prenatal factors, like low socioeconomic status and substance abuse, but not with maternal depression (Barker et al. 2011). Since smoking and drug abuse are themselves risk factors for infant pathology, it would be important in future studies to separate adverse factors associated with stress from depression and drug abuse in determining their impact on child development and behavior.

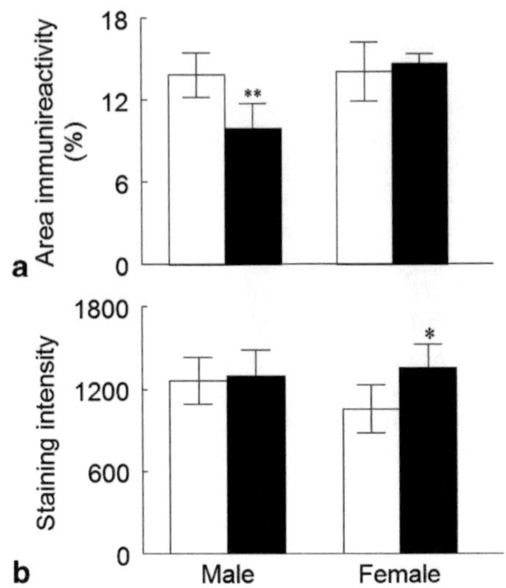

Fig. 1.4 Effect of prenatal stress on neurogenesis and synaptic plasticity in the dentate gyrus of the hippocampus. **a** Represents the % area of immunoreactivity in cells labeled with an antibody to doublecortin. **b** Represents the intensity of staining with an antibody to GAP43. Significantly different from controls; *$p<0.05$; **$p<0.01$

1.7 Experiments Differentiating Pre- and Postnatal Effects of Maternal Stress in Rats

1.7.1 Maternal Behavior

In experimental studies in rats, a clear separation can be made between the influence of prenatal and postnatal factors wih no confounds due to drug intake by the use of cross-fostering of pups from a stressed onto a control mother and vice versa. Thus, we found that chronic variable stress during the last week of pregnancy increased the anxiogenic behavior of stressed mothers in the EPM test measured 2 days after their pups are weaned (Fig. 1.4), and this may reduce their maternal behavior towards their pups (Moore and Power 1986; Power and Moore 1986; Smith et al. 2004). This may have resulted from excess levels of maternal COR released in response to stress since diminished maternal care was also seen when COR was administered during pregnancy (Brummelte and Galea 2010). The presence (Smith et al. 2004) or absence (Poltyrev and Weinstock 1999) of an alteration in maternal care appears to depend on the magnitude of the increase in COR and its duration (Brummelte and Galea 2010).

1.7.2 Cross-fostering

Fostering PS pups onto control mothers reduced their anxiety in the EPM test but increased anxiety in controls pups reared by a stressed mother (Barros et al. 2006). This testifies to the influence of postnatal factors in the etiology of anxiety in the offspring. The fostering procedure of PS rats onto control mothers also normalized the activity of the HPA axis of the males and prevented the downregulation of glucocorticoid receptors (Maccari et al. 1995). In contrast to the amelioration of anxiety, rearing by a control foster mother had no effect on the learning deficits or the reduction in LTP in PS males (Yang et al. 2006). This indicated that the effect on learning and synaptic plasticity is prenatally mediated.

It has been shown that the anxiety and abnormal regulation of the HPA axis, but not the learning deficits, in PS rats are brought about by raised levels of maternal COR (Barbazanges et al. 1996; Salomon et al. 2011). Chronic maternal stress was shown to increase COR in the mothers' milk for periods of up to 3 weeks (Pfister and Muir 1989). Rearing of PS pups by control mothers would also prevent their exposure to such raised levels of COR. The finding that heightened anxiety in PS rats is associated with the quality of postnatal maternal mood, and rearing ability accords well with the findings in human subjects. The closer association between maternal stress during the prenatal period and the cognitive outcome is also in agreement with the human data (Laplante et al. 2008).

1.7.3 Rearing in an Enriched Environment

Environmental enrichment (EE) consists in modifying the rat's housing conditions to provide enhanced sensory motor and cognitive stimulation. This is probably only significant for laboratory rats that are normally housed in small cages with no source of stimulation or little room for movement. EE has been shown to increase total brain weight (Wainwright et al. 1993) and the number of dendritic branches in the hippocampus (Greenough and Volkmar 1973). Housing PS male rats in an EE from weaning restored the response of the HPA axis to stress to that in control rats (Morley-Fletcher et al. 2003) and reduced their anxiety (Li et al. 2012). In contrast to the lack of effect of fostering, an EE also normalized the spatial performance of PS rats in the MWM test, and hippocampal LTP (Lui et al. 2011; Yang et al. 2007). However, the reduction in social interaction in adolescent and adult male PS rats, which was associated with a decrease in oxytocin in the PVN and was restored to that of controls by injection of oxytocin into the amygdala, was not improved by EE (Lee et al. 2007). Others housed the pregnant rats themselves in an EE while they were stressed and found that their male offspring showed less anxiety, performed like controls in the MWM, and had an increase in spine density in the hippocampal CA1 and DG regions (Li et al. 2012). It is not yet known if maternal housing in an EE would prevent the reduction in social interaction induced by prenatal stress. The mechanism by which EE reverses some of the effects of prenatal stress is not clear, but it may occur through stimulation of neurogenesis and formation of more dendritic spines to overcome their loss in PS rats.

1.8 Effect of Antidepressant Treatment

During ontogenesis, 5-HT serves as a developmental signal for both serotonergic neurons and target tissues (Lauder 1990). In the mature brain, 5-HT acts as a neurotransmitter and also modulates neuronal function and plasticity (Lesch 2001). In order to exert its proper function during development, serotonin must be present in various brain regions in optimal concentrations which are controlled through the levels of 5-HT synthesis and metabolism. The level of 5-HT in the brain is largely regulated via its reuptake through the serotonin transporter (SERT) which appears in humans and rats during midgestation (Daws and Gould 2011). Alterations in serotonergic activity are believed to occur in subjects with depression (Ressler and Nemeroff 2000).

1.8.1 Humans

Selective serotonin reuptake inhibitors (SSRIs), fluoxetine, paroxetine, and citalopram, are most frequently prescribed antidepressants for maternal depression because they are generally considered to cause fewer adverse effects than the older tricyclic antidepressants (Cipriani et al. 2005; Westenberg and Sandner 2006). The number of anxious and depressed pregnant women using these medications varies from 5 to 20 % (Marcus 2009; Moses-Kolko and Roth 2004; Nordeng et al. 2012). A higher incidence of preterm births (Hayes et al. 2012; Klieger-Grossmann et al. 2012), autism (Croen et al. 2011), irritability (Thormahlen 2006), and lower psychomotor development index (Casper et al. 2003) has been reported in infants of mothers treated with SSRIs, but these symptoms are also seen in those of untreated subjects with depression (Louik et al. 2007). Others were able to differentiate an influence of SSRIs from that of depression on neuronal function. Infants from untreated depressed mothers had significantly lower attention scores than those of nondepressed mothers, while those of drug-treated mothers had a lower gestational age, more hypertonia, and a higher number of central nervous system stress signs than those of either untreated mothers or controls (Salisbury et al. 2011).

Only two studies have compared the effect on behavior of the children of depressed mothers with and without SSRI treatment. In a relatively small group of women (22) given either paroxetine, sertraline, or fluoxetine, no difference was found in maternal mood or the incidence of behavioral abnormalities in 4-year-old children of treated and untreated mothers (Oberlander et al. 2007). In a larger group, depressed mothers were either untreated or given venlafaxine, sertraline, paroxetine, fluoxetine, or citalopram in the first semester or throughout pregnancy. There was a reduction in the number of women who experienced a depressive episode in the 1st year following childbirth in those receiving venlafaxine but not the other drugs. However, irrespective of drug treatment and maternal outcome, their children aged 3, 6, and 12 years had significantly higher rates of poor neonatal adaptation, problematic externalizing and internalizing behaviors, and lower verbal and

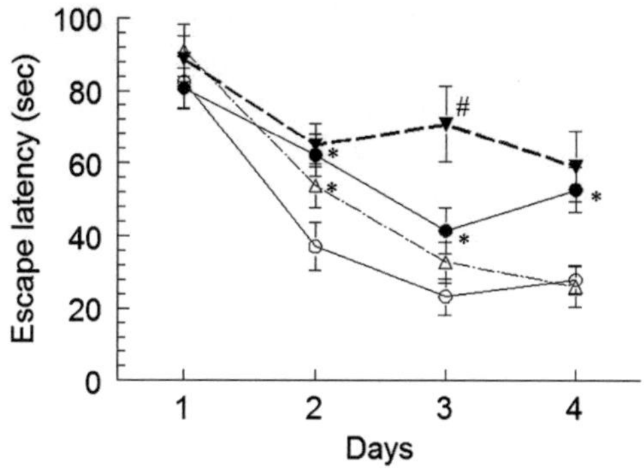

Fig. 1.5 Effect of maternal stress and citalopram treatment on spatial learning in the Morris water maze of adult male offspring. *Open circles:* controls, mothers untreated; *closed circles:* prenatally stressed, mothers untreated; *open triangles:* controls, mothers treated with citalopram (10 mg/kg/day); *closed triangles:* prenatally stressed, mothers treated with citalopram (10 mg/kg/day). Significantly different from controls; $^*p<0.05$; significantly different from prenatally stressed $^{\#}p<0.05$

performance IQs than those of nondepressed mothers (Nulman et al. 2012). Thus, there is no evidence that treatment of pregnant women with these drugs produces any benefit in terms of the behavioral outcome in the children.

1.8.2 Rats

Administration of citalopram (10 mg/kg/day) to pregnant rats from day 7 of gestation 1 week before commencement of stress, until the pups were weaned, reduced the anxiogenic behavior of stressed mothers but had no effect on that of unstressed mothers (Fig. 1.4). However, like SSRI treatment in depressed pregnant women (Nulman et al. 2012), citalopram did not ameliorate anxiogenic behavior of PS male offspring or their spatial learning deficits. When given to control mothers, citalopram induced learning deficits in their offspring (Fig. 1.5). Maternal administration of fluoxetine from gestational day 11 to unstressed mothers resulted in a reduction in social play behavior in their juvenile offspring and anxiety at adulthood (Olivier et al. 2011). However, when fluoxetine was administered to rat mothers after parturition, anxiogenic behavior of the PS male offspring was decreased and there was no increased anxiety in controls (Rayen et al. 2011). In addition, fluoxetine treatment restored the suppressed neurogenesis in the SGZ of the DG of PS males and females to that in controls. Imipramine or fluoxetine given chronically to PS males in adulthood reduced their anxiety in the open field and the levels of COR and glucocorticoid receptor in response to stress (Szymanska et al. 2009). How can one explain the difference in outcome when SSRIs are given pre- or postnatally? Prenatal administration of SSRIs inhibits the SERT during a critical period of neuron development and exposes the brain to excess amounts of 5-HT, as also shown in

SERT knockout mice (Borue et al. 2007). Once the serotoninergic systems are fully developed and neuronal guidance is complete, the actions of SSRIs on abnormal behavior result from readjustment of alterations induced by prenatal stress in pre- and postsynaptic 5-HT receptor activation. Taken together, the data from experiments in human subjects and rats suggest that treatment of pregnant mothers with SSRIs may improve their depressed mood in some subjects. However, SSRIs can adversely affect neuronal guidance and the development of serotoninergic systems in the offspring by inhibiting the SERT at a critical time during development.

Conflict of Interest The author declares no conflicts of interest.

References

Abe H, Hidaka N, Kawagoe C, Odagiri K, Watanabe Y, Ikeda T, Ishizuka Y, Hashiguchi H, Takeda R, Nishimori T, Ishida Y (2007) Prenatal psychological stress causes higher emotionality, depression-like behavior, and elevated activity in the hypothalamo-pituitary-adrenal axis. Neurosci Res 59(2):145–151. doi:10.1016/j.neures.2007.06.1465

Abrous DN, Koehl M, Le Moal M (2005) Adult neurogenesis: from precursors to network and physiology. Physiol Rev 85(2):523–569. doi:10.1152/physrev.00055.2003

Alonso SJ, Arevalo R, Afonso D, Rodriguez M (1991) Effects of maternal stress during pregnancy on forced swimming test behavior of the offspring. Physiol Behav 50(3):511–517

Andrade JP, Madeira MD, Paula-Barbosa MM (2000) Sexual dimorphism in the subiculum of the rat hippocampal formation. Brain Res 875(1-2):125–137

Baibazarova E, van de Beek C, Cohen-Kettenis PT, Buitelaar J, Shelton KH, van Goozen SH (2012) Influence of prenatal maternal stress, maternal plasma cortisol and cortisol in the amniotic fluid on birth outcomes and child temperament at 3 months. Psychoneuroendocrinology 38(6):907–915. doi:10.1016/j.psyneuen.2012.09.015

Baker S, Chebli M, Rees S, Lemarec N, Godbout R, Bielajew C (2008) Effects of gestational stress: 1. Evaluation of maternal and juvenile offspring behavior. Brain Res 1213:98–110. doi:10.1016/j.brainres.2008.03.035

Barbazanges A, Piazza PV, Le Moal M, Maccari S (1996) Maternal glucocorticoid secretion mediates long-term effects of prenatal stress. J Neurosci 16(12):3943–3949

Barker ED, Jaffee SR, Uher R, Maughan B (2011) The contribution of prenatal and postnatal maternal anxiety and depression to child maladjustment. Depress Anxiety 28(8):696–702. doi:10.1002/da.20856

Barros VG, Rodriguez P, Martijena ID, Perez A, Molina VA, Antonelli MC (2006) Prenatal stress and early adoption effects on benzodiazepine receptors and anxiogenic behavior in the adult rat brain. Synapse 60(8):609–618. doi:10.1002/syn.20336

Bayer SA, Altman J, Russo RJ, Zhang X (1993) Timetables of neurogenesis in the human brain based on experimentally determined patterns in the rat. Neurotoxicology 14(1):83–144

Bergman K, Sarkar P, O'Connor TG, Modi N, Glover V (2007) Maternal stress during pregnancy predicts cognitive ability and fearfulness in infancy. J Am Acad Child Adolesc Psychiatry 46(11):1454–1463. doi:10.1097/chi.0b013e31814a62f6

Blokland A, Rutten K, Prickaerts J (2006) Analysis of spatial orientation strategies of male and female Wistar rats in a Morris water escape task. Behav Brain Res 171(2):216–224. doi:10.1016/j.bbr.2006.03.033

Bock J, Murmu MS, Biala Y, Weinstock M, Braun K (2011) Prenatal stress and neonatal handling induce sex specific changes in dendritic complexity and dendritic spine density in hippocampal subregions of prepubertal rats. Neuroscience 193:34–43. doi:10.1016/j.neuroscience.2011.07.048

Borue X, Chen J, Condron BG (2007) Developmental effects of SSRIs: lessons learned from animal studies. Int J Dev Neurosci 25(6):341–347. doi:10.1016/j.ijdevneu.2007.06.003

Brummelte S, Galea LA (2010) Chronic corticosterone during pregnancy and postpartum affects maternal care, cell proliferation and depressive-like behavior in the dam. Horm Behav 58(5):769–779. doi:10.1016/j.yhbeh.2010.07.012

Brunton PJ, Russell JA (2010) Prenatal social stress in the rat programmes neuroendocrine and behavioural responses to stress in the adult offspring: sex-specific effects. J Neuroendocrinol 22(4):258–271. doi:10.1111/j.1365-2826.2010.01969.x

Brunton PJ, Donadio MV, Russell JA (2011) Sex differences in prenatally programmed anxiety behaviour in rats: differential corticotropin-releasing hormone receptor mRNA expression in the amygdaloid complex. Stress 14(6):634–643. doi:10.3109/10253890.2011.604750

Buss C, Davis EP, Shahbaba B, Pruessner JC, Head K, Sandman CA (2012) Maternal cortisol over the course of pregnancy and subsequent child amygdala and hippocampus volumes and affective problems. Proc Natl Acad Sci U S A 109(20):E1312–E1319. doi:10.1073/pnas.1201295109

Casper RC, Fleisher BE, Lee-Ancajas JC, Gilles A, Gaylor E, DeBattista A, Hoyme HE (2003) Follow-up of children of depressed mothers exposed or not exposed to antidepressant drugs during pregnancy. J Pediatr 142(4):402–408. doi:10.1067/mpd.2003.139

Cipriani A, Brambilla P, Furukawa T, Geddes J, Gregis M, Hotopf M, Malvini L, Barbui C (2005) Fluoxetine versus other types of pharmacotherapy for depression. Cochrane Database Syst Rev (4):CD004185. doi:10.1002/14651858.CD004185.pub2

Coe CL, Kramer M, Czeh B, Gould E, Reeves AJ, Kirschbaum C, Fuchs E (2003) Prenatal stress diminishes neurogenesis in the dentate gyrus of juvenile rhesus monkeys. Biol Psychiatry 54(10):1025–1034

Croen LA, Grether JK, Yoshida CK, Odouli R, Hendrick V (2011) Antidepressant use during pregnancy and childhood autism spectrum disorders. Arch Gen Psychiatry 68(11):1104–1112. doi:10.1001/archgenpsychiatry.2011.73

Davis EP, Sandman CA (2010) The timing of prenatal exposure to maternal cortisol and psychosocial stress is associated with human infant cognitive development. Child Dev 81(1):131–148. doi:10.1111/j.1467-8624.2009.01385.x

Davis EP, Sandman CA (2012) Prenatal psychobiological predictors of anxiety risk in preadolescent children. Psychoneuroendocrinology 37(8):1224–1233. doi:10.1016/j.psyneuen.2011.12.016

Daws LC, Gould GG (2011) Ontogeny and regulation of the serotonin transporter: providing insights into human disorders. Pharmacol Ther 131(1):61–79. doi:10.1016/j.pharmthera.2011.03.013

De Bellis MD, Casey BJ, Dahl RE, Birmaher B, Williamson DE, Thomas KM, Axelson DA, Frustaci K, Boring AM, Hall J, Ryan ND (2000) A pilot study of amygdala volumes in pediatric generalized anxiety disorder. Biol Psychiatry 48(1):51–57

de Bruin JP, Moita MP, de Brabander HM, Joosten RN (2001) Place and response learning of rats in a Morris water maze: differential effects of fimbria fornix and medial prefrontal cortex lesions. Neurobiol Learn Mem 75(2):164–178. doi:10.1006/nlme.2000.3962

Delcour M, Russier M, Amin M, Baud O, Paban V, Barbe MF, Coq JO (2012) Impact of prenatal ischemia on behavior, cognitive abilities and neuroanatomy in adult rats with white matter damage. Behav Brain Res 232(1):233–244. doi:10.1016/j.bbr.2012.03.029

DiPietro JA, Novak MF, Costigan KA, Atella LD, Reusing SP (2006) Maternal psychological distress during pregnancy in relation to child development at age two. Child Dev 77(3):573–587. doi:10.1111/j.1467-8624.2006.00891.x

Estanislau C, Morato S (2005) Prenatal stress produces more behavioral alterations than maternal separation in the elevated plus-maze and in the elevated T-maze. Behav Brain Res 163(1):70–77. doi:10.1016/j.bbr.2005.04.003

Evans LM, Myers MM, Monk C (2008) Pregnant women's cortisol is elevated with anxiety and depression-but only when comorbid. Arch Womens Ment Health 11(3):239–248. doi:10.1007/s00737-008-0019-4

Fan JM, Chen XQ, Jin H, Du JZ (2009) Gestational hypoxia alone or combined with restraint sensitizes the hypothalamic-pituitary-adrenal axis and induces anxiety-like behavior in adult male rat offspring. Neuroscience 159(4):1363–1373. doi:10.1016/j.neuroscience.2009.02.009

Fride E, Weinstock M (1988) Prenatal stress increases anxiety related behavior and alters cerebral lateralization of dopamine activity. Life Sci 42(10):1059–1065

Fujioka A, Fujioka T, Ishida Y, Maekawa T, Nakamura S (2006) Differential effects of prenatal stress on the morphological maturation of hippocampal neurons. Neuroscience 141(2):907–915. doi:10.1016/j.neuroscience.2006.04.046

Fujioka T, Fujioka A, Tan N, Chowdhury GM, Mouri H, Sakata Y, Nakamura S (2001) Mild prenatal stress enhances learning performance in the non-adopted rat offspring. Neuroscience 103(2):301–307

Giardinelli L, Innocenti A, Benni L, Stefanini MC, Lino G, Lunardi C, Svelto V, Afshar S, Bovani R, Castellini G, Faravelli C (2012) Depression and anxiety in perinatal period: prevalence and risk factors in an Italian sample. Arch Womens Ment Health 15(1):21–30. doi:10.1007/s00737-011-0249-8

Gotz AA, Stefanski V (2007) Psychosocial maternal stress during pregnancy affects serum corticosterone, blood immune parameters and anxiety behaviour in adult male rat offspring. Physiol Behav 90(1):108–115. doi:10.1016/j.physbeh.2006.09.014

Gould E, Woolley CS, Frankfurt M, McEwen BS (1990) Gonadal steroids regulate dendritic spine density in hippocampal pyramidal cells in adulthood. J Neurosci 10(4):1286–1291

Gray TS, Bingaman EW (1996) The amygdala: corticotropin-releasing factor, steroids, and stress. Crit Rev Neurobiol 10(2):155–168

Greenough WT, Volkmar FR (1973) Pattern of dendritic branching in occipital cortex of rats reared in complex environments. Exp Neurol 40(2):491–504

Hayes RM, Wu P, Shelton RC, Cooper WO, Dupont WD, Mitchel E, Hartert TV (2012) Maternal antidepressant use and adverse outcomes: a cohort study of 228,876 pregnancies. Am J Obstet Gynecol 207(1):49 e41–e49. doi:10.1016/j.ajog.2012.04.028

Holson RR, Gough B, Sullivan P, Badger T, Sheehan DM (1995) Prenatal dexamethasone or stress but not ACTH or corticosterone alter sexual behavior in male rats. Neurotoxicol Teratol 17(4):393–401

Honkaniemi J, Pelto-Huikko M, Rechardt L, Isola J, Lammi A, Fuxe K, Gustafsson JA, Wikstrom AC, Hokfelt T (1992) Colocalization of peptide and glucocorticoid receptor immunoreactivities in rat central amygdaloid nucleus. Neuroendocrinology 55(4):451–459

Hosseini-Sharifabad M, Hadinedoushan H (2007) Prenatal stress induces learning deficits and is associated with a decrease in granules and CA3 cell dendritic tree size in rat hippocampus. Anat Sci Int 82(4):211–217. doi:10.1111/j.1447-073X.2007.00186.x

Huizink AC, Robles de Medina PG, Mulder EJ, Visser GH, Buitelaar JK (2003) Stress during pregnancy is associated with developmental outcome in infancy. J Child Psychol Psychiatry 44(6):810–818

Husain N, Cruickshank K, Husain M, Khan S, Tomenson B, Rahman A (2012) Social stress and depression during pregnancy and in the postnatal period in British Pakistani mothers: a cohort study. J Affect Disord 140(3):268–276. doi:10.1016/j.jad.2012.02.009

Imamura Y, Nakane Y, Ohta Y, Kondo H (1999) Lifetime prevalence of schizophrenia among individuals prenatally exposed to atomic bomb radiation in Nagasaki City. Acta Psychiatr Scand 100(5):344–349

Jansson T, Powell TL (2007) Role of the placenta in fetal programming: underlying mechanisms and potential interventional approaches. Clin Sci (Lond) 113(1):1–13. doi:10.1042/CS20060339

Jensen Pena C, Monk C, Champagne FA (2012) Epigenetic effects of prenatal stress on 11beta-hydroxysteroid dehydrogenase-2 in the placenta and fetal brain. PLoS One 7 (6):e39791. doi:10.1371/journal.pone.0039791

Keck ME (2006) Corticotropin-releasing factor, vasopressin and receptor systems in depression and anxiety. Amino Acids 31(3):241–250. doi:10.1007/s00726-006-0333-y

Khashan AS, Abel KM, McNamee R, Pedersen MG, Webb RT, Baker PN, Kenny LC, Mortensen PB (2008) Higher risk of offspring schizophrenia following antenatal maternal exposure to severe adverse life events. Arch Gen Psychiatry 65(2):146–152. doi:10.1001/archgenpsychiatry.2007.20

Klieger-Grossmann C, Weitzner B, Panchaud A, Pistelli A, Einarson T, Koren G, Einarson A (2012) Pregnancy outcomes following use of escitalopram: a prospective comparative cohort study. J Clin Pharmacol 52(5):766–770. doi:10.1177/0091270011405524

Koenig JI, Kirkpatrick B, Lee P (2002) Glucocorticoid hormones and early brain development in schizophrenia. Neuropsychopharmacology 27(2):309-318. doi:10.1016/S0893-133x(01)00396–7

Koenig JI, Elmer GI, Shepard PD, Lee PR, Mayo C, Joy B, Hercher E, Brady DL (2005) Prenatal exposure to a repeated variable stress paradigm elicits behavioral and neuroendocrinological changes in the adult offspring: potential relevance to schizophrenia. Behav Brain Res 156(2):251–261. doi:10.1016/j.bbr.2004.05.030

Kofman O (2002) The role of prenatal stress in the etiology of developmental behavioural disorders. Neurosci Biobehav Rev 26(4):457–470

Laloux C, Mairesse J, Van Camp G, Giovine A, Branchi I, Bouret S, Morley-Fletcher S, Bergonzelli G, Malagodi M, Gradini R, Nicoletti F, Darnaudery M, Maccari S (2012) Anxiety-like behaviour and associated neurochemical and endocrinological alterations in male pups exposed to prenatal stress. Psychoneuroendocrinology 37(10):1646–1658. doi:10.1016/j.psyneuen.2012.02.010

Laplante DP, Brunet A, Schmitz N, Ciampi A, King S (2008) Project Ice Storm: prenatal maternal stress affects cognitive and linguistic functioning in 5 1/2-year-old children. J Am Acad Child Adolesc Psychiatry 47(9):1063–1072. doi:10.1097/CHI.0b013e31817eec80

Lauder JM (1990) Ontogeny of the serotonergic system in the rat: serotonin as a developmental signal. Ann N Y Acad Sci 600:297–313; discussion 314

Lee PR, Brady DL, Shapiro RA, Dorsa DM, Koenig JI (2007) Prenatal stress generates deficits in rat social behavior: reversal by oxytocin. Brain Res 1156:152–167. doi:10.1016/j.brainres.2007.04.042

Lemaire V, Koehl M, Le Moal M, Abrous DN (2000) Prenatal stress produces learning deficits associated with an inhibition of neurogenesis in the hippocampus. Proc Natl Acad Sci U S A 97(20):11032–11037

Lereya ST, Wolke D (2012) Prenatal family adversity and maternal mental health and vulnerability to peer victimisation at school. J Child Psychol Psychiatry. doi:10.1111/jcpp.12012

Lesch KP (2001) Variation of serotonergic gene expression: neurodevelopment and the complexity of response to psychopharmacologic drugs. Eur Neuropsychopharmacol 11(6):457–474

Li H, Li X, Jia N, Cai Q, Bai Z, Chen R, Song T, Zhu Z, Liu J (2008) NF-kappaB regulates prenatal stress-induced cognitive impairment in offspring rats. Behav Neurosci 122(2):331–339. doi:10.1037/0735-7044.122.2.331

Li M, Wang M, Ding S, Li C, Luo X (2012) Environmental enrichment during gestation improves behavior consequences and synaptic plasticity in hippocampus of prenatal-stressed offspring rats. Acta Histochem Cytochem 45(3):157–166. doi:10.1267/ahc.11054

Loomans EM, van der Stelt O, van Eijsden M, Gemke RJ, Vrijkotte T, den Bergh BR (2011) Antenatal maternal anxiety is associated with problem behaviour at age five. Early Hum Dev 87(8):565–570. doi:10.1016/j.earlhumdev.2011.04.014

Louik C, Lin AE, Werler MM, Hernandez-Diaz S, Mitchell AA (2007) First-trimester use of selective serotonin-reuptake inhibitors and the risk of birth defects. New Engl J Med 356(26):2675–2683. doi:10.1056/NEJMoa067407

Lui CC, Wang JY, Tain YL, Chen YC, Chang KA, Lai MC, Huang LT (2011) Prenatal stress in rat causes long-term spatial memory deficit and hippocampus MRI abnormality: differential effects of postweaning enriched environment. Neurochem Int 58(3):434–441. doi:10.1016/j.neuint.2011.01.002

Maccari S, Piazza PV, Kabbaj M, Barbazanges A, Simon H, Le Moal M (1995) Adoption reverses the long-term impairment in glucocorticoid feedback induced by prenatal stress. J Neurosci 15(1 Pt 1):110–116

Madeira MD, Lieberman AR (1995) Sexual dimorphism in the mammalian limbic system. Prog Neurobiol 45(4):275–333

Marcus SM (2009) Depression during pregnancy: rates, risks and consequences-Motherisk Update 2008. Can J Clin Pharmacol 16(1):e15-e22

Markham JA, Taylor AR, Taylor SB, Bell DB, Koenig JI (2010) Characterization of the cognitive impairments induced by prenatal exposure to stress in the rat. Front Behav Neurosci 4:173. doi:10.3389/fnbeh.2010.00173

Martini J, Knappe S, Beesdo-Baum K, Lieb R, Wittchen HU (2010) Anxiety disorders before birth and self-perceived distress during pregnancy: associations with maternal depression and obstetric, neonatal and early childhood outcomes. Early Hum Dev 86(5):305–310. doi:10.1016/j.earlhumdev.2010.04.004

McCormick CM, Smythe JW, Sharma S, Meaney MJ (1995) Sex-specific effects of prenatal stress on hypothalamic-pituitary-adrenal responses to stress and brain glucocorticoid receptor density in adult rats. Brain Res Dev Brain Res 84(1):55–61

Mennes M, Stiers P, Lagae L, Van den Bergh B (2006) Long-term cognitive sequelae of antenatal maternal anxiety: involvement of the orbitofrontal cortex. Neurosci Biobehav Rev 30(8):1078–1086. doi:10.1016/j.neubiorev.2006.04.003

Merchenthaler I, Vigh S, Petrusz P, Schally AV (1982) Immunocytochemical localization of corticotropin-releasing factor (CRF) in the rat brain. Am J Anat 165(4):385–396. doi:10.1002/aja.1001650404

Meyer G, Schaaps JP, Moreau L, Goffinet AM (2000) Embryonic and early fetal development of the human neocortex. J Neurosci 20(5):1858–1868

Miszkurka M, Goulet L, Zunzunegui MV (2012) Antenatal depressive symptoms among Canadian-born and immigrant women in Quebec: differential exposure and vulnerability to contextual risk factors. Soc Psychiatry Psychiatr Epidemiol 47(10):1639–1648. doi:10.1007/s00127-011-0469-2

Moore CL, Power KL (1986) Prenatal stress affects mother-infant interaction in Norway rats. Dev Psychobiol 19(3):235–245. doi:10.1002/dev.420190309

Morley-Fletcher S, Rea M, Maccari S, Laviola G (2003) Environmental enrichment during adolescence reverses the effects of prenatal stress on play behaviour and HPA axis reactivity in rats. Eur J Neurosci 18(12):3367–3374

Morley-Fletcher S, Darnaudery M, Mocaer E, Froger N, Lanfumey L, Laviola G, Casolini P, Zuena AR, Marzano L, Hamon M, Maccari S (2004) Chronic treatment with imipramine reverses immobility behaviour, hippocampal corticosteroid receptors and cortical 5-HT_{1A} receptor mRNA in prenatally stressed rats. Neuropharmacology 47(6):841–847. doi:10.1016/j.neuropharm.2004.06.011

Moses-Kolko EL, Roth EK (2004) Antepartum and postpartum depression: healthy mom, healthy baby. J Am Med Womens Assoc 59(3):181–191

Munoz-Cueto JA, Garcia-Segura LM, Ruiz-Marcos A (1990) Developmental sex differences and effect of ovariectomy on the number of cortical pyramidal cell dendritic spines. Brain Res 515(1-2):64–68

Newhouse P, Newhouse C, Astur RS (2007) Sex differences in visual-spatial learning using a virtual water maze in pre-pubertal children. Behav Brain Res 183(1):1–7. doi:10.1016/j.bbr.2007.05.011

Nordeng H, van Gelder MM, Spigset O, Koren G, Einarson A, Eberhard-Gran M (2012) Pregnancy outcome after exposure to antidepressants and the role of maternal depression: results from the Norwegian Mother and Child Cohort Study. J Clin Psychopharmacol 32(2):186–194. doi:10.1097/JCP.0b013e3182490eaf

Nulman I, Koren G, Rovet J, Barrera M, Pulver A, Streiner D, Feldman B (2012) Neurodevelopment of children following prenatal exposure to venlafaxine, selective serotonin reuptake inhibitors, or untreated maternal depression. Am J Psychiatry 169(11):1165–1174. doi:10.1176/appi.ajp.2012.11111721

O'Connor TG, Heron J, Golding J, Beveridge M, Glover V (2002) Maternal antenatal anxiety and children's behavioural/emotional problems at 4 years. Report from the Avon longitudinal study of parents and children. Br J Psychiatry 180:502–508

O'Donnell KJ, Bugge Jensen A, Freeman L, Khalife N, O'Connor TG, Glover V (2012) Maternal prenatal anxiety and downregulation of placental 11beta-HSD2. Psychoneuroendocrinology 37(6):818–826. doi:10.1016/j.psyneuen.2011.09.014

Oberlander TF, Reebye P, Misri S, Papsdorf M, Kim J, Grunau RE (2007) Externalizing and attentional behaviors in children of depressed mothers treated with a selective serotonin reuptake inhibitor antidepressant during pregnancy. Arch Pediatr Adolesc Med 161(1):22–29. doi:10.1001/archpedi.161.1.22
Ohkawa T, Takeshita S, Murase T, Okinaga S, Arai K (1991) The effect of an acute stress in late pregnancy on hypothalamic catecholamines of the rat fetus. Nihon Sanka Fujinka Gakkai Zasshi 43(7):783–787
Olivier JD, Valles A, van Heesch F, Afrasiab-Middelman A, Roelofs JJ, Jonkers M, Peeters EJ, Korte-Bouws GA, Dederen JP, Kiliaan AJ, Martens GJ, Schubert D, Homberg JR (2011) Fluoxetine administration to pregnant rats increases anxiety-related behavior in the offspring. Psychopharmacology 217(3):419–432. doi:10.1007/s00213-011-2299-z
Ooishi Y, Kawato S, Hojo Y, Hatanaka Y, Higo S, Murakami G, Komatsuzaki Y, Ogiue-Ikeda M, Kimoto T, Mukai H (2012) Modulation of synaptic plasticity in the hippocampus by hippocampus-derived estrogen and androgen. J Steroid Biochem Mol Biol 131(1-2):37–51. doi:10.1016/j.jsbmb.2011.10.004
Pellow S, File SE (1986) Anxiolytic and anxiogenic drug effects on exploratory activity in an elevated plus-maze: a novel test of anxiety in the rat. Pharmacol Biochem Behav 24(3):525–529
Pereira OC, Bernardi MM, Gerardin DC (2006) Could neonatal testosterone replacement prevent alterations induced by prenatal stress in male rats? Life Sci 78(24):2767–2771. doi:10.1016/j.lfs.2005.10.035
Pfister HP, Muir JL (1989) Psychological stress and administered oxytocin during pregnancy: effect corticosterone and prolactin response in lactating rats. Int J Neurosci 45(1-2):91–99
Phan A, Gabor CS, Favaro KJ, Kaschack S, Armstrong JN, MacLusky NJ, Choleris E (2012) Low doses of 17beta-estradiol rapidly improve learning and increase hippocampal dendritic spines. Neuropsychopharmacology 37(10):2299–2309. doi:10.1038/npp.2012.82
Poltyrev T, Weinstock M (1999) Effect of gestational stress on maternal behavior in response to cage transfer and handling of pups in two strains of rat. Stress 3(1):85–95. doi:10.3109/10253899909001114
Poltyrev T, Gorodetsky E, Bejar C, Schorer-Apelbaum D, Weinstock M (2005) Effect of chronic treatment with ladostigil (TV-3326) on anxiogenic and depressive-like behaviour and on activity of the hypothalamic-pituitary-adrenal axis in male and female prenatally stressed rats. Psychopharmacology 181(1):118–125. doi:10.1007/s00213-005-2229-z
Power KL, Moore CL (1986) Prenatal stress eliminates differential maternal attention to male offspring in Norway rats. Physiol Behav 38(5):667–671
Qu Z, Tian D, Zhang Q, Wang X, He H, Zhang X, Huang L, Xu F (2012) The impact of the catastrophic earthquake in China's Sichuan province on the mental health of pregnant women. J Affect Disord 136(1-2):117–123. doi:10.1016/j.jad.2011.08.021
Rao MS, Shetty AK (2004) Efficacy of doublecortin as a marker to analyse the absolute number and dendritic growth of newly generated neurons in the adult dentate gyrus. Eur J Neurosci 19(2):234–246
Rayen I, van den Hove DL, Prickaerts J, Steinbusch HW, Pawluski JL (2011) Fluoxetine during development reverses the effects of prenatal stress on depressive-like behavior and hippocampal neurogenesis in adolescence. PLoS One 6(9):e24003. doi:10.1371/journal.pone.0024003
Ressler KJ, Nemeroff CB (2000) Role of serotonergic and noradrenergic systems in the pathophysiology of depression and anxiety disorders. Depress Anxiety 12(Suppl 1):2–19. doi:10.1002/1520-6394(2000)12:1+<2::AID-DA2>3.0.CO;2-4
Reul JM, Holsboer F (2002) On the role of corticotropin-releasing hormone receptors in anxiety and depression. Dialogues Clin Neurosci 4(1):31–46
Rice D, Barone S Jr (2000) Critical periods of vulnerability for the developing nervous system: evidence from humans and animal models. Environ Health Perspect 108(Suppl 3):511–533
Richardson HN, Zorrilla EP, Mandyam CD, Rivier CL (2006) Exposure to repetitive versus varied stress during prenatal development generates two distinct anxiogenic and neuroendocrine profiles in adulthood. Endocrinology 147(5):2506–2517. doi:10.1210/en.2005-1054

Rimondini R, Agren G, Borjesson S, Sommer W, Heilig M (2003) Persistent behavioral and autonomic supersensitivity to stress following prenatal stress exposure in rats. Behav Brain Res 140(1-2):75–80

Salisbury AL, Wisner KL, Pearlstein T, Battle CL, Stroud L, Lester BM (2011) Newborn neurobehavioral patterns are differentially related to prenatal maternal major depressive disorder and serotonin reuptake inhibitor treatment. Depress Anxiety 28(11):1008–1019. doi:10.1002/da.20883

Salm AK, Pavelko M, Krouse EM, Webster W, Kraszpulski M, Birkle DL (2004) Lateral amygdaloid nucleus expansion in adult rats is associated with exposure to prenatal stress. Brain Res Dev Brain Res 148(2):159–167. doi:10.1016/j.devbrainres.2003.11.005

Salomon S, Bejar C, Schorer-Apelbaum D, Weinstock M (2011) Corticosterone mediates some but not other behavioural changes induced by prenatal stress in rats. J Neuroendocrinol 23(2):118–128. doi:10.1111/j.1365-2826.2010.02097.x

Sandman CA, Davis EP, Buss C, Glynn LM (2011) Prenatal programming of human neurological function. Int J Pept 2011:837596. doi:10.1155/2011/837596

Sandstrom NJ, Kaufman J, Huettel SA (1998) Males and females use different distal cues in a virtual environment navigation task. Brain Res Cogn Brain Res 6(4):351–360

Schulz KM, Pearson JN, Neeley EW, Berger R, Leonard S, Adams CE, Stevens KE (2011) Maternal stress during pregnancy causes sex-specific alterations in offspring memory performance, social interactions, indices of anxiety, and body mass. Physiol Behav 104(2):340–347. doi:10.1016/j.physbeh.2011.02.021

Schwabe L, Bohbot VD, Wolf OT (2012) Prenatal stress changes learning strategies in adulthood. Hippocampus 22(11):2136–2143. doi:10.1002/hipo.22034

Sidman RL, Rakic P (1973) Neuronal migration, with special reference to developing human brain: a review. Brain Res 62(1):1–35

Smith JW, Seckl JR, Evans AT, Costall B, Smythe JW (2004) Gestational stress induces postpartum depression-like behaviour and alters maternal care in rats. Psychoneuroendocrinology 29(2):227–244

Sneider JT, Sava S, Rogowska J, Yurgelun-Todd DA (2011) A preliminary study of sex differences in brain activation during a spatial navigation task in healthy adults. Percept Mot Skills 113(2):461–480

Szuran TF, Pliska V, Pokorny J, Welzl H (2000) Prenatal stress in rats: effects on plasma corticosterone, hippocampal glucocorticoid receptors, and maze performance. Physiol Behav 71(3-4):353–362

Szymanska M, Budziszewska B, Jaworska-Feil L, Basta-Kaim A, Kubera M, Leskiewicz M, Regulska M, Lason W (2009) The effect of antidepressant drugs on the HPA axis activity, glucocorticoid receptor level and FKBP51 concentration in prenatally stressed rats. Psychoneuroendocrinology 34(6):822–832. doi:10.1016/j.psyneuen.2008.12.012

Takahashi LK, Turner JG, Kalin NH (1998) Prolonged stress-induced elevation in plasma corticosterone during pregnancy in the rat: implications for prenatal stress studies. Psychoneuroendocrinology 23(6):571–581

Tamura M, Sajo M, Kakita A, Matsuki N, Koyama R (2011) Prenatal stress inhibits neuronal maturation through downregulation of mineralocorticoid receptors. J Neurosci 31(32):11505–11514. doi:10.1523/JNEUROSCI.3447-10.2011

Thormahlen GM (2006) Paroxetine use during pregnancy: is it safe? Ann Pharmacother 40(10):1834–1837. doi:10.1345/aph.1H116

Vallee M, Mayo W, Dellu F, Le Moal M, Simon H, Maccari S (1997) Prenatal stress induces high anxiety and postnatal handling induces low anxiety in adult offspring: correlation with stress-induced corticosterone secretion. J Neurosci 17(7):2626–2636

Van den Bergh BR, Marcoen A (2004) High antenatal maternal anxiety is related to ADHD symptoms, externalizing problems, and anxiety in 8- and 9-year-olds. Child Dev 75(4):1085–1097. doi:10.1111/j.1467-8624.2004.00727.x

Van den Bergh BR, Van Calster B, Smits T, Van Huffel S, Lagae L (2008) Antenatal maternal anxiety is related to HPA-axis dysregulation and self-reported depressive symptoms in adolescence: a prospective study on the fetal origins of depressed mood. Neuropsychopharmacology 33(3):536–545. doi:10.1038/sj.npp.1301450

Wadhwa PD, Sandman CA, Porto M, Dunkel-Schetter C, Garite TJ (1993) The association between prenatal stress and infant birth weight and gestational age at birth: a prospective investigation. Am J Obstet Gynecol 169(4):858–865

Wainwright PE, Levesque S, Krempulec L, Bulman-Fleming B, McCutcheon D (1993) Effects of environmental enrichment on cortical depth and Morris-maze performance in B6D2F2 mice exposed prenatally to ethanol. Neurotoxicol Teratol 15(1):11–20

Ward HE, Johnson EA, Salm AK, Birkle DL (2000) Effects of prenatal stress on defensive withdrawal behavior and corticotropin releasing factor systems in rat brain. Physiol Behav 70(3-4):359–366

Weinstock M (1997) Does prenatal stress impair coping and regulation of hypothalamic-pituitary-adrenal axis? Neurosci Biobehav Rev 21(1):1–10

Weinstock M (2001) Alterations induced by gestational stress in brain morphology and behaviour of the offspring. Prog Neurobiol 65(5):427–451

Weinstock M (2005) The potential influence of maternal stress hormones on development and mental health of the offspring. Brain Behav Immun 19(4):296–308. doi:10.1016/j.bbi.2004.09.006

Weinstock M (2008) The long-term behavioural consequences of prenatal stress. Neurosci Biobehav Rev 32(6):1073–1086. doi:10.1016/j.neubiorev.2008.03.002

Weinstock M (2011) Sex-dependent changes induced by prenatal stress in cortical and hippocampal morphology and behaviour in rats: an update. Stress 14(6):604–613. doi:10.3109/10253890.2011.588294

Weinstock M, Fride E, Hertzberg R (1988) Prenatal stress effects on functional development of the offspring. Prog Brain Res 73:319–331. doi:10.1016/S0079-6123(08)60513-0

Weinstock M, Matlina E, Maor GI, Rosen H, McEwen BS (1992) Prenatal stress selectively alters the reactivity of the hypothalamic-pituitary adrenal system in the female rat. Brain Res 595(2):195–200

Weinstock M, Poltyrev T, Schorer-Apelbaum D, Men D, McCarty R (1998) Effect of prenatal stress on plasma corticosterone and catecholamines in response to footshock in rats. Physiol Behav 64(4):439–444

Westenberg HG, Sandner C (2006) Tolerability and safety of fluvoxamine and other antidepressants. Int J Clin Pract 60(4):482–491. doi:10.1111/j.1368-5031.2006.00865.x

Wilson CA, Schade R, Terry AV Jr (2012) Variable prenatal stress results in impairments of sustained attention and inhibitory response control in a 5-choice serial reaction time task in rats. Neuroscience 218:126–137. doi:10.1016/j.neuroscience.2012.05.040

Wu J, Song TB, Li YJ, He KS, Ge L, Wang LR (2007) Prenatal restraint stress impairs learning and memory and hippocampal PKCbeta1 expression and translocation in offspring rats. Brain Res 1141:205–213. doi:10.1016/j.brainres.2007.01.024

Yaka R, Salomon S, Matzner H, Weinstock M (2007) Effect of varied gestational stress on acquisition of spatial memory, hippocampal LTP and synaptic proteins in juvenile male rats. Behav Brain Res 179(1):126–132. doi:10.1016/j.bbr.2007.01.018

Yang J, Han H, Cao J, Li L, Xu L (2006) Prenatal stress modifies hippocampal synaptic plasticity and spatial learning in young rat offspring. Hippocampus 16(5):431–436. doi:10.1002/hipo.20181

Yang J, Hou C, Ma N, Liu J, Zhang Y, Zhou J, Xu L, Li L (2007) Enriched environment treatment restores impaired hippocampal synaptic plasticity and cognitive deficits induced by prenatal chronic stress. Neurobiol Learn Mem 87(2):257–263. doi:10.1016/j.nlm.2006.09.001

Zagron G, Weinstock M (2006) Maternal adrenal hormone secretion mediates behavioural alterations induced by prenatal stress in male and female rats. Behav Brain Res 175(2):323–328. doi:10.1016/j.bbr.2006.09.003

Zohar I, Weinstock M (2011) Differential effect of prenatal stress on the expression of corticotrophin-releasing hormone and its receptors in the hypothalamus and amygdala in male and female rats. J Neuroendocrinol 23(4):320–328. doi:10.1111/j.1365-2826.2011.02117.x

Zuena AR, Mairesse J, Casolini P, Cinque C, Alema GS, Morley-Fletcher S, Chiodi V, Spagnoli LG, Gradini R, Catalani A, Nicoletti F, Maccari S (2008) Prenatal restraint stress generates two distinct behavioral and neurochemical profiles in male and female rats. PLoS One 3(5):e2170. doi:10.1371/journal.pone.0002170

Chapter 2
Sleep in Prenatally Restraint Stressed Rats, a Model of Mixed Anxiety-Depressive Disorder

Jérôme Mairesse, Gilles Van Camp, Eleonora Gatta, Jordan Marrocco, Marie-Line Reynaert, Michol Consolazione, Sara Morley-Fletcher, Ferdinando Nicoletti and Stefania Maccari

Abstract Prenatal restraint stress (PRS) can induce persisting changes in individual's development. PRS increases anxiety and depression-like behaviors and induces changes in the hypothalamo–pituitary–adrenal (HPA) axis in adult PRS rats after exposure to stress. Since adaptive capabilities also depend on temporal organization and synchronization with the external environment, we studied the effects of PRS on circadian rhythms, including the sleep–wake cycle, that are parameters altered in depression. Using a restraint stress during gestation, we showed that PRS induced phase advances in hormonal/behavioral circadian rhythms in adult rats, and an increase in the amount of paradoxical sleep, positively correlated to plasma corticosterone levels. Plasma corticosterone levels were also correlated with immobility in the forced swimming test, indicating a depressive-like profile in the PRS rats. We observed comorbidity with anxiety-like profile on PRS rats that was correlated with a reduced release of glutamate in the ventral hippocampus. Pharmacological approaches aimed at modulating glutamate release may represent a novel therapeutic strategy to treat stress-related disorders. Finally, since depressed patients exhibit changes in HPA axis activity and in circadian rhythmicity as well as in the paradoxical sleep regulation, we suggest that PRS could represent an original animal model of depression.

S. Maccari (✉) · J. Mairesse · G. Van Camp · E. Gatta · M.-L. Reynaert · M. Consolazione · S. Morley-Fletcher
International Associated Laboratory—Prenatal Stress and Neurodegenerative Diseases, Neural Plasticity Team-UMR CNRS/USTL n 8576 Structural and Functional Glycobiology Unit, University of Lille 1, Lille, France
e-mail: stefania.maccari@univ-lille1.fr

J. Marrocco
IRCCS Centro Neurolesi "Bonino-Pulejo", Messina, Italy

F. Nicoletti
International Associated Laboratory—Prenatal Stress and Neurodegenerative Diseases, IRCCS Neuromed, Pozzilli, Italy

F. Nicoletti
International Associated Laboratory—Prenatal Stress and Neurodegenerative Diseases, Department of Pharmacology and Human Physiology, Sapienza University of Rome, Rome, Italy

M. C. Antonelli (ed.), *Perinatal Programming of Neurodevelopment,* Advances in Neurobiology 10, DOI 10.1007/978-1-4939-1372-5_2,

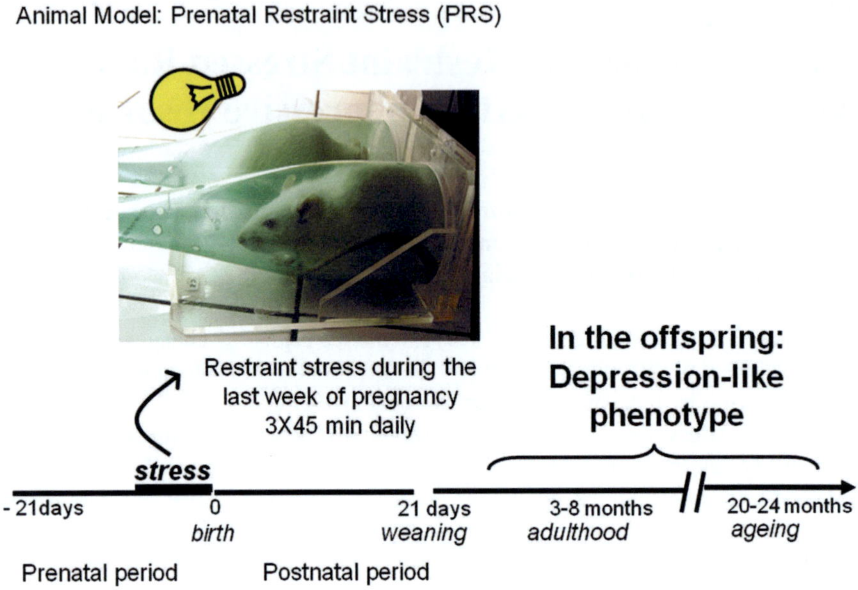

Fig. 2.1 Schematic representation of PRS procedure; during the ten days of gestation, dam is restrained under a strong light for 45 min, 3 times per day. No painful contention is applied, even though dams cannot avoid this situation. The offspring and the mother are then left undisturbed until weaning. The first aim of our group was to investigate the long-term effect of prenatal stress to determine if the offspring of stressed mothers were more vulnerable to affective disorders

2.1 Early Life Events, HPA Axis Activity, and Maternal Factors

Very little is known about the long-term effects of an early adverse experience on the development of response to the rotation of the earth (predictive adaptation). In order to study biological rhythms while avoiding the obvious ethical dilemmas involved in human experimentation, different animal models of perinatal stress have been developed since 1957 (Thompson 1957; Dahlof et al. 1978; Fride and Weinstock 1984; Peters 1982; Ward 1972; Alonso et al. 1991; Takahashi and Kalin 1991; Maccari et al. 1995; Seckl 1998). In particular, during the past 20 years, our group, first in Bordeaux and more recently in Lille and Rome, has studied the long-term outcome of prenatal restraint stress (PRS) on the activity of the HPA axis and the function of the circadian rhythm. PRS animal model in the rat (Fig. 2.1) is a well-documented model of early stress known to induce long-term neurobiological and behavioral alterations after exposure to stressors (Darnaudéry and Maccari 2008). Indeed, prenatal environment exerts profound influences on the development of an organism (Seckl 2004; de Kloet et al. 2005), inducing changes, which extend from early to later life (Vallée et al. 1999). Early environmental triggers or stressors may have a permanent rather than a transient effect on the organism inducing an imprinting of physiological systems known as perinatal "programming" (Barker 1999).

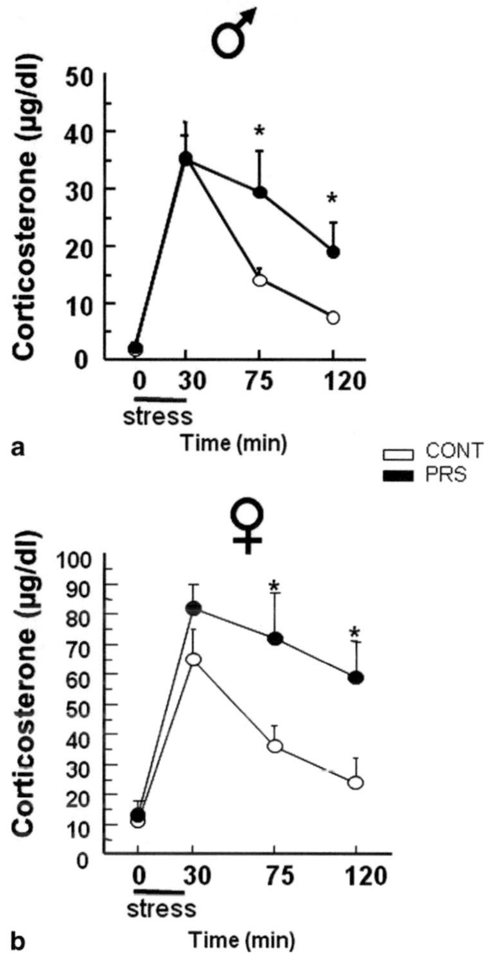

Fig. 2.2 Corticosterone stress response in both male (**a**) and female (**b**) rats is shown. $*p<0.05$ versus controls. *PRS* prenatal restraint stress. (Original data are reported in Koehl et al. 1999)

It has been shown that adult prenatal stress rats (called PRS rats) exhibit abnormalities under stressful conditions, such as increased "anxiety-like behaviors", "hyperemotionality-like behaviors" or "depression-like behavior" (Maccari and Morley-Fletcher 2007). The increased anxious/depressive-like behavior is evidenced by increased ultrasonic vocalizations in infancy (Laloux et al. 2012), reduced social play during adolescence (Morley-Fletcher et al. 2003a), reduced exploration in the elevated-plus maze (EPM) test and open-field test (Vallée 1997, 1999), or increased immobility in the forced swim test during adulthood (Morley-Fletcher et al. 2003b, 2004, 2011).The HPA axis is involved in the ability to cope with stress and represents one of the biological substrates possibly mediating these behavioral disorders. Indeed, PRS induces an increase in stress-induced plasma adrenocorticotropin (ACTH) levels, a prolongation of stress-induced corticosterone secretion both in male and female PRS rats (Fig. 2.2), and a decreased binding capacity of hippocampal corticosteroid receptors (Maccari et al. 1995; Van Waes et al. 2006). A number of neurotransmitters are involved in those hormonal and behavioral responses. For example, PRS increases the levels of 5-HT_2 receptors (Peters 1986, 1988, 1990), induces an increase in the expression of 5HT_{1A} mRNA in the prefrontal cortex (Morley-Fletcher et al. 2004), and increases acetylcholine release in the hippocampus after mild stress or corticotropin-releasing factor (CRF) injection (Day et al. 1998). Recently, we showed that a decrease in glutamate release in the ventral hippocampus was causally related to the increased anxiety observed in PRS rats (Marrocco et al. 2012).

Those long-term neuroendocrinological effects are mediated, at least in part, by stress-induced maternal corticosterone increase during pregnancy (Barbazanges et al. 1996). This is also supported by the recent observation that PRS leads to a decrease in placental 11beta-hydroxysteroid dehydrogenase 2 (11beta-HSD2) activity, and the ensuing increase in maternal corticosterone reaching the fetus (Mairesse et al. 2007b). However, considering that mothers who are themselves stressed bring up these newborns, it is more appropriate to define the stress as perinatal rather than simply prenatal, and to take into account postnatal maternal factors as well. Also, maternal factors other than hormones may contribute to the long-term changes in HPA reactivity in the offspring. Indeed, the state of pregnancy and lactation brings about a range of physiological and behavioral changes in the mother, and while the impact of maternal stress on behavior in the offspring has been well documented, few studies have explored its effect on the pregnant rodents themselves. We have shown that PRS can persistently affect maternal behavior from delivery to weaning (21 days after the end of the stress). Stressed mothers seem to be more anxious and to show impaired coping in inescapable situations. They also exhibit a decrease in the corticosterone response to novelty. The restraint stress procedure results in reduced body weight gain in pregnant females. A similar effect has also been observed in the post-stress period, suggesting that body weight differences between stressed and unstressed dams persist until the weaning of their pups. Interestingly, no prolonged effects of chronic restraint stress have been reported in nulliparous females (Darnaudéry et al. 2004a). These results indicate that pregnancy constitutes a period of high vulnerability to stress, and that the effects of this stress on female behavior are long lasting. This long-lasting effect of PRS is also the reason why it is preferable to refer to the stress as "perinatal" rather than simply "prenatal". The results mentioned above are in agreement with previous data from the literature regarding the impairment of maternal behavior following gestational stress (Fride et al. 1985; Poltyrev and Weinstock 1999; Neumann et al. 1998; Smith et al. 2004; Champagne and Meaney 2006). Such behavioral disturbances affect maternal care during the lactating period and can contribute to the long-term effects of perinatal stress on the offspring. Indeed, a previous report has shown that early adoption, which enhances maternal behavior (Maccari et al. 1995; Darnaudéry et al. 2004b), prevents prenatal-stress-induced impairments in glucocorticoid feedback (Maccari et al. 1995). We have also observed that PRS increases maternal anxiety in dams subjected to such a procedure (Darnaudéry et al. 2004a). Furthermore, postnatal handling of the offspring results in changes in maternal behavior (Bell et al. 1971; Lee and Williams 1975; Francis et al. 1999) and contributes to improved behavioral performances by the handled offspring. Indeed, early postnatal factors seem to play an important role in the impact of PRS, since adoption, postnatal handling, environmental enrichment (both pre- and postnatal), and disruption of maternal behavior all strongly modify it (Maccari et al. 1995; Koo et al. 2010; Lemaire et al. 2006; Weaver et al. 2004; Champagne and Meaney 2006). The notion that epigenetic mechanisms, which exert lasting control over gene expression without altering the genetic code, could mediate stable changes in brain function (Tsankova et al. 2007) provides a possible explanation for these changes.

2.2 Early Life Events and Circadian Systems

The circadian system is essential for the optimal functioning and adaptation of the organism to the dramatic changes in the environment imposed by earth rotation on its axis. Indeed, the circadian system maintains temporal synchronization not only between the organism and daily changes in the physical environment but also between diverse internal physiological processes (Turek and Van Reeth 1988). Most biochemical, physiological, or behavioral processes within the organism fluctuate on a regular basis throughout the 24-hour day. These daily rhythms arise from an internal time-keeping system, the circadian clock, located in the hypothalamic suprachiasmatic nuclei (SCN). In the absence of any environmental input, these rhythms persist with a period of about 24 h and are therefore referred to as "circadian rhythms." For many years, the circadian system was thought to be only sensitive to changes in the light–dark (LD) cycle. However, it has been shown that circadian function and/or sleep patterns can also be reset by neurochemical or behavioral stimuli. Among those stimuli, steroids can have marked effects on the functioning of the circadian system, and chronic stress in adult rats can induce changes in sleep patterns and circadian rhythms (Baker and Driver 2007; Cutolo et al. 2006; Smith et al. 2005; Young et al. 2004). Interestingly, we demonstrated a negative correlation between the corticosterone response to an acute stress and the ability for the circadian clock to adapt to an acute chronobiological stress (jet lag; Weibel et al. 2002). Perinatal events seem to have complex influences on the long-term development of circadian function. For instance, it is known that abnormal development of visual system in anophthalmic mice have clear effects on the integrity of the SCN and on the development of behavioral rhythms. Restricted access to the natural mother has been shown to shift endogenous corticosterone rhythm of rat pups. Despite the fact that postnatal events have complex influences on the long-term development of the circadian clock, very little attention has been given to the possible effects of prenatal stress on circadian clock function. Previous studies reported that sustained hypoxia in adults affects the functioning of the internal clock located in the suprachiasmatic nuclei. We have hypothesized that gestational hypoxia, found in many pathological conditions in pregnant women and naturally associated with life at high altitude, can cause long-lasting consequences for the synchronization of the circadian clock to light. Using gestational hypoxia as a prenatal stress procedure, we demonstrated that, once adult, rats born to hypoxic mothers had significant alterations in their circadian rhythm of locomotor activity (recorded in freely accessible running wheels). Under a regular LD cycle, they showed a phase advance of their rhythm of activity and were less active than control rats. After an abrupt 6-hour phase delay in the LD cycle, rats from the prenatal hypoxic group took significantly more time to resynchronize to the new LD cycle compared to controls. Under constant darkness, prenatal hypoxic group and control rats had a similar period of activity but the response of prenatal hypoxic group rats to a light pulse in the early subjective night was less marked than that of control rats. When submitted to acute restraint stress, prenatal hypoxic group rats had a prolonged secretion of corticosterone compared to controls. These results indicate that prenatal hypoxia is a factor that has long-lasting consequences for the functional output of the biological clock and the hormonal response to stress (Joseph et al. 2002).

Our previous data provide evidence for the long-term effect of a PRS procedure on circadian locomotor activity (Maccari and Morley-Fletcher 2007). More recently, in order to clarify relationships between PRS, gender, and the circadian system, we have monitored the running wheel behavior in male and female adult PRS rats, first under a LD cycle and then after an abrupt 6-h advance shift in the LD cycle. The locomotor activity pattern of both male and female PRS rats was significantly more erratic and fragmented compared to controls. PRS differentially alters the circadian activity rhythm in male and female rats. PRS increases total activity in males and decreases it in females. Also, PRS induced a significant phase advance in the circadian activity rhythm only in male rats and increased the time required to resynchronize the activity rhythm after an abrupt phase advance of the LD cycle but to a larger extent in females than in males (Fig. 2.3). These observations highlight the importance of the circadian system in the gender-specific outcome observed in animals exposed to prenatal stressful events and suggest a gender difference in the response of the circadian clock to light (unpublished data). Concerning hormonal rhythms, we previously have demonstrated that PRS induces an increase in corticosterone secretion at the end of the light period in both males (Fig. 2.4) and females, and hypercorticism throughout the diurnal cycle in females only (Koehl et al. 1997, 1999). Interestingly, most of the PRS consequences seem to be sex specific. PRS-reduced hippocampal neurogenesis, metabotropic glutamate (mGlu) receptors expression, and brain-derived neurotrophic factor (BDNF) only in males (Zuena et al. 2008) and chronic ethanol treatment increased mGlu1 receptors in PRS males (Van Waes et al. 2009) while it reduced it in PRS females (Van Waes et al. 2011). However, prolonged hypothalamo–pituitary–adrenal (HPA) axis response to stress measured as corticosterone secretion is observed in PRS rats of both genders (Bowman et al. 2004; Koehl et al. 1999). These effects might be mediated, at least in part, by a reduction in hippocampal mineralocorticoid receptor (MR)/glucocorticoid receptor (GR) at specific times of the day (Koehl et al. 1999). Results also show that prepartum stress could alter the pattern of corticosterone secretion in pregnant females gestational. As for the response to stress, altered circadian corticosterone activity is associated with disturbed circadian behavioral cycles including increased REM sleep and sleep fragmentation (Dugovic et al. 1999). Male PRS rats exhibited a significant increase in the amount of paradoxical sleep (PS) over the 24-h recording

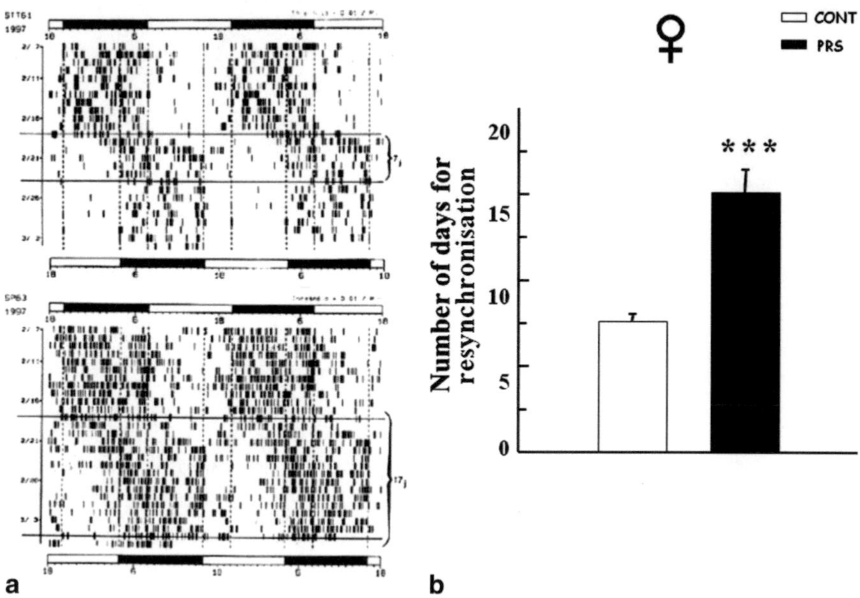

Fig. 2.3 **a** Representative activity records of the circadian running wheel activity from female, control (*upper panel*), and prenatal restraint stress (*lower panel*) (*PRS*) rats subjected to abrupt 6-h phase advanced of the LD cycle (on day 7 on these records). **b** Time required to resynchronize circadian rhythm of running wheel activity to the new LD cycle. $***P<0.001$ versus the respective controls. (Original data are reported in Mairesse et al. 2007a, b; Poster SFN)

session (Fig. 2.5). Other changes include an increased sleep fragmentation and light slow wave sleep (SWS 1) time, and a slight decrease in the percentage of deep slow wave sleep (SWS 2) relative to total sleep time. The present results demonstrate that exposure to PRS can produce long-term and selective changes in both the structure and the continuity of sleep. Although there were preliminary reports of abnormal "sleep-like behaviors" in prenatal stress monkeys and prenatal stress humans, our data provide the first polygraphic demonstration of long-term effects of PRS on the sleep–wake cycle when the animals reach adulthood.

The sleep–wake cycle that is dramatically modified by PRS (Dugovic et al. 1999; Mairesse et al. 2013), with a significant increase in the amount of REM sleep over the 24-hour recording session, positively correlated to plasma corticosterone levels. Remarkably, a number of behavioral alterations (i.e., immobility time in the forced swimming test) observed in PRS are positively correlated to enhanced corticosterone response to stress. Also, we found a high negative correlation between the extent of depolarization-evoked glutamate release in the ventral hippocampus and anxiety-like behavior in both control and PRS rats (Fig. 2.6). Strategies aimed at restoring HPA axis functioning, i.e., environmental enrichment during adolescence (Morley-Fletcher et al. 2003a), or pharmacological manipulations during adulthood (Morley-Fletcher et al. 2011; Marrocco et al. 2012), correct the associated behavioral patterns (Fig. 2.7).This is extremely relevant since it provides evidence that

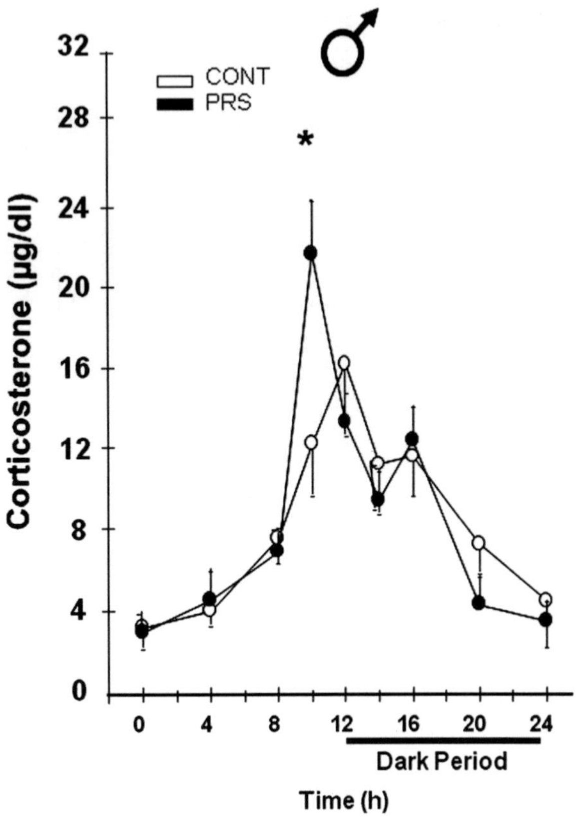

Fig. 2.4 Circadian secretion of corticosterone in male rats is shown. *$p<0.05$ versus the respective controls. *PRS* prenatal restraint stress. (Original data are reported in Koehl et al. 1999)

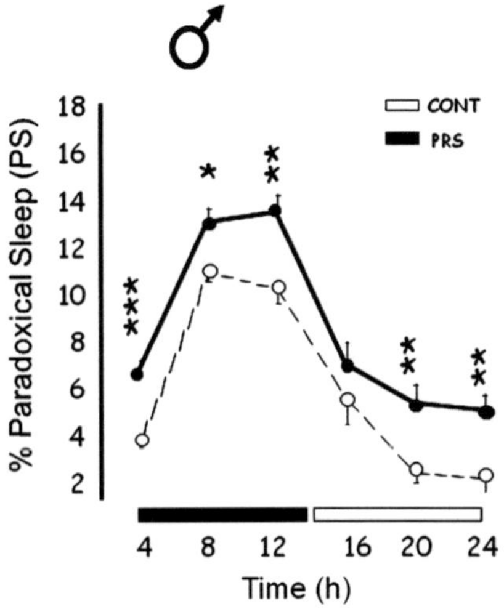

Fig. 2.5 Over 24 h, paradoxical sleep or rapid eyes movements in controls and prenatal restraint stress (PRS) rats. *$p< 0.05$; **$p< 0.01$; ***$p< 0.001$ (two-tailed unpaired Student's t test) for between-group comparisons. (Original data are reported in Dugovic et al. 1999)

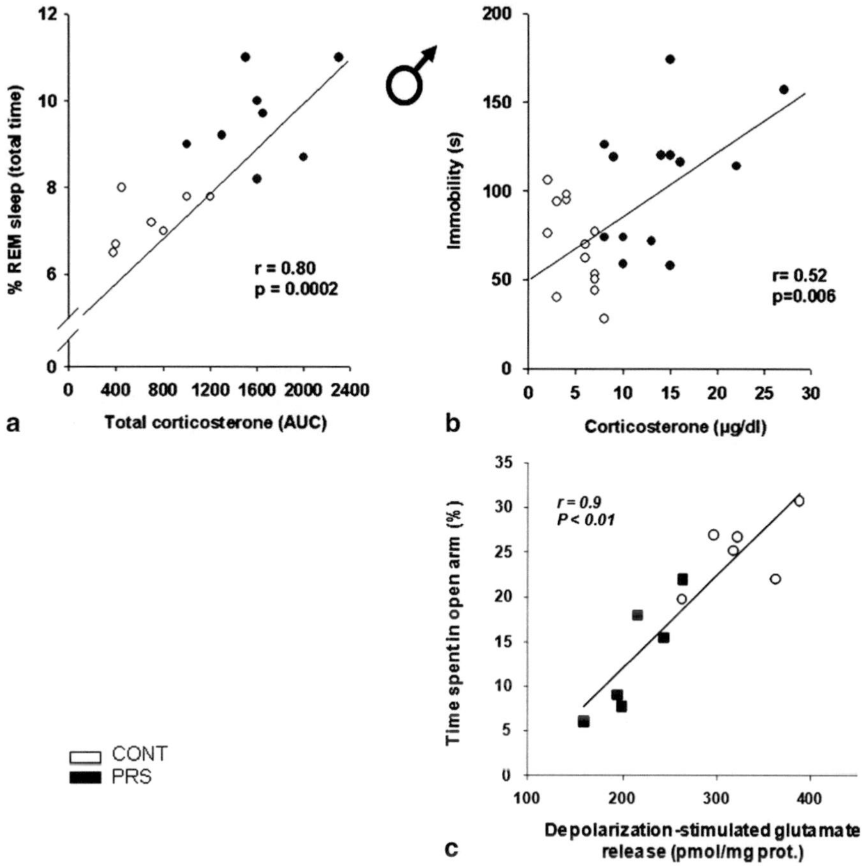

Fig. 2.6 **a** Positive correlation between rapid eye movement (*REM*) sleep and area under curve (*AUC*) of corticosterone secretion after response to stress. **b** Positive correlation between immobility in the forced swimming test (depression-like behavior) and corticosterone secretion after response to stress. **c** Positive correlation between time spent in open arm (anxiety-like behavior) and depolarization-stimulated glutamate release in the ventral hippocampus. r=coefficient of Pearson's correlation analysis. *PRS* prenatal restraint stress. (Original data are reported in Dugovic et al. 1999)

glutamate release in the ventral hippocampus is associated with anxiety, a stress-related disorder (Marrocco et al. 2012)

2.3 PRS, an Animal Model of Depression

These are evidently concerned with the neurobiological symptoms of depression, and not the syndrome of depression, which includes mood aspects that clearly cannot be reproduced in the rat model. In fact, it is quite obvious that the etiology of a

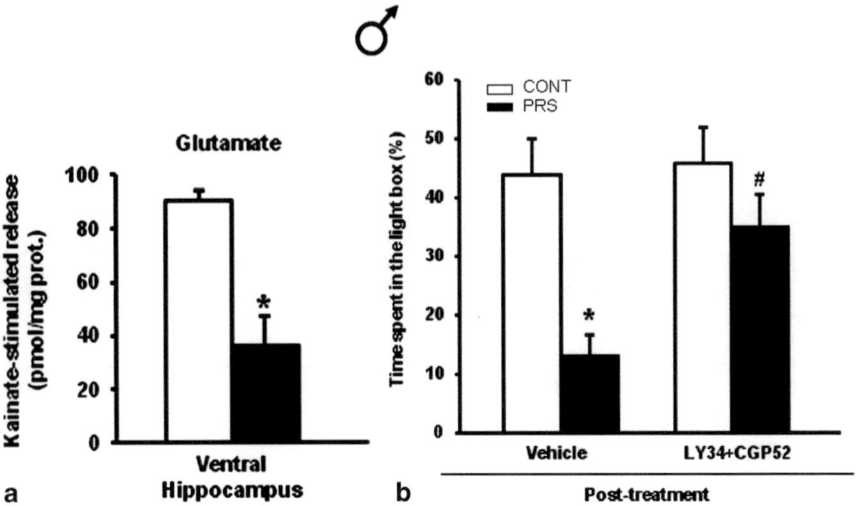

Fig. 2.7 **a** Prenatal restraint stress (*PRS*) rats display reduced depolarization-evoked glutamate release selectively in the ventral hippocampus. **b** Microinfusion in the ventral hippocampus of a cocktail that inhibits both mGluR2/3 receptors and GABAB receptors reduces their anxiety-like behavior in the Elevated-Plus Maze. *$p<0.05$ or $p<0.01$ versus the respective controls. (Original data are reported in Marrocco et al. 2012)

complex psychiatric human disease like depression cannot be reduced merely to a few biological changes exhibited by rats. Despite this consideration, the usefulness of an animal model lies in its capacity to improve our knowledge of the mechanisms underlying the actions of antidepressants on specific symptoms, with the goal of developing more appropriate and better-targeted drugs and to increase their efficacy on those specific symptoms.

As recently reviewed by Popoli et al. (2012), acute exposure to stress or treatment with glucocorticoids enhances glutamate release in the hippocampus, amygdala, and prefrontal cortex, three brain regions that are critically involved in the pathophysiology of psychiatric disorders. Our group has proven increasing evidence of an involvement of the glutamate transmission and machinery in particular at the level of metabotropic glutamate receptors in the neuroplastic programming and anxious phenotype. PRS rats show a reduced expression and function of group-I and group-II metabotropic glutamate receptors in the hippocampus (Zuena et al. 2008; Van Waes et al. 2009; Morley-Fletcher et al. 2011; Laloux et al. 2012). Hippocampal levels of mGlu1 and mGlu5 receptors are already reduced in infant PRS rats at postnatal day 10, whereas expression of mGlu2/3 receptors declined only after weaning (Laloux et al. 2012). We have recently shown that PRS causes a selective impairment of glutamate release in the ventral hippocampus, a brain region that specifically encodes memories related to stress and emotions (Fanselow and Dong 2010). The reduction in the evoked release of glutamate found in the ventral hippocampus of PRS rats was not due to an impaired glutamate synthesis in presynaptic terminals because it was also seen in synaptosomes preloaded with

D-[^{3}H]-aspartate, a nonmetabolizable analogue of glutamate, but was very probably associated to the marked reduction of synaptic vesicle proteins found in the ventral hippocampus (Marrocco et al. 2012). Thus, PRS would lead to an impairment of the intrinsic machinery of exocytotic glutamate release in the ventral hippocampus. This profile of expression of vesicle-associated proteins nicely fits with the finding that glutamate release was reduced in the ventral hippocampus, but not in the dorsal hippocampus of PRS rats. Remarkably, at least two proteins that have been found to be reduced in the ventral hippocampus of PRS rats, i.e., munc-18 and Rab3A are specifically associated to the regulation of glutamate release. Accordingly, munc-18 regulates the size of the readily releasable vesicle pool in glutamatergic but not GABAergic terminals (Augustin et al. 1999), and Rab3A is preferentially, albeit not exclusively, expressed at glutamatergic terminals (Geppert et al. 1994). Therefore, one of the most striking findings of our study is that PRS had profound effects on glutamate release, but it failed to affect gamma-aminobutyric acid (GABA) release. Our data suggest that PRS causes an imbalance between excitatory and inhibitory neurotransmission in the ventral hippocampus, an effect that might perturb cognitive functions related to stress and emotions (reviewed by Bannerman et al. 2004; Fanselow and Dong 2010). Presynaptic alterations in the glutamate/GABA balance have been associated with anxiety, depressive-like behavior, and memory impairment (Tordera et al. 2007; Garcia-Garcia et al. 2009; Chen et al. 2010). Thus, the imbalance between excitatory and inhibitory neurotransmission in the ventral hippocampus might contribute to explain the anxious/depressive phenotype of PRS rats (Vallée et al. 1997; Zuena et al. 2008; Morley-Fletcher et al. 2011). The lack of changes in glutamate release in the dorsal hippocampus is in agreement with the finding that PRS rats do not show abnormalities in spatial memory unless they are of >10 months of age (Vallée et al. 1999; Van Waes et al. 2009). At this old age, PRS rats show changes in the hippocampal expression and activity of group-I metabotropic glutamate receptors (Van Waes et al. 2009), which are localized postsynaptically and are involved in the regulation of learning and memory (reviewed by Nicoletti et al. 2011). We cannot exclude that changes in postsynaptic glutamate receptors in the hippocampus contribute to the behavioral phenotype of PRS rats particularly during ageing. The involvement of postsynaptic glutamate machinery in the dorsal hippocampus in the alterations of spatial memory has been reported in other models of stress during prenatal or juvenile/adult stage (Yaka et al. 2007; Schmidt et al. 2010). The value of PRS rats as a model of depression is strengthened by the finding that these animals show a persistent deficit in hippocampal neurogenesis and abnormalities in transcription factors and surface receptors that have been related to the pathophysiology of depression and anxiety (Darnaudéry et al. 2006; Lemaire et al. 2000; Schmitz et al. 2002; Zuena et al. 2008).

PRS rats display biobehavioral alterations that can parallel to some extent indices in human depression research, thus becoming a useful tool for the design and testing of new pharmacologic strategies in mood and sleep disorders. The criteria proposed by Willner and Mitchell (2002) require that animal models of depression exhibit face, predictive, and construct validity. Face validity refers to the phenomenological similarity, whereas predictive validity refers to the accuracy of a model in forecasting the course and outcome of a human syndrome. Finally, construct validity represents the degree to which both the human syndrome and the animal model are unambiguously defined such that a rational theory can be constructed to explain the pathophysiology of a disorder. However, because mental disorder is a human pathology, the perfect homology of an animal model to a human psychiatric condition cannot be absolutely demonstrated. In contrast, it is possible to use animal models to highlight some similar symptoms and develop new pharmacological strategies. Various clinical observations in humans suggest a possible pathophysiological link between depression and disturbances in hypothalamus–pituitary axis, circadian rhythmicity, body temperature fluctuations, various peripheral hormone concentrations, and urinary levels of neurotransmitter metabolites (Holsboer 2001). Added to our previous findings in PRS rats of high anxiety and emotionality, dysfunction of the HPA axis, and circadian timing abnormalities, the observation of long-term changes in their sleep structure (Dugovic et al. 1999; Mairesse et al. 2013) supports the validity of the PRS model as a valid animal model of anxiety/depression.

Our group has provided increasing evidence for the predictive validity of PRS model by means of chronic treatment with different classes of antidepressants in adult rats (Morley-Fletcher et al. 2003b, 2004, 2011). Indeed, imipramine (tricyclic) and tianeptine (a selective serotonin reuptake enhancer, structurally similar to the tricyclic antidepressants) reverse several PRS-induced alterations at the behavioral, neurochemical, and neuroanatomical level. Thus, after antidepressant treatment, PRS rats displayed reduced immobility behavior in the forced swim test, increased exploration of the open arm in the Elevated-Plus Maze (EPM), enhanced mineralocorticoid and glucocorticoid receptors densities in the hippocampus, and modified 5-HT_{1A} mRNA expression (Morley-Fletcher et al. 2003b, 2004). We have also tested the therapeutic efficacy of agomelatine, a novel antidepressant that behaves as a mixed MT1/MT2 melatonin receptor agonist/5-HT_{2c} serotonin receptor antagonist (Morley-Fletcher et al. 2011). Agomelatine treatment corrected *all* abnormalities displayed by PRS rats, suggesting that the drug impacts mechanisms that lie at the core of the maladaptive programming induced by PRS (Fig. 2.8). We wish to highlight that agomelatine had no effect on any of the parameters we have tested in control rats. This suggests that agomelatine, at least in PRS rats, acts as an etiopathogenetic drug and its action is specific to the pathological state (i.e., agomelatine behaves as a "disease-dependent" drug). In response to agomelatine, the ventral hippocampus of PRS rats may reacquire the ability to link stress-related events that occur simultaneously or close in time and to separate recent and remote stress-related memories, a mechanism that may critically affect resilience to stress.

Because of its peculiar pharmacological profile, agomelatine caters the potential to correct the abnormalities of circadian rhythms associated with mood disorders, including abnormalities of the sleep/wake cycle. Hence, we examined the effect of chronic agomelatine treatment on sleep architecture and circadian rhythms of motor activity in PRS male rats (Mairesse et al. 2013). We found a reduced duration of Slow-Wave Sleep (SWS), an increased duration of rapid eye movement (REM) sleep, an increased number of REM sleep events, and an increase in motor activity before the beginning of the dark phase of the LD cycle. In addition, adult PRS rats showed an increased

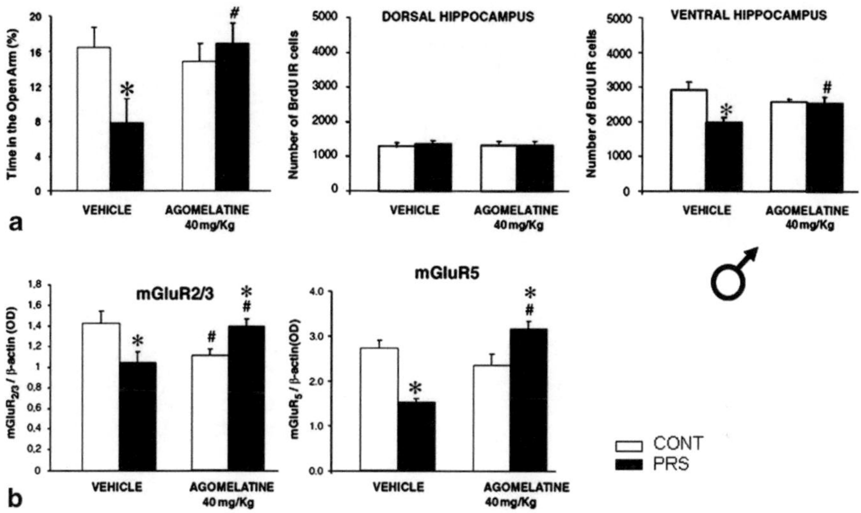

Fig. 2.8 **a** A chronic treatment with the antidepressant agomelatine reduces anxiety-like behavior of prenatal restraint stress (*PRS*) rats in the elevated-plus maze (EPM) and reverses *PRS*-induced deficit in hippocampal neurogenesis specifically in the ventral hippocampus, a key region involved in anxiety regulation. **b** Chronic treatment with agomelatine increases mGluR5 and mGluR2/3 in the hippocampus of adult *PRS* rats. $*p<0.05$ versus the respective controls or # versus *PRS* rats treated with vehicle. (Original data are reported in Morley-Fletcher et al. 2011)

expression of the transcript of the primary response gene, c-Fos, in the hippocampus just prior to the beginning of the dark phase. All these changes were reversed by a chronic oral treatment with agomelatine (2000 ppm in the diet; Fig. 2.9). The e.9ct of agomelatine on sleep was largely attenuated by treatment with the MT1/MT2 melatonin receptor antagonist S22153, which caused PRS-like sleep disturbances on its own (Mairesse et al. 2013). These data provided the first evidence that agomelatine corrects sleep architecture and restores circadian homeostasis in a preclinical model of depression and supports the value of agomelatine as a novel antidepressant that resynchronizes circadian rhythms under pathological conditions.

2.4 Conclusions

These results suggest that PRS leads to long-term disturbances both in predictive adaptations to the earth's rotation (circadian rhythms) and especially in the adaptation to stress (stress response); interestingly, these alterations persist all along the life span. In other words, PRS rats present an "allostatic load" or a high cost for their adaptation in the face of challenges, expressed by increased anxiety, depression-like behavior, vulnerability to drugs and impaired learning during ageing. The studies conducted so far by our group and others indicate that the face as well as the predictive value of the PRS model is high, since several abnormalities observed in the PRS rats

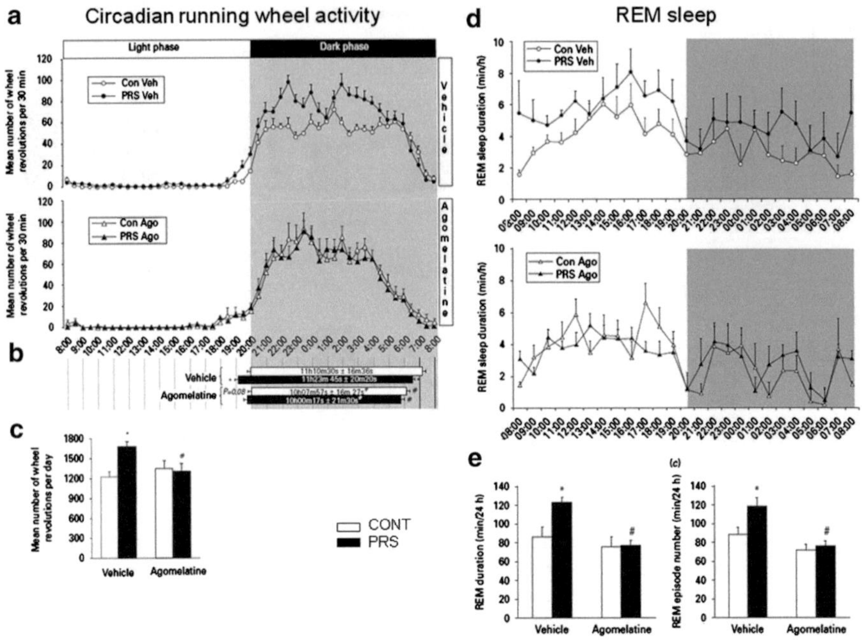

Fig. 2.9 Agomelatine restores circadian homeostasis and corrects sleep architecture in a preclinical model of depression, prenatally restraint stressed rats. **a** Circadian profile of running wheel activity (mean from 8 consecutive days), for control (*Con*) and prenatal restraint stress (*PRS*) animals, after treatment with virgin food (*upper panel*) or with food containing agomelatine (2000 ppm; *lower panel*). **b** Circadian characteristics of the running wheel activity including the onset, the offset of the activity, and the duration of the period of activity. **c** Levels of total activity per days. Rapid eye movement (*REM*) sleep profile over 24 h, *left panel*. **d** *REM* sleep circadian profile. **e** *REM* sleep duration per day. **f** *REM* sleep episode numbers per day. Values are means ¡S.E.M. of six rats per group. *$p<0.05$, PRS versus Con rats; $^{\#}p<0.05$, agomelatine (*Ago*) versus virgin food (*Veh*). (Original data are reported in Mairesse et al. 2013)

parallel those found in human depression. Indeed, like depressed patients, PRS rats escape from the feedback inhibition responsible for returning corticosterone secretion to basal levels after stress. Abnormalities in a variety of overt circadian rhythms, including the cortisol rhythm, have been documented in depressed patients. Also, alteration in the sleep–wake cycle is a hallmark of human depression, with changes including a shortened PS latency, an increase in the amount and frequency of PS during the first part of the night, increased sleep fragmentation, and a decrease in the amount of SWS. In contrast to other stress-related animal models of depression, the persistence of all induced abnormalities after stressor removal in PRS rats could be seen as particularly advantageous for the design and testing of new therapeutical strategies in mood and sleep disorders. Finally, considering that maternal mediators act during a critical period in development, it could be postulated that the high maternal corticosterone levels and altered maternal behavior following PRS contribute to the long-term effects described in the offspring, through epigenetic modifications.

Conflict of Interest The authors declare no conflicts of interest.

References

Alonso SJ, Arevalo R, Afonso D, Rodriguez M (1991) Effects of maternal stress during pregnancy on forced swimming test behavior of the offspring. Physiol Behav 50:511–517

Augustin I, Rosenmund C, Südhof TC, Brose N (1999) Munc13-1 is essential for fusion competence of glutamatergic synaptic vesicles. Nature 400(6743):457–461

Baker FC, Driver HS (2007) Circadian rhythms, sleep, and the menstrual cycle. Sleep Med 8(6):613–622

Bannerman DM, Rawlins JN, McHugh SB, Deacon RM, Yee BK, Bast T, Zhang WN, Pothuizen HH, Feldon J (2004) Regional dissociations within the hippocampus-memory and anxiety. Neurosci Biobehav Rev 28(3):273–283

Barbazanges A, Piazza PV, Le Moal M, Maccari S (1996) Maternal glucocorticoid secretion mediates long-term effects of prenatal stress. J Neurosci 16:3943–3949

Barker DJ (1999) Fetal origins of cardiovascular disease. Ann Med 31(Suppl 1):3–6

Bell RW, Nitschke W, Gorry TH, Zachman TA (1971) Infantile stimulation and ultrasonic signaling: a possible mediator of early handling phenomena. Dev Psychobiol 4(2):181–191

Bowman RE, MacLusky NJ, Sarmiento Y, Frankfurt M, Gordon M, Luine VN (2004) Sexually dimorphic effects of prenatal stress on cognition, hormonal responses, and central neurotransmitters. Endocrinology 145:3778–3787

Champagne FA, Meaney MJ (2006) Stress during gestation alters postpartum maternal care and the development of the offspring in a rodent model. Biol Psychiatry 59:1227–1235

Chen G, Henter ID, Manji HK (2010) Presynaptic glutamatergic dysfunction in bipolar disorder. Biol Psychiatry 67(11):1007–1009

Cutolo M, Sulli A, Pizzorni C, Secchi ME, Soldano S, Seriolo B, Straub RH, Otsa K, Maestroni GJ (2006) Circadian rhythms: glucocorticoids and arthritis. Ann N Y Acad Sci 1069:289–299.

Dahlof LG, Hard E, Larsson K (1978) Influence of maternal stress on the development of the fetal genital system. Physiol Behav 20:193–195

Darnaudéry M, Maccari S (2008) Epigenetic programming of the stress response in male and female rats by prenatal restraint stress. Brain Res Rev 57(2):571–585

Darnaudéry M, Dutriez I, Viltart O, Morley-Fletcher S, Maccari S (2004a) Stress during gestation induces lasting effects on emotional reactivity of the dam rat. Behav Brain Res 153:211–216

Darnaudéry M, Koehl M, Barbazanges A, Cabib S, Le Moal M, Maccari S (2004b) Early and later adoptions differently modify mother-pup interactions. Behav Neurosci 118:590–596

Darnaudéry M, Perez-Martin M, Belizaire G, Maccari S, Garcia-Segura LM (2006) Insulin-like growth factor 1 reduces age-related disorders induced by prenatal stress in female rats. Neurobiol Aging 27:119–127

Day JC, Koehl M, Deroche V, Le Moal M, Maccari S (1998) Prenatal stress enhances stress- and corticotropin-releasing factor-induced stimulation of hippocampal acetylcholine release in adult rats. J Neurosci 18:1886–1892

de Kloet ER, Sibug RM, Helmerhorst FM, Schmidt M (2005) Stress, genes and the mechanism of programming the brain for later life. Neurosci Biobehav Rev 29:271–281

Dugovic C, Maccari S, Weibel L, Turek FW, Van Reeth O (1999) High corticosterone levels in prenatally stressed rats predict persistent paradoxical sleep alterations. J Neurosci 19:8656–8664

Fanselow MS, Dong HW (2010) Are the dorsal and ventral hippocampus functionally distinct structures? Neuron 65(1):7–19

Francis DP, Davies LC, Coats AJ (1999) Handling complexity in oxygen delivery in the univentricular circulation. Circulation 100(2):213–214

Fride E, Weinstock M (1984) The effects of prenatal exposure to predictable or unpredictable stress on early development in the rat. Dev Psychobiol 17:651–660

Fride E, Dan Y, Gavish M, Weinstock M (1985) Prenatal stress impairs maternal behavior in a conflict situation and reduces hippocampal benzodiazepine receptors. Life Sci 36:2103–2109
Garcia-Garcia AL, Elizalde N, Matrov D, Harro J, Wojcik SM, Venzala E, Ramírez MJ, Del Rio J, Tordera RM (2009) Increased vulnerability to depressive-like behavior of mice with decreased expression of VGLUT1. Biol Psychiatry 66(3):275–282
Geppert M, Bolshakov VY, Siegelbaum SA, Takei K, De Camilli P, Hammer RE, Südhof TC (1994) The role of Rab3A in neurotransmitter release. Nature 369(6480):493–497
Holsboer F (2001) Stress, hypercortisolism and corticosteroid receptors in depression: implications for therapy. J Affect Disord 62(1–2):77–91
Joseph V, Mamet J, Lee F, Dalmaz Y, Van RO (2002) Prenatal hypoxia impairs circadian synchronisation and response of the biological clock to light in adult rats. J Physiol 543:387–395
Koehl M, Barbazanges A, Le Moal M, Maccari S (1997) Prenatal stress induces a phase advance of circadian corticosterone rhythm in adult rats which is prevented by postnatal stress. Brain Res 759:317–320
Koehl M, Darnaudéry M, Dulluc J, Van Reeth O, Le Moal M, Maccari S (1999) Prenatal stress alters circadian activity of hypothalamo-pituitary-adrenal axis and hippocampal corticosteroid receptors in adult rats of both gender. J Neurobiol 40:302–315
Koo JW, Russo SJ, Ferguson D, Nestler EJ et al (2010) Nuclear factor-kappaB is a critical mediator of stress-impaired neurogenesis and depressive behavior. Proc Natl Acad Sci U S A 107:2669–2674
Laloux C, Mairesse J, Van Camp G, Giovine A, Branchi I, Bouret S, Morley-Fletcher S, Bergonzelli G, Malagodi M, Gradini R, Nicoletti F, Darnaudéry M, Maccari S (2012) Anxiety-like behaviour and associated neurochemical andendocrinological alterations in male pups exposed to prenatal stress. Psychoneuroendocrinology 37(10):1646–1658
Lee MH, Williams DI (1975) Long term changes in nest condition and pup grouping following handling of rat litters. Dev Psychobiol 8:91–95
Lemaire V, Koehl M, Le Moal M, Abrous DN (2000) Prenatal stress produces learning deficits associated with an inhibition of neurogenesis in the hippocampus. Proc Natl Acad Sci U S A 97:11032–11037
Lemaire V, Lamarque S, Le Moal M, Piazza PV, Abrous DN (2006) Postnatal stimulation of the pups counteracts prenatal stress-induced deficits in hippocampal neurogenesis. Biol Psychiatry 59:786–792
Maccari S, Morley-Fletcher S (2007) Effects of prenatal restraint stress on the hypothalamus-pituitary-adrenal axis and related behavioural and neurobiological alterations. Psychoneuroendocrinology 32(Suppl 1):S10–S15
Maccari S, Piazza PV, Kabbaj M, Barbazanges A, Simon H, Le Moal M (1995) Adoption reverses the long-term impairment in glucocorticoid feedback induced by prenatal stress. J Neurosci 15:110–116
Mairesse J, Viltart O, Salome N, Giuliani A, Catalani A, Casolini P, Morley-Fletcher S, Nicoletti F, Maccari S (2007a) Prenatal stress alters the negative correlation between neuronal activation in limbic regions and behavioral responses in rats exposed to high and low anxiogenic environments. Psychoneuroendocrinology 32:765–776
Mairesse J, Lesage J, Breton C, Breant B, Hahn T, Darnaudéry M, Dickson SL, Seckl J, Blondeau B, Vieau D, Maccari S, Viltart O (2007b) Maternal stress alters endocrine function of the fetoplacental unit in rats. Am J Physiol Endocrinol Metab 292(6):E1526–1533
Mairesse J, Silletti V, Laloux C, Zuena AR, Giovine A, Consolazione M, Van Camp G, Malagodi M, Gaetani S, Cianci S, Catalani A, Mennuni G, Mazzetta A, van Reeth O, Gabriel C, Moca-Laloux C, Zuena AR, Giovine A, Consolazione M, (2013) Chronic agomelatine treatment corrects the abnormalities in the circadian rhythm of motor activity and sleep/wake cycle induced by prenatal restraint stress in adult rats. Int J Neuropsychopharmacol 16(2):323–338
Marrocco J, Mairesse J, Ngomba RT, Silletti V, Van Camp G, Bouwalerh H, Summa M, Pittaluga A, Nicoletti F, Maccari S, Morley-Fletcher S (2012) Anxiety like behavior of prenatally stressed rats is associated with a selective reduction of glutamate release in the ventral hippocampus. J Neurosci 32(48):17143–17154

Morley-Fletcher S, Rea M, Maccari S, Laviola G (2003a) Environmental enrichment during adolescence reverses the effects of prenatal stress on play behaviour and HPA axis reactivity in rats. Eur J Neurosci 18:3367–3374

Morley-Fletcher S, Darnaudéry M, Koehl M, Casolini P, Van Reeth O, Maccari S (2003b) Prenatal stress in rats predicts immobility behavior in the forced swim test. Effects of a chronic treatment with tianeptine. Brain Res 989:246–251

Morley-Fletcher S, Darnaudéry M, Mocaer E, Froger N, Lanfumey L, Laviola G, Casolini P, Zuena AR, Marzano L, Hamon M, Maccari S (2004) Chronic treatment with imipramine reverses immobility behaviour, hippocampal corticosteroid receptors and cortical 5-HT(1) receptor mRNA in prenatally stressed rats. Neuropharmacology 47:841–847

Morley-Fletcher S, Mairesse J, Soumier A, Banasr M, Fagioli F, Gabriel C, Mocaer E, Daszuta A, McEwen B, Nicoletti F, Maccari S (2011) Chronic agomelatine treatment corrects behavioral, cellular, and biochemical abnormalities induced by prenatal stress in rats. Psychopharmacology 217(3):301–313

Neumann ID, Wigger A, Liebsch G, Holsboer F, Landgraf R (1998) Increased basal activity of the hypothalamo-pituitary-adrenal axis during pregnancy in rats bred for high anxiety-related behaviour. Psychoneuroendocrinology 23(5):449–463

Nicoletti F, Bockaert J, Collingridge GL, Conn PJ, Ferraguti F, Schoepp DD, Wroblewski JT, Pin JP (2011) Metabotropic glutamate receptors: from the workbench to the bedside. Neuropharmacology 60(7–8):1017–1041

Peters DA (1982) Prenatal stress: effects on brain biogenic amine and plasma corticosterone levels. Pharmacol Biochem Behav 17:721–725

Peters DA (1986) Prenatal stress: effect on development of rat brain serotonergic neurons. Pharmacol Biochem Behav 24:1377–1382

Peters DA (1988) Effects of maternal stress during different gestational periods on the serotonergic system in adult rat offspring. Pharmacol Biochem Behav 31:839–843

Peters DA (1990) Maternal stress increases fetal brain and neonatal cerebral cortex 5-hydroxytryptamine synthesis in rats: a possible mechanism by which stress influences brain development. Pharmacol Biochem Behav 35:943–947

Poltyrev T, Weinstock M (1999) Effect of gestational stress on maternal behavior in response to cage transfer and handling of pups in two strains of rat. Stress 3(1):85–95

Popoli M, Yan Z, McEwen BS, Sanacora G (2012) The stressed synapse: the impact of stress and glucocorticoids on glutamate transmission. Nat Rev Neurosci 13(1):22–37

Schmidt MV, Trümbach D, Weber P, Wagner K, Scharf SH, Liebl C, Datson N, Namendorf C, Gerlach T, Kühne C, Uhr M, Deussing JM, Wurst W, Binder EB, Holsboer F, Müller MB (2010) Individual stress vulnerability is predicted by short-term memory and AMPA receptor subunit ratio in the hippocampus. J Neurosci 30(50):16949–16958

Schmitz C, Rhodes ME, Bludau M, Kaplan S et al (2002) Depression: reduced number of granule cells in the hippocampus of female, but not male, rats due to prenatal restraint stress. Mol Psychiatry 7:810–813

Seckl JR (1998) Physiologic programming of the fetus. Clin Perinatol 25:939–962, vii

Seckl JR (2004) Prenatal glucocorticoids and long-term programming. Eur J Endocrinol 151(Suppl 3):U49–U62

Smith JW, Seckl JR, Evans AT, Costall B, Smythe JW (2004) Gestational stress induces postpartum depression-like behaviour and alters maternal care in rats. Psychoneuroendocrinology 29:227–244

Smith RG, Betancourt L, Sun Y (2005) Molecular endocrinology and physiology of the aging central nervous system. Endocr Rev 26(2):203–250 (Epub 23 Nov 2004. Review)

Takahashi LK, Kalin NH (1991) Early developmental and temporal characteristics of stress-induced secretion of pituitary-adrenal hormones in prenatally stressed rat pups. Brain Res 558:75–78

Thompson WR (1957) Influence of prenatal maternal anxiety on emotionality in young rats. Science 125:698–699

Tordera RM, Totterdell S, Wojcik SM, Brose N, Elizalde N, Lasheras B, Del Rio J (2007) Enhanced anxiety, depressive-like behavior and impaired recognition memory in mice with reduced expression of the vesicular glutamate transporter 1 (VGLUT1). Eur J Neurosci 25(1):281–290

Tsankova N, Renthal W, Kumar A, Nestler EJ (2007) Epigenetic regulation in psychiatric disorders. Nat Rev Neurosci 8:355–367

Turek FW, Van Reeth O (1988) Altering the mammalian circadian clock with the short-acting benzodiazepine triazolam. Trends Neurosci 11(12):535–541.

Vallée M, Mayo W, Dellu F, Le Moal M, Simon H, Maccari S (1997) Prenatal stress induces high anxiety and postnatal handling induces low anxiety in adult offspring: correlation with stress-induced corticosterone secretion. J Neurosci 17(7):2626–2636

Vallée M, Maccari S, Dellu F, Simon H, Le Moal M, Mayo W (1999) Long-term effects of prenatal stress and postnatal handling on age-related glucocorticoid secretion and cognitive performance: a longitudinal study in the rat. Eur J Neurosci 11:2906–2916

Van Waes V, Enache M, Dutriez I, Lesage J, Morley-Fletcher S, Vinner E, Lhermitte M, Vieau D, Maccari S, Darnaudéry M (2006) Hypo-response of the hypothalamic-pituitary-adrenocortical axis after an ethanol challenge in prenatally stressed adolescent male rats. Eur J Neurosci 24:1193–1200

Van Waes V, Enache M, Zuena A, Mairesse J, Nicoletti F, Vinner E, Lhermitte M, Maccari S, Darnaudéry M (2009) Ethanol attenuates spatial memory deficits and increases mGlu1a receptor expression in the hippocampus of rats exposed to prenatal stress. Alcohol Clin Exp Res 33(8):1346–1354

Van Waes V, Enache M, Berton O, Vinner E, Lhermitte M, Maccari S, Darnaudéry M (2011) Effect of prenatal stress on alcohol preference and sensitivity to chronic alcohol exposure in male rats. Psychopharmacology (Berl). Mar;214(1):197–208

Ward IL (1972) Prenatal stress feminizes and demasculinizes the behavior of males. Science 175:82–84

Weaver IC, Cervoni N, Champagne FA, D'Alessio AC, Sharma S, Seckl JR, Dymov S, Szyf M, Meaney MJ (2004) Epigenetic programming by maternal behavior. Nat Neurosci 7:847–854

Weibel L, Maccari S, Van RO (2002) Circadian clock functioning is linked to acute stress reactivity in rats. J Biol Rhythms 17:438–446

Willner P, Mitchell PJ (2002) The validity of animal models of predisposition to depression. Behav Pharmacol 13(3):169–188

Yaka R, Salomon S, Matzner H, Weinstock M (2007) Effect of varied gestational stress on acquisition of spatial memory, hippocampal LTP and synaptic proteins in juvenile male rats. Behav Brain Res 179(1):126–132

Young EA, Abelson J, Lightman SL (2004) Cortisol pulsatility and its role in stress regulation and health. Front Neuroendocrinol 25(2):69–76

Zuena AR, Mairesse J, Casolini P, Cinque C et al (2008) Prenatal restraint stress generates two distinct behavioural and neurochemical profiles in male and female rats. PLoS ONE 3(5):e2170

Chapter 3
Hormonal Modulation of Catecholaminergic Neurotransmission in a Prenatal Stress Model

María Eugenia Pallarés and Marta C. Antonelli

Abstract Our laboratory has a long-standing interest in the effects of *prenatal stress* (*PS*) on various neurotransmitter pathways and the morphology of the developing brain as well as in behavioural aspects of the offspring. Employing a commonly used PS paradigm in which the dams were subjected to restraint stress during the last week of gestation, we observed that several of these pathways were altered in the offspring brain. In this chapter, we will summarize and discuss the results obtained with the main catecholaminergic pathways, namely *dopamine* (*DA*) and *norepinephrine* (*NE*). In our hands, PS produces an increase in dopamine D2-type receptors in limbic areas, a decreased DA release after amphetamine stimulation in *prefrontal cortex* (*PFC*) and an increase in NE release in the same area of the adult offspring brain. In addition, DA uptake is altered at prepubertal stages that persist through adulthood. However, the expression of the step-limiting enzyme of the DA synthesis, *tyrosine hydroxylase* (*TH*), is only impaired at early stages of development after PS in the neuronal bodies. At the nuclear regulation level, dopaminergic transcription factors Nurr1 and Ptx3 showed a high vulnerability to PS showing changes along the lifespan. It was striking to observe that many impairments observed in most of these pathways differed depending on whether they were tested before or after puberty indicating a particular sensitivity of the systems to variations in gonadal hormones peaks. In fact, we observed that PS induced long-term effects on the male offspring reproductive system and spermatogenesis development, particularly by inducing a long-term imbalance of circulating sexual hormone levels. Our findings suggest that PS exerts long-term effects on various neurotransmitter pathways altering the normal connectivity between brain areas. Since the developing forebrain was shown to be influenced by androgen exposure, and PS was shown to disrupt prenatal testosterone surges, our results suggest that prenatal insults might be affecting the *organizational* role of androgens during brain development and differentially modulating their *activational* role during pubertal brain maturation.

M. C. Antonelli (✉) · M. E. Pallarés
Instituto de Biología Celular y Neurociencias, Facultad de Medicina,
Universidad de Buenos Aires, Buenos Aires, Argentina
e-mail: mca@fmed.uba.ar

M. C. Antonelli (ed.), *Perinatal Programming of Neurodevelopment*,
Advances in Neurobiology 10, DOI 10.1007/978-1-4939-1372-5_3,

3.1 Introduction

In the past, several reviews have summarized data that clearly demonstrate that exposure to different stressful events during the last week of pregnancy in rats interferes with the correct progeny development showing delays in motor development, impaired adaptation to stressful conditions, altered sexual behaviour and learning deficits (Weinstock 2001, 2008; Huizink et al. 2004; Darnaudery and Maccari 2008). In addition, the offspring display anomalies in neuronal development and brain morphology, as well as changes in cerebral asymmetry that persist into adulthood (Fride and Weinstock 1989). Evidences provided by animal research, as well as retrospective studies in humans, pointed out that exposure to adverse events in early life can alter adult behaviours and neurochemical indicators of several neurotransmitter pathways in the central nervous system.

In this chapter, we will concentrate our efforts in discussing the information available on the effect of prenatal stress (PS) on the catecholaminergic neurotransmission pathways bringing into attention the age-specific variations observed and the participation of sexual hormones in these variations. We will summarize the data available on the effects of PS on the dopaminergic and norepinephrinergic pathways but we will concentrate on the dopaminergic system when describing the interactions with the male hypothalamic–pituitary–gonadal axis.

3.2 Effects of Prenatal Stress on Central Catecholaminergic Pathways

3.2.1 Prenatal Stress and the Dopaminergic System

The dopaminergic metabolism is a well-studied system in brain that is mainly divided into four pathways whose neurons synthesize dopamine (DA) through a short synthetic sequence limited by the enzyme tyrosine hydroxylase (TH). DA is released in the terminals and exerts its action over two main subgroups of DA receptor: D1-like and D2-like. The D1-like receptor family contains the D1 and D5 receptor subtypes and the D2-like family contains the D2, D3 and D4 subtypes (for details see Baier et al. 2012).

Henry et al. (1995) were the first to evaluate the levels of DA receptors in the adult offspring after restraint PS and found that it increases the number of D2-like receptors while D3 receptors decrease in *nucleus accumbens* (*NAcc*). In our laboratory, we found that prenatal restraint stress enhances D2-like receptors in cortical areas and NAcc while no differences were found in D2-like receptors levels in motor areas (Berger et al. 2002). However, the changes in D2 limbic receptors produced by PS that persist until adulthood can be reverted when the progeny is manipulated early in life. We have shown that adoption of a prenatally stressed offspring by a control mother during the first postnatal days can reverse the increase of DA receptors (Barros et al. 2004).

Several reports have supported the notion that DA neurotransmission in the prefrontal cortex (PFC) controls subcortical DA. Moreover, there is now a large body of evidence showing that mesocortical DA exerts a tonic inhibitory influence on subcortical DA function. Brake et al. (2000) have shown that perinatal stress induces a mesocortical DA deficit and a subcortical DA hyperfunction. In line with these observations, we have reported that PS produces a decreased DA release after amphetamine stimulation in PFC of adult offspring (Carboni et al. 2010), suggesting that this cortical dopaminergic deficit might be triggering a NAcc hyperfunction and an overall dopaminergic imbalance in the prenatally stressed brain.

Since asymmetrical levels of DA have been reported in several rat brain areas (Rosen et al. 1984) and a right dominance is lost in prenatally stressed animals (Weinstock 2001), we analyzed D2 receptors in left and right brain hemispheres of adult male rats exposed to PS. Loss of asymmetries due to PS induces a higher rate of DA turnover in the right PFC and reduced DA activity in the right NAcc and left caudate, which were related to a reduced directional preference in prenatally stressed adult rats in response to amphetamine (Fride and Weinstock 1988). Hemispheric asymmetries are found throughout the brain but their correlation with function and behaviour are not well defined (LeMay 1999). D2 receptors show a left–right asymmetry in NAcc core (*NAcc-C*) that was selectively lost in PS offspring (Adrover et al. 2007). The asymmetrical and higher distribution of D2 receptors in right versus left sides of NAcc-C might reflect a constitutive left-biased difference in DA content between sides generating different levels of dopaminergic D2 receptors.

The precise anatomical localization and functional differentiation of dopaminergic neurons in the mammalian brain is a complex and multistep process. Significant progress has been made in trying to identify some transcriptional determinants which are regulating early developmental events such as fate specification, differentiation, migration, neurite growth, guidance pruning and synapse formation such as Nurr 1, Pitx3, engrailed 1 and 2 (En-1 and En-2) and Lmx1b (Smidt and Burbach 2007). Among these, Nurr1 and Pitx3 have been the most extensively studied *transcription factors* (*TF*). Nurr1 regulates proteins required for DA synthesis and transport (Zetterstrom et al. 1997; Saucedo-Cardenas et al. 1998; Wallen et al. 1999; Smits et al. 2003; Le et al. 1999; Baffi et al. 1999; Smidt and Burbach 2007; Weidong et al. 2009). Pitx3 influences the development of this specific *mesodiencephalic dopaminergic* (*mdDA*) subpopulation and seems to regulate the expression of TH.

In our hands, the expression of both Nurr1 and Pitx3 increased in prenatally stressed adult offspring in the *ventral tegmental area* (*VTA*), whereas no changes were observed in *substantia nigra* (*SN*) (Katunar et al. 2009). Pitx3 expression in the control group shows a significant decrease with age, whereas Nurr1 shows an increase that is significant between *postnatal days* (*PND*) 28 and 60. The different expression levels described for Pitx3 and Nurr1 in this study might be supporting the notion that Pitx3 has a prominent role at early stages in the postnatal development of the mdDA system, whereas Nurr1 plays a crucial role in adulthood probably in the maintenance of dopaminergic metabolism through the regulation of the expression of its key enzymes and transporters (Katunar et al. 2010). Growing evidence has lately emerged showing that Pitx3 and Nurr1 pathways are interconnected at a functional level (Jacobs et al. 2009).

Altogether, these results allow us to confirm that the changes observed in key TF as Nurr1 and Pitx3 are selectively found in the VTA area of the mesencephalon, indicating a major vulnerability of this limbic region to PS in accordance with the limbic selectivity found for the D2 receptors in prenatally stressed offspring (Berger et al. 2002).

3.2.2 Prenatal Stress and the Noradrenergic System

The noradrenergic system that utilizes NE as a neurotransmitter mainly originates in a relatively small number of cells located in the locus coeruleus and projects to the entire neuroaxis, from olfactory bulb to spinal cord. NE mainly exerts modulatory effects in several brain circuits and has been shown to be mediated, via different transduction mechanisms, by both alpha-1 and beta adrenergic receptors (Aghajanian and Rogawski 1983; Jiang et al. 1996; Waterhouse et al. 1990; Woodward et al. 1991). Dysregulation of noradrenergic neurotransmission has been implicated in stress-related psychiatric diseases such as depression, post-traumatic stress disorder and other anxiety disorders (Schatzberg and Schildkraut 1995; Southwick et al. 1993; Sullivan et al. 1999).

Relative few studies have addressed the effect of PS on the noradrenergic system. One of the earliest studies showed that a relatively mild protocol of PS exerted an increase in NE levels in various brain areas of infant rats (Peters 1982). Concomitantly, the same group showed that the alpha1 and beta noradrenergic receptors were reduced in cerebral cortex of PS offspring at PND 16, the same age that showed an elevation of NE levels (Peters 1984). The hypothalamus showed persistent changes throughout postnatal life (Peters 1982). Takahashi et al. (1992) found that PS adult rats showed reduced levels of NE in cerebral cortex but higher turnover rates in the same area. Moreover, Reznikov et al. (2003), found that PS eliminates sex-related differences in NE levels, increasing NE in preoptic hypothalamic areas of male PS brains to levels of intact females. Wilson et al. (2012) and Wilson and Terry (2013) reported that certain behavioural signs of PS adult rats can be improved by atomoxetine, an NE reuptake inhibitor.

In our hands, basal levels of NE in *NAcc Shell* (*NAcc-S*) is lower than in controls in prepubertal but not in PS adults rats (Silvagni et al. 2008). Moreover, basal levels of NE in PFC are lower, both in prepubertal and control offspring (Carboni et al. 2010). After an amphetamine challenge, levels of NE increase only in adult PNS rats in NAcc-S and PFC.

Taken together, all these studies indicate that PS significantly alters brain noradrenergic systems in an area- and age-specific manner.

3.3 Age-Specific Effects of Prenatal Stress on Catecholaminergic Pathways

As summarized in the above sections, both main catecholamine neurotransmitter systems show impairments in the brain of offspring subjected to PS. A detailed analysis of the results show that the impairments are evident all along the postnatal development of the offspring but the period of adolescence seems to be a pivotal stage after which some impairments become evident or are reversed. In agreement, it has been reported that several behavioural and biochemical alterations exerted by PS were seen only after puberty (Diaz et al. 1997; Henry et al. 1995). Since the hormonal changes are the hallmark of the adolescent period, sexual hormones become the first suspects of the differential effects of PS. In fact, the Pitx3 expression profile in prenatally stressed rats could be interpreted as a consequence of the gonadal hormone surge that might be exerting important challenges to the dopaminergic system during the pubertal period. When evaluating Nurr1 expression in the VTA of prenatally stressed rats at PND 28, we found a decline that was recovered after puberty, probably indicating an altered vulnerability to gonadal hormones (Katunar et al. 2010). In agreement with other authors, our hypothesis is that the perinatal events might render the catecholaminergic circuitries more vulnerable to puberty variation of the hormonal circulating levels. In turn, PS might exert a misbalanced hormonal milieu by altering the hypothalamic–pituitary–gonadal axis of the offspring. Catecholaminergic pathways are programmed during foetal stages, but we suggest that PS can modify this program that is further altered under the influence of the hormonal pattern in turn modified by the prenatal insult.

3.4 Adolescence and Neural Development

Adolescence and puberty are commonly used as similar concepts but they refer to different events of life. Although the timing of these periods overlaps, *adolescence* is defined as the gradual period of transition from childhood to adulthood whereas *puberty* refers to the precise moment where an individual physiologically matures to its reproductive stage. Therefore, adolescence is a longer period of life even though its precise onset and offset are difficult to determine with precision (Spear 2000).

The physiological changes that take place during adolescence are directed by the sexual hormones that regulate several neuroendocrine systems, modulate behaviours and stimulate (or suppress) the differentiation, as well as the plasticity of several neural populations. Hence, on peripheral tissues, they induce the appearance of secondary sexual characteristics which are necessary to achieve the reproductive maturation (Nussey and Whitehead 2001; Sato et al. 2008). On the other hand, their effects on the brain orchestrate several morphological and neurochemical modifications which are necessary for the adolescent brain to mature to an adult form (Spear 2000; Sato et al. 2008; Paus et al. 2008; MacLusky et al. 2006; Andersen

2003). During this phase, the brain awakens to pleasure, risk and other behavioural features that are common among adolescents of a variety of mammal species in order to provide to the individual, experience and information from the external environment which are necessary for its conception of the "adult world" (Spear 2000; Andersen 2003; Paus et al. 2008; Sato et al. 2008). Hormonal actions on several neurotransmitter pathways are very important for the expression of such behaviours. Sexual hormones were shown to regulate the synthesis and the release of neurotransmitters, and also to direct the remodelling of several synaptic circuits (Kuppers et al. 2000; Alonso-Solis et al. 1996). Especially, it was observed that they are directly implicated with the development of cognitive, motor and motivational processes related to dopaminergic neurotransmission, probably by inducing modifications at transcriptional and maturational levels (Purves-Tyson et al. 2012). In this respect, it was demonstrated in rats that the densities of both type 1 and 2 DA receptors in PFC of adult males increase from birth until PND 40, where a maximum is achieved. Thereafter the pruning process takes place, during which the number of receptors is diminished (58–75 %) to reach adult levels (Teicher et al. 1995). In humans, a similar phenomenon was reported (Huttunen and Niskanen 1978; Seeman et al. 1987). The changes in the number of DA receptors during adolescence parallels the increase and the decrease in the symptoms of *attention deficit hyperactivity disorder* (*ADHD*) and *Tourette syndrome* (*TS*). In addition, they also parallel with the adolescent androgen surges.

Therefore, adolescence is a dynamic period of neural development when behavioural circuits are refined and many of the developmental processes that occur during perinatal brain development are repeated during this phase of life: Processes such as neurogenesis, programmed cell death, pruning of dendritic arborizations and synapses and myelination and sexual differentiation take place (Sato et al. 2008). Perturbations in the timing of pubertal hormone influences on the adolescent brain would be predicted to have long-lasting consequences for adult behaviour, since adolescence is also a pivotal time on the aetiology of certain psychopathologies.

3.5 Gonadal Hormones in Brain Development

Physiologically, testosterone is mainly released to circulation by testes. However, minor concentrations are also released by the adrenal glands. The biological availability of testosterone in blood is regulated by the specific binding to globulins and albumin. Therefore, only a little percentage of the total testosterone released is available (0.5–2 %) to exert its effects on specific tissues (Bialek et al. 2004). Because of its steroid nature, testosterone is able to cross the hematoencephalic barrier to reach the brain. Conversely, it could be locally synthesised de novo from cholesterol (Celotti et al. 1997; Tsuruo 2005). Brain-specific steroidogenic enzymes, such as 5α-reductase or aromatase, catalyse testosterone local conversion to more potent metabolites such as *dihydrotestosterone* (*DHT*) or oestradiol. In addition, testosterone and its metabolites regulate the expression of the steroidogenic enzymes, which vary in their availability and brain distribution according to the developmental stage and sex of the individual (Tsuruo 2005).

Steroids exert their action by binding to two different types of specific receptors: *membrane receptors* mediate non-genomic fast mechanisms such as affection on specific membrane conductance or by executing second messenger cascades; or *intracellular nuclear receptors* that can act as ligand-dependent TF, by regulating gene expression which in turn modulate neurotransmitter synthesis and release and postsynaptic receptor sensitivities (Wendler et al. 2010; McCarthy 2011). Testosterone and DHT bind to *androgen receptors* (*ARs*) whereas oestrogens bind to *oestrogen receptors* (*ERs*) which are available in two different isoforms: ERα and ERβ. The brain distribution of both AR and ER receptors was described by biochemical and immunohistochemical methods. Sex-specific variations were reported: Higher AR protein levels were found in male than in female rats and its distribution includes hypothalamic nucleus, limbic regions (such as amygdala, cortex and hippocampus) and NAcc (Xiao and Jordan 2002; Bialek et al. 2004; Sato et al. 2008; Phillips-Farfan and Fernandez-Guasti 2009). In mesencephalic structures, AR presence was reported on the SN and VTA (Kritzer 1997). Brain distribution of both ER isoforms basically overlaps since they were both mainly found on the hypothalamic nucleus, the amygdala and less abundantly on PFC and hippocampal areas (McEwen and Alves 1999).

The role of oestrogens as neurotrophic, protector and antioxidant agents on brain was explored. Oestrogens promote the development of the hypothalamus, the hippocampus, several mesencephalic structures and the cortex and modulate apoptotic processes as well as synaptic formations (McEwen and Alves 1999). Moreover, the protective effects of oestrogens on neuronal substrates were studied in several neurodegenerative disease models. The deleterious effects induced by 6-hidroxydopamine (6-OHDA) on the nigrostriatal dopaminergic pathway, widely used as a model of Parkinson's disease, were reduced when the administration of the toxin was co-administered with oestrogens. Particularly, oestrogen modulates the release and the uptake of DA in the NAcc (Creutz and Kritzer 2002). Interestingly, Kuppers et al. (2000) and Dluzen (2000) found that neither testosterone nor DHT induced the same protective oestrogenic effect after 6-OHDA injection.

On the other hand, evidences from clinical studies support the notion that testosterone could also induce neuroprotective effects since it was demonstrated that the maintenance of physiological androgen level is important for reducing the incidence of depressive and neurodegenerative disorders such as Parkinson's or Alzheimer's diseases (Bialek et al. 2004; MacLusky et al. 2006). Testosterone neurotrophic effects were also reported: They promote and increase the neural differentiation of neuronal cultures (Beyer and Hutchison 1997). Androgens were shown to reduce neurotoxin-induced cellular death and they promote regenerative faculties on adult neurones which suffer damage (DonCarlos et al. 2003).

The modulation of dopaminergic pathways by gonadal hormones has long been demonstrated; and in this regard, Yang and Shieh (2007) postulated that the nigrostriatal and mesolimbic systems are modulated by circulating gonadal steroids. Moreover, according to Creutz and Kritzer (2004) the soma projections of midbrain dopaminergic systems contain high number of oestrogens and androgen receptors. One of the most important sexual dimorphic brain nuclei is the preoptic area of the hypothalamus where the distribution of dopaminergic neurons and fibres are also sex dependent (Beyer et al. 1991). Kawashima and Takagi (1994) studied the effects of testosterone and oestradiol on survival, process growth and dopaminergic function on cultured cells derived from neonatal rat preoptic area. These authors found that steroid administration stimulated basal level of DA synthesis and release in the hypothalamic culture cells. Moreover, neuronal survival was enhanced in the group treated with testosterone, as well as the length of the total processes and the number of branches (Kawashima and Takagi 1994). Several studies have investigated the regulation exerted by sex hormones on biosynthetic enzyme gene expression related to dopaminergic systems. Regulation of TH gene expression is an important mechanism underlying modulation of catecholamine biosynthesis and homeostasis. The action of steroid hormones on its expression is not well elucidated yet, and there is some discrepancy on its effects, but it was demonstrated that both ARs and ERs control TH gene expression by acting at the promoter level in a ligand-dependent manner or indirectly by changing levels of trophic factors (Maharjan et al. 2005; Jeong et al. 2006). Interestingly, other physiological signals, such as cold stress or glucocorticoids, also regulate its expression (Adler et al. 1999).

3.6 Effect of Prenatal Stress on the Male Hypothalamic–Pituitary–Gonadal Axis

As exposed above, gonadal hormones participate on both windows of brain maturation: (a) the *organizational* phase that takes place during critical periods of foetal development. In this phase, sex steroids promote cellular and molecular events which will determine sexual differentiation of the brain and (b) the *activational* phase during the onset of puberty: Oscillations in their circulating levels affect a wide variety of neuronal phenomena, ranging from cyclic remodelling of synaptic circuitry to transsynaptic modulation of neurotransmission (Alonso and Lopez-Coviella 1998).

The process of brain sexual differentiation in the rat is a complex phenomenon in which the androgens have a crucial role during specific developmental stages (see Chap. 18, Olvera-Hernández and Fernández-Guasti). During the brain differentiation period, perinatal testosterone surges are responsible for the *masculinisation* and *defeminisation* of the sexual behaviour, the establishment of gonadotropin secretion patterns and the constitution of central, as well as peripheral, sexual morphological indexes which are characteristic for each sex. It was demonstrated that if testosterone surges are disrupted, the morphology and physiology of dimorphic structures are *feminised*. Various studies have indicated that stress during critical windows of foetal brain development reduces the fertility and fecundity of such individual, inducing alterations in the sexual behaviour (Weinstock 2001; Pereira et al. 2006). In male rats, exposure to stress during gestation diminished the rate of copula and reduced the number of ejaculations. Moreover, feminine sexual behaviours such as "lordosis" and enhanced male partner preferences over receptive female were also observed (Shono and Suita 2003; Gerardin et al. 2005; Kapoor and Matthews 2011). Also, PS was shown to alter the tonic pattern of gonadotropin secretion inducing long-term modifications on the physiological concentrations in the serum of testosterone and *luteinising hormone* (*LH*) (Shono and Suita 2003; Gerardin et al. 2005; Rodriguez et al. 2007). As mentioned above, PS affects dimorphic brain sexual structures: The number of neurons at the bulbocavernosus nucleus and the *c-fos* activity at the preoptic area of males are reduced, similar to what is physiologically found in females. Furthermore, Viltart and Vanbesien-Mailliot (2007) reported an increase in neonatal mRNA levels of oestrogen receptors and diminished activity of the aromatase enzyme in the hypothalamus of male rats. Several parameters of genitalia morphology such as the anogenital distance and the time in which both testes descend on the scrotal sac was shown to be modified by maternal stress during gestation (Shono and Suita 2003; Barros et al. 2004; Gerardin et al. 2005; Rodriguez et al. 2007). These modifications were related to a malfunction of the foetal reproductive axis, since insufficient concentrations of androgens during the last week of gestation due to the administration of an anti-androgen (e.g. flutamide) feminised brain development and deregulated reproductive axis endocrine function at adulthood (Huhtaniemi 1995; Knickmeyer and Baron-Cohen 2006). In this sense, gestational stress, as well as prenatal administration of dexamethasone during the last week of gestation, diminished testosterone foetal surges which are fundamental for the expression and the maintenance of sexual differentiation (Ward and Weisz 1984; Scott et al. 2009). In agreement with this observation, Pereira et al. (2006) found that neonatal administration of testosterone propionate to prenatally stressed offspring reverses PS consequences on male anogenital distance, testosterone concentrations in serum in adult offspring and reproductive behaviour.

Increased glucocorticoid concentration exerts disruptive effects on the foetal capacity to synthesise testosterone by inducing apoptosis to Leydig cells. Moreover, glucocorticoids suppress the steroidogenic enzyme synthesis and activity (Gao et al. 2002; Hardy et al. 2005; Scott et al. 2009).

In our hands, PS induced a long-term effect on the male offspring reproductive system and spermatogenesis development. We found that the pituitary LH and follicle-stimulating hormone were decreased at PND 28 prenatally stressed rats. On the other hand, the analysis of testicular androgen concentrations in serum of PS rats revealed that total androgens (testosterone and its potent metabolite DHT) were increased from prepubertal to adult stages. The concentrations of the testosterone metabolite *5α-androstane-3b-17β diol (Diol),* increased in serum of PS offspring at PNDs 28 and 45. However, we found a reduction of specific testosterone concentrations in serum at PND 75 offspring. The increase of total androgen concentrations correlates with the increase in the expression of the enzyme 5α Reductase-1 mRNA levels found in PS prepubertal rats. Moreover, PS accelerated the spermatogenesis rate at PNDs 35 and 60 and was accompanied by an increase in the mean diameter of the seminiferous tubules in pubertal offspring. A reduction in Leydig cell number was observed at PNDs 35 and 60. Finally, AR expression was evaluated showing that PND 35 offspring had increased numbers of immunopositive Sertoli cells (Pallares et al. 2013). Taken together, we hypothesise that the untimely increase of Diol and total androgens together with the increase in AR levels in Sertoli cells could be responsible for the acceleration of the spermatogenesis process in pubertal PS rats. The spermatogenesis acceleration was unexpected in this hormonal context but in agreement with Schopper et al. (2012) we propose that advancing the reproductive maturity might well be a strategy to cope with a possible unfavourable environment anticipated by the stress suffered in utero.

3.7 Sexual Hormones, Prenatal Stress and Brain Development

Brain development is a complex process in which different maturational events are well regulated and take place in a chronological order: After neural progenitor cells are born they migrate to specific brain areas where, according to their fate, differentiate and develop specific morphological patterns. Dendrite and axon production, elongation and branching processes initiate inside the womb but continue after delivery and along the whole lifespan of the individual. Stimuli coming from the external environment modulate the morphology of such connections. Therefore, the brain is prepared to receive, decode and deliver information belonging to the outside world and to adjust its structure to the necessities that are perceived from the external life. In mammals, there are two delimitated dynamical periods of neural development that redefine behavioural circuits by pruning synaptic contacts and by reducing the number of neurotransmitter receptors. Those periods are: (a) *the perinatal life* (which in rodents includes the period from the last week of gestation to the first postnatal week) and (b) *the adolescence*. Because of the magnitude and the speed where such structural changes take place, both temporal windows of brain development are vulnerable to the presence of perturbations or insults that could induce long-term consequences to the individual by modifications in the brain architecture and neuronal functioning (Spear 2000; Andersen 2003; Paus et al. 2008; Sato et al. 2008; Champagne 2012). Early-life stress or adversity can modulate the programming typical neuronal functioning that may be maintained throughout an organism lifespan. Some of the final consequences of early *genetic × environmental* interactions is to shape brain development at the epigenetic level that can profoundly alter the overall health and well-being of an individual (Auger and Auger 2013). The data reviewed in this chapter demonstrate that individual differences in early-life experiences may underlie risk and resilience to disease. Another contributing risk factor is gender since it seems to play an important role in influencing the occurrence, severity and age of onset of a number of mental health disorders. Steroid hormones can elicit transient actions at the genome, but they can also induce lasting changes in cell number, migration patterns, phenotypical differentiation, as well as morphological differentiation of brain cells between the sexes. Therefore, steroid hormone concentrations during critical windows of brain development should be very important on the *genetic × environmental* modulation of brain maturation.

Midbrain dopaminergic system regulates diverse behavioural and cognitive functions which are critical for integrating mammalian responses and adaptations to the environment. Moreover, it is considered to be of particular interest for the pathophysiology of idiopathic psychiatric disorders (Biederman 2005). Considering the importance of dopaminergic neurotransmission in the mesocorticolimbic pathways and its relation to cognition, emotion, positive reinforcement, food intake and decision making, it is tempting to hypothesize that if the alterations observed in prenatally stressed animals models could be extrapolated to the effects of PS in humans, vulnerable limbic areas in conjunction with *genetic × environmental* factors might facilitate the development of schizophrenia, ADHD or drug abuse later in life. The age of onset of many of the DA-related cognitive pathologies (before puberty for ADHD and young adult for schizophrenia) might be supporting the notion that prenatal programming of a vulnerable limbic dopaminergic system might be incapable of managing the hormonal variations during puberty.

Despite the fact that the consequences of PS on brain development have been extensively explored lately, the mechanisms by which early stressful events differentially programme the brain are still unclear. However, the emerging studies of the *organizational/activational* role of sex steroids on the programming of brain development open new insights into the mechanism involved in the disruptive actions of early event exposure. Abnormal androgen concentrations during critical windows of brain development differentially programme the brain, increasing the incidence of psychopathologics later in life.

Conflict of Interest The authors declare no conflicts of interest.

References

Adler V, Yin Z, Tew KD, Ronai Z (1999) Role of redox potential and reactive oxygen species in stress signaling. Oncogene 18(45):6104–6111. doi:10.1038/sj.onc.1203128

Adrover E, Berger MA, Perez AA, Tarazi FI, Antonelli MC (2007) Effects of prenatal stress on dopamine D2 receptor asymmetry in rat brain. Synapse 61(6):459–462

Aghajanian GK, Rogawski MA (1983) The physiological role of alpha-adrenoceptors in the CNS: new concepts from single-cell studies. Trends Pharmacol Sci 4:315–317

Alonso R, Lopez-Coviella I (1998) Gonadal steroids and neuronal function. Neurochem Res 23(5):675–688

Alonso-Solis R, Abreu P, Lopez-Coviella I, Hernandez G, Fajardo N, Hernandez-Diaz F, Diaz-Cruz A, Hernandez A (1996) Gonadal steroid modulation of neuroendocrine transduction: a transynaptic view. Cell Mol Neurobiol 16(3):357–382

Andersen SL (2003) Trajectories of brain development: point of vulnerability or window of opportunity? Neurosci Biobehav Rev 27(1-2):3–18

Auger CJ, Auger AP (2013) Permanent and plastic epigenesis in neuroendocrine systems. Front Neuroendocrinol 34(3):190–197. doi:10.1016/j.yfrne.2013.05.003

Baffi JS, Palkovits M, Castillo SO, Mezey E, Nikodem VM (1999) Differential expression of tyrosine hydroxylase in catecholaminergic neurons of neonatal wild-type and Nurr1-deficient mice. Neuroscience 93(2):631–642. doi:10.1016/S0306-4522(99)00124-4

Baier CJ, Katunar MR, Adrover E, Pallares ME, Antonelli MC (2012) Gestational restraint stress and the developing dopaminergic system: an overview. Neurotox Res 22(1):16–32. doi:10.1007/s12640-011-9305-4

Barros VG, Berger MA, Martijena ID, Sarchi MI, Perez AA, Molina VA, Tarazi FI, Antonelli MC (2004) Early adoption modifies the effects of prenatal stress on dopamine and glutamate receptors in adult rat brain. J Neurosci Res 76(4):488–496

Berger MA, Barros VG, Sarchi MI, Tarazi FI, Antonelli MC (2002) Long-term effects of prenatal stress on dopamine and glutamate receptors in adult rat brain. Neurochem Res 27(11):1525–1533

Beyer C, Hutchison JB (1997) Androgens stimulate the morphological maturation of embryonic hypothalamic aromatase-immunoreactive neurons in the mouse. Brain Res Dev Brain Res 98(1):74–81. doi:10.1016/S0165-3806(96)00170-8

Beyer C, Pilgrim C, Reisert I (1991) Dopamine content and metabolism in mesencephalic and diencephalic cell cultures: sex differences and effects of sex steroids. J Neurosci 11(5):1325–1333

Bialek M, Zaremba P, Borowicz KK, Czuczwar SJ (2004) Neuroprotective role of testosterone in the nervous system. Pol J Pharmacol 56(5):509–518

Biederman J (2005) Attention-deficit/hyperactivity disorder: a selective overview. Biol Psychiatry 57(11):1215–1220. doi:10.1016/j.biopsych.2004.10.020

Brake WG, Sullivan RM, Gratton A (2000) Perinatal distress leads to lateralized medial prefrontal cortical dopamine hypofunction in adult rats. J Neurosci 20(14):5538–5543

Carboni E, Barros VG, Ibba M, Silvagni A, Mura C, Antonelli MC (2010) Prenatal restraint stress: an in vivo microdialysis study on catecholamine release in the rat prefrontal cortex. Neuroscience 168(1):156–166. doi:10.1016/j.neuroscience. 2010.03.046

Celotti F, Negri-Cesi P, Poletti A (1997) Steroid metabolism in the mammalian brain: 5alpha-reduction and aromatization. Brain Res Bull 44(4):365–375. doi:10.1016/S0361-9230(97)00216-5

Champagne FA (2012) Interplay between social experiences and the genome: epigenetic consequences for behavior. Adv Genet 77:33-57. doi:10.1016/B978-0-12-387687-400002-7

Creutz LM, Kritzer MF (2002) Estrogen receptor-beta immunoreactivity in the midbrain of adult rats: regional, subregional, and cellular localization in the A10, A9, and A8 dopamine cell groups. J Comp Neurol 446(3):288–300. doi:10.1002/cne.10207

Creutz LM, Kritzer MF (2004) Mesostriatal and mesolimbic projections of midbrain neurons immunoreactive for estrogen receptor beta or androgen receptors in rats. J Comp Neurol 476(4):348–362. doi:10.1002/cne.20229

Darnaudery M, Maccari S (2008) Epigenetic programming of the stress response in male and female rats by prenatal restraint stress. Brain Res Rev 57(2):571–585

Diaz R, Fuxe K, Ogren SO (1997) Prenatal corticosterone treatment induces long-term changes in spontaneous and apomorphine-mediated motor activity in male and female rats. Neuroscience 81(1):129–140

Dluzen DE (2000) Neuroprotective effects of estrogen upon the nigrostriatal dopaminergic system. J Neurocytol 29(5-6):387–399

DonCarlos LL, Garcia-Ovejero D, Sarkey S, Garcia-Segura LM, Azcoitia I (2003) Androgen receptor immunoreactivity in forebrain axons and dendrites in the rat. Endocrinology 144(8):3632–3638

Fride E, Weinstock M (1988) Prenatal stress increases anxiety related behavior and alters cerebral lateralization of dopamine activity. Life Sci 42(10):1059–1065

Fride E, Weinstock M (1989) Alterations in behavioral and striatal dopamine asymmetries induced by prenatal stress. Pharmacol Biochem Behav 32(2):425–430

Gao HB, Tong MH, Hu YQ, Guo QS, Ge R, Hardy MP (2002) Glucocorticoid induces apoptosis in rat leydig cells. Endocrinology 143(1):130–138

Gerardin DC, Pereira OC, Kempinas WG, Florio JC, Moreira EG, Bernardi MM (2005) Sexual behavior, neuroendocrine, and neurochemical aspects in male rats exposed prenatally to stress. Physiol Behav 84(1):97–104. doi:10.1016/j.physbeh. 2004.10.014

Hardy MP, Gao HB, Dong Q, Ge R, Wang Q, Chai WR, Feng X, Sottas C (2005) Stress hormone and male reproductive function. Cell Tissue Res 322(1):147–153. doi:10.1007/s00441-005-0006-2

Henry C, Guegant G, Cador M, Arnauld E, Arsaut J, Le Moal M, Demotes-Mainard J (1995) Prenatal stress in rats facilitates amphetamine-induced sensitization and induces long-lasting changes in dopamine receptors in the nucleus accumbens. Brain Res 685(1-2):179–186

Huhtaniemi I (1995) Molecular aspects of the ontogeny of the pituitary-gonadal axis. Reprod Fertil Dev 7(5):1025–1035

Huizink AC, Mulder EJ, Buitelaar JK (2004) Prenatal stress and risk for psychopathology: specific effects or induction of general susceptibility? Psychol Bull 130(1):115–142

Huttunen MO, Niskanen P (1978) Prenatal loss of father and psychiatric disorders. Arch Gen Psychiatry 35(4):429–431

Jacobs FM, van Erp S, van der Linden AJ, von Oerthel L, Burbach JP, Smidt MP (2009) Pitx3 potentiates Nurr1 in dopamine neuron terminal differentiation through release of SMRT-mediated repression. Development 136(4):531–540. doi:10.1242/dev.029769

Jeong JY, Einhorn Z, Mercurio S, Lee S, Lau B, Mione M, Wilson SW, Guo S (2006) Neurogenin1 is a determinant of zebrafish basal forebrain dopaminergic neurons and is regulated by the conserved zinc finger protein Tof/Fezl. Proc Natl Acad Sci U S A 103(13):5143–5148. doi:-10.1073/pnas.0600337103

Jiang M, Griff ER, Ennis M, Zimmer LA, Shipley MT (1996) Activation of locus coeruleus enhances the responses of olfactory bulb mitral cells to weak olfactory nerve input. J Neurosci 16(19):6319–6329

Kapoor A, Matthews SG (2011) Testosterone is involved in mediating the effects of prenatal stress in male guinea pig offspring. J Physiol 589(Pt 3):755–766. doi:10.1113/jphysiol.2010.200543

Katunar M, Saez T, Brusco A, Antonelli MC (2009) Inmunocytochemical expression of dopamine-related transcription factors Pitx3 and Nurr1 in prenatally stressed adult rats. J Neurosci Res 87:1014–1022

Katunar M, Saez T, Brusco A, Antonelli MC (2010) Ontogenetic expression of dopamine-related transcription factors and tyrosine hydroxylase in prenatally stressed rats. Neurotox Res 18(1):69–81

Kawashima S, Takagi K (1994) Role of sex steroids on the survival, neuritic outgrowth of neurons, and dopamine neurons in cultured preoptic area and hypothalamus. Horm Behav 28(4):305–312. doi:10.1006/hbeh.1994.1026

Knickmeyer RC, Baron-Cohen S (2006) Fetal testosterone and sex differences in typical social development and in autism. J Child Neurol 21(10):825–845

Kritzer MF (1997) Selective colocalization of immunoreactivity for intracellular gonadal hormone receptors and tyrosine hydroxylase in the ventral tegmental area, substantia nigra, and retrorubral fields in the rat. J Comp Neurol 379(2):247–260. doi:10.1002/(SICI)1096-9861(19970310)379:2<247::AID-CNE6>3.0.CO;2–3

Kuppers E, Sabolek M, Anders U, Pilgrim C, Beyer C (2000) Developmental regulation of glutamic acid decarboxylase mRNA expression and splicing in the rat striatum by dopamine. Brain Res Mol Brain Res 81(1-2):19–28. doi:10.1016/S0169-328X(00)00156-X

Le W, Conneely OM, He Y, Jankovic J, Appel SH (1999) Reduced Nurr1 expression increases the vulnerability of mesencephalic dopamine neurons to MPTP-induced injury. J Neurochem 73(5):2218–2221

LeMay M (1999) Functional and anatomical asymmetries of the human brain. Eur J Neurol 6(1):79–85

MacLusky NJ, Hajszan T, Prange-Kiel J, Leranth C (2006) Androgen modulation of hippocampal synaptic plasticity. Neuroscience 138(3):957–965. doi:10.1016/j.neuroscience.2005.12.054

Maharjan S, Serova L, Sabban EL (2005) Transcriptional regulation of tyrosine hydroxylase by estrogen: opposite effects with estrogen receptors alpha and beta and interactions with cyclic AMP. J Neurochem 93(6):1502–1514. doi:10.1111/j.1471-4159.2005.03142.x

McCarthy M (ed) (2011) Sex and the developing brain. Morgan & Claypool Life Sciences, San Rafael

McEwen BS, Alves SE (1999) Estrogen actions in the central nervous system. Endocr Rev 20(3):279–307

Nussey S, Whitehead S (2001) Endocrinology, an integrated approach. Bios Scientific Pulishers, Oxford

Pallares ME, Adrover E, Baier CJ, Bourguignon NS, Monteleone MC, Brocco MA, Gonzalez-Calvar SI, Antonelli MC (2013) Prenatal maternal restraint stress exposure alters the reproductive hormone profile and testis development of the rat male offspring. Stress 16(4):429–440. doi:10.3109/10253890.2012.761195

Paus T, Keshavan M, Giedd JN (2008) Why do many psychiatric disorders emerge during adolescence? Nat Rev Neurosci 9(12):947–957. doi:10.1038/nrn2513

Pereira OC, Bernardi MM, Gerardin DC (2006) Could neonatal testosterone replacement prevent alterations induced by prenatal stress in male rats? Life Sci 78(24):2767–2771. doi:10.1016/j.lfs.2005.10.035

Peters DA (1982) Prenatal stress: effects on brain biogenic amine and plasma corticosterone levels. Pharmacol Biochem Behav 17(4):721–725

Peters DA (1984) Prenatal stress: effect on development of rat brain adrenergic receptors. Pharmacol Biochem Behav 21(3):417–422

Phillips-Farfan BV, Fernandez-Guasti A (2009) Endocrine, neural and pharmacological aspects of sexual satiety in male rats. Neurosci Biobehav Rev 33(3):442–455. doi:10.1016/j.neubiorev.2008.11.003

Purves-Tyson TD, Handelsman DJ, Double KL, Owens SJ, Bustamante S, Weickert CS (2012) Testosterone regulation of sex steroid-related mRNAs and dopamine-related mRNAs in adolescent male rat substantia nigra. BMC Neurosci 13:95. doi:10.1186/1471-2202-13-95

Reznikov AG1, Nosenko ND, Tarasenko LV (2003) Participation of endogenous opioids in pathogenesis of early neuroendocrine manifestations of prenatal stress syndrome. Bull Exp Biol Med. May;135(5):421–423

Rodriguez N, Mayer N, Gauna HF (2007) Effects of prenatal stress on male offspring sexual maturity. Biocell 31(1):67–74

Rosen GD, Finklestein S, Stoll AL, Yutzey DA, Denenberg VH (1984) Neurochemical asymmetries in the albino rat's cortex, striatum, and nucleus accumbens. Life Sci 34(12):1143–1148

Sato S, Yin C, Teramoto A, Sakuma Y, Kato M (2008) Sexually dimorphic modulation of GABA(A) receptor currents by melatonin in rat gonadotropin-releasing hormone neurons. J Physiol Sci 58(5):317–322. doi:10.2170/physiolsci.RP006208

Saucedo-Cardenas O, Quintana-Hau JD, Le WD, Smidt MP, Cox JJ, De Mayo F, Burbach JP, Conneely OM (1998) Nurr1 is essential for the induction of the dopaminergic phenotype and the survival of ventral mesencephalic late dopaminergic precursor neurons. Proc Natl Acad Sci U S A 95(7):4013–4018

Schatzberg AF, Schildkraut JJ (1995) Recent studies on norepinephrine systems in mood disorders. In: Bloom FE, Kupfer DJ (ed) Psychopharmacology: the fourth generation of progress. Raven, New York, pp 911–920

Schopper H, Klaus T, Palme R, Ruf T, Huber S (2012) Sex-specific impact of prenatal stress on growth and reproductive parameters of guinea pigs. J Comp Physiol B 182(8):1117–1127. doi:10.1007/s00360-012-0680-9

Scott HM, Mason JI, Sharpe RM (2009) Steroidogenesis in the fetal testis and its susceptibility to disruption by exogenous compounds. Endocr Rev 30(7):883–925. doi:10.1210/er.2009-0016

Seeman P, Bzowej NH, Guan HC, Bergeron C, Becker LE, Reynolds GP, Bird ED, Riederer P, Jellinger K, Watanabe S et al (1987) Human brain dopamine receptors in children and aging adults. Synapse 1(5):399–404. doi:10.1002/syn.890010503

Shono T, Suita S (2003) Disturbed pituitary-testicular axis inhibits testicular descent in the prenatal rat. BJU Int 92(6):641–643. doi:10.1046/j.1464-410X.2003.04436.x

Silvagni A, Barros V, Mura M, Antonelli MC, Carboni E (2008) Prenatal restrain stress differentially modifies basal and stimulated dopamine and noradrenaline release in the rat nucleus accumbens shell: an "in vivo" microdialysis study. Eur J Neurosci 28(4):744–758

Smidt MP, Burbach JP (2007) How to make a mesodiencephalic dopaminergic neuron. Nat Rev Neurosci 8(1):21–32

Smits SM, Ponnio T, Conneely OM, Burbach JP, Smidt MP (2003) Involvement of Nurr1 in specifying the neurotransmitter identity of ventral midbrain dopaminergic neurons. Eur J Neurosci 18(7):1731–1738. doi:10.1046/j.1460-9568.2003.02885.x

Southwick SM, Krystal JH, Morgan CA, Johnson D, Nagy LM, Nicolaou A, Heninger GR, Charney DS (1993) Abnormal noradrenergic function in posttraumatic stress disorder. Arch Gen Psychiatry 50(4):266–274

Spear LP (2000) The adolescent brain and age-related behavioral manifestations. Neurosci Biobehav Rev 24(4):417–463. doi:10.1016/S0149-7634(00)00014-2

Sullivan GM, Coplan JD, Kent JM, Gorman JM (1999) The noradrenergic system in pathological anxiety: a focus on panic with relevance to generalized anxiety and phobias. Biol Psychiatry 46(9):1205–1218

Takahashi LK, Turner JG, Kalin NH (1992) Prenatal stress alters brain catecholaminergic activity and potentiates stress-induced behavior in adult rats. Brain Res 574(1-2):131–137

Teicher MH, Andersen SL, Hostetter JC Jr (1995) Evidence for dopamine receptor pruning between adolescence and adulthood in striatum but not nucleus accumbens. Brain Res Dev Brain Res 89(2):167–172

Tsuruo Y (2005) Topography and function of androgen-metabolizing enzymes in the central nervous system. Anat Sci Int 80(1):1–11. doi:10.1111/j.1447-073x.2005.00098.x

Viltart O, Vanbesien-Mailliot CC (2007) Impact of prenatal stress on neuroendocrine programming. ScientificWorldJournal 7:1493–1537. doi:10.1100/tsw.2007.204

Wallen A, Zetterstrom RH, Solomin L, Arvidsson M, Olson L, Perlmann T (1999) Fate of mesencephalic AHD2-expressing dopamine progenitor cells in NURR1 mutant mice. Exp Cell Res 253(2):737–746. doi:10.1006/excr.1999.4691

Ward IL, Weisz J (1984) Differential effects of maternal stress on circulating levels of corticosterone, progesterone, and testosterone in male and female rat fetuses and their mothers. Endocrinology 114(5):1635–1644

Waterhouse BD, Azizi SA, Burne RA, Woodward DJ (1990) Modulation of rat cortical area 17 neuronal responses to moving visual stimuli during norepinephrine and serotonin microiontophoresis. Brain Res 514(2):276–292

Weidong L, Shen C, Jankovic J (2009) Etiopathogenesis of Parkinson disease: a new beginning? Neuroscientist 15(1):28–35. doi:10.1177/1073858408319974

Weinstock M (2001) Alterations induced by gestational stress in brain morphology and behaviour of the offspring. Prog Neurobiol 65(5):427–451

Weinstock M (2008) The long-term behavioural consequences of prenatal stress. Neurosci Biobehav Rev 32(6):1073–1086

Wendler A, Baldi E, Harvey BJ, Nadal A, Norman A, Wehling M (2010) Position paper: rapid responses to steroids: current status and future prospects. Eur J Endocrinol 162(5):825–830. doi:10.1530/EJE-09-1072

Wilson CA, Terry AV Jr (2013) Variable maternal stress in rats alters locomotor activity, social behavior, and recognition memory in the adult offspring. Pharmacol Biochem Behav 104:47–61. doi:10.1016/j.pbb.2012.12.015

Wilson CA, Schade R, Terry AV Jr (2012) Variable prenatal stress results in impairments of sustained attention and inhibitory response control in a 5-choice serial reaction time task in rats. Neuroscience 218:126–137. doi:10.1016/j.neuroscience.2012.05.040

Woodward DJ, Moises HC, Waterhouse BD, Yeh HH, Cheun JE (1991) The cerebellar norepinephrine system: inhibition, modulation, and gating. Prog Brain Res 88:331–341

Xiao L, Jordan CL (2002) Sex differences, laterality, and hormonal regulation of androgen receptor immunoreactivity in rat hippocampus. Horm Behav 42(3):327–336. doi:10.1006/hbeh.2002.1822

Yang SC, Shieh KR (2007) Gonadal hormones-mediated effects on the stimulation of dopamine turnover in mesolimbic and nigrostriatal systems by cocaine- and amphetamine-regulated transcript (CART) peptide in male rats. Neuropharmacology 53(7):801–809. doi:10.1016/j.neuropharm.2007.08.007

Zetterstrom RH, Solomin L, Jansson L, Hoffer BJ, Olson L, Perlmann T (1997) Dopamine neuron agenesis in Nurr1-deficient mice. Science 276(5310):248–250

Chapter 4
Involvement of Nitric Oxide, Neurotrophins and HPA Axis in Neurobehavioural Alterations Induced by Prenatal Stress

Damian G. Maur, Cecilia G. Pascuan, Ana M. Genaro and Maria A. Zorrilla-Zubilete

Abstract Several studies suggest that negative emotions during pregnancy generate adverse effects on the cognitive, behavioural and emotional development of the descendants. The psychoneuroendocrine pathways involve the transplacentary passage of maternal glucocorticoids in order to influence directly on fetal growth and brain development.

Nitric oxide is a gaseous neurotransmitter that plays an important role in the control of neural activity by diffusing into neurons and participates in learning and memory processes. It has been demonstrated that nitric oxide is involved in the regulation of corticosterone secretion. Thus, it has been found that the neuronal isoform of nitric oxide synthase (nNOS) is an endogenous inhibitor of glucocorticoid receptor (GR) in the hippocampus and that nNOS in the hippocampus may participate in the modulation of hypothalamic–pituitary–adrenal axis activity via GR.

Neurotrophins are a family of secreted growth factors consisting of nerve growth factor (NGF), brain-derived neurotrophic factor (BDNF), neurotrophin 3 (NT3) and NT4. Although initially described in the nervous system, they regulate processes such as cell survival, proliferation and differentiation in several other compartments. It has been demonstrated that the NO–citrulline cycle acts together with BDNF in maintaining the progress of neural differentiation.

In the present chapter, we explore the interrelation between nitric oxide, glucocorticoids and neurotrophins in brain areas that are key structures in learning and memory processes. The participation of this interrelation in the behavioural and cognitive alterations induced in the offspring by maternal stress is also addressed.

M. A. Zorrilla-Zubilete (✉) · D. G. Maur · C. G. Pascuan · A. M. Genaro
Department of Pharmacology, School of Medicine, University of Buenos Aires, Buenos Aires, Argentina
e-mail: mariazorrillaz@gmail.com

D. G. Maur · C. G. Pascuan · A. M. Genaro · M. A. Zorrilla-Zubilete
Center of Pharmacological and Botanical Studies (CEFYBO-CONICET-UBA), Paraguay 2155, 15th Floor, C1121ABG, Buenos Aires, Argentina

M. C. Antonelli (ed.), *Perinatal Programming of Neurodevelopment*, Advances in Neurobiology 10, DOI 10.1007/978-1-4939-1372-5_4,

4.1 Introduction

Stress is defined as any situation capable of perturbing physiological or psychological homeostasis. Exposure to adverse situations affects an important number of aspects of our daily life. The stress response is thus referred to as 'allostasis', meaning 'maintaining stability through change' (McEwen 1998). Important interrelations between the hypothalamus–pituitary–adrenal axis, the sympathetic–parasympathetic efferent pathways and the chemical messengers (hormones, neurotransmitters, interleukins, neurotrophins) which make possible multiple communication for maintaining the homeostatic balances have been described. Prolonged stress can lead to disruption of these interactions and have several negative repercussions, ranging from impairments in learning and memory to enhanced neuronal cell death (McEwen 1999, 2007).

Exposure to adverse events early in life may profoundly affect brain development, leading to long-lasting effects on neuronal structures and behaviour and playing a role in the etiology of mood and anxiety disorders (de Lima et al. 2011). Recent research in humans has related psychosocial work stress during pregnancy to postnatal consequences, namely the decrease in birth weight and gestational time (Lee et al. 2011). Moreover, animal and human studies have linked prenatal stress (PS) with alterations in development and behavioural disorders, and suggest that the effects of PS can persist long after birth (Vallee et al. 1997; Zuena et al. 2008). In fact, PS can have long-lasting effects on the offspring immune system and in neurocognitive functions (Ruiz and Avant 2005). Many studies have associated different models of PS with deficits in offspring's development. These studies suggest that PS could predispose rats to behavioural abnormalities such as increased anxiety, greater tendency to drug addiction and depressive-like behaviour (Weinstock 2001; Maccari et al. 2003; Mueller and Bale 2006; Vallee et al. 1999). There has also been reported an alteration in circadian rhythms (Van Reeth et al. 1998). Behavioural studies have shown that PS offspring exhibit increased exploratory behaviour and increased locomotor activity (Vallee et al. 1997, Van den Hove et al. 2005) as well as impairments in learning and memory performance (Wu et al. 2007, Maur et al. 2007). Deficits in social behaviour are found in several neuropsychiatric disorders with a presumed developmental origin (Patin et al. 2005). The differential effect of PS on learning and memory ability of males and females probably has a physiological basis that is determined in intrauterine development. Not surprisingly, stress exerted in this period can differentially affect males and females since during intrauterine development there is a brain and neuronal differentiation characteristic of each sex.

The psychoneuroendocrine pathways involve the transplacentary passage of maternal glucocorticoids in order to influence directly on fetal growth and brain development. Stress and adaptation to stress require numerous homeostatic adjustments; among them the balance between oxidants and antioxidants seems to play a critical role (Liu and Mori 1999). Nitric oxide, a gaseous neurotransmitter that plays an important role in the control of neural activity, has an important role in the regulation of oxidative state, it has been found that NO has both pro-oxidant and antioxidant actions (Lipton 1999; Wink et al. 1999). Interestingly, NO is involved in the regulation of corticosterone secretion (Lopez-Figueroa et al. 1998).

Neurotrophins are a family of secreted growth factors consisting of nerve growth factor (NGF), brain-derived neurotrophic factor (BDNF), neurotrophin 3 (NT3) and NT4 that play important functions in the nervous system. It has been demonstrated that the NO–citrulline cycle acts together with BDNF for maintaining the progress of neural differentiation.

In the present chapter we explore the interrelation between nitric oxide, glucocorticoids and neurotrophins in brain areas that are key structures in learning and memory processes. The participation of this interrelation in the behavioural and cognitive alterations induced in the offspring by maternal stress is also addressed.

4.2 Nitric Oxide

Nitric oxide (NO) is the main product of the reaction catalysed by nitric oxide synthase (NOS) that converts L-arginine to L-citrulline (Moncada and Higgs 1993). The family of NOS, the enzymes that produce NO, consists of two different classes: constitutive and inducible isoforms (Dawson and Snyder 1994; Marletta 1994). Neuronal NOS (nNOS) and endothelial NOS (eNOS) are constitutive (cNOS) isoforms dependent on the transient influx of Ca^{2+} to activate calmodulin, which binds to the isoform to elicit enzyme activity. The inducible NOS (iNOS) is permanently bound to calmodulin, so it is not regulated by calcium concentration but is regulated transcriptionally (Nathan and Xie 1994; Di Monte et al. 1997). NOS is abundant in brain tissue (Marletta 1994; Araki et al. 1999; Muramatsu et al. 2003), nNOS primarily being expressed in neurons, eNOS in pyramidal cells and endothelial cells and iNOS in astrocytes, microglia and inflammatory cells (Moncada et al. 1991). NO is an important brain messenger released upon activation of the glutamate N-methyl-D-aspartate (NMDA) receptor and the subsequent Ca^{2+}-dependent activation of nNOS (Kiss and Vizi 2001). Moreover, this molecule also has important neuromodulatory roles as a retrograde intracellular messenger mediating cell-to-cell interactions in the brain including cell-mediated immune system, cerebral smooth muscle relaxation, inhibition of platelet aggregation, synaptic plasticity and more complex processes like learning (Bredt and Snyder 1994; Schulz et al. 1995). It has been demonstrated that NO is a retrograde signaling molecule that might be involved in learning and memory processes conserved through evolution (Edwards and Rickard 2007; Sullivan et al. 1997). NO participates in long-term potentiation (LTP) (Hopper and Garthwaite 2006) and other forms of synaptic plasticity in many different brain areas but where it is synthesized and how it acts remains controversial (for review see Feil and Kleppisch 2008). NO's signal transduction generally occurs through binding to soluble guanylyl cyclase (sGC)-coupled NO receptors and thus stimulating the synthesis of cGMP, with cGMP-dependent protein kinase being one of the downstream effectors leading to changes in synaptic strength (Feil et al. 2005).

In addition, a number of experimental studies have also demonstrated that NO and NO donors can enhance the basal release of several neurotransmitters in the mammalian brain, including dopamine, glutamate and acetylcholine (Lonart and Johanson 1992; Nathan and Xie 1994).

The production of NO might lead to either toxicity or neuroprotection depending on the level of NO, the location of NO production, the extent of oxidative stress and the type of neurodegenerative process. It has been found that NO has both pro-oxidant and antioxidant actions (Colasanti and Susuki 2000).

In general, it is accepted that a normal pathophysiologic response of the damaged tissue may involve controlled NO production and the inhibition of this response may interfere with the normal repair process. In a previous report we analysed the participation of NOS activity in the morphological and learning alterations induced by chronic mild stress (CMS) exposure in BALB/c mice (Palumbo et al. 2007). Our results indicate that CMS induces a reduction in NO production by nNOS. NO production by iNOS isoform was not detectable. The magnitude of oxidative stress, evaluated by reactive oxygen species production after excitotoxic levels of NMDA was increased in hippocampus of CMS mice. Moreover, ROS formation in the presence of nNOS inhibitor was higher in both control and CMS mice. Finally, treatment of mice with nNOS inhibitors resulted in behavioural alterations similar to those observed in CMS animals and in an important increment in ROS formation. These findings suggest a role for nNOS as a protection against injury that trigger tissue toxicity leading to memory impairment. Nevertheless, it has been suggested that a high production of NO after the induction of iNOS expression participates in neurodestructive events either via formation of the toxic species peroxynitrite as a result of the reaction of NO with superoxide anion or by S-nitrosylation of regulatory protein groups (Miranda et al. 2000; Zorrilla-Zubilete et al. 2010).

It has been demonstrated that prenatal restraint stress leads to offspring's cognitive impairment accompanied by a decrease in neuron number, and by increases in other parameters including nNOS expression, oxidative damage to mitochondrial DNA and damage to the hippocampal antioxidant system (Feil et al. 2005; Feil et al. 2008). In addition our group found that PS induces impairments in spatial memory and territory discrimination in adult rats. PS offspring also displayed alterations in cerebellar NOS expression and activity. Moreover, a correlation between spatial memory deficits and the increase in NOS activity was found, pointing to a role of cerebellar NO in the behavioural alterations induced by stress during early development stages. It has been described that PS disrupts the normal sequence of sexual differentiation reducing adult social behaviours, such as aggression and copulation. Miller et al. (1999) found a desregulated nNOS distribution in medial preoptic area and basolateral amygdala that could be involved in the demasculinization of male rats induced by PS.

Moreover, it has been described that NO is an important molecule contributing to the regulation of HPA axis (Stern 2004) which plays an important role in the mechanisms underlying behavioural alterations due to PS.

4.3 HPA Axis

The HPA axis plays a vital role in adaptation of the organism to homeostatic challenge. The activation of the HPA axis leads to rapid secretion of corticotropin-releasing hormone (CRH) and arginine-vasopressin (AVP) from the hypothalamic paraventricular nucleus (PVN), which in turn activates the secretion of adrenocorticotrophic hormone (ACTH) from the pituitary, which finally stimulates the secretion of glucocorticoids (cortisol in primates and corticosterone in rodents) from the adrenal glands. CRH is essentially synthesized in the parvocellular portion of the PVN where it is colocalized with AVP (Whitnall 1993). The activated HPA axis not only regulates body peripheral functions such as metabolism and immunity but also has profound effects on the brain. For example, glucocorticoids regulate neuronal survival, neurogenesis, learning and memories (Wong and Herbert 2006).

A bidirectional interaction between HPA axis and NO production has been shown. Evidence suggests that NO may modulate the release of the stress hormones ACTH and corticosterone, and NOS activity and transcription is increased in the limbic HPA axis following various stressful stimuli. Furthermore, following activation of the stress axis, glucocorticoids are thought to downregulate the transcription and activity of NOS via a feedback mechanism (Lopez-Figueroa et al. 1998).

4.3.1 Nitric Oxide in the Stress Axis

Lee et al. (2003) showed that the ability of increased NO levels in the brain to release ACTH and stimulate PVN neuronal activity is enhanced in adult male rats exposed to alcohol prenatally, supporting the hypothesis that alterations in HPA axis activity in adult offspring of alcohol-exposed dams may be related to changes in hypothalamic responsiveness to NO (Lee et al. 2003).

Activation of the HPA axis in response to psychological stress in the pregnant female could play an important role in the regulation of the stress response in the offspring in later life and may be responsible for some of their behavioural pathology.

There are several pieces of evidence that stress and anxiety during human pregnancy result in activation of the HPA axis: suggest that significantly higher plasma ACTH, β-endorphin and cortisol concentrations, particularly if the stress was of a chronic rather than episodic nature (Demyttenaiere et al. 1989; Wadhwa et al. 1996). Gestational stress not only activates the maternal HPA axis but can also cause an increase in the release of CRH from the placenta by catecholamines and cortisol (Petraglia et al. 1996) as well as by fetal hypoxia (Sug-Tang et al. 1992; see Fig. 4.1). In contrast to the negative feedback control that it exerts on the release of hypothalamic CRH, cortisol stimulates the release of the peptide from the placenta resulting in a positive feedback control. CRH can also cause the release of β-endorphin from the placenta (Chrousos et al. 1998), which may influence opioid activity in the fetus. Weinstock's group found that when PS and control rats were repeatedly exposed to the same open field, the controls no longer released significant amounts

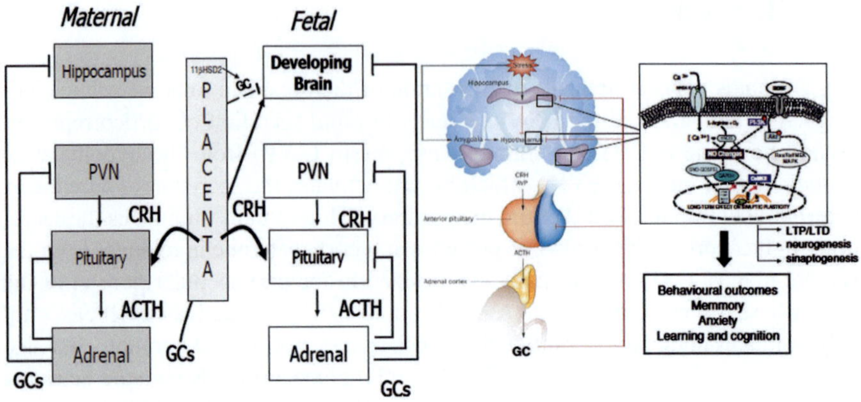

Fig. 4.1 Interaction between maternal and fetal HPA axis. *Left* panel shows a schematic representation of the interrelation between maternal and fetal HPA axis. Activation of the maternal HPA axis leads to an increase in circulating glucocorticoids which stimulate placental CRH release and also enter into fetal bloodstream and thus affects fetal brain development. Placental CRH acts directly on fetal pituitary modulating fetal HPA axis. *Central* panel shows an anatomic representation of the areas involved in the stress response in the adult prenatally stressed offspring. *Right* panel shows some of the molecular pathways altered by prenatal stress which would lead to changes in plasticity in the limbic regions involved in learning, memory and anxiety

of corticosterone after the fourth exposure, but the PS rats continued to release high amounts even after 8 consecutive days (Fride et al. 1985). This indicates that gestational stress could interfere with the normal adaptive process to mild novelty stress. There is a fast feedback by circulating glucocorticoids that inhibit the further release of CRH and ACTH through specific glucocorticoid receptors, designated type I (MR) and type II (GR), in the hippocampus and other brain regions. There are also slower feedback mechanisms, which involve suprahypothalamic neuronal systems and suppression of gene expression by a corticoid–receptor complex.

Rats exposed to stress during the last week of gestation have significantly decreased dendritic spine density in the anterior cingulate gyrus and orbitofrontal cortex (Murmu et al. 2006). Furthermore, prenatal exposure to glucocorticoids leads to increased adult CRH levels in the central nucleus of the amygdala, a key region in the regulation of fear and anxiety (Cratty et al. 1995). Exposure to PS has three major effects on adult behaviour: learning impairments, especially in aging rats (Valle et al. 1999; Maur et al. 2012); and increases in anxiety- and depression-related behaviours (Vallee et al. 1997).

4.3.2 *Neurotrophins*

Neurotrophic factors are peptides that promote neuronal development and differentiation, and are expressed in a particular period and region. Neurotrophic factors act in autocrine or paracrine manner and can be secreted by neurons that innervate tissues and by glial cells (Korsching 1993). The most important members of neurotrophic factors family are the neurotrophins. This family is composed of nerve growth factor (NGF), brain-derived neurotrophic factor (BDNF), neurotrophin 3 (NT3) and neurotrophin 4 (NT4) among others. The first neurotrophin to be characterized in vertebrates was NGF. Later BDNF, NT3 and NT4 were discovered. It is a fairly conserved group of proteins that share 50% homology (Binder and Scharfman 2004). Neurotrophins are synthesized as pro-hormones which are then cleaved giving rise to the 14-kDa mature form. Finally, two of them combine to form a homodimer by noncovalent bonds (Blum and Konnerth 2005). NGF acts in a retrograde manner, while NT3 and BDNF act both retrogade and anterograde (Korsching 1993).

Neurotrophins bind to a receptor called p75NTR (also known as TNFRSF16) and to one of the three tropomyosin-related kinase (TRK; also known as NTRK) receptors—NGF binds to TRKA, BDNF and NT4 bind to TRKB and NT3 binds to TRKC.

Through the differential expression and cellular localization of their receptors, neurotrophins can elicit diverse cellular functions in different types of neurons and at different cellular loci.

The TRK receptors are distributed all along the cell body and neurites. Binding of neurotrophins to the receptor induces conformational changes leading to receptor dimerization and activation (Blum and Konnerth 2005). It can trigger signaling cascades from the membrane or may be endocytosed and activate signaling cascades from the vesicle and then being retrogradely transported to the soma. The signaling pathways that can be triggered are (Huang and Reichardt 2001 and Van den Hove et al. 2012):

- The MAP kinase cascade, which is responsible for cell survival induced by neurotrophins and is vital for protection of neurons from oxidative stress and damage by cytokines. This cascade can be activated by growth factors or transcription factors, and may also be involved in the LTP process.
- The phospholipase C (PLCgamma) signalling pathway which leads to the depletion of Ca^{2+} from intracellular reservoirs and the subsequent activation of kinases. This pathway is also involved in the LTP process.
- Phosphatidylinositol 3′-kinase (PI3K) pathway which is largely responsible for NGF-induced survival. This phenomenon is involved in apoptotic cascade, inhibiting procaspase-9 and stimulating the synthesis of pro-apoptotic factor Bcl-2, among others.

An important finding in the neurotrophin's function was the discovery that the synthesis of neurotrophins in the brain is increased by both seizure activity and sensory stimulation. Indeed, active research over the past two decades has shown that neurotrophins regulate nearly all aspects of neural circuit development and function, including cell proliferation and differentiation, axon and dendrite growth, synaptogenesis and synaptic function and plasticity.

Both BDNF and NT3 play a role in the synapse. This is a very important function of BDNF, as a promoter of LTP. Hippocampal slices treated with BDNF or NGF and NT3 show a persistent increase in postsynaptic potential amplitude, similar to what happens with the induction of LTP (Kang and Schuman 1995). Conversely, in BDNF knockout mice tetanic stimulation fails to induce LTP, and this is reversed by the addition of exogenous BDNF (Korte et al. 1995; Patterson et al. 1996). Furthermore, electrical stimulations induce LTP and increase the expression of BDNF and NGF (Patterson et al. 1992). Some experiments demonstrate that the effect of BDNF on LTP would be at the presynapsis affecting neurotransmitter release, but there is also strong evidence of postsynaptic effects (McKay et al. 1999, Binder and Scharfman 2004). In postsynapsis, BDNF modulates the NMDA receptor and GABA channels that activate voltage-dependent sodium channels leading to the release of neurotransmitters from the dendrites (Blum and Konnerth 2005). It has been demonstrated that induction of the expression of BDNF in the hippocampus during contextual learning improves spatial learning (Binder and Scharfman 2004; Mizuno et al. 2000).

It has been demonstrated that BDNF is able to directly affect HPA axis activity (Tapia-Arancibia 2004; Jeanneteau et al. 2012). Moreover genetic disruption of GR in the PVN disinhibited both the HPA axis and the expression of hypothalamic BDNF (Jeanneteau et al. 2012). It was supported that BDNF has a key position in integrating neural, immune and endocrine responses to stress (Capoccia et al. 2013).

The neurotrophin gene BDNF is associated with several neuropsychiatric disorders, including depression, autism, bipolar disorder and schizophrenia, and it is an important gene of interest in stress research (Correia et al. 2010; Angelucci et al. 2005). BDNF, as previously mentioned, induces the survival, development, and function (Huang and Reichardt 2001) of selected neuronal populations of the peripheral and central nervous systems and participates in the modulation of dendritic growth and morphology (Binder and Scharfman 2004; Bibel and Barde 2000). In the later stages of nervous system development and in the adult BDNF regulates synaptic transmission and plasticity and acts as a central modulator of pain (Pezet and McMahon 2006).

BDNF regulates synaptic plasticity in neuronal networks involved in depressive behaviours (Pittenger and Duman 2007; Schinder and Poo 2000). Regulation of BDNF may reverse stress-induced deficits in structural and synaptic plasticity in the adult brain, resulting in cognitive flexibility and, subsequently, an increased ability to adapt/cope with environmental challenges that may precipitate or exacerbate depressive episodes (Schmidt et al. 2011).

A review hypothesizes that, during gestation, neurotrophins may play numerous roles in angiogenesis, energy homoeostasis, regulation of growth factor's actions

and development and maturation of the feto-placental unit (Mayeur et al. 2011).

Neurotrophins such as BDNF potentiate the placental development and play an important role in cytotrophoblast differentiation, proliferation and survival (Kawamura et al. 2009, 2011). Abnormal brain development in a compromised prenatal and/or early postnatal environment is thought to be a risk factor for several neurological disorders (Rehn and Rees 2005; Numakawa et al. 2010; Dhobale et al. 2012). Reports indicate that maternal BDNF reaches the fetal brain through utero-placental barrier and might contribute to its development (Kodomari et al. 2009). It has been previously shown that PS in the last week of fetal gestation can produce multiple endophenotypic changes related to schizophrenia in the Sprague-Dawley rat strain (Kinnunen et al. 2003; Koenig et al. 2005; Lee et al. 2007; Taylor et al. 2002).

Some mechanisms have been found to regulate BDNF levels during the gestation.

Thus, levels of BDNF are known to be regulated by omega-3 fatty acids (docosahexaenoic acid: DHA; Wu et al. 2004).

DHA is a structural component of the plasma membrane and reductions in DHA can have a direct influence on the function of the membrane. Disruptions in membrane fluidity due to DHA deficiency have been suggested to lower BDNF and signalling through TrkB in the rat brain (Bhatia et al. 2011). Dhobale et al. (2012) found a negative association between levels of placental TrkB and docosahexaenoic acid (DHA) and a negative association between maternal plasma BDNF levels and placental weight.

Other mechanism proposed for the modulation of BDNF during pregnancy is the regulation by glucocorticoids. It has been reported that GR (glucocorticoid receptor) interacts with receptor tyrosine kinase for BDNF (TrkB; Numakawa 2010). In addition Numakawa et al. propose that TrkB–GR interaction plays a critical role in the BDNF-stimulated PLC pathway, which is required for glutamate release, and the decrease in TrkB–GR interaction caused by chronic exposure to glucocorticoids results in the suppression of BDNF-mediated neurotransmitter release via a glutamate transporter.

Finally, an epigenetic regulation of BDNF gene has been described. BDNF gene has 11 exons and nine functional promoters that are tissue-specific and have activity-dependent regulation across development and in adulthood (Nair et al. 2007; Aid et al. 2007; Liu et al. 2006; Sathanoori et al. 2004). It is mainly expressed in the central nervous system, particularly in the cortex, hippocampus and forebrain but it is also found in peripheral tissues (Pruunsild et al. 2007).

Clinical and experimental studies indicate that the prefrontal cortex (PFC) and hippocampus might play a pivotal role in the cognitive deficits and aberrant emotional behaviours originated from early-life adversity (De Bellis 2005; Kaffman and Meaney 2007; Fumagalli et al. 2007; Noble et al. 2005; Teicher et al. 2003).

Dennis and Levitt (2005) have determined that methylation of exon IV was a likely mechanism mediating BDNF gene expression during development and thus susceptible to environmental insults. Thus, Roth et al. (2009) have evaluated offspring exposed to a stressed-abusive mother (maltreatment) or to a normal maternal care or positive caregiving mother (cross-fostered care). They demonstrate that early experiences of maltreatment increase BDNF DNA methylation at both exons IV and IX within the PFC that persist into adulthood, highlighting an epigenetic molecular mechanism potentially underlying lifelong and transgenerational perpetuation of changes in gene expression and behaviour incited by early abuse and neglect.

4.4 Concluding Remarks

There is a clear interrelation between glucocorticoids, NO and BDNF in many different processes, such as the neuroendocrine modulation of the stress response, brain plasticity, learning and memory formation. As described here, NO participates in the neuroendocrine signaling involved in glucocorticoid synthesis and is also necessary for BDNF increases during memory formation, as shown in knockouts and with pharmacological interventions. Moreover, there is a regulation of BDNF through GR.

Finally, we conclude that in addition to the classical description of a disruption of HPA axis, NO and BDNF are relevant intracellular mediators of brain plasticity and its participation in the different processes induced by PS should be taken into account for the development of new therapeutic approaches.

Conflict of Interest The authors declare no conflicts of interest.

References

Aid T, Kazantseva A, Piirsoo M, Palm K, Timmusk T (2007) Mouse and rat BDNF gene structure and expression revisted. J Neurosci Res 85:525–535

Angelucci F, Brene S, Mathe AA (2005) BDNF in schizophrenia, depression and corresponding animal models. Mol Psychiatry 10:345–352

Araki T, Tanji H, Fujihara K et al (1999) Increases in [3H]FK-506 and [3H]L-NG-nitroarginine binding in the rat brain after nigrostriatal dopaminergic denervation. Metab Brain Dis 14:21–31

Bibel M, Barde Y-A (2000) Neurotrophins: key regulators of cell fate and cell shape in the vertebrate nervous system. Genes Dev 14:2919–2937

Binder DK, Scharfman HE (2004) Brain-derived neurotrophic factor. Growth Factors 22(3):123–131

Blum R, Konnerth A (2005) Neurotrophin-mediated rapid signaling in the central nervous system: mechanisms and functions. Physiology (Bethesda) 20:70–78

Bredt DS, Snyder SH (1994) Nitric oxide: a physiologic messenger molecule. Ann Rev Biochem 63:175–195

Capoccia S, Berry A, Bellisario V, Vacirca D, Ortona E, Alleva E, Cirulli F Quality and timing of stressors differentially impact on brain plasticity and neuroendocrine-immune function in mice. Neural Plast. 2013; 2013:971817. doi: 10.1155/2013/971817

Chrousos GP, Torpy DJ, Gold PW (1998) Interactions between the hypothalamic-pituitary-adrenal axis and the female reproductive system: clinical implications. Ann Intern Med 129:229–240

Colasanti M, Suzuki H (2000 Jul) The dual personality of NO. Trends Pharmacol Sci 21(7):249–252

Correia CT, Coutinho AM, Sequeira AF, Sousa IG, Venda LLO, Almeida JP, Abreu RL, Lobo C, Miguel TS, Conroy J, Cochrane L, Gallagher L, Gill M, Ennis S, Oliveira GG, Vicente AM (2010) Increased BDNF levels and NTRK2 gene association suggest a disruption of BDNF/TrkB signaling in autism. Genes Brain Behav 9:841–848

Cratty MS, Ward HE, Johnson EA, Azzaro AJ, Birkle DL (1995) Prenatal stress increases corticotropin releasing factor (CRF) content and release in rat amygdala minces. Brain Res 675:297–302

Dawson TM, Snyder SH (1994) Gases as biological messengers: nitric oxide and carbon monoxide in the brain. J Neurosci 14:5147–5159

De Bellis MD (2005) The psychobiology of neglect. Child Maltreat 10:150–172
de Lima MN, Presti-Torres J, Vedana G, Alcalde LA, Stertz L, Fries GR, Roesler R, Andersen ML, Quevedo J, Kapczinski F, Schröder N (2011) Early life stress decreases hippocampal BDNF content and exacerbates recognition memory deficits induced by repeated D-amphetamine exposure. Behav Brain Res 224(1):100–106
Demyttenaiere K, Nijs P, Evers-Kiebooms F, Konincky PR (1989) The effect of a specific emotion stressor on prolactin, cortisol, and testosterone concentrations in women varies with their trait anxiety. Fertil Steril 52:942–948
Dennis KE, Levitt P (2005) Regional expression of brain derived neurotrophic factor (BDNF) is correlated with dynamic patterns of promoter methylation in the developing mouse forebrain. Mol Brain Res 140:1–9
Dhobale M, Mehendale S, Pisal V, D'Souza V, Joshi S (2012)Association of brain-derived neurotrophic factor and tyrosine kinase B receptor in pregnancy. Neuroscience 216:31–37
Di Monte DAR, Anderson A et al (1997) Inhibition of monoamine oxidase contributes to the protective effect of 7-nitroindazole against MPTP neurotoxicity. J Neurochem 69:1771–1773
Edwards TM, Rickard NS (2007) New perspectives on the mechanisms through which nitric oxide may affect learning and memory processes. Neurosci Biobehav Rev 31(3):413–425. (Epub 2006 Dec 26)
Feil R, Hofmann F, Kleppisch T (2005) Function of cGMP-dependent protein kinases in the nervous system. Rev Neurosci 16:23–41
Feil R, Kleppisch T (2008) NO/cGMP-dependent modulation of synaptic transmission. Handb Exp Pharmacol 184:529–605
Fride E, Dan Y, Gavish M, Weinstock M (1985) Prenatal stress impairs maternal behavior in a conflict situation and reduces hippocampal benzodiazepine receptors. Life Sci 36(22):2103–2109
Fumagalli F, Molteni R, Racagni G, Riva MA (2007) Stress during development: impact on neuroplasticity and relevance to psychopathology. Prog Neurobiol 81:197–217
Hopper RA, Garthwaite J (2006) Tonic and phasic nitric oxide signals in hippocampal long-term potentiation. J Neurosci 26:11513–11521
Huang EJ, Reichardt LF (2001) Neurotrophins: roles in neuronal development and function. Annu Rev. Neurosci 24:677–736
Jeanneteau FD, Lambert WM, Ismaili N, Bath KG, Lee FS, Garabedian MJ, Chao MV (2012) BDNF and glucocorticoids regulate corticotrophin-releasing hormone (CRH) homeostasis in the hypothalamus. Proc Natl Acad Sci U S A 109:1305–1310
Kaffman A, Meaney MJ (2007) Neurodevelopmental sequelae of postnatal maternal care in rodents. Clinical and research implications of molecular insights. J Child Psychol Psychiatry 48:224–244
Kang H, Schuman EM (1995) Long-lasting neurotrophin-induced enhancement of synaptic transmission in the adult hippocampus. Science 267(5204):1658–1662
Kawamura K, Kawamura N, Kumazawa Y, Kumagai J, Fujimoto T, Tanaka T (2011 Mar) Brain-derived neurotrophic factor/tyrosine kinase B signaling regulates human trophoblast growth in an in vivo animal model of ectopic pregnancy. Endocrinology 152(3):1090–1100. doi: 10.1210/en.2010-1124. (Epub 2011 Jan 14)
Kawamura K, Kawamura N, Sato W, Fukuda J, Kumagai J, Tanaka T (2009 Aug) Brain-derived neurotrophic factor promotes implantation and subsequent placental development by stimulating trophoblast cell growth and survival. Endocrinology 150(8):3774–3782. doi: 10.1210/en.2009-0213. (Epub 2009 Apr 16)
Kinnunen AK, Koenig JI, Bilbe G (2003 Aug) Repeated variable prenatal stress alters pre- and postsynaptic gene expression in the rat frontal pole. J Neurochem 86(3):736–748
Kiss JP, Vizi ES (2001) Nitric oxide: a novel link between synaptic and nonsynaptic transmission. Trends Neurosci 24:211–215
Kodomari I, Wada E, Nakamura S, Wada K (2009 Feb) Maternal supply of BDNF to mouse fetal brain through the placenta. Neurochem Int 54(2):95–98
Koenig JI, Elmer GI, Shepard PD, Lee PR, Mayo C, Joy B, Hercher E, Brady DL (2005) Prenatal exposure to a repeated variable stress paradigm elicits behavioral and neuroendocrinological

changes in the adult offspring: potential relevance to schizophrenia. Behav Brain Res 156(2):251–261
Korsching S (1993) The neurotrophic factor concept: a reexamination. J Neurosci 13(7):2739–2748
Korte M, Carroll P, Wolf E, Brem G, Thoenen H, Bonhoeffer T (1995 Sep 12) Hippocampal long-term potentiation is impaired in mice lacking brain-derived neurotrophic factor. Proc Natl Acad Sci U S A 92(19):8856–8860
Lee BE, Ha M, Park H, Hong YC, Kim Y, Kim YJ, Ha EH (2011) Psychosocial work stress during pregnancy and birthweight. Paediatr Perinat Epidemiol 25(3):246–254
Lee PR, Brady DL, Shapiro RA, Dorsa DM, Koenig JI (2007) Prenatal stress generates deficits in rat social behavior: reversal by oxytocin. Brain Res 2 1156:152–167
Lee S, Blanton CA, Rivier C (2003 Jun) Prenatal ethanol exposure alters the responsiveness of the rat hypothalamic-pituitary-adrenal axis to nitric oxide. Alcohol Clin Exp Res 27(6):962–969
Lipton SA (1999) Neuronal protection and destruction by NO. Cell Death Differ 6:943–951
Liu J, Mori A (1999) Stress, aging, and brain oxidative damage. Neurochem Res 24:1479–1497
Liu Q-R, Lu L, Zhu X-G, Gong J-P, Shaham Y, Uhl GR (2006) Rodent BDNF genes, novel promoters, novel splice variants and regulation by cocaine. Brain Res 1067:1–12
Lonart G, Johanson KM (1992) Inhibitory effects of nitric oxide on the uptake of [3H]dopamine and [3H] glutamate by striatal synaptosomes. J Neurochem 63:2108–2117
Lopez-Figueroa MO, Itoi K, Watson SJ (1998) Regulation of nitric oxide synthase messenger RNA expression in the rat hippocampus by glucocorticoids. Neuroscience 87:439–446
Maccari S, Darnaudery M, Morley-Fletcher S, Zuena AR, Cinque C, Van Reeth O (2003) Prenatal stress and long-term consequences: implications of glucocorticoid hormones. Neurosci Biobehav Rev 27(1–2):119–127
Marletta MA (1994) Nitric oxide synthase: aspects concerning structure and catalysis. Cell 78:927–930
Maur D, Romero C, Palumbo M, Genaro A, Zorrilla-Zubilete M (2007) Alterations in behavioral learning due to prenatal stress in rats. Biocell 31:100
Maur DG, Romero CB, Burdet B, Palumbo ML, Zorrilla-Zubilete MA (2012 Dec) Prenatal stress induces alterations in cerebellar nitric oxide that are correlated with deficits in spatial memory in rat's offspring. Neurochem Int 61(8):1294–1301
Mayeur S, Lukaszewski MA, Breton C, Storme L, Junien C, Vieau D, Lesage J (2011) BDNF in feto-placental development. Med Sci (Paris) 27(3):251–252
McEwen BS (1998) Stress, adaptation, and disease. Allostasis and allostatic load. Ann N Y Acad Sci 1;840:33–44
McEwen BS (1999) Stress and hippocampal plasticity. Annu Rev Neurosci 22:105–122. (Review)
McEwen BS (2007 Jul) Physiology and neurobiology of stress and adaptation: central role of the brain. Physiol Rev 87(3):873–904. (Review)
McKay SE, Purcell AL, Carew TJ (1999) Regulation of synaptic function by neurotrophic factors in vertebrates and invertebrates: implications for development and learning. Learn Mem 6(3):193–215
Miller SD, Mueller E, Gifford GW, Kinsley CH (1999 Jun 29) Prenatal stress-induced modifications of neuronal nitric oxide synthase in amygdala and medial preoptic area. Ann N Y Acad Sci 877:760–763
Miranda KM, Espey MG, Wink DA (2000) A discussion of the chemistry of oxidative and nitrosative stress in cytotoxicity. J Inorg Biochem 79:237–240
Mizuno M, Yamada K, Olariu A, Nawa H, Nabeshima T (2000) Involvement of brain-derived neurotrophic factor in spatial memory formation and maintenance in a radial arm maze test in rats. J Neurosci 20(18):7116–7121
Moncada S, Higgs A (1993) The L-arginine-nitric oxide pathway. New Engl J Med 329:2002–2012
Moncada S, Palmer RM, Higgs EA (1991) Nitric oxide: physiology, pathophysiology and pharmacology. Pharmacol Rev 43:109–142
Mueller BR, Bale TL (2006) Impact of prenatal stress on long term body weight is dependent on timing and maternal sensitivity. Physiol Behav 88:605–614

Muramatsu Y, Kurosaki R, Watanabe H et al (2003) Cerebral alterations in a MPTP-mouse model of Parkinson's disease-an immunocytochemical study. J Neural Transm 110:1129–1144
Murmu MS et al (2006) Changes of spine density and dendritic complexity in the prefrontal cortex in offspring of mothers exposed to stress during pregnancy. Eur J Neurosci 24:1477–1487
Nair A, Vadodaria KC, Banerjee SB, Benekareddy M, Dias BG, Duman RS, Vaidya VA (2007) Stressor-specific regulation of distinct brain-derived neurotrophic factor transcripts and cyclic AMP response element-binding protein expression in the postnatal and adult rat hippocampus. Neurophsychopharmacol 32:1504–1519
Nathan C, Xie QW (1994) Nitric oxide synthases: roles, tolls, and controls. Cell 78:915–918
Noble KG, Tottenham N, Casey BJ (2005) Neuroscience perspectives on disparities in school readiness and cognitive achievement. Future Child 15:71–89
Numakawa T, Kumamaru E, Adachi N, Yagasaki Y, Izumi A, Kunugi H (2010) Glucocorticoid receptor interaction with TrkB promotes BDNF-triggered PLC-gamma signaling for glutamate release via a glutamate transporter. Proc Natl Acad Sci U S A 106(2):647–652
Palumbo ML, Fosser NS, Rios H, Zorrilla-Zubilete MA, Guelman LR, Cremaschi GA, Genaro AM (2007) Loss of hippocampal neuronal nitric oxide synthase contributes to the stress-related deficit in learning and memory. J Neurochem 102(1):261–274
Patin V, Lordi B, Vincent A, Caston J (2005) Effects of prenatal stress on anxiety and social interactions in adult rats. Brain Res Dev Brain Res 160(2):265–274
Patterson SL, Abel T, Deuel TA, Martin KC, Rose JC, Kandel ER (1996) Recombinant BDNF rescues deficits in basal synaptic transmission and hippocampal LTP in BDNF knockout mice. Neuron 16(6):1137–1145
Patterson SL, Grover LM, Schwartzkroin PA, Bothwell M (1992) Neurotrophin expression in rat hippocampal slices: a stimulus paradigm inducing LTP in CA1 evokes increases in BDNF and NT-3 mRNAs. Neuron 9(6):1081–1088
Petraglia F, Florio P, Nappi C, Genazzani AR (1996) Peptide signaling in human placenta and membranes: autocrine, paracrine, and endocrine mechanisms. Endocr Rev 17:156–186
Pezet S, McMahon SB (2006) Neurotrophins: mediators and modulators of pain. Annu Rev Neurosci 29:507–538
Pittenger C, Duman RS (2007) Stress, depression, and neuroplasticity: a convergence of mechanisms. Neuropsychopharmacology 33:88–109
Pruunsild P, Kazantseva A, Aid T, Palm K, Timmusk T (2007) Dissecting the human BDNF locus: Bidirectional transcription, complex splicing, and multiple promoters. Genomics 90(3):397–406
Rehn AE, Rees SM (2005) Investigating the neurodevelopmental hypothesis of schizophrenia. Clin Exp Pharmacol Physiol 32(9):687–696. (Review)
Roth TL, Lubin FD, Funk AJ, Sweat D (2009) Lasting epigenetic influence of early-life adversity on the BDNF Gene. Biol Pshychiatry 65:760–769
Ruiz R, Avant C (2005) Effects of maternal prenatal stress on infant outcomes. ANS Adv Nurs Sci 28:345–355
Sathanoori M, Dias BG, Nair AR, Banjerjee SB, Tole S, Vaidya VA (2004) Differential regulation of multiple brain-derived neurotrophic factor transcripts in the postnatal and adult rat in hippocampus during development, and in response to kitinate administration. Brain Res Mol Brain Res 130:170–177
Schinder AF, Poo M (2000) The neurotrophin hypothesis for synaptic plasticity. Trends Neurosci 23:639–645
Schmidt HD, Shelton RC, Duman RS (2011) Functional biomarkers of depression: diagnosis, treatment, and pathophysiology. Neuropsychopharmacology 36(12):2375–2394
Schulz JB, Matthews RT, Muqit MMK (1995) Inhibition of neuronal nitric oxide synthase by 7-nitroindazole protects against MPTP induced neurotoxicity in mice. J Neurochem 64:936–939
Stern JE (2004 Feb–Apr) Nitric oxide and homeostatic control: an intercellular signalling molecule contributing to autonomic and neuroendocrine integration? Prog Biophys Mol Biol 84(2–3):197–215

Sug-Tang A, Bocking AD, Brooks AN, Hooper S, White SE, Jacobs RA, Fraher LJ, Challis JR (1992) Effects of restricting uteroplacental blood flow on concentrations of corticotrophin-releasing hormone, adrenocorticotrophin, cortisol, and prostaglandin E2 in the sheep fetus during late pregnancy. Can J Physiol Pharmacol 70:1396–1402

Sullivan BM, Wong S, Schuman EM (1997) Modification of hippocampal synaptic proteins by nitric oxide-stimulated ADP ribosylation. Learn Mem 3:414–424

Tapia-Arancibia L, Rage F, Givalois L, Arancibia S (2004) Physiology of BDNF: focus on hypothalamic function. Front Neuroendocrin 25:77–107

Taylor JP, Hardy J, Fischbeck KH (2002) Toxic proteins in neurodegenerative disease. Science 296(5575):1991–1995

Teicher MH, Andersen SL, Polcari A, Anderson CM, Navalta CP, Kim DM (2003) The neurobiological consequences of early stress and childhood maltreatment. Neurosci Biobehav Rev 87:873–904

Vallee M et al (1997) Prenatal stress induces high anxiety and postnatal handling induces low anxiety in adult offspring: correlation with stress-induced corticosterone secretion. J Neurosci 17:2626–2636

Vallee M et al (1999) Long-term effects of prenatal stress and postnatal handling on age-related glucocorticoid secretion and cognitive performance: a longitudinal study in the rat. Eur J Neurosci 11:2906–2916

Van den Hove DL, Steinbusch H, Scheepens A, Van deBW, Kooiman L, Boosten B, Prickaerts J, Blanco C (2005) Prenatal stress and neonatal rat brain development. Dev Neurosci 27(5):313–320

Van den Hove DL, Kenis G, Brass A, Opstelten R, Rutten BP, Bruschettini M, Blanco CE, Lesch KP, Steinbusch HW, Prickaerts J (2012) Vulnerability versus resilience to prenatal stress in male and female rats; implications from gene expression profiles in the hippocampus and frontal cortex. Eur Neuropsychopharmacol. doi:10.1016/j.euroneuro.2012.09.011

Van Reeth O, Koehl M, Weibel L, Le Moal M, Maccari S (1998) Effects of prenatal stress on circadian synchronization in adult rats. J Sleep Res 7(Suppl. 2):287

Wadhwa PD, Dunkel-Schetter C, Chicz-DeMet A, Porto M, Sandman CA (1996) Prenatal psychosocial factors and the neuroendocrine axis in human pregnancy. Psychosomat Med 58:432–446

Weinstock M (2001) Alterations induced by gestational stress in brain morphology and behaviour of the offspring. Prog Neurobiol 65(5):427–451

Whitnall MH (1993 May) Regulation of the hypothalamic corticotropin-releasing hormone neurosecretory system. Prog Neurobiol 40(5):573–629. (Review)

Wink DA, Vodovotz Y, Grisham MB et al (1999) Antioxidant effects of nitric oxide. Methods Enzymol 301:413–424

Wong EY, Herbert J (2006) Raised circulating corticosterone inhibits neuronal differentiation of progenitor cells in the adult hippocampus. Neuroscience 137(1):83–92. (Epub 2005 Nov 10)

Wu A, Ying Z, Gomez-Pinilla F (2004) Dietary omega-3 fatty acids normalize BDNF levels, reduce oxidative damage, and counteract learning disability after traumatic brain injury in rats. J Neurotrauma 21(10):1457–1467

Wu J, Song TB, Li YJ, He KS, Ge L, Wang LR (2007) Prenatal restraint stress impairs learning and memory and hippocampal PKCbeta1 expression and translocation in offspring rats. Brain Res 13(1141):205–213

Zorrilla-Zubilete MA, Maur DG, Palumbo ML, Genaro AM (2010) Role of nitric oxide signaling pathways in brain injuries. Curr Chem Biol 4:250–261. (Bentham Science Publishers Ltd.)

Zuena AR, Mairesse J, Casolini P, Cinque C, Alemà GS, Morley-Fletcher S, Chiodi V, Spagnoli LG, Gradini R, Catalani A, Nicoletti F, Maccari S (2008) Prenatal restraint stress generates two distinct behavioral and neurochemical profiles in male and female rats. PLoS One 3(5):e2170

Chapter 5
Prenatal Stress and Adult Drug-Seeking Behavior: Interactions with Genes and Relation to Nondrug-Related Behavior

Tod E. Kippin, Jaonnalee C. Campbell, Kyle Ploense, Chris P. Knight and Jared Bagley

Abstract Addiction inflicts large personal, social, and economic burdens, yet its etiology is poorly defined and effective treatments are lacking. As with other neuropsychiatric disorders, addiction is characterized by a core set of symptoms and behaviors that are believed to be influenced by complex gene–environment interactions. Our group focuses on the interaction between early stress and genetic background in determining addiction vulnerability. Prior work by our group and others has indicated that a history of prenatal stress (PNS) in rodents elevates adult drug seeking in a number of behavioral paradigms. The focus of the present chapter is to summarize work in the area of PNS and addiction models as well as our recent studies of PNS on drug seeking in different strains of mice as a strategy to dissect gene–environment interactions underlying cocaine addiction vulnerability. These studies indicate that ability of PNS to elevate adult cocaine seeking is strain dependent. Further, PNS also alters other nondrug behaviors in a fashion that is dependent on different strains and independent from the strain dependence of drug seeking. Thus, it appears that the ability of PNS to alter behavior related to different psychiatric conditions is orthogonal, with similar nonspecific susceptibility to prenatal stress across genetic backgrounds but with the genetic background determining the specific nature of the PNS effects. Finally, the advent of recombinant inbred mouse strains is allowing us to determine the genetic bases of these gene–environment interactions. Understanding these effects will have broad implications to determining the nature of vulnerability to addiction and perhaps other disorders.

T. E. Kippin (✉) · J. C. Campbell · K. Ploense · C. P. Knight · J. Bagley
Department of Psychological and Brain Sciences, University of California,
Santa Barbara, CA 93106-9660, USA
e-mail: kippin@psych.ucsb.edu

T. E. Kippin
Neuroscience Research Institute, University of California,
Santa Barbara, CA 93106-9660, USA

The Institute for Collaborative Biotechnology, University of California,
Santa Barbara, CA 93106-9660, USA

J. C. Campbell
Center for Neuropharmacology and Neuroscience, Albany Medical College,
MC-136, 47 New Scotland Avenue, Albany, NY 12208, USA

M. C. Antonelli (ed.), *Perinatal Programming of Neurodevelopment*,
Advances in Neurobiology 10, DOI 10.1007/978-1-4939-1372-5_5,

5.1 Introduction

Addiction to drugs of abuse, including cocaine, is a leading health problem in Western society. The prevalence of substance-use disorders among the US general population is estimated to be approximately 9.2% (Aldworth et al. 2007). In the USA, approximately 2% of the general population has used cocaine at least once in their life time, of which 2 million are current users and 1.5 million are cocaine dependent (NIDA 2004). Addiction produces significant costs to both the individual addicts, their family, and friends, as well as economic and social burdens to society in general. The addiction-associated costs resulting from productivity loss, traffic accidents, related diseases (e.g., spread through infected needles), and drug-related crime, including domestic violence, are estimated to be up to 3.5% of the gross domestic product (GDP) in Western countries (Pouletty 2002). Accordingly, given the US$ 13.86 trillion estimated GDP of the USA in 2007 (CIA 2008), the cost of drug addiction to the country is around US$ 485 billion.

The reasons why some individuals become addicted to cocaine or other drugs, while others do not, are poorly understood, but it appears that cocaine addiction results from a complex interaction between genetic and environmental factors (for reviews see, e.g., Crabbe 2002; Enoch 2006; Kendler et al. 2007; Kreek 2001). For instance, Ellenbroek and colleagues (Ellenbroek et al. 2002; van der Kam et al. 2005) argue that addiction, as with other neuropsychiatric disorders, is mediated by the interaction of genetic polymorphisms, early environment, and current environment. In both humans and animals, there are large individual differences on a wide range of measures following acute and repeated exposure to and/or level of intake of psychostimulants such as cocaine and d-amphetamine (DeWit et al. 1986, Cohen 1990). Thus, it is presumed that there are polygenetic backgrounds which interact with specific experiences to determine cocaine responsiveness, which will make an individual more or less vulnerable to developing cocaine addiction. Our ultimate goal is to elucidate genes that modulate the ability of early environmental experiences to alter adult responsiveness to cocaine. Recent research has focused on delineating specific early environmental conditions that increase cocaine responsiveness. In particular, maternal stress during gestation or neonatal isolation produces an adult phenotype that has increased sensitivity to cocaine, greater cocaine intake during cocaine self-administration, and higher cocaine seeking during extinction and reinstatement procedures (Kippin et al. 2008; Thomas et al. 2009; Kosten et al. 2000, 2004, 2006; Lynch et al. 2005; Zhang et al. 2005; see also Deminiere et al. 1992), suggesting that a history of early life stress increases vulnerability to cocaine addiction. Conversely, genetic background, namely gene polymorphism, also plays a role in determination of cocaine responsiveness. For example, substantial differences exist between inbred mouse strains regarding the behavioral and physiological effects of cocaine (Crawley et al. 1997; Crabbe et al. 1999). Further, the advent of recombinant inbred (RI) mouse strains, in conjunction with the mapping of the mouse genome, now allows quantitative analyses of the contribution of specific genetic loci to traits (Belknap and Crabbe 1992; Chesler et al. 2003) and these analyses have been performed for several behavioral responses to acute or repeated cocaine exposure, including cocaine-induced convulsions and locomotion (see for reviews, e.g., Phillips and Belknap 2002; Crabbe 2002). However, despite emerging evidence that gene–environment interactions (G × E) are critical determinants of cocaine responsiveness, there has been almost no systematic investigation that would allow determination of the specific genetic loci, and ultimately genes located within these loci, that facilitate or impede the capacity of early environmental experience to influence subsequent cocaine responsiveness. Accordingly, a major impediment to our ability to predict individual cocaine responsiveness, and potentially individual vulnerability to addiction, is a poor understanding of the interactions between genes and environment. The overall goal of the present review is to provide a broad analysis of the evidence implicating prenatal stress (PNS) in drug responsiveness and drug-seeking behavior, then to more narrowly review studies of the genetic bases of drug responsiveness/seeking behavior, followed by a description of our approach to provide a genetic analysis of the effects of early environmental experience, namely PNS, upon cocaine addiction vulnerability employing a mouse model of cocaine reward. We will also summarize what is known about the neurobiology of PNS-induced alterations in drug responsiveness and, finally, the relation of addiction vulnerability to vulnerability to other neuropsychiatric disorders will be addressed.

5.2 Early Environmental Stress and Neuropsychiatric Disease

Environmental stimuli during the embryonic and early postnatal periods have profound effects upon developmental processes, resulting in permanent alterations in nervous system structure and function. In humans, children of mothers experiencing stress during gestation show alterations in early motor development, anomalies in brain morphology, and behavioral abnormalities such as attention deficit hyperactivity disorder (ADHD), sleep disturbances, cognitive dysfunction, increased anxiety, and are at an increased risk of a variety of neuropsychiatric disorders, including higher incidence of substance abuse (reviewed in Huizink et al. 2004; Kofman 2002; Weinstock 2001). Stress during gestation or PNS can be modeled in laboratory rodents using a number of procedures, of which a widely employed one involves repeated bouts of restraint stress to a pregnant dam during the last week of gestation. The data from animal studies using this PNS model are consistent with the epidemiological and clinical findings, but are far more extensive. In animals, PNS sequelae include (but are not limited to) increased anxiety, learning and memory impairments, altered circadian rhythm function, impaired sexual function, and increased psychomotor responsiveness to, as well as increased propensity to self-administer, drugs of abuse (for recent reviews see, e.g., Darnaudery and Maccari 2008; Weinstock 2008; Matthews 2002). Thus, there is a remarkable parallel between the human literature and that derived from animal models regarding the long-term behavioral consequences resulting from stressful experiences during development.

Although the detailed mechanisms producing the PNS syndrome are not known, the intermediate cause appears to be maternal secretion of glucocorticoids during stressful events. The hypothalamic-pituitary-adrenal (HPA) axis is a primary element in the nervous system response to environmental stress and is used extensively as an index of nervous system stress output in animal models. Evidence from animal models has shown that maternal stress during gestation elevates glucocorticoids in the fetal brain in a variety of species (Lephart et al. 1997; Montano et al. 1991). Removal of the adrenal glands, the maternal source of circulating glucocorticoids, prevents the development of the PNS syndrome (Barbazanges et al. 1996). Systemic administration of exogenous glucocorticoids to the mother during gestation mimics the PNS syndrome (reviewed in Maccari et al. 2003; Takahashi 1998). Similarly, in humans, exposure to glucocorticoids during development is associated with increased risk for psychiatric disorders (reviewed in, e.g., Bertram and Hanson 2002; Matthews et al. 2002; Sekyl and Meaney 2004). Accordingly, it has been widely proposed that embryonic nervous system programming is highly sensitive to glucocorticoids with lifelong consequences that may compose a significant component of individual vulnerability to neuropsychiatric conditions, including drug addiction.

5.3 Early Environmental Stress Contributes to Adult Drug and Alcohol Responsiveness

In order to address the potential role of early environmental stress in addiction vulnerability, a number of groups have examined the role of PNS or prenatal glucocorticoid exposure (PNG) in modulating the responsiveness to cocaine and other drugs of abuse in adulthood. The common findings across studies are a general elevation in drug responsiveness as well as the motivation to take drugs, with these effects spanning psychostimulants, depressants/opiates, and alcohol.

5.3.1 PNS Alters Response to and the Propensity to Take Stimulant Drugs

Psychostimulants are the class of drugs where the effects of PNS on the responsiveness to and the seeking or consumption of drugs of abuse has been most examined. These studies have formed the strongest basis for the linkage between PNS and changes in drug-use patterns. For amphetamine, Deminiere et al. (1992) found that while administration of a 0.3 mg/kg (intraperitoneal, IP) dose of amphetamine increased subsequent motor activity in all rats, PNS male rats given amphetamine produced double the locomotor counts of control males during the first 10 min after administration. Further, when rats self-administered amphetamine (30 µl/infusion; intravenous, IV), the PNS rats had higher intakes compared to controls on all trials after the first day (Deminiere et al. 1992). While others have not always found initial differences in the locomotor response to amphetamine (Henry et al. 1995; 1 mg/kg dose, IP), they have seen greater sensitization of locomotor behavior in male PNS rats (Henry et al. 1995). Additionally, PNS female rats have increased rotational behavior (left turns) after amphetamine administration, indicating PNS asymmetries in motor system sensitivity to stimulants (Weinstock and Fride 1989). There are also changes in the response to amphetamine after corticosterone is used to mimic a hormonal aspect of PNS. Diaz et al. (1995) implanted pregnant females with corticosterone pellets during late gestation, and the resulting corticosterone-exposed (prenatal corticosterone administration, PNC) weanlings were assessed for locomotor activity after amphetamine. Amphetamine increased locomotor activity in both normal and PNC offspring, but again there was an enhancement of movement in PNC animals. There was also a sex difference, in that male PNC pups increased all types of movement, while female PNC pups displayed increased rearing behavior only (Diaz et al. 1995), indicating a more robust effect in males. PNS status also interacts with behavior on cocaine. Kippin et al. (2008) found that a 15 mg/kg (IP) dose of cocaine enhanced locomotion in adult male PNS rats compared to adult male controls with similar levels of cocaine experience. On the other hand, Thomas et al. (2009) did not see pronounced differences between PNS and control males in locomotion after either acute or repeated cocaine exposure, but did find enhanced locomotor sensitization in PNS females. This change in female locomotor sensitization did not carry over to self-administration, where female PNS and control rats self-administered similar amounts of cocaine (Thomas et al. 2009). In males, there were no differences on a constant (0.2 mg/kg/infusion, IV) cocaine dose, but on a weekly escalating (0.3–0.5 mg/kg/infusion, IV) schedule PNS males had significantly higher total intake (Thomas et al. 2009). In comparison, a study using a higher but constant dose (1.0 mg/kg) reported a trend towards PNS males self-administering more cocaine than controls (Kippin et al. 2008). Yet during extinction, these same PNS males had significantly greater initial responding, took longer to reach extinction criterion, and exhibited greater cocaine-primed reinstatement (Kippin et al. 2008). Interestingly, noncontingent cocaine injections produced an extinction deficit following removal of brain self-stimulation in PNS but not control rats (Gao et al. 2011) suggesting that adult psychostimulant exposure may produce generalized cognitive deficits dependent upon early environmental stress.

PNS also increases responsiveness to other stimulant drugs. PNS increases the sensitivity to nicotine and the magnitude of nicotine-induced locomotor activity (Kochl et al. 2000). For caffeine, adult PNS male rats displayed greater rearing and corner activity in an open field compared to controls, including at a dose where there was no difference in plasma caffeine concentrations between control and PNS animals (10 mg/kg, IP; Pohorecky et al. 1989). For 3,4-Methylenedioxymethamphetamine

(MDMA, "ecstasy"; 5 mg/kg, IP), adolescent PNS female rats had fewer periods of inactivity while also demonstrating more slips and body twists on a runway, and had higher blood levels of MDMA after the same dose was administered compared to controls (Morley-Fletcher et al. 2004). In summary, PNS increases responsiveness to and self-administration of a variety of psychostimulant drugs, and this response interacts with sex and task type. We will later discuss the PNS interaction with genetic background on measures thought to probe cocaine reward.

5.3.2 PNS Alters Response to and Preference for Morphine, and Response to Diazepam

PNS can also impact the response to therapeutics with abuse potential such as morphine and diazepam. PNS male rats are less sensitive than control males to the analgesic effects of morphine (5 mg/kg, IP) in the tail-flick test; conversely, PNS females are more sensitive than female controls to the analgesic effects of morphine, with longer latencies to remove the tail (Kinsley et al. 1988). For both PNS and brief postnatal separations from the dam in mice, adult mice displayed greater morphine-induced analgesia 30 min after administration, with similar patterns with subtle sex-specific alterations after 60 min (Sternberg and Ridgeway 2003). PNS male rats also show a difference in conditioned place preference (CPP) to morphine. While all male rats given morphine (10 mg/kg) showed a significant place preference for the morphine-paired side, the group males only exposed to PNS showed a significantly greater preference than the control group (Yang et al. 2006).

PNS can also alter the response to diazepam. Male PNS rats responded to a low dose of diazepam (1 mg/kg, IP) that did not impact control behavior in the open field (Pohorecky and Roberts 1991). Additionally, it has also been shown that PNS rats are hyperresponsive to bright-light-potentiated acoustic startle, and startle levels are normalized with diazepam treatment (3.2 mg/kg; Tazumi et al. 2005).

5.3.3 PNS Alters Response to and the Propensity to Take Alcohol

There are several changes in the physiological response to alcohol after PNS. Adult PNS rats given alcohol in the 1–2 g/kg dose range display initial blunting of alcohol-induced hypothermia, a reduction in alcohol-induced corticosterone release, reduced free fatty acid release, and decreased alcohol-induced startle deficits, and they retain better motor control on the rotorod but less capability in a swim test (DeTurck and Pohorecky 1987). Adolescent male PNS rats also exhibit this HPA axis blunting, such that a 1.5 g/kg dose of alcohol does not produce the sustained increase in serum adrenocorticotropic hormone (ACTH) and corticosterone levels in PNS rats versus controls (Van Waes et al. 2006).

Displaying a higher initial tolerance to alcohol increases the chance of later alcohol dependence in humans (for review, Schuckit 2009), suggesting that PNS could predispose an individual to initial high alcohol consumption and increasing the risk of later addiction.

PNS also appears to impact voluntary alcohol consumption in animal models. In the two-bottle choice task, where animals are allowed access to a bottle of water and another bottle of alcohol solution, Darnaudery et al. (2007) found that a subpopulation of adult female rats showed a change due to PNS after a stressor in adulthood. After 2 weeks of two-bottle choice, these female rats were designated as high or low drinkers and they were subjected to an inescapable shock. Post-stress, PNS and control low-drinking females did not differ from each other, and high-drinking control females decreased their alcohol consumption from that point onwards, but high-drinking PNS females returned to their previous high level of consumption in the weeks following the stress (Darnaudery et al. 2007). PNS also increases the preference for a sweetened alcohol solution versus water in adolescent female rats (Van Waes et al. 2011a), although PNS does not change two-bottle choice behavior in adolescent or adult male rats (Van Waes et al. 2011b).

Interestingly, there are effects of PNS on alcohol consumption which only emerge after alcohol exposure in the adult environment. Our studies indicate that there is not an effect of PNS on adult alcohol consumption under free-access conditions, but there is an impact of that free-access alcohol consumption on subsequent motivation for alcohol (Campbell et al., under review). Briefly, PNS and control males and females without adult exposure to alcohol exhibited equivalent operant behavior for reinforcement with sucrose (15 %, 20 μl/infusion), sucrose–alcohol fading (decreasing sucrose from 15–0 % with concurrent increasing of alcohol from 0 to 10 %), or alcohol alone (10 %, 20 μl. infusion) even on a response schedule with increasing demand. In contrast, control males allowed prior continuous access to alcohol for 8 weeks exhibited a substantial decline in operant behavior for alcohol—a finding consistent with those indicating that alcohol experience by either injection (Carrara-Nascimento et al. 2012) or voluntary consumption (our own unpublished observations) reduces alcohol reward as measured in CPP. However, PNS males allowed alcohol access exhibited the same level of intake as controls but this alcohol exposure failed to diminish subsequent operant responsiveness for alcohol. PNS females only exhibited this difference from control females for the 10 % alcohol solution on the FR1 schedule. Again, these male and female C57BL/6J mice with either a PNS or control history exhibited the same level of alcohol consumption during continuous free access to alcohol in a two-bottle choice task over a period of 8 weeks, but then PNS males in particular did not show the expected behavior on the operant task. Accordingly, the ability of PNS to alter motivation for alcohol appears to be more subtle than the effects observed for psychostimulants as it is dependent upon fairly extensive exposure to alcohol in adulthood. Nonetheless, the altered alcohol-induced motivational plasticity of PNS-exposed individuals is consistent with the general findings of enhanced drug-seeking behavior following early environmental stress exposure.

In summary, PNS is known to change the response to a diverse number of drugs of abuse and increase seeking and consumptive behaviors towards drugs such as psychostimulants, morphine, and alcohol, with the behavioral alterations in many cases modulated by sex. However, these studies have largely occurred in parallel with separate lines of research indicating that genetic background modulates drug responsiveness and seeking behavior. To address the issue of genetic background influence, as well as interactions with early environment, on drug responsiveness and potential addiction vulnerability, the remainder of the chapter will focus more narrowly on cocaine.

5.4 Hereditability of Cocaine Responsiveness: Mouse Strain Differences

Studies using humans and animal models have suggested that genetic factors are important in substance abuse (DeWit et al. 1986; Cohen 1990; Crabbe et al. 1994). Inbreeding has been performed in both rats and mice to reduce the allelic diversity between close genetic relatives and to produce a strain of animal in which same-sex members are almost monozygotic (identical) twins of all other members. There are currently over 100 inbred strains of mice available from the Jackson Laboratory and the stability of strain differences over decades of research has provided an extensive database concerning groups of genes that influence a particular phenotypic trait. Comparisons between different inbred mouse strains have provided substantial evidence for genetic contributions to an individual's responsiveness to drugs of abuse, including cocaine. In this respect, strain differences between the C57BL/6J (B6) and DBA/2J (D2) strains have been the most thoroughly studied (see for a review, Crabbe et al. 1999; Crawley et al. 1997; Wahlsten et al. 2003). In most studies, the D2 strain exhibits greater acute cocaine-induced locomotion and/or greater cocaine-induced locomotor sensitization upon repeated treatment (Cunningham et al. 1999; de Jong et al. 2007, 2008; Orsini et al. 2005; Tolliver and Carney 1994a, b, 1995; but see also Kafkafi and Elmer 2005). Conversely, a strain comparison of the reinforcing effects of cocaine indicates that B6 mice exhibit higher cocaine intake than D2 mice during both IV cocaine self-administration under operant conditions (van der Veen et al. 2007, 2008) and access to cocaine in drinking water (George and Goldberg 1989; Seale and Carney 1991). Similarly, CPP has also been used to compare cocaine reward among these strains and B6 mice show a cocaine CPP with lower doses of cocaine, shorter conditioning trials, and/or a greater magnitude preference than do D2 (Cunningham et al. 1999; Seale and Carney 1991; Orsini et al. 2005, 2008). Accordingly, comparison of the genetic differences between these strains is an approach to understanding individual differences in both the psychomotor stimulant responsiveness to drug exposure and the motivational components of drug-seeking behavior.

5.5 Search for Genetic Substrates Mediating Hereditability of Cocaine Responsiveness

Determining the genetic influences that contribute to hereditability of cocaine responsiveness has proven difficult because individual differences in behavioral responses to drugs are highly complex. In addition to substantial nongenetic factors, including potential intergenerational epigenetic hereditability, the range of drug responsiveness across different genotypes tends to exhibit a continuous distribution (see, e.g., Miner and Marley 1995a; Phillips et al. 1998). This indicates a polygenetic pattern of inheritance involving many genes rather than one or two major genes. Quantitative trait loci (QTL) analyses have high utility for determining the relation between genetic influences that account for relatively small amounts of variation in a given behavior (Crabbe et al. 1999). One strategy that has been employed in the search for QTLs influencing cocaine responsiveness is the employment of RI strains of mice derived from the systematic brother–sister matings of the F2 generation of two progenitor inbred mouse strains, which results in a panel of new inbred strains having unique recombinations of parental chromosomal material distinct from the two progenitor strains and all other RI strains (Belknap and Crabbe 1992; Complex Trait Consortium 2003). RI strains have allowed significant advances in behavioral genetics because these mice have fixed genomes and genotypes of known genetic markers are already stored in a database; thus, they can be used for behavioral tests and then QTL can be performed without additional molecular analysis facilitating genetic mapping of behavioral traits (Belknap 1998). The BXD and DXB panels of RI strains are derived from the B6 and D2 inbred parental strains and are ideal for genetically dissecting responsiveness to cocaine and other abused drugs because of the distinct phenotypes exhibited by the original strains (see, e.g., Chesler et al. 2003). Accordingly, this panel has been employed to genetically dissect several distinct cocaine-related behaviors which include the acute locomotor stimulant effect of cocaine (Miner and Marley 1995b; Phillips et al. 1998; Tolliver et al. 1994), sensitization to the locomotor stimulant effect of cocaine with repeated treatment (Phillips et al. 1998; Tolliver et al. 1994), cocaine-induced seizures (Miner and Marley 1995a), cocaine-induced CPP (Philip et al. 2010), and even cocaine self-administration in a small subset of available strains (Cervantes et al. 2013). As the available RI strains are unable to resolve individual gene polymorphisms, additional studies need to employ fine mapping strategies utilizing loci-directed production of additional RI strains, accompanied by nucleotide sequences, to determine the specific genes within a loci that exhibit polymorphisms within RI panels that could contribute to phenotypic variability (e.g., Fehr et al. 2002; Hood et al. 2006). Thus, it is possible to utilize rodent models to obtain fine-detail genetic analysis of specific aspects of cocaine responsiveness. Critically, the approaches outlined above are amenable to investigation of the impact of early environmental factors and can be used as a basis for forward and reverse genetic approaches to determine the nature of G × Es in cocaine-addiction-related behavior.

Of relevance to such approaches to studying interactions between early environmental stress and strain differences are strain differences in the stress response during adulthood. Although B6 mice are often characterized as stress resistant, both B6 and D2 mice show substantial changes in plasma corticosterone levels in response to psychological stressors (e.g., Prakash et al. 2006; Belzung et al. 2001; Crawley et al. 1997) and show stress-related modulation of programming in early development (reviewed in, e.g., Weinstock 2001; Zhang et al. 2006)—notably, we have recently compared changes in stress-induced glucocorticoid levels during pregnancy (i.e., during application of the PNS procedure) and both strains appear to have similar HPA activation during the PNS procedure indicating the differences in PNS-induced changes in adult behavior between these strains are unlikely to be due to stress responses of the mother during gestation. Be that as it may, RI strains may have different response profiles than either of the parental lines (e.g., Phillips et al. 2002); thus, the magnitude and duration of stress-induced changes in plasma corticosterone will be monitored during the PNS procedure in all strains allowing determination of its relation to adult behavior changes induced by PNS.

5.6 Interactions Between PNS and Genetic Background on Drug Seeking

With respect to studies of genetic background–environment interactions, D2 mice show greater changes in cocaine responsiveness following adult stress exposure whereas B6 mice appear to be relatively insensitive to adult stress-induced changes in cocaine responsiveness (Badiani et al. 1992; Cabib and Bonaverntura 1997; Cabib et al. 2000; van der Veen et al. 2007). As discussed above, PNS procedures that enhance cocaine responsiveness in rats (Kippin et al. 2008; Thomas et al. 2009) also increase the level of responding for, and consumption of, alcohol during operant self-administration procedures in B6 mice with extensive alcohol experience (Campbell et al., under review). Accordingly, our studies investigating the impact of PNS on cocaine responsiveness in B6 and D2 mice are summarized below. These studies employ standard PNS procedures of repeated restraint stress during late gestation and CPP that we have standardized between strain to be optimally sensitive to PNS–gene interactions.

5.6.1 Impact of Conditioning Parameters on Cocaine-Induced CPP in Male and Female B6 and D2 Mice

Prior work indicates that strain differences (between B6 and D2 mice) in cocaine-induced CPP exist and are sensitive to the duration of conditioning trials (Cunningham et al. 1999);, however, this evidence employed discreet conditioning stimuli (distinct tactile flooring) and did not examine females. Accordingly, we evaluated the impact of conditioning session duration on cocaine-induced CPP magnitude employing contextual conditioning stimuli in both B6 and D2 males and females as these stimuli were employed in our preliminary studies demonstrating PNS-induced potentiation of CPP. This experiment was a necessary first step to our investigation into the impact of PNS on cocaine seeking because it determined the procedures that allow assessment of cocaine-induced CPP in both male and female mice in the B6 and D2 parental strains. Specifically, both male and female B6 and D2 mice (obtained from Jackson Laboratories, Bar Harbor, ME) were tested on an unbiased CPP procedure that it is sensitive to psychomotor stimulant responsiveness and cocaine reward, extinction and reinstatement of CPP (e.g., Szumlinski et al. 2008). Briefly, these experiments used a two-compartment chamber with distinct compartments (based on wall/flooring pattern and texture) that can be divided using an insert and supports acquisition and expression CPP as measured by video tracking. Mice were first allowed to explore the entire apparatus without the divider insert allowing assessment of novelty-induced locomotion. Next, each mouse received injections of cocaine (10 mg/kg, IP) and saline on alternating days and then were placed into one of the two compartments (with the divider insert in place) such that each compartment was repeatedly paired with cocaine or saline. Next, mice were allowed access to the entire apparatus on a "posttest" in order to determine direction and magnitude of place conditioning produced by the cocaine injections relative to the saline ones. In order to examine the rate of extinction, mice were subsequently placed into the undivided apparatus for 15-min sequential sessions and finally subjected to a "primed test" during which they first received a saline injection and CPP was reassessed for 15 min, and finally a cocaine injection (same dose as given during conditioning) and CPP was reassessed for 15 min. We have completed behavioral testing of both male and female B6 and D2 mice at session durations of 15, 20, and 30 min. The results of this experiment revealed a significant sex × strain interaction on the magnitude of CPP (exhibited during the initial posttest) with females exhibiting a larger CPP than males in the D2 strain but a smaller CPP than males in the B6 strain. Further, there was a significant strain effect (independent of sex) on extinction with B6 mice exhibiting rapid extinction (within two to three 15-min sessions) but D2 mice failing to exhibit significant extinction (even after 12 × 15-min sessions), thus, the absence of extinction in one of the strains is unfortunate as it complicates examination of PNS–gene interactions on this measure. However, neither a strain × session duration interaction nor strain effect was detected on the magnitude of CPP under these conditions. These findings indicate that strain effects on acquisition of CPP are minor when contextual stimuli (in contract to discreet stimuli) are employed and, accordingly, the subsequent PNS–gene interaction experiments employ a 15-min conditioning duration in the above-described apparatus.

5.6.2 *Impact of PNS and Cocaine Dose on Cocaine-Induced CPP—PNS × Genetic Background Effect but Not Sex Effects*

We used the procedures described above (with a 15-min conditioning session duration) to determine the impact of PNS on CPP induced by 3, 10, or 30 mg/kg (IP) in male and female B6 and D2 mice. Briefly, B6 and D2 mice (obtained from the same source as above) were mated and then pregnant dams were subject to either repeated restraint stress (3 × 1 h per day from E14 to delivery) or left alone for the entire gestation period. Offspring were weaned at 3 weeks of age and housed into same-sex groups until testing at 8–10 weeks of age. The results indicate that PNS potentiates cocaine-induced CPP at all employed doses in B6, but not D2, mice. Specifically, there was a significant effect of cocaine dose with all groups exhibiting increased CPP magnitude with increasing cocaine dose. Moreover, there was an interaction between PNS and strain with B6 mice subject to PNS exhibiting greater CPP than control B6 mice (independent of sex or dose) but no differences between PNS and control D2 mice (in either sex or at any dose). The finding that PNS increases CPP in both males and females was somewhat surprising as it contrasts with other research indicating greater sensitivity to PNS potentiation of drug seeking in males relative to females in experiments employing operant self-administration of cocaine in rats (Thomas et al. 2009) as well as our own research employing operant self-administration of alcohol in B6 mice (Campbell et al. 2009). Accordingly, we replicated this experiment (at the 10 mg/kg dose) and produced the same results in a second cohort of PNS and control B6 males and females. Thus, we are highly confident in this effect and, consistent with other measures of drug seeking, PNS does increase the magnitude of cocaine-induced CPP in B6, but not D2, mice indicating that our approach is sufficient to detect gene–early environment interactions in adult drug responsiveness. Moreover, this strain × PNS interaction indicates that utilization of inbred strains, including RI strains, will be a feasible approach to determine the genetic factors which permit early environmental stress to increase drug-seeking behavior.

5.6.3 *Impact of PNS in B6 and D2 Strains on Alcohol-Induced CPP—PNS × Genetic Background × Sex Effects*

As described above, it was surprising that both male and female B6 mice exhibited greater CPP following PNS because sex-dependent effects of PNS are observed on the rate of operant drug self-administration. In order to determine whether the effects of PNS are consistent between CPP and operant self-administration models of drug-seeking, we examined the impact of PNS on alcohol-induced CPP in male and female B6 and D2 mice. Mice were generated as above and then conditioning was performed using the same apparatus and procedures except that each mouse received eight alcohol (2 g/kg, i.p.) and eight vehicle pairings prior to the posttest. The results of this experiment mirror those of our operant alcohol self-administration study. Specifically, we observed a sex × strain × PNS interaction with male B6 mice exhibiting greater alcohol-induced CPP than control B6 mice but PNS and control differences were not observed in other sex/genotype conditions. Accordingly, these results indicate that PNS male, but not female, B6 mice exhibit greater alcohol-seeking behavior, that the impact of PNS on drug seeking is specific to the drug abuse examined, and that the effects of PNS on drug seeking are consistent across operant self-administration and CPP models.

Mouse strains appear to be ideal for elucidating gene–early environment interactions for several reasons. First, the impact of early environmental stress, namely PNS, appears to be consistent across species, including inbred mouse strains, as well as a variety of measures of drug-seeking behavior (i.e., CPP and operant). Second, strain differences are generally stable across laboratories. Third, the availability of RI strains allows detailed dissections of gene–early environment interactions and is an approach currently being employed in our laboratory. Fourth, the availability of a host of well-established behavioral models for a variety of indices of neuropsychiatric disorders which can be utilized for investigation into the relation between heightened cocaine responsiveness and alterations in nondrug behaviors which will further our understanding of nature of comorbid neuropsychiatric vulnerability with addiction.

5.7 Interactions Between PNS and Genetic Background on Nondrug Behaviors

Direct evidence for interactions between genetic background and early environmental stress is observed in a small number of studies that have examined the impact of PNS in different strains of rodents. PNS increases aggression in male and female B6 mice, but decreases this measure in female D2 mice and does not affect aggression in male D2 mice (Kinsley and Svare 1987). PNS increases exploratory behavior in a novel environment in B6 mice but decreases it in BALB/c mice (DeFries 1964; DeFries et al. 1967; Weir and DeFries 1964). Similarly, strain specificity in the effects of PNS upon the behavior of adult rats also appears to be strain specific. For instance, PNS effects upon active avoidance and forced swim-induced immobility are observed in Lewis, but not Fisher 344, strains of rats (Stöhr et al. 1998). Further, PNS produces different changes in emotionality and/or neuroendocrine responsiveness in “high” versus “low” novelty-seeking substrains of Sprague-Dawley rats (Clinton et al. 2008) or “high” versus “low” anxiety substrains of Wistar rats (Bosch et al. 2006; Neumann et al. 2005). Further, analyses of the impact of PNS in B6 × BALB/c F1 progeny generated by bidirectional crossing of males and females revealed contributions of both maternal and paternal genetic background to the expression of PNS-induced changes in open-field behavior (DeFries 1964; DeFries et al. 1967). Importantly, these data indicate that maternal genes are important for, but cannot account for all of, the gene–environment effects observed in PNS-induced changes across genetic background. Such data further the notion that G × Es are complex and likely involve polygenetic influences.

Given widely reported comorbidity of addiction with heritable traits (e.g., novelty seeking) as well as other neuropsychiatric disorders (e.g., schizophrenia), we conducted experiments to assess open-field behavior and prepulse inhibition as measures of reactivity to novel environments and of altered sensory-motor gating observed in schizophrenics. B6 and D2 mice were mated and then pregnant dams were either subject to repeated restraint stress (3 × 1 h per day from E14 to delivery) or left alone for the entire gestation period. Adult offspring were tested for acoustic startle and prepulse inhibition (PPI) and then 3–5 days later were tested for open-field behavior (as reported previously by our group; see Szumlinski et al. 2005). All subjects exhibited robust acoustic startle which was not altered by PNS. However, the ability of a sub-startle acoustic prepulse to attenuate response to a startle tone was altered by PNS in a strain-specific fashion. Importantly, on this measure the D2, not the B6, strain exhibited attenuated PPI following a history of PNS. Thus, these findings with those of our CPP experiments indicate that both B6 and D2 strains are sensitive to the enduring effects of PNS but that the different genetic backgrounds produced sensitivity to PNS on different behavioral measures. Similarly, open-field behavior was altered in a strain-specific fashion by PNS, with PNS B6 mice exhibiting increased explorations and reduced episodes of immobility compared to their controls whereas PNS D2 mice exhibited increased immobility compared to their controls. Accordingly, our findings suggest that the ability of PNS to modify neuropsychiatric disease vulnerability interacts with genetic background such that specific outcomes are altered in a genetic background-specific fashion, rather than simply having some backgrounds sensitive and others resilient to the effects of PNS. We believe that this finding is critical for proper conceptualization of how the interaction between genetic vulnerability and environment interact across disease vulnerability as well as the potential comorbidity of a disease.

5.8 Towards a Defined Neurobiology of PNS-Induced Alterations in Drug Seeking

In addition to establishing the pattern of interaction between genes and environment, another major benefit of animal studies of drug responsiveness is the ability to elucidate the underlying neurobiology of that altered responsiveness. Although the impact of drugs of abuse on brain function has been extensively studied, the way in which PNS alters these functions have not been well elucidated. Moreover, there has been even more limited attention on how gene–PNS interactions may be mediated at a neurobiological level. The following section will review what is known about PNS-induced changes in the neurobiology, particularly at a neurochemical level, and how PNS modulates neurochemical responses to drugs, as well as introduce our approach to further this literature by examination of PNS across mouse strains towards an understanding of G × E neurobiology.

Drugs of abuse, like natural rewards, interact with the mesocorticolimbic dopamine system. This system is involved with providing incentive salience to stimuli and inducing the performance of goal-directed behavior. Dopamine cell bodies in the ventral tegmental area (VTA) project to area including the amygdala, hippocampus, nucleus accumbens (NAc), and areas of the prefrontal cortex (Feltenstein and See 2008). Interestingly, in male and female PNS rats there is reduced dopamine turnover in the left dorsal striatum (Weinstock and Fride 1989). There are also higher dopamine levels in dorsal striatum (Gerardin et al. 2005) and the ventral striatum (vSTR) of PNS rats (Alonso et al. 1994; McArthur et al. 2005). This neurochemical change is similar to that of prenatal exposure to corticosterone levels, which increases basal dopamine metabolism in the dorsal and ventral striatum of both male and female offspring (Diaz et al. 1995). Also, adolescent and adult male PNS rats show higher basal levels of extracellular dopamine in the NAc (Kippin et al. 2008; Silvagni et al. 2008). There are changes in the expression and/or binding potential of dopamine D1 and D2 receptors, the dopamine transporter (DAT), and changes in the number of tyrosine hydroxylase-positive cells within the mesocorticolimbic dopamine system (i.e., NAc, medial PFC, hippocampal subregions) in PNS animals (Alonso et al. 1994; Berger et al. 2002; Henry et al. 1995; McArthur et al. 2005). Furthermore, PNS reduces spine density of medium spiny cells of the NAc in adult male rats, but not in preadolescents (Martinez-Tellez et al. 2009). These baseline changes in the dopamine system could contribute to reactions to drugs of abuse which may facilitate overuse, as discussed later. The dopamine systems of PNS animals react differently from controls to some drugs of abuse. Amphetamine stimulates greater dopamine output in the NAc of PNS adolescent and adult rats compared to controls (Silvagni et al. 2008). Conversely, amphetamine-stimulated dopamine output was blunted in the PFC in PNS animals (Carboni et al. 2010). PNS males also show higher dopamine output in the NAc after first cocaine exposure, while cocaine-experienced PNS rats exhibit increased PFC dopamine at baseline with enhanced NAc and PFC dopamine output following cocaine administration (Kippin et al. 2008).

Neurotransmitter systems beyond dopamine also contribute to drug seeking (Koob and Volkow 2010), and these systems show differences in PNS animals at baseline and after drug administration (Baier et al. 2012). In the NAc, basal levels of norepinephrine are lower in adolescent PNS rats, although this change is not present in adulthood (Silvagni et al. 2008). Both adult and adolescent rats have decreased basal PFC norepinephrine output (Carboni et al. 2010). When either amphetamine or nicotine is given, Nac norepinephrine output is increased in adult PNS rats (Silvagni et al. 2008). In the PFC, PNS increases amphetamine-stimulated norepinephrine output and PNS decreased nicotine-stimulated norepinephrine output in adults (Carboni et al. 2010). In contrast, there are no basal differences in hippocampal acetylcholine release between PNS and control male and female rats, although when exposed to injection stress or intracerebroventricular corticotropin-releasing factor (CRF) administration PNS rats have higher acetylcholine release compared to controls (Day et al. 1998). As for CRF, PNS animals do have differences in that system, with PNS rats displaying higher amygdala levels of CRF and increased CRF release in response to depolarization (Cratty et al. 1995), and increased CRF receptor binding (Ward et al. 2000). The serotonin system also changes in PNS animals. PNS male rats show reduced baseline serotonin in the NAc (Kippin et al. 2008). PNS also alters 5-HT1A receptor binding, with males showing significantly lower binding compared to controls in the ventral hippocampus, with a similar trend for PNS females (Van den Hove et al. 2006). Considering the effects of prenatal corticosterone administration alone, autoradiographs of the dorsal hippocampus found significant decreases of the 5-HT1A receptor system in the hippocampal CA1 region (Meerlo et al. 2001). Male PNS rats also show higher levels of serotonin and 5-HIAA in the dorsal striatum than controls (Gerardin et al. 2005).

The glutamate system and opiate system are also altered by PNS. PNS male rats show reductions in basal glutamate in the NAc compared to controls (Kippin et al. 2008). PNS also increases in N-methyl-D-aspartate (NMDA) receptors in several regions, including the medial PFC, dorsal frontal cortex (DFC), the CA1 region of the hippocampus, the medial striatum, and the NAc, with increases in group III metabotropic glutamate receptors in both the medial PFC and DFC (Berger et al. 2002). PNS male rats also show reduced mGluR1/5 activity in the ventral hippocampus, while female PNS rats show increased mGluR1/5 activity in both the dorsal and ventral hippocampus (Zuena et al. 2008). Further, the distribution of Homer proteins which are glutamatergic scaffolding proteins, e.g., are also altered throughout limbo-corticostriatal regions (Ary et al. 2007). Additionally, on the first administration of cocaine adult male PNS rats show enhanced glutamate neurotransmission, while cocaine-experienced PNS rats show reduced NAc glutamate at baseline and enhanced NAc glutamate following a cocaine challenge (Kippin et al. 2008). Finally, there are differences to note in the opiate system. Receptor autoradiography has shown that male and female PNS rats show decreased μ opiate receptor binding in the striatum, specifically in the anterior striatum, the NAc, the lateral amygdala, and the endopiriform cortex (Insel et al. 1990).

PNS also produces morphological changes in brain structures associated with drug reward. In the amygdala, some subnuclei diverge between PNS and control males across several measures (Krazpulski et al. 2006). By early adulthood, though, PNS individuals begin to match controls on obvious features such as nuclear volume, length, and neuron and glial cell number (Krazpulski et al. 2006). PNS animals may even surpass control volume in the lateral nucleus in later adulthood (Salm et al. 2004), highlighting the importance of developmental time point on the effects of PNS. Hippocampal morphology changes over time as well, with preadolescent PNS males displaying increased spine density in the CA1 subregion and decreased density in the CA3 subregion (Martinez-Tellez et al. 2009). By adulthood, PNS males display reductions in dendritic spines in both the CA1 and CA3 subregions (Martinez-Tellez et al. 2009). PNS or prenatal corticosterone treatment causes a decrease in synaptophysin and an increase in GAP-43 and phosphorylated GAP-43 in the hippocampus, possibly contributing to the alteration in synapse formation in the hippocampus seen after PNS (Afadlal et al. 2010). As we have shown, there are many ways in which PNS is known to change brain morphology, the basal state of neurotransmitter systems, and the reactions of these systems to drugs of abuse. Any of these changes, or even PNS changes not identified yet, might contribute to a greater individual predisposition towards addiction, and further research is needed to determine which of these changes are critical to the addiction process.

PNS also adversely impacts the ability of the adult brain to produce new neurons. New neurons are produced in high numbers within the adult olfactory bulbs and hippocampus with the ultimate precursors for these new cells being the proliferative and multipotential neural stem cells (Taupin and Gage 2002; Weiss and van der Kooy 1998). PNS reduces the number of neural stem cells in the lateral ventricle subventricular zone (SVZ) across the lifespan (Kippin et al. 2004). PNS also reduces overall proliferation in the dentate gyrus (DG; Fujioka et al. 2006; Kawamura et al. 2006; Lemaire et al. 2000, 2006; Odagiri et al. 2008; Rayen et al. 2011) and in the SVZ (Kippin et al. 2004) as well survival of new born precursors is reduced following PNS (Koo et al. 2003) in adolescent and adult rodents with similar findings reported in PNS adult primates (Coe et al. 2003). Alternatively, a wide variety of drugs of abuse negatively impact adult neurogenesis (for reviews, see, e.g., Canales 2010; Eisch and Harburg 2006) and one study reported elevated cocaine self-administration following ablation of the hippocampal neurogenesis (Noonan et al. 2010) suggesting that PNS-induced changes in neurogenesis may be critically involved in PNS-induced changes in drug-seeking behavior. Other differences in the neurogenic system emerge between the sexes following PNS (Koehl et al. 2009; Schmitz et al. 2002; Biala et al. 2011; Mandyam et al. 2008; Zuena et al. 2008) and these effects may be relevant to sex-differentiated response to PNS modulation in responsiveness to drugs of abuse (Thomas et al. 2009).

Given our observed PNS × genetic background effects on behavior, we have initiated a project to identify associated molecular epigenetic changes. In our first experiment, we specifically targeted a region of the brain that plays an important role in emotional learning and memory, the ventromedial prefrontal cortex (mPFC; i.e., infralimbic and prelimbic cortices), and determined whether epigenetic mechanisms are impacted by PNS × genetic background interactions. To this end, genome-wide DNA methylation profiling was performed using a methyl-capture microarray approach on mPFC tissue derived from PNS and control mice of B6 and D2 strains (that received no further experimental manipulations, i.e., drug naïve). Indeed, widely varied differences in DNA methylation were observed in non-PNS DBA/2J and C57BL/6J adult offspring, suggesting that underlying genetic background influences the epigenotype, which was further impacted by PNS. Interestingly, many of the genes displaying differential DNA methylation are well known to play a role in the neurodevelopmental process and neuroplasticity, and altered epigenetic regulation of these genes might therefore be responsible for the distinct behavioral phenotypes observed in these offspring in later life. To verify the results of the genome-wide assay, select genes are being examined employing bisulfite conversion and mass spectrophotometry as well as quantitative polymerase chain reaction (PCR) to determine mRNA levels. Although this project is still in its infancy, we believe that it will be instrumental in explaining the ability of prenatal perturbations to alter adult brain function, both in terms of drug responsiveness as well as other types of behaviors, such as PPI deficits. Further, the potentially fortuitous finding that PNS effects are genetically dissociable on drug-induced CPP versus PPI allows for built-in controls for elucidating PNS-induced molecular changes that are relevant to specific neuropsychiatric outcomes, although these will need to be functionally verified.

5.9 Summary and Future Goals

Clinical and preclinical studies not only indicate that the genetic inheritance of cocaine addiction ranges from 0.30 to 0.70 (reviewed in, e.g., Agrawal and Lynskey 2008) but also have implicated early environmental factors as a major risk factor for cocaine addiction (see, e.g., Ellenbroek et al. 2005). However, there is limited research on the impact of early environmental factors, such as stress, and genetic background, which comprises a large gap in our understanding of the determinants of individual vulnerability to cocaine addiction. For instance, contributions of individual genes for addiction vulnerability determined by genome-wide association studies (GWAS) typically account for a very small amount of variability (e.g., Visscher et al. 2012). However, such analyses generally neglect and/or are insensitive to environmental contributing factors that may interact with specific genetic backgrounds. Evidence from our studies and others indicate that early environmental programming can modulate genetic background differences (i.e., across strains), and may be a reason for the weak contributions determined by GWAS studies. That is by looking at genetic contribution across environmental histories, the G × E contributions may confound such approaches and dramatically underestimate the actual importance of specific genes to neuropsychiatric vulnerability. Accordingly, one of the overall benefits of examining the effects of PNS across various genetic background in a parametric fashion (i.e., RI strains) is that it will enable a way to compare genetic alone versus G × E contributions to a phenotype for specific genetic factors (i.e., loci or ideally specific gene).

Another major benefit of our current studies is that it is providing fundamental insight into the nature and relation of G × E effects on indices of multiple neuropsychiatric diseases. The changes produced by PNS in behavior and neural function have been argued to be relevant to addiction (Campbell et al. 2009; Koehl et al. 2002) as well as many neuropsychiatric diseases, including schizophrenia, ADHD, depression, and anxiety disorders (see, e.g., Koenig et al. 2002; Kofman 2002; Van den Bergh et al. 2008; Glover 2011). Further, there is a high comorbidity between addiction and other neuropsychiatric conditions (e.g., Siegfried 1998; Cuffel 1996). Accordingly, it was somewhat surprising that we were able to find strain-independent G × E effects on drug seeking (in CPP) versus psychoses (PPI;

response to novelty) following PNS in B6 and D2 mice, respectively, as this suggests independent genetic vulnerability for each disorder based on environmental history. Although this finding indicates that the role of PNS in different disorders is distinct, an extension of this suggests that disease comorbidity may be related to presence of multiple G × E effects which may exacerbate each other. Thus, when individuals have elevated drug responsiveness/reward/seeking with other disruptions, addiction may be more pronounced (e.g., exhibiting faster transition from recreational use to need for clinical treatment or being resistant to available clinical interventions). We feel understanding the relation of G × E effects on behaviors of relevance to different neuropsychiatric disorders will be a critical milestone in for identifying individuals at risk.

Acknowledgments This work was supported by National Institute on Drug Abuse (NIDA) grant DA-027525 and a National Alliance for Research on Schizophrenia and Depression (NARSAD) Young Investigator Award to TEK.

Conflict of Interest The authors declare no conflicts of interest.

References

Afadlal S, Polaboon N, Surakul P, Govitrapong P, Jutapakdeegul N (2010) Prenatal stress alters presynaptic marker proteins in the hippocampus of rat pups. Neurosci Lett 470(1):24–27

Agrawal A, Lynskey MT (2008) Are there genetic influences on addiction: evidence from family, adoption and twin studies. Addiction 103(7):1069–1081

Aldworth J, Ault K, Bishop E (2007) Results from the 2006 National Survey on Drug Use and Health: National Findings: Office of Applied Studies, NSDUH Series H-32, DHHS Publication No. SMA 07-4293, Rockville, MD

Alonso SJ, Navarro E, Rodriguez M (1994) Permanent dopaminergic alterations in then accumbens after prenatal stress. Pharmacol Biochem Behav 49:353–358

Ary AW, Aguilar VR, Szumlinski KK, Kippin TE (2007) Prenatal stress alters limbo-corticostriatal Homer protein expression. Synapse 61(11):938–941

Badiani A, Cabib S, Puglisi-Allegra S (1992) Chronic stress induces strain-dependent sensitization to the behavioral effects of amphetamine in the mouse. Pharmacol Biochem Behav 43(1):53–60

Baier CJ, Katunar MR, Adrover E, Pallarés ME, Antonelli MC (2012) Gestational restraint stress and the developing dopaminergic system: an overview. Neurotox Res 22(1):16–32

Barbazanges A, Piazza PV, Le Moal M, Maccari S (1996) Maternal glucocorticoid secretion mediates long-term effects of prenatal stress. J Neurosci 16(12):3943–3949

Belknap JK (1998) Effect of within-strain sample size on QTL detection and mapping using recombinant inbred mouse strains. Behav Genet 28(1):29–38

Belknap JK, Crabbe JC (1992) Chromosome mapping of gene loci affecting morphine and amphetamine responses in BXD recombinant inbred mice. Ann N Y Acad Sci 654:311–323

Belzung C, El Hage W, Moindrot N, Griebel G (2001) Behavioral and neurochemical changes following predatory stress in mice. Neuropharmacology 41(3):400–408

Berger MA, Barros VG, Sarchi MI, Tarazi FI, Antonelli MC (2002) Long-term effects of prenatal stress on dopamine and glutamate receptors in adult rat brain. Neurochem Res 27(11):1525–1533

Bertram CE, Hanson MA (2002) Prenatal programming of postnatal endocrine responses by glucocorticoids. Reproduction 124(4):459–467

Biala YN, Bogoch Y, Bejar C, Linial M, Weinstock M (2011) Prenatal stress diminishes gender differences in behavior and in expression of hippocampal synaptic genes and proteins in rats. Hippocampus 21(10):1114–1125

Bosch OJ, Krömer SA, Neumann ID (2006) Prenatal stress: opposite effects on anxiety and hypothalamic expression of vasopressin and corticotropin-releasing hormone in rats selectively bred for high and low anxiety. Eur J Neurosci 23(2):541–551

Cabib S, Bonaventura N (1997) Parallel strain-dependent susceptibility to environmentally-induced stereotypies and stress-induced behavioral sensitization in mice. Physiol Behav 61(4):499–506

Cabib S, Orsini C, Le Moal M, Piazza PV (2000) Abolition and reversal of strain differences in behavioral responses to drugs of abuse after a brief experience. Science 289(5478):463–465

Campbell JC, Dunn CP, Nayak RR, Cornateuanu, RJ, Kippin TE (under review) Prenatal stress produces sex-specific effects on alcohol-induced motivational plasticity in C57Bl/6 J mice. Physiol Behav (submitted)

Campbell JC, Szumlinski KK, Kippin TE (2009) Contribution of early environmental stress to alcoholism vulnerability. Alcohol 43(7):547–554

Canales JJ (2010) Comparative neuroscience of stimulant-induced memory dysfunction: role for neurogenesis in the adult hippocampus. Behav Pharmacol 21(5–6):379–393

Carboni E, Barros VG, Ibba M, Silvagni A, Mura C, Antonelli MC (2010) Prenatal restraint stress: an in vivo microdialysis study on catecholamine release in the rat prefrontal cortex. Neuroscience 168(1):156–166

Carrara-Nascimento PF, Olive MF, Camarini R (2012) Ethanol pre-exposure during adolescence or adulthood increases ethanol intake but ethanol-induced conditioned place preference is enhanced only when pre-exposure occurs in adolescence. Dev Psychobiol (epub ahead of print)

Cervantes MC, Laughlin RE, Jentsch JD (2013) Cocaine self-administration behavior in inbred mouse lines segregating different capacities for inhibitory control. Psychopharmacology (Berl) (Epub ahead of print)

Chesler EJ, Wang J, Lu L, Qu Y, Manly KF, Williams RW (2003) Genetic correlates of gene expression in recombinant inbred strains: a relational model system to explore neurobehavioral phenotypes. Neuroinformatics 1(4):343–357

CIA (2008) The world factbook: office of public affairs, CIA, Washington, DC

Clinton S, Miller S, Watson SJ, Akil H (2008) Prenatal stress does not alter innate novelty-seeking behavioral traits, but differentially affects individual differences in neuroendocrine stress responsivity. Psychoneuroendocrinology 33(2):162–177

Coe CL, Kramer M, Czeh B, Gould E, Reeves AJ, Kirschbaum C, Fuchs E (2003) Prenatal stress diminishes neurogenesis in the dentate gyrus of juvenile Rhesus monkeys. Biol Psychiatry 54(10):1025–1034

Cohen P (1990) Desires for cocaine. In: Warburton DM (ed) Addiction controversies. Chur, Harwood, 212–222

Complex Trait Consortium (2003) The nature and identification of quantitative trait loci: a community's view. Nat Rev Genet 4(11):911–916

Crabbe JC (2002) Genetic contributions to addiction. Annu Rev Psychol 53:435–462

Crabbe JC, Belknap JK, Buck KJ, Metten P (1994) Use of recombinant inbred strains for studying genetic determinants of responses to alcohol. Alcohol Alcohol Suppl 2:67–71

Crabbe JC, Phillips TJ, Buck KJ, Cunningham CL, Belknap JK (1999) Identifying genes for alcohol and drug sensitivity: recent progress and future directions. Trends Neurosci 22(4):173–179

Cratty MS, Ward HE, Johnson EA, Azzaro AJ, Birkle DL (1995) Prenatal stress increases corticotropin-releasing factor (CRF) content and release in rat amygdala minces. Brain Res, 675(1–2):297–302

Crawley JN, Belknap JK, Collins A, Crabbe JC, Frankel W, Henderson N, Hitzemann RJ, Maxson SC, Miner LL, Silva AJ, Wehner JM, Wynshaw-Boris A, Paylor R (1997) Behavioral phenotypes of inbred mouse strains: implications and recommendations for molecular studies. Psychopharmacology (Berl) 132(2):107–124

Cuffel BJ (1996) Comorbid substance use disorder: prevalence, patterns of use, and course. New Dir Ment Health Serv 70:93–105

Cunningham CL, Dickinson SD, Grahame NJ, Okorn DM, McMullin CS (1999) Genetic differences in cocaine-induced conditioned place preference in mice depend on conditioning trial duration. Psychopharmacology (Berl) 146(1):73–80

Darnaudéry M, Maccari S (2008) Epigenetic programming of the stress response in male and female rats by prenatal restrain stress. Brain Res Rev 57(2):571–585

Darnaudery M, Louvart H, Defrance L, Leonhardt M, Morley-Fletcher S, Gruber SH et al (2007) Impact of an intense stress on ethanol consumption in female rats characterized by their pre-stress preference: modulation by prenatal stress. Brain Res 1131:181–186

Day JC, Koehl M, Deroche V, Le Moal M, Maccari S (1998) Prenatal stress enhances stress- and corticotropin-releasing factor-induced stimulation of hippocampal acetylcholine release in adult rats. J Neurosci 18(5):1886–1892

DeFries JC (1964) Prenatal maternal stress in mice: differential effects on behavior. J Hered 55:289–925

DeFries JC, Weir MW, Hegmann JP (1967) Differential effects of prenatal stress on offspring behavior in mice as a function of genotype and stress. J Comp Physiol Psychol 63(2):332–334

de Jong IE, Oitzl MS, de Kloet ER (2007) Adrenalectomy prevents behavioural sensitization of mice to cocaine in a genotype-dependent manner. Behav Brain Res 177(2):329–339

de Jong IE, Steenbergen PJ, de Kloet ER (2008) Strain differences in the effects of adrenalectomy on the midbrain dopamine system: implication for behavioral sensitization to cocaine. Neuroscience 153(3):594–604

Deminiere JM, Piazza PV, Guegan G, Abrous N, Maccari S, Le Moal M et al (1992) Increased locomotor response to novelty and propensity to intravenous amphetamine self-administration in adult offspring of stressed mothers. Brain Res 586:135–139

DeTurck KH, Pohorecky LA (1987) Ethanol sensitivity in rats: effect of prenatal stress. Physiol Behav 40:407–410

DeWit J, Uhlenhuth EH, Johanson CE (1986) Individual differences in the reinforcing and subjective effects of amphetamine and diazepam. Drug Alcohol Depend 16:341–360

Diaz R, Ogren SO, Blum M, Fuxe K (1995) Prenatal corticosterone increases spontaneous and d-amphetamine induced locomotor activity and brain dopamine metabolism in prepubertal male and female rats. Neuroscience 66:467–473

Eisch AJ, Harburg GC (2006) Opiates, psychostimulants, and adult hippocampal neurogenesis: insights for addiction and stem cell biology. Hippocampus 16(3):271–286

Ellenbroek BA, van der Kam EL, van der Elst MC, Cools AR (2005) Individual differences in drug dependence in rats: the role of genetic factors and life events. Eur J Pharmacol 526(1–3):251–258

Enoch MA (2006) Genetic and environmental influences on the development of alcoholism: resilience vs. risk. Ann N Y Acad Sci, 1094:193–201

Fehr C, Shirley RL, Belknap JK, Crabbe JC, Buck KJ (2002) Congenic mapping of alcohol and pentobarbital withdrawal liability loci to a <1 centimorgan interval of murine chromosome 4: identification of Mpdz as a candidate gene. J Neurosci 22:3730–3738

Feltenstein MW, See RE (2008) The neurocircuitry of addiction: an overview. Br J Pharmacol 154(2)261–274

Fujioka A, Fujioka T, Ishida Y, Maekawa T, Nakamura S (2006) Differential effects of prenatal stress on the morphological maturation of hippocampal neurons. Neuroscience 141(2), 907–915

Gao S, Suenaga T, Oki Y, Yukie M, Nakahara D (2011) Cocaine enhances resistance to extinction of responding for brain-stimulation reward in adult prenatally stressed rats. Behav Brain Res 223(2):411–416

George FR, Goldberg SR (1989) Genetic approaches to the analysis of addiction processes. Trends Pharmacol Sci 10(2):78–83

Gerardin DC, Pereira OC, Kempinas WG, Florio JC, Moreira EG, Bernardi MM (2005) Sexual behavior, neuroendocrine, and neurochemical aspects in male rats exposed prenatally to stress. Physiol Behav 84(1):97–104

Glover V (2011) Annual research review: prenatal stress and the origins of psychopathology: an evolutionary perspective. J Child Psychol Psychiatry 52(4):356–367

Henry C, Guegant G, Cador M, Arnauld E, Arsaut J, Le Moal M et al (1995) Prenatal stress in rats facilitates amphetamine-induced sensitization and induces long-lasting changes in dopamine receptors in the nucleus accumbens. Brain Res 685 179–186

Hood HM, Metten P, Crabbe JC, Buck KJ (2006) Fine mapping of a sedative-hypnotic drug withdrawal locus on mouse chromosome 11. Genes Brain Behav 5(1):1–10
Huizink AC, Mulder EJ, Buitelaar JK (2004) Prenatal stress and risk for psychopathology: specific effects or induction of general susceptibility? Psychol Bull 130:115–142
Insel TR, Kinsley CH, Mann PE, Bridges RS (1990) Prenatal stress has long-term effects on brain opiate receptors. Brain Res 511:93–97
Kafkafi N, Elmer GI (2005) Activity density in the open field: a measure for differentiating the effect of psychostimulants. Pharmacol Biochem Behav 80(2):239–249
Kawamura T, Chen J, Takahashi T, Ichitani Y, Nakahara D (2006) Prenatal stress suppresses cell proliferation in the early developing brain. Neuroreport 17(14):1515–1518
Kendler KS, Myers J, Prescott CA (2007) Specificity of genetic and environmental risk factors for symptoms of cannabis, cocaine, alcohol, caffeine, and nicotine dependence. Arch Gen Psychiatry 64(11):1313–1320
Kinsley C, Svare B (1987) Genotype modulates prenatal stress effects on aggression in male and female mice. Behav Neural Biol 47(2):138–150
Kinsley CH, Mann PE, Bridges RS (1988) Prenatal stress alters morphine- and stress-induced analgesia in male and female rats. Pharmacol Biochem Behav 30:123–128
Kippin TE, Cain SW, Masum Z, Ralph MR (2004) Neural stem cells show bidirectional experience-dependent plasticity in the perinatal mammalian brain. J Neurosci 24:2832–2836
Kippin TE, Szumlinski KK, Kasparova Z, Rezner B, See RE (2008) Prenatal stress enhances responsiveness to cocaine. Neuropsychopharmacology 33(4):769–782
Koehl M, Bjijou Y, Le Moal M, Cador M (2000) Nicotine-induced locomotor activity is increased by preexposure of rats to prenatal stress. Brain Res 882(1–2):196–200
Koehl M, Lemaire V, Mayo W, Abrous DN, Maccari S, Piazza PV, Le Moal M, Vallée M (2002) Individual vulnerability to substance abuse and affective disorders: role of early environmental influences. Neurotox Res 4(4):281–296
Koenig JI, Kirkpatrick B, Lee P (2002) Glucocorticoid hormones and early brain development in schizophrenia. Neuropsychopharmacology 27(2):309–318
Kofman O (2002) The role of prenatal stress in the etiology of developmental behavioural disorders. Neurosci Biobehav Rev 26(4):457–470
Koo JW, Park CH, Choi SH, Kim NJ, Kim HS, Choe JC, Suh YH (2003) The postnatal environment can counteract prenatal effects on cognitive ability, cell proliferation, and synaptic protein expression. FASEB J 11:1556–1558
Koob GF, Volkow ND (2010) Neurocircuitry of addiction. Neuropsychopharmacology 35:217–238
Koehl M, Lemaire V, Le Moal M, Abrous DN (2009) Age-dependent effect of prenatal stress on hippocampal cell proliferation in female rats. Eur J Neurosci 29:635–640
Kosten TA, Miserendino MJ, Kehoe P (2000) Enhanced acquisition of cocaine self-administration in adult rats with neonatal isolation stress experience. Brain Res 875(1–2):44–50
Kosten TA, Sanchez H, Zhang XY, Kehoe P (2004) Neonatal isolation enhances acquisition of cocaine self-administration and food responding in female rats. Behav Brain Res 151(1–2):137–149
Kosten TA, Zhang XY, Kehoe P (2006) Heightened cocaine and food self-administration in female rats with neonatal isolation experience. Neuropsychopharmacology 31(1):70–76
Krazpulski M, Dickerson PA, Salm AK (2006) Prenatal stress affects the developmental trajectory of the rat amygdala. Stress 9(2):85–95
Kreek MJ (2001) Drug addictions: molecular and cellular endpoints. Ann N Y Acad Sci 937:27–49
Lemaire V, Koehl M, Le Moal M, Abrous DN (2000) Prenatal stress produces learning deficits associated with an inhibition of neurogenesis in the hippocampus Proc Natl Acad Sci U S A 97(20):11032–11037
Lemaire V, Lamarque S, Le Moal M, Piazza PV, Abrous DN (2006) Postnatal stimulation of the pups counteracts prenatal stress-induced deficits in hippocampal neurogenesis. Biol Psychiatry 59(9):786–792

Lephart ED, Watson MA, Rhees RW, Ladle DR, Jacobson NA (1997) Developmental expression of calretinin in the medial basal hypothalamus and amygdala from male and female rats. Neurosci Res 28(3):269–273
Lynch WJ, Mangini LD, Taylor JR (2005) Neonatal isolation stress potentiates cocaine seeking behavior in adult male and female rats. Neuropsychopharmacology 30:322–329
Maccari S, Darnaudery M, Morley-Fletcher S, Zuena AR, Cinque C, Van Reeth O (2003) Prenatal stress and long-term consequences: implications of glucocorticoid hormones. Neurosci Biobehav Rev 27(1–2):119–127
Mandyam CD, Crawford EF, Eisch AJ, Rivier CL, Richardson HN (2008) Stress experiences in utero reduces sexual dichotomies in neurogenesis, microenvironment, and cell death in the adult rat hippocampus. Dev Neurobio 68(5):575–589
Martinez-Tellez RI, Hernandez-Torres E, Gamboa C, Flores G (2009) Prenatal stress alters spine density and dendritic length of nucleus accumbens and hippocampus neurons in rat offspring. Synapse 63(9):794–804
Matthews SG (2002) Early programming of the hypothalamo-pituitary-adrenal axis. Trends Endocrinol Metab 13(9):373–380
McArthur S, McHale E, Dalley JW, Buckingham JC, Gillies GE (2005) Altered mesencephalic dopaminergic populations in adulthood as a consequence of brief perinatal glucocorticoid exposure. J Neuroendocrinol 17:475–482
Meerlo P, Horvath KM, Luiten PG, Angelucci L, Catalani A, Koolhaas JM (2001) Increased maternal corticosterone levels in rats: effects on brain 5-HT1A receptors and behavioral coping with stress in adult offspring. Behav Neurosci 115(5):1111–1117
Miner LL, Marley RJ (1995a) Chromosomal mapping of loci influencing sensitivity to cocaine-induced seizures in BXD recombinant inbred strains of mice. Psychopharmacology (Berl) 117(1):62–66
Miner LL, Marley RJ (1995b) Chromosomal mapping of the psychomotor stimulant effects of cocaine in BXD recombinant inbred mice. Psychopharmacology (Berl) 122(3):209–214
Montano MM, Wang MH, Even MD, vom Saal FS (1991) Serum corticosterone in fetal mice: sex differences, circadian changes, and effect of maternal stress. Physiol Behav 50(2):323–329
Morley-Fletcher S, Puopolo M, Gentili S, Gerra G, Macchia T, Laviola G (2004) Prenatal stress affects 3,4-methylenedioxymethamphetamine pharmacokinetics and drug-induced motor alterations in adolescent female rats. Eur J Pharmacol 489(1–2):89–92
Neumann ID, Krömer SA, Bosch OJ (2005) Effects of psycho-social stress during pregnancy on neuroendocrine and behavioural parameters in lactation depend on the genetically determined stress vulnerability. Psychoneuroendocrinology 30(8):791–806
Noonan MA, Bulin SE, Fuller DC, Eisch AJ (2010) Reduction of adult hippocampal neurogenesis confers vulnerability in an animal model of cocaine addiction. J Neurosci 30(1):304–315
Odagiri K, Abe H, Kawagoe C, Takeda R, Ikeda T, Matsuo H, Nonaka H, Ebihara K, Nishimori T, Ishizuka Y, Hashiguchi H, Ishida Y (2008) Psychological prenatal stress reduced the number of BrdU immunopositive cells in the dorsal hippocampus without affecting the open field behavior of male and female rats at one month of age. Neurosci Lett 446(1):25–29
Orsini C, Bonito-Oliva A, Conversi D, Cabib S (2005) Susceptibility to conditioned place preference induced by addictive drugs in mice of the C57BL/6 and DBA/2 inbred strains. Psychopharmacology (Berl) 181(2):327–336
Orsini C, Bonito-Oliva A, Conversi D, Cabib S (2008) Genetic liability increases propensity to prime-induced reinstatement of conditioned place preference in mice exposed to low cocaine. Psychopharmacology (Berl) 198(2):287–296
Philip VM, Duvvuru S, Gomero B, Ansah TA, Blaha CD, Cook MN, Hamre KM, Lariviere WR, Matthews DB, Mittleman G, Goldowitz D, Chesler EJ (2010) High-throughput behavioral phenotyping in the expanded panel of BXD recombinant inbred strains. Genes Brain Behav 9(2):129–159
Phillips TJ, Belknap JK (2002) Complex-trait genetics: emergence of multivariate strategies. Nat Rev Neurosci 3(6):478–485

Phillips TJ, Huson MG, McKinnon CS (1998) Localization of genes mediating acute and sensitized locomotor responses to cocaine in BXD/Ty recombinant inbred mice. J Neurosci 18(8):3023–3034

Phillips TJ, Belknap JK, Hitzemann RJ, Buck KJ, Cunningham CL, Crabbe JC (2002) Harnessing the mouse to unravel the genetics of human disease. Genes Brain Behav 1(1):14–26

Pohorecky LA, Roberts P (1991) Activity in a modified open-field apparatus: effect of diazepam and prenatal stress. Neurotoxicol Teratol 13:129–133

Pohorecky LA, Roberts P, Cotler S, Carbone JJ (1989) Alteration of the effects of caffeine by prenatal stress. Pharmacol Biochem Behav 33(1):55–62

Pouletty P (2002) Drug addictions: towards socially accepted and medically treatable diseases. Nat Rev Drug Discov 1(9):731–736

Prakash P, Merali Z, Kolajova M, Tannenbaum BM, Anisman H (2006) Maternal factors and monoamine changes in stress-resilient and susceptible mice: cross-fostering effects. Brain Res 1111(1):122–133

Rayen I, Van den Hove DL, Prickaerts J, Steinbusch HW, Pawluski JL (2011) Fluoxetine during development reverses the effects of prenatal stress on depressive-like behavior and hippocampal neurogenesis in adolescence. PLoS ONE 6(9):e24003

Salm AK, Pavelko M, Krouse EM, Webster W, Kraszpulski M, Birkle D (2004) Lateral amygdaloid nucleus expansion in adult rats is associated with exposure to prenatal stress. Dev Brain Res 148:159–167

Schmitz C, Rhodes ME, Bludau M, Kaplan S, Ong P, Ueffing I, Vehoff J, Korr H, Frye CA (2002) Depression: reduced number of granule cells in the hippocampus of female, but not male, rats due to prenatal restraint stress. Mol Psychiatry 7(7):810–813

Schuckit MA (2009) An overview of genetic influences in alcoholism. J Subst Abuse Treat 36(1):S5–S14

Seale TW, Carney JM (1991) Genetic determinants of susceptibility to the rewarding and other behavioral actions of cocaine. J Addict Dis 10(1–2):141–162

Seckl JR, Meaney MJ (2004) Glucocorticoid programming. Ann N Y Acad Sci 1032:63–84

Siegfried N (1998) A review of comorbidity: major mental illness and problematic substance use. Aust N Z J Psychiatry 32:707–717

Silvagni A, Barros VG, Mura C, Antonelli MC, Carboni E (2008) Prenatal restraint stress differentially modifies basal and stimulated dopamine and noradrenaline release in the nucleus accumbens shell: an 'in vivo' microdialysis study in adolescent and young adult rats. Eur J Neurosci 28(4):744–758

Sternberg WF, Ridgeway CG (2003) Effects of gestational stress and neonatal handling on pain, analgesia, and stress behavior of adult mice. Physiol Behav 78:375–383

Stöhr T, Schulte Wermeling D, Szuran T, Pliska V, Domeney A, Welzl H, Weiner I, Feldon J (1998) Differential effects of prenatal stress in two inbred strains of rats. Pharmacol Biochem Behav 59(4):799–805

Szumlinski KK, Lominac KD, Kleschen M, Oleson EB, Dehoff M, Schwartz M, Seeberg P, Worley PF, Kalivas PW (2005) Behavioural and neurochemical phenotyping of *Homer1* mutant mice: possible implications for schizophrenia. Genes Brain Behav 4:273–288

Szumlinski KK, Ary AW, Lominac KD, Klugmann M, Kippin TE (2008) Accumbens Homer2 over-expression facilitates alcohol-induced neuroplasticity in C57BL/6 J mice. Neuropsychopharmacology 33(6):1365–1378

Takahashi LK, Turner JB, Kalin NH (1998) Prolonged stress-induced elevation in plasma corticosterone during pregnancy in the rat: implications for prenatal stress studies. Psychoneuroendocrinology 23(6):571–581

Taupin P, Gage FH (2002) Adult neurogenesis and neural stem cells of the central nervous system in mammals. J Neurosci Res 69(6):745–749

Tazumi T, Hori E, Uwano T, Umeno K, Tanebe K, Tabuchi E, Ono T, Nishijo H (2005) Effects of prenatal maternal stress by repeated cold environment on behavioral and emotional development in the rat offspring. Behav Brain Res 162:153–160

Thomas MB, Hu M, Lee TM, Bhatnagar S, Becker JB (2009) Sex-specific susceptibility to cocaine in rats with a history of prenatal stress. Physiol Behav 97(2):270–277
Tolliver BK, Carney JM (1994a) Comparison of cocaine and GBR 12935: effects on locomotor activity and stereotypy in two inbred mouse strains. Pharmacol Biochem Behav 48(3):733–739
Tolliver BK, Carney JM (1994b) Sensitization to stereotypy in DBA/2 J but not C57BL/6 J mice with repeated cocaine. Pharmacol Biochem Behav 48(1):169–173
Tolliver BK, Carney JM (1995) Locomotor stimulant effects of cocaine and novel cocaine analogs in DBA/2 J and C57BL/6 J inbred mice. Pharmacol Biochem Behav 50(2):163–169
Tolliver BK, Belknap JK, Woods WE, Carney JM (1994) Genetic analysis of sensitization and tolerance to cocaine. J Pharmacol Exp Ther 270(3):1230–1238
Van den Bergh BR, Van Calster B, Smits T, Van Huffel S, Lagae L (2008) Antenatal maternal anxiety is related to HPA-axis dysregulation and self-reported depressive symptoms in adolescence: a prospective study on the fetal origins of depressed mood. Neuropsychopharmacology 33(3):536–545
Van den Hove DL, Lauder JM, Scheepens A, Prickaerts J, Blanco CE, Steinbusch HW (2006) Prenatal stress in the rat alters 5-HT1A receptor binding in the ventral hippocampus. Brain Res 1090(1):29–34
van der Kam EL, Ellenbroek BA, Cools AR (2005) Gene-environment interactions determine the individual variability in cocaine self-administration. Neuropharmacology 48(5):685–695
van der Veen R, Piazza PV, Deroche-Gamonet V (2007) Gene-environment interactions in vulnerability to cocaine intravenous self-administration: a brief social experience affects intake in DBA/2 J but not in C57BL/6 J mice. Psychopharmacology (Berl) 193(2):179–186
van der Veen R, Koehl M, Abrous DN, de Kloet ER, Piazza PV, Deroche-Gamonet V (2008) Maternal environment influences cocaine intake in adulthood in a genotype-dependent manner. PLoS ONE 3(5):e2245
Van Waes V, Enache M, Dutriez I, Lesage J, Morley-Fletcher S, Vinner E et al (2006) Hypo-response of the hypothalamic-pituitaryadrenocortical axis after an ethanol challenge in prenatally stressed adolescent male rats. Eur J Neurosci 24:1193–1200
Van Waes V, Darnaudery M, Marrocco J, Gruber SH, Talavera E, Mairesse J, Van Camp G, Casolla B, Nicoletti F, Mathe AA, Maccari S, Morley-Fletcher S (2011a) Impact of early life stress on alcohol consumption and on the short- and long-term responses to alcohol in adolescent female rats. Behav Brain Res 221(1):43–49
Van Waes V, Enache M, Berton O, Vinner E, Lhermitte M, Maccari S, Darnaudery M (2011b) Effects of prenatal stress on alcohol preference and sensitivity to chronic alcohol exposure in male rats. Psychopharmacology (Berl) 214(1):197–208
Visscher PM, Brown MA, McCarthy MI, Yang J (2012) Five years of GWAS discovery. Am J Hum Genet 90(1):7–24
Wahlsten D, Metten P, Phillips TJ, Boehm SL 2nd, Burkhart-Kasch S, Dorow J, Doerksen S, Downing C, Fogarty J, Rodd-Henricks K, Hen R, McKinnon CS, Merrill CM, Nolte C, Schalomon M, Schlumbohm JP, Sibert JR, Wenger CD, Dudek BC, Crabbe JC (2003) Different data from different labs: lessons from studies of gene-environment interaction. J Neurobiol 54(1):283–311
Ward HE, Johnson EA, Salm AK, Birkle DL (2000) Effects of prenatal stress on defensive withdrawal behavior and corticotropin releasing factor systems in rat brain. Physiol Behav 70(3–4):359–366
Weinstock M (2001) Alterations induced by gestational stress in brain morphology and behaviour of the offspring. Prog Neurobiol 65:427–451
Weinstock M (2008) The long-term behavioural consequences of prenatal stress. Neurosci Biobehav Rev 32:1073–1086
Weinstock M, Fride E (1989) Alterations in behavioral and striatal dopamine asymmetries induced by prenatal stress. Pharmacol Biochem Behav 32(2):425–430
Weir MW, DeFries JC (1964) Prenatal maternal influence of behavior in mice: evidence of a genetic basis. J Comp Physiol Psychol 58:412–417

Weiss S, van der Kooy D (1998) CNS stem cells: where's the biology (a.k.a. beef)? J Neurobiol 36(2):307–314

Yang J, Li W, Liu X, Li Z, Li H, Yang G, Xu L, Li L (2006) Enriched environment treatment counteracts enhanced addictive and depressive-like behavior induced by prenatal chronic stress. Brain Res 1125:192–197

Zhang XY, Sanchez H, Kehoe P, Kosten TA (2005) Neonatal isolation enhances maintenance but not reinstatement of cocaine self-administration in adult male rats. Psychopharmacology (Berl) 177(4):391–399

Zhang TY, Bagot R, Parent C, Nesbitt C, Bredy TW, Caldji C, Fish E, Anisman H, Szyf M, Meaney MJ (2006) Maternal programming of defensive responses through sustained effects on gene expression. Biol Psychol 73(1):72–89

Zuena AR, Mairesse J, Casolini P, Cinque C, Alema GS, Morley-Fletcher S, Chiodi V, Spagnoli LG, Gradini R, Catalani A, Nicoletti F, Maccari S (2008) Prenatal restraint stress generates two distinct behavioral and neurochemical profiles in male and female rats. PLoS ONE 3(5):e2170

Chapter 6
A Self-Medication Hypothesis for Increased Vulnerability to Drug Abuse in Prenatally Restraint Stressed Rats

Marie-Line Reynaert, Jordan Marrocco, Eleonora Gatta, Jérôme Mairesse, Gilles Van Camp, Francesca Fagioli, Stefania Maccari, Ferdinando Nicoletti and Sara Morley-Fletcher

Abstract Stress-related events that occur in the perinatal period can permanently change brain and behavior of the developing individual and there is increasing evidence that early-life adversity is a contributing factor in the etiology of drug abuse and mood disorders. Neural adaptations resulting from early-life stress may mediate individual differences in novelty responsiveness and in turn contribute to drug abuse vulnerability. Prenatal restraint stress (PRS) in rats is a well-documented model of early stress known to induce long-lasting neurobiological and behavioral alterations including impaired feedback mechanisms of the HPA axis, enhanced novelty seeking, and increased sensitiveness to psychostimulants as well as anxiety/depression-like behavior. Together with the HPA axis, functional alterations of the mesolimbic dopamine system and of the metabotropic glutamate receptors system appear to be involved in the addiction-like profile of PRS rats.

Note: Authors Ferdinando Nicoletti and Sara Morley-Fletcher are co-last authors.

S. Maccari (✉) · M.-L. Reynaert · E. Gatta · J. Mairesse · G. Van Camp · S. Morley-Fletcher · F. Nicoletti
International Associated Laboratory (LIA) "Prenatal Stress and Neurodegenerative Diseases", UMR8576 University Lille 1/CNRS, Villeneuve d'Ascq, France and Sapienza University of Rome/IRCCS Neuromed, Pozzilli, Italy
e-mail: stefania.maccari@univ-lille1.fr

J. Marrocco
IRCCS Centro Neurolesi "Bonino-Pulejo", Messina, Italy

F. Fagioli
Department of Psychiatry, Sapienza University of Rome, Roma, Italy

AziendaSanitaria Locale, RM.E. Unità Operativa Complessa Adolescent, Rome, Italy

M. C. Antonelli (ed.), *Perinatal Programming of Neurodevelopment,*
Advances in Neurobiology 10, DOI 10.1007/978-1-4939-1372-5_6,

6.1 Programming of a Phenotype by Stress: Validity of the PRS Rat as a Model of "Addiction"

Stress-related events that occur in the perinatal period can permanently change brain and behavior of the developing individual and, there is increasing evidence that early-life adversity is a contributing factor in the etiology of drug abuse and mood disorders (Charmandari et al. 2003; Seckl 2008; McEwen 2012). Stress plays a key role in modulating the development and the expression of addictive behavior, and is moreover a major cause of relapse following periods of abstinence (Ungless et al. 2010; Shaham et al. 2003). In animals, a greater behavioral reactivity to a mild stress, such as exposure to a novel environment, is an index of the vulnerability to acquire amphetamine self-administration (Maccari et al. 1991a, b; Piazza et al. 1991). Biological responses to stress as well as behavioral reactivity to novelty may predict such vulnerability. Indeed, previous work have shown that repeated corticosterone administration in drinking water increased the locomotor response to amphetamine, suggesting that corticosterone secretion may be one of the mechanisms by which repeated stress increases the behavioral responses to amphetamine (for a review, see Shaham et al. 2003).

The hypothalamus–pituitary–adrenal (HPA) axis modulates adaptive behavior and there are compelling evidences about a key role of the stress system in shaping individual vulnerability to drug abuse. Individual predisposition to develop amphetamine self-administration is associated with impairment in corticosteroid negative feedback mechanisms (Maccari et al. 1991a; Fig. 6.1a). Indeed, a decrease of corticosteroid receptors, type I (mineralocorticoid receptor, MR) and type II (glucocorticoid receptor, GR), in hippocampus is associated to higher amphetamine self-administration in high-responding rats. Moreover, repeated exposure to stress sensitizes motor and addictive effects of drugs of abuse, and stress-induced behavioral sensitization depends on the secretion of glucocorticoids (Maccari et al. 1991b; Piazza et al. 1991). Rats with a longer duration of corticosterone secretion after exposure to novelty showed facilitation of acquisition of amphetamine self-administration and, corticosterone administration in nonpredisposed individuals increases the reinforcing value of the drug and facilitates the acquisition of amphetamine self-administration (Piazza et al. 1991; Fig. 6.1b). These results clearly indicate that the stress-related activity of the HPA axis may play a role in the pathogenesis of psychostimulant addiction and suggest that pharmacological manipulations of HPA axis system could reveal new therapeutic strategies for drug abuse.

During the past 20 years, we have studied the influences of a prenatal restraint stress (PRS) in a rat animal model. The prenatal stress procedure we have used consisted in restraining the pregnant rat—in a transparent Plexiglas cylinder, three times/day for 45 min under bright light—at the day 11 of pregnancy until delivery at 21–22 days (Maccari et al. 1995; 2003; Morley-Fletcher et al. 2003a; Fig. 6.2). The HPA axis of the PRS offspring is long-term impaired with a prolonged corticosterone stress response during life span (Maccari et al. 1995, 2003; Vallée et al. 1997; Koehl et al. 1999; Morley-Fletcher et al. 2003b) and reduced levels of both

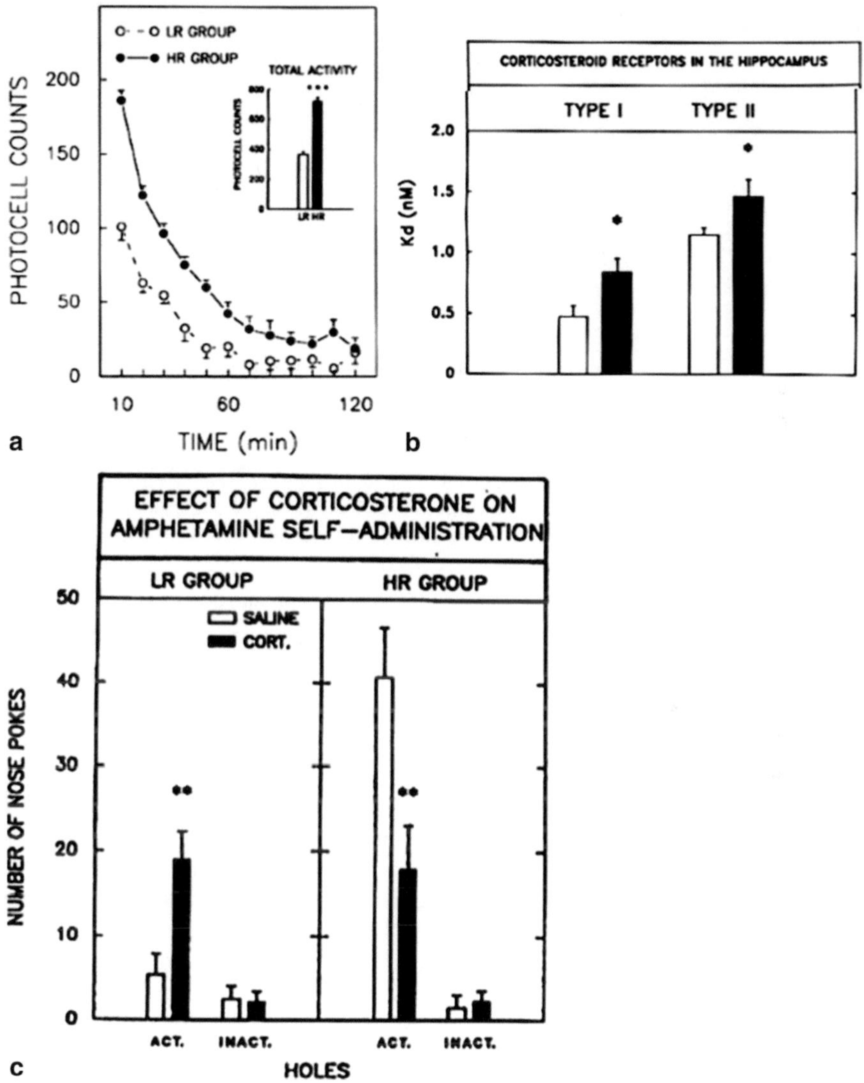

Fig. 6.1 Impact of corticosterone (*CORT*) on amphetamine self-administration. **a** HR (High responding) rats are characterized by an increased reactivity to novelty, decreased affinities of hippocampal type I and type II receptors and **b** increased amphetamine self-administration respect to LR (Low responding) animals. **c** Intravenous infusion of *CORT* increased the nose pokes in LR animals while it decreased it in HR rats. $*P<0.05$ versus LR rats and $**P<0.05$ versus *CORT*-treated group. (Original data are reported in Maccari et al. 1991a (**a**, **b**) and Piazza et al. 1991 (**c**))

type I (MR) and type II (GR) corticosteroid receptors in the hippocampus at the adolescent and adult stage (Henry et al. 1994; Maccari et al. 1995; Van Waes et al. 2006; Fig. 6.3). The age-related HPA axis dysfunctions are enhanced by PRS. Indeed, the HPA axis period of hyporesponsiveness was abolished in newborn PRS

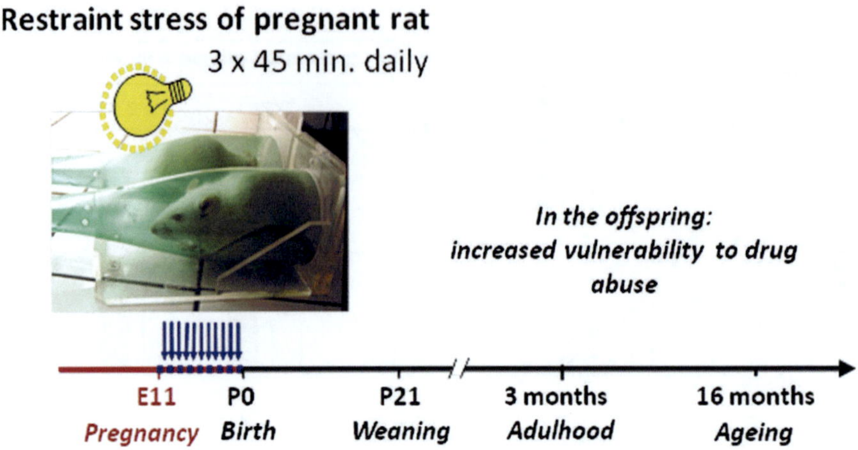

Fig. 6.2 The prenatal restraint stress model in the rat. During the last week of gestation, the dam is restrained under a strong light for 45 min, three times per day. No painful contention is applied, but she cannot escape to this situation. The offspring and the mother are then left undisturbed until weaning (*P21*)

rats (Henry et al. 1994) and circulating glucocorticoids levels of PRS middle-aged animals were similar to those found in old nonstressed animals (Vallée et al. 1999). We have reported several behavioral disturbances in PRS rats that are associated with the impairment of HPA axis activity such as increased anxiety (Vallée et al. 1997, 1999; Laloux et al. 2012; Marrocco et al. 2012), reduced social play during adolescence (Morley-Fletcher et al. 2003b), or increased immobility in the forced swim test during adulthood (Morley-Fletcher et al. 2003a, 2004a, 2011; for review, see Darnaudery and Maccari 2008).

PRS thus induces persistent behavioral and neurobiological alterations leading to enhanced response to drug abuse. Those effects are evident in both male and female PRS animals and in response to different reinforcing stimuli. Indeed, PRS rats express a higher locomotor response to novelty and enhanced amphetamine self-administration (Deminière et al. 1992; Fig. 6.4a), and increased sensitiveness to nicotine-induced locomotor activity (Koehl et al. 2000; Fig. 6.4b). Cocaine-naive PRS rats exhibit increased locomotor activity in response to both a novel environment and noncontingent cocaine injections (Kippin et al. 2008). Moreover, PRS rats with a history of cocaine self-administration exhibit increased cocaine seeking (as measured by unreinforced active lever presses) during extinction training and cocaine-primed reinstatement following operant self- administration of cocaine (Kippin et al. 2008). PRS enhances also sensitivity to 3,4-methylenedioxy-N-methylamphetamine (MDMA) “ecstasy” motor alterations and MDMA pharmacokinetics in adolescent female rats (30 days; Morley-Fletcher et al. 2004b; Fig. 6.4c). We found that PRS increased levels of plasmatic MDMA during the kinetic assessment with respect to controls and induced a higher frequency of altered motor coordination following MDMA administration, thus indicating a strong consistency between drug blood levels and behavior. Moreover, the blood concentrations of MDMA in adolescent

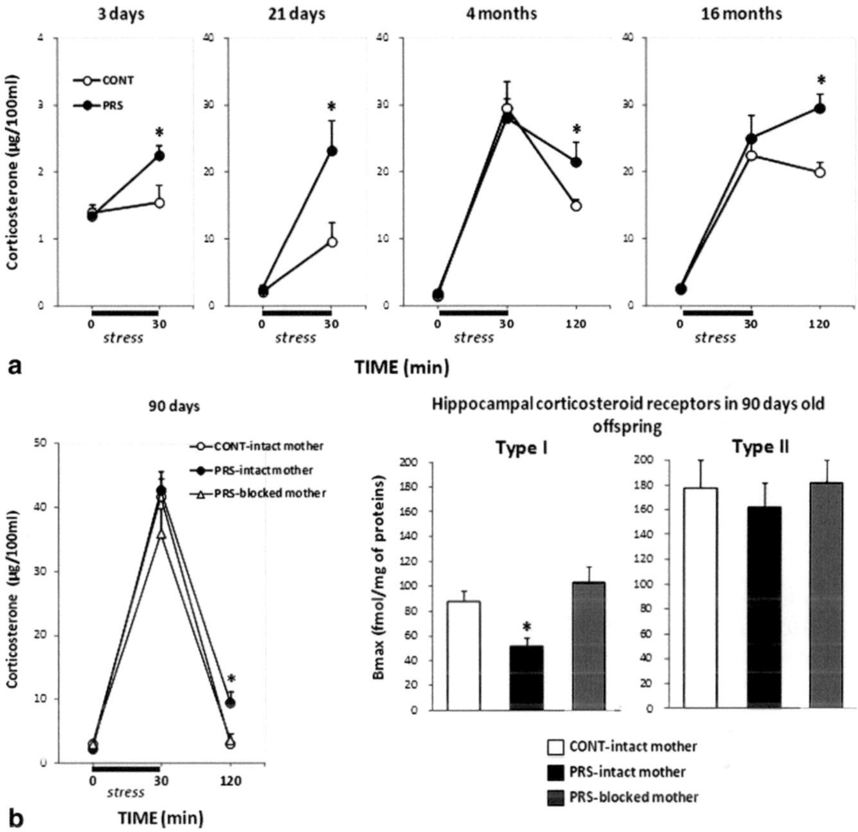

Fig. 6.3 HPA axis alterations in prenatal restraint stress (*PRS*) rats. **a** *PRS* enhances corticosterone response to stress during life span. **b** Blockade of the mother's stress-induced glucocorticoid secretion suppresses the prolonged stress-induced corticosteroid response and the decrease in type I hippocampal corticosteroid receptors usually observed in *PRS* adults. $^*P<0.05$ and $^{**}P<0.01$ *versus* controls. (Original data are reported in Henry et al. 1995 and Vallée et al. 1999 (**a**) and Barbazanges et al. 1996 (**b**))

rats were already within the range reported following a single MDMA administration in humans (Helmlin et al. 1996), thereby supporting the periadolescent rodent as a valid animal model to be used in the assessment of the vulnerability to psychostimulants (Laviola et al. 1999). The constant recovery of MDMA in the prenatal stress group seems to point towards an inhibition of MDMA metabolism. Differences in the metabolism of MDMA have been reported in humans (Hiramatsu 1990) as well as in animals (Malpass et al. 1999). Although enzymatic assays are required to validate this hypothesis, the different degradation route observed in the prenatal stress group could lead to suppose an altered enzymatic activity. Given the importance of drug metabolism in determining the magnitude of the effects of a drug, the differences in MDMA metabolism observed may impact the likelihood of adverse consequences in PRS rats. One direct consequence would be the development of

Fig. 6.4 Response to psychostimulants in prenatal restraint stress (*PRS*) male rats. **a** Amphetamine self-administration behavior exhibited by *PRS* and control male rats. *PRS* rats exhibited greater responsiveness (number of injections) to amphetamine than control rats over sessions. **b** Effects of acute injections of different doses of nicotine in *PRS* and control male rats. The locomotor activating effect of nicotine is increased in *PRS* male rats. **c** Pharmacokinetic examination of 3,4-methylenedioxy-N-methylamphetamine (*MDMA*) blood levels shown by adolescent prenatally stressed and control female rats (30 days of age). $*P<0.05$ and $^{\#}P<0.05$ versus previous dose. (Original data are reported in Deminière et al. 1992 (**a**), Koehl et al. 2000 (**b**), and Morley-Fletcher et al. 2004 (**c**))

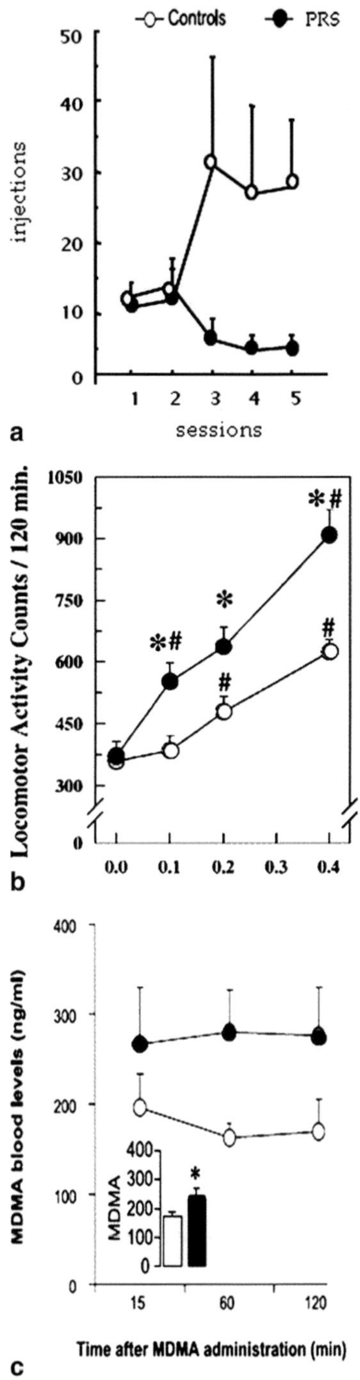

acute toxicity at moderate doses of MDMA because the drug would accumulate in the body instead of being metabolized and inactivated. It has been suggested that in addition to serotonergic system (Ricaurte et al. 1988), the HPA axis plays a role in neurotoxic effects of MDMA. Thus, MDMA stimulates corticosterone release (Nash et al. 1988) and alters corticoid receptor gene expression (Yau et al. 1994). Together with the aforementioned alterations in the feedback inhibition of the HPA axis, PRS induces an impairment of the serotonergic system (Peters 1982, 1990; Morley-Fletcher et al. 2004a). These data strongly suggest that HPA axis as well as serotonergic system activity play a key role in the enhanced response to MDMA found in PRS female rats.

In addition to psychostimulants, PRS also modulates response to alcohol. We have conducted several studies on the effect of chronic ethanol treatment on PRS rats and measured the effects of alcohol exposure in both male and female PRS rats during adolescence, adulthood, or ageing. Male adolescent PRS rats show a reduced activation of the HPA axis in response to acute alcohol administration (Van Waes et al. 2006), a phenomenon observed in heavy drinkers and their relatives, as well as in alcohol-dependent rats. We examined the impact of chronic ethanol treatment on spatial recognition performance in middle-aged (8- to 9-month old) PRS rats. Prolonged ethanol consumption is associated with cognitive disorders, especially related to memory, in humans (Oscar-Berman et al. 2004). However, studies of the cognitive outcome associated with moderate chronic ethanol consumption have yielded complex and heterogeneous results in humans as well as in rodents, leading to a consideration of vulnerable phenotypes. Indeed, memory performance has variously been reported to be impaired (Farr et al. 2005; Matthews and Morrow 2000), unaffected (Fadda et al. 1999; Gál and Bárdos 1994; Homewood et al. 1997), or even improved (Krazem et al. 2003a, b; Robles and Sabriá 2008; Steigerwald and Miller 1997) in experimental animals following chronic ethanol exposure. Several factors such as the dose, the duration of ethanol exposure, or the mode of treatment (withdrawal episodes) determine the impact of ethanol on memory. However, little is known about the repercussions of factors linked to the history of an individual on the effects of alcohol in the long term.

We found that chronic ethanol consumption induced memory impairment in control animals, whereas it attenuated the memory deficit observed in PRS rats (Fig. 6.5a). Several variables, including the presence and extent of a withdrawal period, can strongly influence the effect of ethanol consumption on cognitive function. For example, Lukoyanov et al. (1999) have shown that rats chronically exposed to ethanol for 13 months develop cognitive impairment only after 6 weeks of withdrawal and, Krazem and colleagues have found a "bidirectional effect" of chronic ethanol exposure on spatial memory, depending on the age (Krazem et al. 2003a, b). Despite an absence of withdrawal period in our study, we found a significant impairment of spatial memory in the alcoholized control rats. Unexpectedly, however, a similar chronic exposure to ethanol in PRS rats attenuated their spatial memory deficits. In a previous experiment, we have shown that the kinetic of blood ethanol levels after ethanol administration was not affected by PRS, and thus it can be assumed that the opposite behavioral changes observed here do not reflect

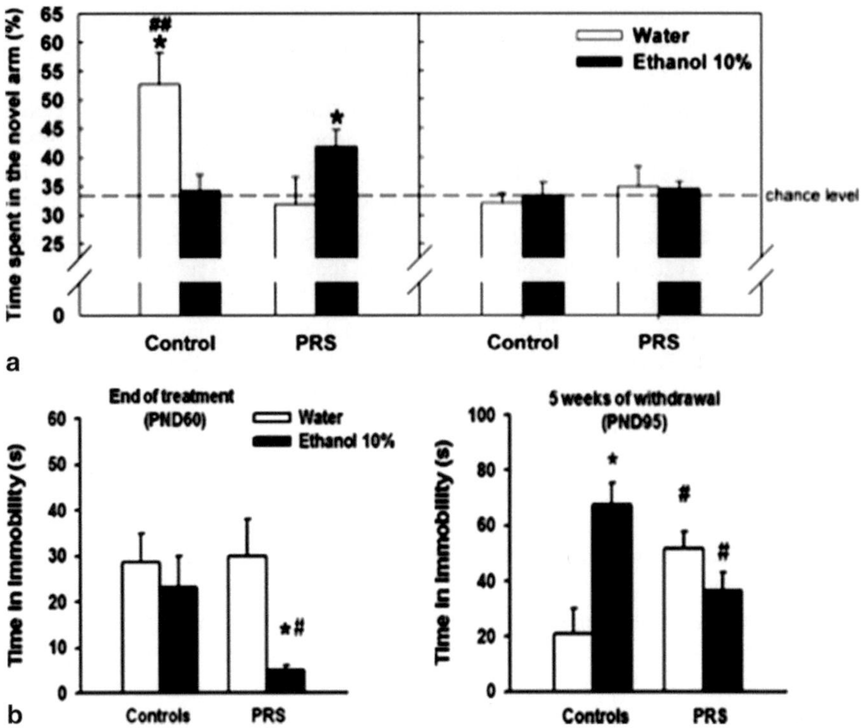

Fig. 6.5 Response to ethanol exposure in prenatal restraint stress (*PRS*) male and female rats. **a** Attenuation of spatial memory deficit by ethanol in *PRS* male rats after a forced consumption of 7 months of a 10 % ethanol solution. Time spent in the novel arm of the Y maze after 6 h intertrial intervals in *PRS* male rats. Ethanol-induced memory deficit in control animals. *PRS* rats presented constitutive spatial memory impairment but this deleterious effect was reduced by ethanol treatment. **b** Antidepressant-like effect of ethanol treatment during adolescence in *PRS* female rats in the forced swim test and after a withdrawal period. $^{*}P < 0.05$ *versus* controls, $^{\#}P < 0.05$ versus water-treated group. (Original data are reported in Van Waes et al. 2009 (**a**) and Van Waes et al. 2011b (**b**))

varying levels of circulating ethanol (Van Waes et al. 2006). Furthermore, a recent work demonstrates that the deleterious impact of chronic ethanol consumption on learning and memory in rodent is not related to changes in caloric intake (Farr et al. 2005).

Studies in nonhuman primates and rodents have reported alterations in the structure and function of the hippocampus as a consequence of prenatal stress (Coe et al. 2003; Son et al. 2006) and several works have shown learning and/or memory impairment in offspring of dams stressed during pregnancy (Mueller and Bale 2007; Zuena et al. 2008). However, the timing of the prenatal stress (early vs. mid- or late-gestation; e.g., Mueller and Bale 2007), the type and/or the severity of the prenatal stress and the sex of the offspring (e.g., Zuena et al. 2008), and the age of the offspring at time of memory assessment (juvenile, adult or aged) appear to be critical. PRS during the late gestation has generally few effects on performance in young male adults (Vallée et al. 1999; Zuena et al. 2008), but exacerbates the memory disorders observed during aging (Darnaudéry et al. 2006; Vallée et al. 1999). Accordingly, we observed a decrease in spatial memory performance in 9-month-old male PRS rats. As opposed to control animals, PRS rats did not differentiate between the novel arm and the two other arms after an intertrial interval of 6 h. In conclusion, the same chronic ethanol treatment produced differential effects in control and PRS animals on memory in the Y maze, thereby suggesting that chronic exposure to ethanol might be detrimental in subjects with "normal" cognitive function, but become beneficial if memory is impaired, as occurs in aged or PRS animals. The damaging effects of ethanol on memory capabilities are thus masked by early-life stress.

Interestingly, spontaneous alcohol consumption is not affected by PRS (Van Waes et al. 2009), although PRS influences mechanisms of neuroadaptation induced by ethanol in the reward pathway, since it exacerbates induction of ΔFosB in the nucleus accumbens (Nac) following chronic exposure to alcohol in PRS males (Van Waes et al. 2011a). To our knowledge, this was the first demonstration that early environmental manipulations can interact with the molecular effect of alcohol later in life. Other data also indicate that environmental enrichment during early stages of life is able to change the consequences of repeated cocaine administration on striatal ΔFosB levels. Indeed, ΔFosB levels are upregulated by cocaine in mice reared in an enriched environment but downregulated after the same treatment in the ones reared in a standard environment (Solinas et al. 2009). Transgenic mice that overexpress ΔFosB display augmented locomotor responses to cocaine and sensitivity to the rewarding effects of cocaine and morphine in place conditioning test (Kelz et al. 1999; Zachariou et al. 2006). Moreover, mice expressing ΔFosB self-administer more cocaine in a progressive ratio procedure, suggesting that ΔFosB may sensitize animals to the incentive motivational properties of this drug (Colby et al. 2003). The link between ΔFosB levels in the Nac and the rewarding effect of alcohol seems more complex. Indeed, dissociation between alcohol preference and saccharin preference has been reported in FosB knockout mice. The permanent elimination of FosB gene products does not alter alcohol intake but enhances the preference for sweet solution in mice (Korkosz et al. 2004). Several experimental procedures, such as a lesion of the subthalamic nucleus, have been shown to increase motivation for alcohol in a self-administration paradigm without impacting alcohol intake (Lardeux and Baunez 2008). In this context, it would be important to examine the motivation for alcohol in an operant paradigm (using a progressive ratio schedule) in PRS animals (Campbell et al. 2009). It remains to determine whether the exacerbated ΔFosB upregulation we observed in EtOh-treated PRS rats could be generalized in the same animal model to other reinforcing stimuli such as psychostimulants or even nondrug reinforces (Olausson et al. 2006).

We also examined the effect of PRS on the neurobehavioral outcome of ethanol exposure during adolescence in female rats (Fig. 6.5b). First, we studied the impact of PRS on ethanol preference during adolescence. PRS slightly increased ethanol preference *per se,* but abolished the effect of social isolation on ethanol preference. We then studied the impact of PRS on short- and long-term responses to ethanol focusing on behavioral and neurochemical parameters related to depression/anxiety.

PRS or unstressed adolescent female rats received 10% ethanol in the drinking water for 4 weeks. At P60 (i.e., at the end of the treatment), the immobility time in the forced-swim test did not differ between PRS and unstressed rats receiving water alone. Ethanol consumption had no effect in unstressed rats, but significantly reduced the immobility time in PRS rats. In contrast, the exposure to ethanol during adolescence induced in control unstressed rats a substantial increase in the immobility time in the forced swim test after five weeks of withdrawal, when female rats had become adult. The forced swim test has "pharmacological validity" as a model of mood disorders because it is responsive to antidepressant drugs (Porsolt et al. 1978). Thus, female nonstressed rats became "depressed" in the adult life after having passively consumed ethanol during adolescence whereas PRS had a strong protective effect on the behavioral outcome of ethanol consumption during adolescence. Thus, surprisingly, early-life stress prevented the detrimental effects of ethanol on mood regulation, and, the other way around, ethanol consumption during adolescence prevented the depressive phenotype induced by PRS in the adult life. We specifically choose to perform this study in female PRS rats because of the higher prevalence of major depression in women. PRS rats represent indeed an etiological model of depression, in which the behavioral and neurochemical abnormalities seen in the adult life likely reflect the development of a pathological epigenetic programming triggered by early-life stress (Maccari and Morley-Fletcher 2007). Thus, present data suggest that the epigenetic program induced by PRS interacts with alcohol consumption during adolescence in shaping the vulnerability to mood disorders later in life.

6.1.1 Neural Substrates Implicated in PRS-Induced Addiction-Like Behavior

An extensive literature has emerged examining the neural circuits and cellular mechanisms through which stress modulates addictive behavior (Koob 2008). Sensitization of the dopaminergic response to drugs is considered the neural substrate of behavioral sensitization and has been implicated in vulnerability to drug abuse. The HPA axis plays a key role in the development of sensitization to psychostimulants. This idea is further supported by the presence of glucocorticoids receptors in the dopaminergic neurons of the ventral tegmental area (VTA) projecting in the Nac (Härfstrand et al. 1986). Glucocorticoids promote sensitization to psychostimulants in rats (Rivet et al. 1989; Maccari et al. 1991b; Piazza et al. 1991; Deroche et al. 1995), while administration of psychostimulants has been shown to activate the HPA axis (Swerdlow et al. 1993). Glucocorticoids control stress-induced sensitization by changing the sensitivity of the mesencephalic dopaminergic transmission to drugs of abuse. There is also evidence for control of corticosteroid receptors by dopamine (DA), since DA inhibits the expression of corticosteroid receptors in the anterior pituitary (Antakly et al. 1987; Casolini et al. 1993), while administration of amphetamine decreased the concentration of corticosteroid receptors (Lowy 1990). Several studies conducted on PRS rats have evidenced functional alterations in mesolimbic DA system. In particular, PRS induces a significant increase in DRD2 receptor binding and mRNA in the core region of the Nac (Henry et al. 1995; Berger et al. 2002; Rodrigues et al. 2012), and a marked decrease in DRD3 receptor binding in both the shell and the core of the NAc (Henry et al. 1995). Moreover, basal and amphetamine-stimulated DA output in adolescent and adult PRS rats is higher than in control unstressed rats when measured by microdialysis in the shell region of the Nac (Silvagni et al. 2008). Recently, Hausknecht et al. (2013) have shown that a PRS paradigm that leads to increased locomotor activity to novelty and enhanced response to amphetamine also causes a persistent reduction in the spontaneous activity of VTA DA neurons recorded in adult animals. Such reduction of neural activity can be reversed by acute apomorphine that normally inhibits the impulse activity of DA neurons. Furthermore, the reduced number of spontaneously active VTA DA neurons can be also reversed by acute psychostimulants (e.g., amphetamine, cocaine), which in control rats inhibited the activity of VTA DA neurons. These findings lead to suppose that the reversal effect on VTA DA neurons observed in PRS animals represents an actual increase in the impulse activity. This effect could contribute to the increased responding to psychostimulants and mediate the increased addiction risk after prenatal stress (Hausknecht et al. 2013).

The administration of drugs of abuse, including cocaine and nicotine, also increases glutamatergic neurotransmission in brain structures implicated in the regulation of reward processes, such as the dorsal striatum (McKee and Meshul 2005), NAc (Pierce et al. 1996) and VTA (Kalivas and Duffy 1995, 1998; Schilstrom et al. 2000). Moreover, drug-induced adaptations in glutamatergic neurotransmission have been suggested to be involved in the development of drug dependence (Nicoletti et al. 1996; Battaglia et al. 2002; Kalivas and Volkow 2005; Liechti and Markou 2008). Interestingly, Kippin et al. (2008) have shown that PRS rats have a heightened corticolimbic DA and glutamate response to cocaine. Indeed, cocaine-naive PRS rats exhibited increased NAc DA and reduced NAc serotonin and glutamate, while cocaine-experienced PRS rats exhibited enhanced NAc glutamate and DA and PFC DA neurotransmission. Recently, we have increasing evidence of an involvement of the glutamate transmission and glutamate machinery in particular at the level of metabotropic glutamate (mGlu) receptors in the neuroplastic programming and behavioral phenotype induced by PRS. We have shown that PRS induces a selective reduction of glutamate release in the ventral hippocampus that is causally related with the enhanced anxiety-like behavior displayed by PRS animals, since pharmacological enhancement of glutamate release in the ventral hippocampus abolished this behavioral pattern (Marrocco et al. 2012).Remarkably, most of the neuroplastic alterations induced by PRS in males occur prominently in the ventral portion of the hippocampus a key region in the regulation of stress, emotions (Kjelstrup et al. 2002; Fanselow and Dong 2010). PRS rats also show a reduced expression and function of group-I and group-II mGlu hippocampal receptors, with a marked reduction of mGlu1 and mGlu5 selectively for males and mGlu2/3 for both males and females (Zuena et al. 2008; Van Waes et al. 2009; Laloux et al. 2012; Fig. 6.6a). The alterations in mGlu receptors induced by PRS in the hippocampus are detectable already at infancy with mGlu5 and mGlu1 receptors reduced in infant PRS rats at postnatal day 10, whereas expression of mGlu2/3 receptors declined

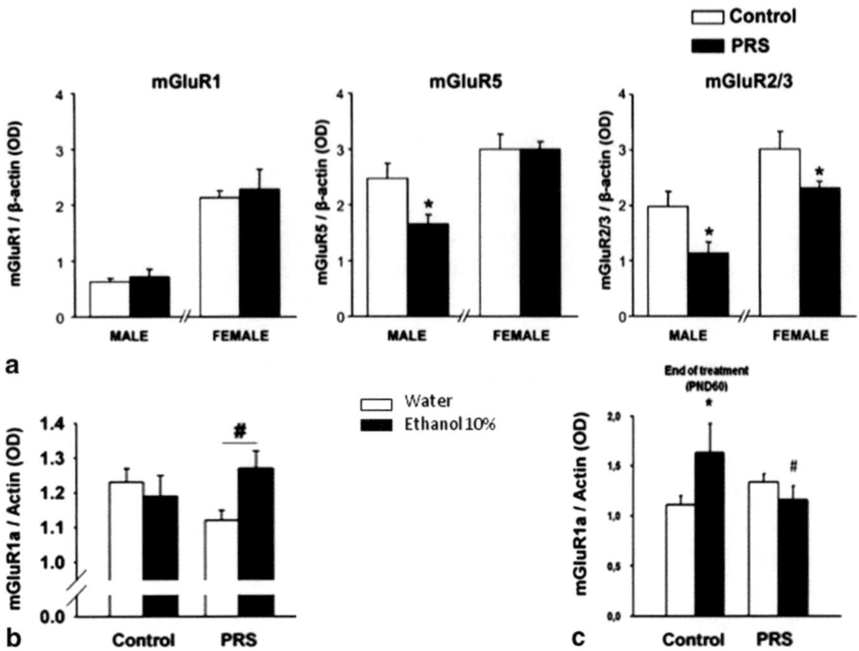

Fig. 6.6 Impact of prenatal restraint stress (*PRS*) on metabotropic glutamate receptors in the hippocampus and influence of ethanol treatment. **a** *PRS* reduced mGlu2/3 receptor expression in the hippocampus of both male and female rats while it reduces mGlu5 receptors in males only. *PRS* has no effect on the expression of mGlu1a receptors. Interestingly, Chronic EtOH treatment modifies the expression of mGlu1a receptors in a sex-dimorphic manner with an increase in *PRS* male rats (**b**) and a decrease in *PRS* females (**c**). $^{*}P<0.05$ and versus controls and $^{\#}P<0.05$ versus water-treated group. (Original data are reported in Zuena et al. 2008 (**a**), Van Waes et al. 2009 (**b**), and Van Waes et al. 2011b (**c**))

only after weaning (Laloux et al. 2012). MGlu receptors are clearly involved in the reinforcing and hyperlocomotor effects of several drugs of abuse. Mice lacking mGlu5 receptor do not self-administer cocaine or display cocaine-induced hyperlocomotion (Chiamulera et al. 2001). Antagonists for mGlu5 receptor reduce drug self-administration for cocaine (Kenny et al. 2003, 2005), nicotine (Paterson and Markou 2005), and ethanol (Bäckström et al. 2004). Localization studies have indicated a high abundance of mGlu5 receptors in brain areas involved in reward processes, including the striatum and NAc further supporting the involvement of mGlu5 receptors in brain reward function.

Chronic treatment with antidepressants correct mGlu5 and mGlu2/3 expressions in the hippocampus of PRS male rats (Morley-Fletcher et al. 2011) and, ethanol treatment modulates hippocampal mGlu1a expression and related behavioral changes in both male and female PRS rats (Van Waes et al. 2009, 2011b). In particular, chronic ethanol treatment had no effect on hippocampal mGlu5 or mGlu2/3 expressions in PRS males while it increased mGlu1a receptor levels and had an effect on improving memory. On the other hand, chronic ethanol reduced mGlu1a and induced memory impairment in control unstressed male rats (Van Waes et al. 2009;

Fig. 6.6b). In females, chronic exposure to ethanol during adolescence reduced mGlu1a receptor levels in PRS females while it increased it in control unstressed females. Those effects were still evident after 5 weeks of ethanol withdrawal (Van Waes et al. 2011b; Fig. 6.6c), whereas control rats also displayed depressive-like behavior. Since mGlu1 receptor antagonists show antidepressant-like effects in the forced swim test (Belozertseva et al. 2007), the increase in mGlu1a receptor levels we have seen in the hippocampus of unstressed female rats after 5 weeks of ethanol withdrawal is in agreement with the depressive-like behavior in the forced swim test. Taken collectively, these data suggest a potential use of mGlu1 receptor antagonists in the treatment of depressive symptoms associated with alcoholism. This expands the possible applications of mGlu1 receptor antagonists in human disorders. It remains to be determined whether changes in mGlu receptors in PRS rats following psychostimulants administration are also observed for mGlu5 and mGlu2/3 subtypes in the hippocampus and in reward-related brain regions.

6.2 PRS Rats as a Model of Comorbidity Between Addiction and Depression

From an epidemiological perspective, there is a higher degree of comorbidity between depression and drug dependence, indicating that the rates of depression among drug abusers and the rates of drug abuse among depressed patients are substantially higher than expected from the individual rates of these disorders (for review, Markou et al. 1998). This leads to suppose that the anhedonic symptoms of depression, which constitute the core feature of this illness, would be due to a dysfunctional brain reward system. Thus, alterations in reward and motivational processes at both the behavioral and neurobiological levels may constitute the defining characteristics of both depression and drug dependence. Nevertheless, it is not clear if drug dependence and depression are different behavioral expressions of the same neurobiological abnormalities, or whether one psychiatric disorder leads the other. An extension of this concept made by Markou et al. (1998) is known as the "self-medication hypothesis" and takes into account the possibility that drug dependence may involve self-medication to reverse some of the abnormalities associated with depression. Hence, it is possible that through the simultaneous use of multiple drugs, people determine the drug or drug combinations that best normalize their behavior.

For these individuals, the need to control their depressive symptomatology through self-medication would play an important role in the maintenance of drug dependence. For example, several works indicate that acute administration of psychostimulants such as opioids or amphetamine can temporarily reverse potential neurochemical deficits that may be found in depressed individuals (Tremblay et al. 2002; for review, Markou et al. 1998). However, none of these drugs of abuse are considered clinically effective as antidepressants by clinicians (Naranjo et al. 2001). In any case, the possibility remains that simultaneous or sequential use of various drugs as self-prescribed by the emotional needs of the drug-using individual leads to an adequate antidepressant effect, but at the same time pushes the individual in an active state of drug dependence. The best clinical support for this self-medication hypothesis is provided by the evidence that antidepressant treatment is significantly more effective in reducing drug use in depressed drug abusers than in nondepressed abusers (Ziedonis and Kosten 1991a, b; Nunes et al. 1993, 1995). Independently of whether the depression was present before the drug abuse or whether it was drug-induced, the reduction of drug use observed with antidepressants suggest that, when there is alleviation of depressive symptomatology through the use of antidepressant compounds, the need for self-medication with drugs of abuse diminishes (Markou et al. 1998). Thus, there are several aspects suggesting that these two psychiatric disorders may be linked by some shared neurobiology as well as common epigenetic mechanisms (Nestler 2002; Renthal and Nestler 2009).

We have shown that the pathological epigenetic programming triggered by early-life stress predispose to drug abuse disorders and anxiety/depression-like behavior (for review, Maccari and Morley-Fletcher 2007; Darnaudery and Maccari 2008). Thus, the known comorbidity between anxiety/depressive disorders and addiction should be interpreted within a context that takes into account the complex interaction between early-life experiences and stressful event occurring during adolescence and or adulthood. As example, we have shown that ethanol intake during adolescence induced depression-like effects in the adult life. These effects were observed in rats subjected to PRS (Van Waes et al. 2011b). Remarkably, the reverse was also true, i.e., ethanol intake during adolescence prevented the "depressive behavior" otherwise seen in adult PRS rats. Perhaps the increase in ethanol preference observed in adolescent PRS female rats might reflect a strategy of self-medication aimed at preventing the onset of depressive disorders later in life.

Therefore, the PRS model in the rat should be considered as a good model for the investigation of *multiple interrelated pathologies*. Since its effects are persistent through life span, it represents also a more advantageous animal model with respect to other models that present the same pattern of coexistence of these diseases (chronic mild stress), but have transitory effects (effects observable up to 2 months after termination of stress procedure).

As a whole, the existence of comorbidity in depression and drug abuse underlines the importance of the adoption of an integrated approach in the treatment of these disorders, where the brain reward system could be considered also as a potentially important therapeutic target. An elucidation of the neurobiological and behavioral mechanisms mediating this comorbidity would not only lead to the development of better treatments for these two psychiatric disorders but also enhance our understanding of the mechanisms subserving motivational and affective processes in both healthy and diseased individuals. Finally, considering that all psychiatric disorders, including depression and drug dependence, involve primarily behavioral symptoms that reflect underlying neurobiological abnormalities, progress in understanding these diseases at any level of analysis would certainly involve a multidisciplinary research approach. In this context, it would be interesting to explore hypothesis generated in the field of depression in animal models of drug dependence and vice versa. Furthermore, exploration of the self-medication hypothesis should be aimed at testing whether various drugs of abuse reverse depressive symptomatology in animal models of depression.

6.3 Conclusions

In an animal model of early stress, it has been shown that stress-related events that occur during the fetal and early postnatal period may have lifelong programming effects on HPA axis functioning and different body functions, with a considerable impact on disease susceptibility. The study of animal models involving early-life environmental manipulations will allow the study of individual vulnerability applied to stress-related disorders and contribute to improve drug abuse treatment by developing new therapeutic strategies. Those studies highlight the importance of early-life events in shaping the neurobehavioral adaptations to environmental challenges in both sexes, and supports the emerging consensus on early origins of drug vulnerability and depression and as potentially interdependent stress-related disorders.

Conflict of Interest The authors declare no conflicts of interest.

References

Antakly T, Mercille S, Côté JP (1987) Tissue-specific dopaminergic regulation of the glucocorticoid receptor in the rat pituitary. Endocrinology 120(4):1558–1562

Bäckström P, Bachteler D, Koch S, Hyytiä P, Spanagel R (2004) mGluR5 antagonist MPEP reduces ethanol-seeking and relapse behavior. Neuropsychopharmacology 29(5):921–928

Barbazanges A, Piazza PV, Le Moal M, Maccari S (1996) Maternal glucocorticoid secretion mediates long-term effects of prenatal stress. J Neurosci 16(12):3943–3949

Battaglia G, Fornai F, Busceti CL, Aloisi G, Cerrito F, Blasi AD, Nicoletti F (2002) Selective blockade of mglu5 metabotropic glutamate receptors is protective against methamphetamine neurotoxicity. J Neurosci 22(6):2135–2141

Belozertseva IV, Kos T, Popik P, Danysz W, Bespalov AY (2007) Antidepressant-like effects of mGluR1 and mGluR5 antagonists in the rat forced swim and the mouse tail suspension tests. Eur Neuropsychopharmacol 17(3):172–179

Berger MA, Barros VG, Sarchi MI, Tarazi FI., Antonelli MC (2002) Long-term effects of prenatal stress on dopamine and glutamate receptors in adult rat brain. Neurochem Res 27(11):1525–1533

Campbell JC, Szumlinski KK, Kippin TE (2009) Contribution of early environmental stress to alcoholism vulnerability. Alcohol 43(7):547–554

Casolini P, Piazza PV, Kabbaj M, Leprat F, Angelucci L, Simon H, Maccari S (1993) The mesolimbic dopaminergic system exerts an inhibitory influence on brain corticosteroid receptor affinities. Neuroscience 55(2):429–434

Charmandari E, Kino T, Souvatzoglou E, Chrousos GP (2003) Pediatric stress: hormonal mediators and human development. Horm Res 59(4):161–179

Chiamulera C, Epping-Jordan MP, Zocchi A, Marcon C, Cottiny C, Tacconi S, Conquet F (2001) Reinforcing and locomotor stimulant effects of cocaine are absent in mGluR5 null mutant mice. Nat Neurosci 4(9):873–874

Coe CL, Kramer M, Czéh B, Gould E, Reeves AJ, Kirschbaum C, Fuchs E (2003) Prenatal stress diminishes neurogenesis in the dentate gyrus of juvenile rhesus monkeys. Biol Psychiatry 54(10):1025–1034

Colby CR, Whisler K, Steffen C, Nestler EJ, Self DW (2003) Striatal cell type-specific overexpression of DeltaFosB enhances incentive for cocaine. J Neurosci 23(6):2488–2493

Darnaudéry M, Maccari S (2008) Epigenetic programming of the stress response in male and female rats by prenatal restraint stress. Brain Res Rev 57(2):571–585

Darnaudéry M, Perez-Martin M, Bélizaire G, Maccari S, Garcia-Segura LM (2006) Insulin-like growth factor 1 reduces age-related disorders induced by prenatal stress in female rats. Neurobiol Aging 27(1):119–127

Deminière JM, Piazza PV, Guegan G, Abrous N, Maccari S, Le Moal M, Simon H (1992) Increased locomotor response to novelty and propensity to intravenous amphetamine self-administration in adult offspring of stressed mothers. Brain Res 586(1):135–139

Deroche V, Marinelli M, Maccari S, Le Moal M, Simon H, Piazza PV (1995) Stress-induced sensitization and glucocorticoids. I. Sensitization of dopamine-dependent locomotor effects of amphetamine and morphine depends on stress-induced corticosterone secretion. J Neurosci 15(11):7181–7188

Fadda F, Cocco S, Stancampiano R, Rossetti ZL (1999) Long-term voluntary ethanol consumption affects neither spatial nor passive avoidance learning, nor hippocampal acetylcholine release in alcohol-preferring rats. Behav Brain Res 103(1):71–76

Fanselow MS, Dong H-W (2010) Are the dorsal and ventral hippocampus functionally distinct structures? Neuron 65(1):7–19

Farr SA, Scherrer JF, Banks WA, Flood JF, Morley JE (2005) Chronic ethanol consumption impairs learning and memory after cessation of ethanol. Alcohol Clin Exp Res 29(6):971–982

Gál K, Bárdos G (1994) The effect of chronic alcohol treatment on the radial maze performance of rats. Neuroreport 5(4):421–424

Härfstrand A, Fuxe K, Cintra A, Agnati LF, Zini I, Wikström AC, Steinbusch H (1986) Glucocorticoid receptor immunoreactivity in monoaminergic neurons of rat brain. Proc Natl Acad Sci U S A 83(24):9779–9783

Hausknecht K, Haj-Dahmane S, Shen R-Y (2013) Prenatal stress exposure increases the excitation of dopamine neurons in the ventral tegmental area and alters their responses to psychostimulants. Neuropsychopharmacology 38(2):293–301

Helmlin HJ, Bracher K, Bourquin D, Vonlanthen D, Brenneisen R (1996) Analysis of 3,4-methylenedioxymethamphetamine (MDMA) and its metabolites in plasma and urine by HPLC-DAD and GC-MS. J Anal Toxicol 20(6):432–440

Henry C, Kabbaj M, Simon H, Le Moal M, Maccari S (1994) Prenatal stress increases the hypothalamo-pituitary-adrenal axis response in young and adult rats. J Neuroendocrinol 6(3):341–345

Henry C, Guegant G, Cador M, Arnauld E, Arsaut J, Le Moal M, Demotes-Mainard J (1995) Prenatal stress in rats facilitates amphetamine-induced sensitization and induces long-lasting changes in dopamine receptors in the nucleus accumbens. Brain Res 685:179–186

Hiramatsu M, Kumagai Y, Unger SE, Cho AK (1990) Metabolism of methylenedioxymethamphetamine: formation of dihydroxymethamphetamine and a quinone identified as its glutathione adduct. J Pharmacol Exp Ther 254(2):521–527

Homewood J, Bond NW, MacKenzie A (1997) The effects of single and repeated episodes of thiamin deficiency on memory in alcohol-consuming rats. Alcohol 14(1):81–91

Kalivas PW, Duffy P (1995) D1 receptors modulate glutamate transmission in the ventral tegmental area. J Neurosci 15(7 Pt 2):5379–5388

Kalivas PW, Duffy P (1998) Repeated cocaine administration alters extracellular glutamate in the ventral tegmental area. J Neurochem 70(4):1497–1502

Kalivas Peter W, Volkow ND (2005) The neural basis of addiction: a pathology of motivation and choice. Am J Psychiatry 162(8):1403–1413. doi:10.1176/appi.ajp. 162.8.1403

Kelz MB, Chen J, Carlezon WA Jr, Whisler K, Gilden L, Beckmann AM, Nestler EJ (1999) Expression of the transcription factor DeltaFosB in the brain controls sensitivity to cocaine. Nature 401(6750):272–276

Kenny PJ, Paterson NE, Boutrel B, Semenova S, Harrison AA, Gasparini F, Markou A (2003) Metabotropic glutamate 5 receptor antagonist MPEP decreased nicotine and cocaine self-administration but not nicotine and cocaine-induced facilitation of brain reward function in rats. Ann N Y Acad Sci 1003:415–418

Kenny PJ, Boutrel B, Gasparini F, Koob GF, Markou A (2005) Metabotropic glutamate 5 receptor blockade may attenuate cocaine self-administration by decreasing brain reward function in rats. Psychopharmacology (Berl) 179(1):247–254

Kippin TE, Szumlinski KK, Kapasova Z, Rezner B, See RE (2008) Prenatal stress enhances responsiveness to cocaine. Neuropsychopharmacology 33(4):769–782

Kjelstrup KG, Tuvnes FA, Steffenach H-A, Murison R, Moser EI, Moser M-B (2002) Reduced fear expression after lesions of the ventral hippocampus. Proc Natl Acad Sci U S A 99(16):10825–10830

Koehl M, Darnaudéry M, Dulluc J, Van Reeth O, Le Moal M, Maccari S (1999) Prenatal stress alters circadian activity of hypothalamo-pituitary-adrenal axis and hippocampal corticosteroid receptors in adult rats of both gender. J Neurobiol 40(3):302–315

Koehl M, Bjijou Y, Le Moal M, Cador M (2000) Nicotine-induced locomotor activity is increased by preexposure of rats to prenatal stress. Brain Res 882(1–2):196–200

Koob GF (2008) A role for brain stress systems in addiction. Neuron 59(1):11–34

Korkosz A, Kolomanska P, Kowalska K, Rogowski A, Radwanska K, Kaczmarek L, Bienkowski P (2004) Dissociation of ethanol and saccharin preference in fosB knockout mice. Physiol Behav 82(2–3):391–395

Krazem A, Marighetto A, Higueret P, Jaffard R (2003a) Age-dependent effects of moderate chronic ethanol administration on different forms of memory expression in mice. Behav Brain Res 147(1–2):17–29

Krazem A, Mons N, Higueret P, Jaffard R (2003b) Chronic ethanol consumption restores the age-related decrease in neurogranin mRNA level in the hippocampus of mice. Neurosci Lett 338(1):62–66

Laloux C, Mairesse J, Van Camp G, Giovine A, Branchi I, Bouret S, Maccari S (2012) Anxiety-like behaviour and associated neurochemical and endocrinological alterations in male pups exposed to prenatal stress. Psychoneuroendocrinology 37(10):1646–1658

Lardeux S, Baunez C (2008) Alcohol preference influences the subthalamic nucleus control on motivation for alcohol in rats. Neuropsychopharmacology 33(3):634–642

Laviola G, Adriani W, Terranova ML, Gerra G (1999) Psychobiological risk factors for vulnerability to psychostimulants in human adolescents and animal models. Neurosci Biobehav Rev 23(7):993–1010

Liechti ME, Markou A (2008) Role of the glutamatergic system in nicotine dependence: implications for the discovery and development of new pharmacological smoking cessation therapies. CNS Drugs 22(9):705–724

Lowy MT (1990) MK-801 antagonizes methamphetamine-induced decreases in hippocampal and striatal corticosteroid receptors. Brain Res 533(2):348–352

Lukoyanov NV, Madeira MD, Paula-Barbosa MM (1999) Behavioral and neuroanatomical consequences of chronic ethanol intake and withdrawal. Physiol Behav 66(2):337–346

Maccari, S, Morley-Fletcher S (2007) Effects of prenatal restraint stress on the hypothalamus-pituitary-adrenal axis and related behavioural and neurobiological alterations. Psychoneuroendocrinology 32(Suppl 1):S10–S15

Maccari S, Piazza PV, Deminière JM, Angelucci L, Simon H, Le Moal M (1991a) Hippocampal type I and type II corticosteroid receptor affinities are reduced in rats predisposed to develop amphetamine self-administration. Brain Res 548(1–2):305–309

Maccari S, Piazza PV, Deminière JM, Lemaire V, Mormède P, Simon H, Le Moal M (1991b) Life events-induced decrease of corticosteroid type I receptors is associated with reduced corticosterone feedback and enhanced vulnerability to amphetamine self-administration. Brain Res 547(1):7–12

Maccari S, Piazza PV, Kabbaj M, Barbazanges A, Simon H, Le Moal M (1995) Adoption reverses the long-term impairment in glucocorticoid feedback induced by prenatal stress. J Neurosci 15(1 Pt 1):110–116

Maccari S, Darnaudery M, Morley-Fletcher S, Zuena AR, Cinque C, Van Reeth O (2003) Prenatal stress and long-term consequences: implications of glucocorticoid hormones. Neurosci Biobehav Rev 27(1–2):119–127

Malpass A, White JM, Irvine RJ, Somogyi AA, Bochner F (1999) Acute toxicity of 3,4-methylenedioxymethamphetamine (MDMA) in Sprague-Dawley and Dark Agouti rats. Pharmacol Biochem Behav 64(1):29–34
Markou A, Kosten TR, Koob GF (1998) Neurobiological similarities in depression and drug dependence: a self-medication hypothesis. Neuropsychopharmacology 18(3):135–174
Marrocco J, Mairesse J, Ngomba RT, Silletti V, Van Camp G, Bouwalerh H, Morley-Fletcher S (2012) Anxiety-like behavior of prenatally stressed rats is associated with a selective reduction of glutamate release in the ventral hippocampus. J Neurosci 32(48):17143–17154
Matthews DB, Morrow AL (2000) Effects of acute and chronic ethanol exposure on spatial cognitive processing and hippocampal function in the rat. Hippocampus 10(1):122–130
McEwen BS (2012) The ever-changing brain: cellular and molecular mechanisms for the effects of stressful experiences. Dev Neurobiol 72(6):878–890
McKee BL, Meshul CK (2005) Time-dependent changes in extracellular glutamate in the rat dorsolateral striatum following a single cocaine injection. Neuroscience 133(2):605–613
Morley-Fletcher S, Darnaudery M, Koehl M, Casolini P, Van Reeth O, Maccari S (2003a) Prenatal stress in rats predicts immobility behavior in the forced swim test. Effects of a chronic treatment with tianeptine. Brain Res 989(2):246–251
Morley-Fletcher S, Rea M, Maccari S, Laviola G (2003b) Environmental enrichment during adolescence reverses the effects of prenatal stress on play behaviour and HPA axis reactivity in rats. Eur J Neurosci 18(12):3367–3374
Morley-Fletcher S, Puopolo M, Gentili S, Gerra G, Macchia T, Laviola G (2004a) Prenatal stress affects 3,4-methylenedioxymethamphetamine pharmacokinetics and drug-induced motor alterations in adolescent female rats. Eur J Pharmacol 489(1–2):89–92
Morley-Fletcher S, Darnaudéry M, Mocaer E, Froger N, Lanfumey L, Laviola G Maccari S et al (2004b) Chronic treatment with imipramine reverses immobility behaviour, hippocampal corticosteroid receptors and cortical 5-HT(1 A) receptor mRNA in prenatally stressed rats. Neuropharmacology 47(6):841–847
Morley-Fletcher S, Mairesse J, Soumier A, Banasr M, Fagioli F, Gabriel C, Maccari S (2011) Chronic agomelatine treatment corrects behavioral, cellular, and biochemical abnormalities induced by prenatal stress in rats. Psychopharmacology 217(3):301–313
Mueller BR, Bale TL (2007) Early prenatal stress impact on coping strategies and learning performance is sex dependent. Physiol Behav 91(1):55–65
Naranjo CA, Tremblay LK, Busto UE (2001) The role of the brain reward system in depression. Prog Neuro Psychopharmacol Biol Psychiatry 25(4):781–823
Nash JF Jr, Meltzer HY, Gudelsky GA (1988) Elevation of serum prolactin and corticosterone concentrations in the rat after the administration of 3,4-methylenedioxymethamphetamine. J Pharmacol Exp Ther 245(3):873–879
Nestler EJ (2002) Common molecular and cellular substrates of addiction and memory. Neurobiol Learn Mem 78(3):637–647
Nicoletti F, Bruno V, Copani A, Casabona G, Knöpfel T (1996) Metabotropic glutamate receptors: a new target for the therapy of neurodegenerative disorders? Trends Neurosci 19(7):267–271
Nunes EV, McGrath PJ, Quitkin FM, Stewart JP, Harrison W, Tricamo E, Ocepek-Welikson K (1993) Imipramine treatment of alcoholism with comorbid depression. Am J Psychiatry 150(6):963–965
Olausson P, Jentsch JD, Tronson N, Neve RL, Nestler EJ, Taylor JR (2006) DeltaFosB in the nucleus accumbens regulates food-reinforced instrumental behavior and motivation. J Neurosci 26(36):9196–9204
Oscar-Berman M, Kirkley SM, Gansler DA, Couture A (2004) Comparisons of Korsakoff and non-Korsakoff alcoholics on neuropsychological tests of prefrontal brain functioning. Alcohol Clin Exp Res 28(4):667–675
Paterson NE, Markou A (2005) The metabotropic glutamate receptor 5 antagonist MPEP decreased break points for nicotine, cocaine and food in rats. Psychopharmacology 179(1):255–261
Peters DA (1982) Prenatal stress: effects on brain biogenic amine and plasma corticosterone levels. Pharmacol Biochem Behav 17(4):721–725

Peters DA (1990) Maternal stress increases fetal brain and neonatal cerebral cortex 5-hydroxytryptamine synthesis in rats: a possible mechanism by which stress influences brain development. Pharmacol Biochem Behav 35(4):943–947

Piazza PV, Maccari S, Deminière JM, Le Moal M, Mormède P, Simon H (1991) Corticosterone levels determine individual vulnerability to amphetamine self-administration. Proc Natl Acad Sci U S A 88(6):2088–2092

Pierce RC, Bell K, Duffy P, Kalivas PW (1996) Repeated cocaine augments excitatory amino acid transmission in the nucleus accumbens only in rats having developed behavioral sensitization. J Neurosci 16(4):1550–1560

Porsolt RD, Anton G, Blavet N, Jalfre M (1978) Behavioural despair in rats: a new model sensitive to antidepressant treatments. Eur J Pharmacol 47(4):379–391

Renthal W, Nestler EJ (2009) Chromatin regulation in drug addiction and depression. Dialogues Clin Neurosci 11(3):257–268

Ricaurte GA, DeLanney LE, Irwin I, Langston JW (1988) Toxic effects of MDMA on central serotonergic neurons in the primate: importance of route and frequency of drug administration. Brain Res 446(1):165–168

Rivet JM, Stinus L, LeMoal M, Mormède P (1989) Behavioral sensitization to amphetamine is dependent on corticosteroid receptor activation. Brain Res 498(1):149–153

Robles N, Sabriá J (2008) Effects of moderate chronic ethanol consumption on hippocampal nicotinic receptors and associative learning. Neurobiol Learn Mem 89(4):497–503

Rodrigues AJ, Leão P, Pêgo JM, Cardona D, Carvalho MM, Oliveira M, Sousa N (2012) Mechanisms of initiation and reversal of drug-seeking behavior induced by prenatal exposure to glucocorticoids. Mol Psychiatry 17(12):1295–1305

Seckl JR (2008) Glucocorticoids, developmental "programming" and the risk of affective dysfunction. Prog Brain Res 167:17–34

Schilström B, Fagerquist MV, Zhang X, Hertel P, Panagis G, Nomikos GG, Svensson TH (2000) Putative role of presynaptic alpha7* nicotinic receptors in nicotine stimulated increases of extracellular levels of glutamate and aspartate in the ventral tegmental area. Synapse 38(4):375–383

Shaham Y, Shalev U, Lu L, De Wit H, Stewart J (2003) The reinstatement model of drug relapse: history, methodology and major findings. Psychopharmacology 168(1–2):31–41. doi:10.1007/s00213-002-1224-x

Silvagni A, Barros VG, Mura C, Antonelli MC, Carboni E (2008) Prenatal restraint stress differentially modifies basal and stimulated dopamine and noradrenaline release in the nucleus accumbens shell: an "in vivo" microdialysis study in adolescent and young adult rats. Eur J Neurosci 28(4):744–758

Solinas M, Thiriet N, El Rawas R, Lardeux V, Jaber M (2009) Environmental enrichment during early stages of life reduces the behavioral, neurochemical, and molecular effects of cocaine. Neuropsychopharmacology 34(5):1102–1111. doi:10.1038/npp. 2008.51

Son GH, Geum D, Chung S, Kim EJ, Jo J-H, Kim C-M, Kim K (2006) Maternal stress produces learning deficits associated with impairment of NMDA receptor-mediated synaptic plasticity. J Neurosci 26(12):3309–3318

Steigerwald ES, Miller MW (1997) Performance by adult rats in sensory-mediated radial arm maze tasks is not impaired and may be transiently enhanced by chronic exposure to ethanol. Alcohol Clin Exp Res 21(9):1553–1559

Swerdlow NR, Koob GF, Cador M, Lorang M, Hauger RL (1993) Pituitary-adrenal axis responses to acute amphetamine in the rat. Pharmacol Biochem Behav 45(3):629–637

Tremblay LK, Naranjo CA, Cardenas L, Herrmann N, Busto UE (2002) Probing brain reward system function in major depressive disorder: altered response to dextroamphetamine. Arch Gen Psychiatry 59(5):409–416

Ungless MA, Argilli E, Bonci A (2010) Effects of stress and aversion on dopamine neurons: implications for addiction. Neurosci Biobehav Rev 35(2):151–156

Vallée M, Mayo W, Dellu F, Moal ML, Simon H, Maccari S (1997) Prenatal stress induces high anxiety and postnatal handling induces low anxiety in adult offspring: correlation with stress-induced corticosterone secretion. J Neurosci 17(7):2626–2636

Vallée M, MacCari S, Dellu F, Simon H, Le Moal M, Mayo W (1999) Long-term effects of prenatal stress and postnatal handling on age-related glucocorticoid secretion and cognitive performance: a longitudinal study in the rat. Eur J Neurosci 11(8):2906–2916

Van Waes V, Enache M, Dutriez I, Lesage J, Morley-Fletcher S, Vinner E, Darnaudéry M (2006) Hypo-response of the hypothalamic-pituitary-adrenocortical axis after an ethanol challenge in prenatally stressed adolescent male rats. Eur J Neurosci 24(4):1193–1200

Van Waes V, Enache M, Zuena A, Mairesse J, Nicoletti F, Vinner E et al (2009) Ethanol attenuates spatial memory deficits and increases mGlu1a receptor expression in the hippocampus of rats exposed to prenatal stress. Alcohol Clin Exp Res 33(8):1346–1354

Van Waes V, Enache M, Berton O, Vinner E, Lhermitte M, Maccari S, Darnaudéry M (2011a) Effect of prenatal stress on alcohol preference and sensitivity to chronic alcohol exposure in male rats. Psychopharmacology 214(1):197–208

Van Waes V, Darnaudéry M, Marrocco J, Gruber SH, Talavera E, Mairesse J et al (2011b) Impact of early life stress on alcohol consumption and on the short- and long-term responses to alcohol in adolescent female rats. Behav Brain Res 221(1):43–49

Yau JL, Kelly PA, Sharkey J, Seckl JR (1994) Chronic 3,4-methylenedioxymethamphetamine administration decreases glucocorticoid and mineralocorticoid receptor, but increases 5-hydroxytryptamine1 C receptor gene expression in the rat hippocampus. Neuroscience 61(1):31–40

Zachariou V, Sgambato-Faure V, Sasaki T, Svenningsson P, Berton O, Fienberg AA, Nestler EJ (2006) Phosphorylation of DARPP-32 at Threonine-34 is required for cocaine action. Neuropsychopharmacology 31(3):555–562

Ziedonis DM, Kosten TR (1991a) Depression as a prognostic factor for pharmacological treatment of cocaine dependence. Psychopharmacol Bull 27(3):337–343

Ziedonis DM, Kosten TR (1991b) Pharmacotherapy improves treatment outcome in depressed cocaine addicts. J Psychoactive Drugs 23(4):417–425

Zuena AR, Mairesse J, Casolini P, Cinque C, Alemà GS, Morley-Fletcher S, Maccari S (2008) Prenatal restraint stress generates two distinct behavioral and neurochemical profiles in male and female rats. PloS ONE 3(5):e2170

Chapter 7
How Postnatal Insults May Program Development: Studies in Animal Models

Carla Dalmaz, Cristie Noschang, Rachel Krolow, Charlis Raineki and Aldo. B. Lucion

Abstract During the postnatal period, the nervous system is modified and shaped by experience, in order to adjust it to the particular environment in which the animal will live. This plasticity, one of the most remarkable characteristics of the nervous system, promotes adaptive changes, but it also makes brain more vulnerable to insults. This chapter will focus on the effects of interventions during the postnatal development in animal models of neonatal handling (usually up to 15 min of handling) and maternal separation (usually at least for 3 h). Sex-specific changes and effects of prepubertal stress such as social isolation later on in life were also considered. These interventions during development induce long-lasting traces in the pups' nervous system, which will be reflected in changes in neuroendocrine functions, including the hypothalamus–pituitary–adrenal and hypothalamus–pituitary–gonadal axes; anxiety and cognitive performance; and feeding, sexual, and social behavior. These enduring changes may be adaptive or maladaptive, depending on the environment in which the animal will live. The challenge researchers facing now is to determine how to reverse the deleterious effects that may result from early-life stress exposure.

7.1 Introduction

During the postnatal period, the nervous system is modified and shaped by experience. The environment to which the animals are exposed during the postnatal period influences brain development. This plasticity, one of the most remarkable characteristics of the nervous system, has the advantage of promoting adaptive changes, but it also makes brain more vulnerable to insults.

A. B. Lucion (✉) · C. Raineki
Department of Physiology, Institute of Basic Health Sciences, Universidade Federal do Rio Grande do Sul, Porte Alegre, RS, Brazil
e-mail: alucion@ufrgs.br

C. Dalmaz · C. Noschang · R. Krolow
Department of Biochemistry, Institute of Basic Health Sciences, Universidade Federal do Rio Grande do Sul, Porte Alegre, RS, Brazil

M. C. Antonelli (ed.), *Perinatal Programming of Neurodevelopment*,
Advances in Neurobiology 10, DOI 10.1007/978-1-4939-1372-5_7,

Earlier studies of brain plasticity during the postnatal development and the functional and structural changes were performed by Hubel and Wiesel (see Q&A in Wiesel and Hubel 2012), who used visual deprivation to investigate plasticity in the visual pathway. In addition, other early studies by Harlow using young monkeys (e.g., Harlow et al. 1965) showed that early social deprivation leads to severe behavioral changes: The animal socially deprived never learns to interact with others.

Besides these marked changes in the environment, exposure to stress or small interventions may also lead to changes in behavior, physiology, or neurochemistry later in life. This chapter will focus on the long-lasting effects of these types of interventions during the neonatal or the prepubertal periods, based on results from animal models.

7.2 The Neonatal Period: Early-Life Stress

There is a continuous interest in how early-life events influence development throughout life (Levine 1957; Meaney et al. 1996; Padoin et al. 2001). The neonatal period is a developmental stage where genetic- and environment-dependent processes interact to establish functional characteristics that may persist until adulthood (Crews et al. 2007). Following birth, the brain continues to develop, and the changes induced by the environment may be adaptive, or not. Therefore, the brain becomes vulnerable to external influences, and these environmental changes may lead to persistent behavioral, endocrine, and neurochemical alterations (Caldji et al. 2000; Sánchez et al. 2001).

During the first weeks of life, rodent pups are completely dependent of their mothers for survival, and a suitable environment is vital for a healthy development. In rats, mother–pup contact primarily occurs within the context of a nest bout, which begins when the mother approaches the litter, licks, grooms, and nurses her pups (Weaver 2007). Observational studies provide evidences for stable individual differences in two forms of maternal behavior, licking/grooming and arched-back nursing posture, over the first weeks of lactation (Liu et al. 1997; Caldji et al. 1998). It is important to highlight that interventions during the neonatal period can influence the relationship between mother and pups. In this sense, maternal care may increase or decrease, depending on the environmental conditions (Liu et al. 1997; Ivy et al. 2008), and this variation is related to future outcomes in adulthood (Liu et al. 1997; Dalle Molle et al. 2012).

Different models for studying early-life stress effects on physiology and behavior have been used, and interesting results came from models that consist of separating the pups from their mothers. A large variety of protocols have been applied and many controversial results between different research groups have been attributed to the method used to separate the pups from their mothers.

Neonatal handling and maternal separation are the terminology frequently used to nominate these procedures. Usually, neonatal early or simply handling consists of removing the pups during 1–15 min and returning them to the dam (Pryce et al.

2005a Noschang et al. 2010; Todeschin et al. 2009). Earlier studies have handled the animals usually during the first 3 postnatal weeks (Levine et al. 1967; Meaney et al. 1985). However, protocols handling the animals for smaller periods have also been used (Noschang et al. 2012; Madruga et al. 2006; Stevenson et al. 2009; Fenoglio et al. 2005). Maternal separation, on the other hand, consists of separating the pups from the dam for several hours (a period longer than that used by the dam when leaving the nest to find food, for example; Pryce et al. 2005a). Multiple protocols of the maternal separation have been applied by different laboratories, with different lengths (1–24 h) and numbers (1–14 days during the first 2 postnatal weeks) of separation episodes (Holmes et al. 2005). It is important to note that the term maternal separation has also been used in the literature referring to handling procedure (Banihashemi et al. 2011; Boufleur et al. 2012).

7.2.1 Early-Life Stress and the HPA Axis

Activation of the stress system leads to behavioral and peripheral changes that improve the ability of the organism to adjust homeostasis and increase its chances for survival (Chrousos and Gold 1992). One of the main endocrine mechanisms is the activation of the hypothalamic–pituitary–adrenal (HPA) axis (Aguilera 1994). This activation begins with the release of corticotropin-releasing hormone (CRH) from the paraventricular nucleus of the hypothalamus (PVN) into the portal system to induce the release of adrenocorticotropic hormone (ACTH) from the anterior pituitary gland. The ACTH release stimulates the production and secretion of the glucocorticoids (GCs) hormones (corticosterone in rats) from the adrenal cortex. Additionally, glucocorticoid receptors (GR) mediate the negative feedback of GCs on the HPA axis following stress. These receptors are localized in distinct brain structures, including the hippocampus, prefrontal cortex, and hypothalamus (Lupien et al. 2005; McEwen 1994; McEwen et al. 1968; De Kloet et al. 1975).

The HPA axis is moderately responsive to stress at birth, but the responsiveness to ACTH has been known to decrease during the neonatal period (Yoshimura et al. 2003). In rats, the late fetal stage, when corticosterone is highly secreted, is followed by a quiescent period until the end of the second postnatal week, known as the stress hyporesponsive period (Sapolsky and Meaney 1986; Yoshimura et al. 2003). The adaptive value of this period can be understood in terms of the effects of GCs on central nervous system (CNS) development: Excessively high or low corticoid levels are associated with abnormal neural and behavioral development (Sapolsky and Meaney 1986). During the stress hyporesponsive period, HPA axis activity is reduced in response to noxious stimulus (Haltmeyer et al. 1966; Bartova 1968). The GC negative-feedback system plays a major role in mediating the stress hyporesponsive period (Yoshimura et al. 2003).

The effects of neonatal handling on the HPA axis and the stress response have been widely described (Pryce and Feldon 2003; Meerlo et al. 1999; Fenoglio et al. 2006; Panagiotaropoulos et al. 2004; Meaney et al. 1991; Severino et al. 2004).

Several laboratories have shown that handling reduces the stress response. In fact, the negative feedback is more effective rendering a smaller HPA axis activation and consequently a shorter increase in peripheral stress hormones. The reduction in stress response is probably caused by the increased number in GR in the hippocampus, which would increase the effectiveness of the stress negative feedback (Meaney et al. 1989, 1996; Avishai-Eliner et al. 2001). The stress activation of noradrenaline release also decreases in adult rats that underwent neonatal handling. They show a decreased noradrenaline activity in the hypothalamus, which would contribute to the reduced stress response (Liu et al. 2000).

The consequences of early interventions on the HPA axis depend on the type of stress that animals are exposed to at the beginning of their lives. Experiments performed in the late 1950s and 1960s revealed a permanent effect of neonatal handling on stress response in rats (Levine 1957; Levine et al. 1967). In the 1980s, Meaney et al. (1989) showed a dramatically altered adrenocortical response to stress in rats handled for the first 3 weeks of life. Those animals secreted less ACTH and corticosterone during and following the stress when compared to non-handled controls. There was also an increased expression of hippocampal GR expression and reduced hypothalamic CRH-messenger RNA (mRNA) levels, as well as decreased stress-induced GC release in adult rats that were handled early in life (Avishai-Eliner et al. 2001). These effects of early-life handling result from increased maternal licking and grooming upon returning handled pups to their home cages (Avishai-Eliner et al. 2001).

GC receptor concentrations in the rat brain are low at birth and increase towards the adult levels during the first 2 weeks of life. Briefly, daily handling of pups for the first 21 postnatal days has been found to permanently increase hippocampal GR concentrations (Meaney and Aitken 1985), and enhance negative-feedback effects of corticosterone on further activation of the HPA axis (Meaney et al. 1989). The handling effect on hippocampal GR concentrations is apparent as early as postnatal day 7. Handling on postnatal days 1–7 is as effective in altering GR concentrations as handling on the first 3 weeks of life; handling on postnatal days 8–14 is somewhat less effective and handling on postnatal days 15–21 has no effect (Meaney and Aitken 1985).

Although the early-life handling has been proposed to decrease HPA axis tone and reduce the response to stress in adulthood, the maternal separation could lead to long-life hyperactivity of the HPA axis (Nishi et al. 2013; Plotsky and Meaney 1993; Wigger and Neumann 1999). Yet, the effects of maternal separation on HPA axis in adult rats are not consistent among studies (Slotten et al. 2006; Ogawa et al. 1994). Adult rats that underwent maternal separation for 180 min from postnatal day 2 to 14 had augmented plasma ACTH and corticosterone concentrations following an acute stressor, compared to rats separated from their mothers for 15 min from postnatal day 2 to 14 and to normal animal facility-reared animals (Lippmann et al. 2007). Maternal separation for 180 min from postnatal day 2 to 14 resulted in elevated plasma corticosterone levels when compared to animals separated from their mothers for 15 min at postnatal day 7, a time when rat pups are normally hyporesponsive to stressors and show limited pituitary–adrenal responses (Huot et al.

2002). The stress hyperresponsiveness observed in maternal separated rats during adulthood could be attributed, at least in part, to impaired GC feedback sensitivity mediated by hippocampal GR (Ladd et al. 2004; Aisa et al. 2008). However, changes in GR expression may be secondary to an increased GC tone (Aisa et al. 2008). Maternal separation is associated with decreased GR levels in the hypothalamus, hippocampus, and frontal cortex, while neonatal handling increases GR expression in the hippocampus and frontal cortex (Meaney et al. 1996; Liu et al. 2000; Ladd et al. 2004). Regarding CRH, adult maternal separated rats showed increased hypothalamic CRH mRNA levels compared with non-handled rats, while CRH mRNA levels in handled rats were significantly lower than either maternal separated or non-handled animals (Plotsky and Meaney 1993). In that study, a 20-min period of restraint stress produced significant CRH depletion in all groups, while the percentage of depletion was significantly lower in handled animals compared with either maternal separated or non-handled animals. Additionally, restraint stress produced significant increase in plasma corticosterone in maternal separated and non-handled when compared to handled animals (Plotsky and Meaney 1993). Noteworthy, a review written by Jahng (2011) showed that the effects of maternal separation on the HPA axis activation after exposure to a stressor later in life appears to vary depending on the type of stressor used.

7.2.2 Early-Life Stress and Anxiety-like Behavior

Besides reduced stress response, handled rats show reduced anxiety-like behaviors (Roman et al. 2006; Meerlo et al. 1999). Several tests have indicated that the neonatal handling procedure decreases fear in novel environments, even when exposed to predators (Padoin et al. 2001), which could be interpreted as an impulsive behavior rather than fear reduction. Thus, it is inferred that the neonatal handling procedure prepares the animal to face adversities later in life by decreasing stress response and fear.

A commonly held belief is that brief manipulations decrease while prolonged manipulations enhance anxiety-like behavior (Meerlo et al. 1999; Núñez et al. 1995; Vallée et al. 1997; see Kosten et al. 2012 for a review). It was shown that pups handled for 1 min from postnatal day 1 to 10 displayed lower anxiety-like behavior in adulthood, both, in males and females, as observed by an increase in the percentage of time spent in the open arms of the elevated plus maze (Severino et al. 2004), a well-established task to evaluate anxiety-like behavior in rats. With a different protocol of neonatal handling, where rats of both sexes were handled for 15 min from postnatal day 1 to 22, decrease in anxiety-like behavior was also observed in the elevated plus maze, compared to non-handled animals (Kiosterakis et al. 2009). On the other hand, no difference between neonatal handled and non-handled animals was observed in anxiety-like behavior in a study developed by Silveira et al. (2005). Maternal separated rats from postnatal day 2 to 14 (180 min of separation) spent less time in the open arms of an elevated plus maze (a behavior compatible with increased anxiety) than neonatal handled animals (15 min of separation between dam and pups) and normal animal facility rearing (Huot et al. 2001).

Other studies have also shown that maternal separation leads to an increased anxiety-like behavior in adulthood. Some studies using the protocol of 180 min of maternal separation from postnatal day 3 to 10 have observed increased anxiety-like behavior in males (Makena et al. 2012) or in both sexes (Wigger and Neumann 1999). It is important to note that differences in the protocol used can contribute to different outcomes considering prolonged manipulations. For example, in a review by Kosten et al. (2012), the authors observed that, when the neonatal procedure was performed for 3 h repeatedly during the neonatal period, an increased anxiety-like behavior was observed (Aisa et al. 2007; Wigger and Neumann 1999), while no effect was reported in animals that were separated from their dams only once for 24 h on days 3, 4, 9, or 18 (Oomen et al. 2011; Lehmann et al. 1999). However, it was reported that animals separated from their mothers on day 11 present increased anxiety-like behavior (Barbosa-Neto et al. 2012). In general, these studies support the notion that early-life adverse experience is a major risk factor for anxiety development (Benekareddy et al. 2011).

Serotonergic neurotransmission is known to be related to anxiety (Gross and Hen 2004), and serotonin type 2 (5-HT2) receptors have been implicated as possible targets to modulate anxiety behavior (Weisstaub et al. 2006). Considering possible mechanisms underlying the anxiety-like behavior presented by early-life maternal separated rats, it was shown that pharmacological blockade of 5-HT2 receptors prevents the emergence of enhanced anxiety in maternal separated animals. It also prevents the upregulation of specific genes linked to 5-HT2 receptor signaling in the prefrontal cortex of adult animals that underwent maternal separation early in life (Benekareddy et al. 2011).

Diehl et al. (2007) evaluated anxiety-like behavior in maternal separation of male and female animals using the time spent in the central area of the open field. They showed an aggravation of the anxiety-like index in maternal separated male rats, when exposed to a stressor, while no effect was observed in females. Sex-specific effects were also shown in mice maternal separated for 24 h on postnatal day 9: Following maternal separation, anxiety was decreased in C57BL/6J and DBA/2J male mice but increased in DBA/2J females (Kember et al. 2012). Therefore, this effect of maternal separation appears to depend on sex and strain of the animals.

Therefore, some, but certainly not all, studies have shown that maternal separation produces increased anxiety-like behavior. The same is true for neonatal handling regarding its anxiety-decreasing effect.

7.2.3 *Early-Life Stress and Feeding Behavior*

Two complementary pathways regulate food intake: homeostatic and hedonic pathways. The homeostatic pathway controls energy balance by increasing the motivation to eat following depletion of energy stores. On the other hand, hedonic regulation can override the homeostatic pathway during periods of relative abundant energy by increasing the desire for highly palatable foods (Lutter and Nestler 2009).

Studies have shown that early-life events can influence feeding behavior and may contribute to feeding disorders. McIntosh et al. (1999) published a study where rat pups were exposed daily to brief handling or maternal separation during the first 3 weeks postpartum: they showed that the intake of palatable snack was higher in handled animals of both sexes and females of the maternal separated group than the non-handled animals. Silveira et al. (2004) showed that adult rats that were handled during the neonatal period (10 min/day during the first 10 days of life) consumed more sweet or savory snacks than controls, an effect observed in males and females. In the case of sweet food, the appetite was independent of the hunger condition (the effect was evident with the animals under food restriction but was particularly present when they were fed ad libitum). However, no difference was observed on standard laboratory chow diet. These authors also suggested that early handling leads to a particular response to positive boosters, such as higher ingestion of palatable food but lower hedonic impact, with decreased dopaminergic metabolism in the nucleus accumbens (Silveira et al. 2010). Regarding the effects of neonatal handling on metabolic changes in adulthood, it was shown that this intervention led to decreased plasma levels of ghrelin without changes in insulin, leptin, glucose, abdominal fat, or body weight (Silveira et al. 2006).

Other studies have evaluated the influence of maternal separation on feeding behavior. Using the 180-min protocol of maternal separation from postnatal day 1 to 14, no difference was observed in food intake (chow) between maternal separated and non-separated rats (Ryu et al. 2009). On the other hand, in an earlier study this same group (Ryu et al. 2008) showed that neonatal maternal separation may lead to an exaggerated feeding response to repeated fasting/refeeding challenges at adolescence in male rats, and suggested that this could be attributed to increased responsiveness of the HPA axis. With a different protocol of maternal separation (6 h of separation from postnatal day 1 to 21), Iwasaki et al. (2000) showed no difference in normal daily food intake between groups, but observed that food intake during rebound hyperphagia was significantly increased in 6–9-week-old female maternally separated rats, compared to control rats, with no difference observed in males.

Therefore, neonatal handling appears to induce increased intake of sweet or savory foods in adult life, and the mechanisms underlying these effects probably involve the reward system of the brain. However, maternal separation during the neonatal period leads to a different pattern, with increased rebound hyperphagia.

7.2.4 *Early-Life Stress and Memory*

There are several methods used to evaluate different types of memory that will not be discussed here (for a review, see Quillfeldt 2010 and Kosten et al. 2012). In addition, many articles about early-life events and memory have been published, and different outcomes have been observed depending on the task and the protocol used for early-life stress (see Kosten et al. 2012, for a recent review). The outcome is also dependent on factors such as the animal's age and condition (stressed or not; Meaney et al. 1991; Garoflos et al. 2005). Here, we will highlight some studies that evaluated sex differences concerning early-life stress and memory.

Considering neonatal handling, spatial learning and memory were evaluated using the water maze task, and the results show that male neonatal handled rats did not differ from non-handled animals, but female neonatal handled rats had a worse performance than the non-handled females. Additionally, female neonatal handled rats showed a decrease in nitric oxide levels compared to non-handled females (Noschang et al. 2010). Nitric oxide seems to be involved in learning and memory formation by facilitating mainly the acquisition process (Qiang et al. 1997). Using another memory task, the *Y* maze, male neonatal handled rats had a worse performance in reversal learning than non-handled males, while female neonatal handled rats had a better performance in the learning phase of this task when compared to non-handled females (Noschang et al. 2012). In the two studies described above (Noschang et al. 2010, 2012), the neonatal handling procedure used lasted for 10 min from postnatal day 1 to 10 and the memory evaluation was performed when the rats reached adulthood. Others have observed that neonatal handling resulted in impairment of inhibitory avoidance learning (involves foot shock), with no effect on circular maze learning (involves escape from bright light) and an enhancement of object recognition memory (presumably not involving aversive stimuli) in both adult male and female rats (Kosten et al. 2007).

Sex-specific effects on memory have also been shown with maternal separation protocol. Adult female rats that were separated from their mothers for 1 h per day from postnatal day 2 to 9 had a tendency to enhance the context-induced fear, while it was reduced in male rats (Kosten et al. 2005). In another study, using adult animals that were handled (15 min from postnatal day 1 to 21), or underwent maternal separation (3 h from postnatal day 1 to 21) in the neonatal period, and animals that were left undisturbed during the same period showed that neonatal handling impaired context- and cue-conditioning fear in both sexes of rats, while maternal separation only impaired context fear in female rats (Kosten et al. 2006). Another study applying a different maternal separation protocol, where pups were separated from their mothers for 24 h, showed that, in adulthood, using the active avoidance paradigm, only males displayed an effect of maternal separation, a severe deficit when separated from their dams on postnatal day 4. Animals with maternal separation on postnatal day 9 presented an improvement. Females did not differ significantly between groups (Lehmann et al. 1999). As the dentate gyrus of the hippocampus is basically formed in the first postnatal days (Altman and Bayer 1990), the structural plasticity of this brain structure is possibly highly susceptible to interventions applied early in life. That could explain the different outcomes that result from different protocols of maternal separation on memory.

Therefore, effects of early-life stress on memory appear to depend on type of memory evaluated and sex. For example, only females had impaired spatial memory.

It is interesting that neonatal handling appears to impair fear-related memory; however, it is important to consider the possibility of decreased fear in these animals, with lower impact on the learning tasks.

7.2.5 Early-Life Stress and the HPG Axis

As mentioned above, changes in maternal behavior can be the possible cause of the long-lasting outcomes induced by the neonatal handling intervention (Francis and Meaney 1999; McLeod et al. 2007). Indeed, the mother appears to mediate the impact of the environment on the offspring. Environmental events apparently would alter the mother–infant relationship and these changes would convey information about the environment, known as the maternal hypothesis (Caldji et al. 1998). The neonatal handling procedure increases maternal behavior after the pups return to the nest, inducing a rupture of maternal care which would exert a behavioral barrage of maternal behavior. Since the neonatal handling reduces stress response and anxiety-like behavior, the increased maternal behavior after the pups return to the nest could explain the "positive" effects of the handling procedure.

The quality of the infant environment plays a crucial role in the establishment of later reproductive strategies during adulthood (Cameron et al. 2005; Carlson and Earls 1997). Individuals exposed to early-life stress may develop alternative mechanisms to overcome this adversity; however, the cost of this alternative strategy can be later reduced reproduction (Cameron et al. 2005; Bateson et al. 2004; Lummaa and Clutton-Brock 2002). Studies using neonatal handling have consistently demonstrated that changes in the environment during infancy can affect several aspects of the reproductive system in both male and female rats. In males, neonatal handling can result in a significant reduction in sexual behavior, as exemplified by drastic reductions in the frequency of mounts with intromission and in the number of animals that perform sexual behavior (Padoin et al. 2001). Besides changes in the consummatory sexual behavior, neonatal handling also induces deficits in motivational aspects of the sexual behavior by reducing the time spent investigating the sexual partner in the partner preference test (Raineki et al. 2013). Additionally, neonatal handling also induces morphophysiological changes in the reproductive system of male rats such as reduced testicular weight, smaller seminiferous tubule diameter, and germinal epithelium thickness, all contributing to the reduction in daily sperm production and in the number of mature spermatids (Mazaro and Lamano-Carvalho 2006).

In females, handling-induced changes in reproduction first emerge as a delay in the pubertal onset as observed in protracted vaginal opening when compared to non-handled rats (Sieck and Ramaley 1975). In adulthood, female rats handled during infancy showed a reduction in sexual behavior (Gomes et al. 2006; Raineki et al. 2008), including reduced lordosis behaviors, resulting in a reduced lordosis quotient when compared to non-handled rats (Gomes et al. 2006; Padoin et al. 2001; Raineki et al. 2008). Moreover, the plasma progesterone surge following sexual behavior necessary for blastocyst implantation and successful pregnancy (Adler 1968; Adler et al. 1970) is reduced in neonatally handled females (Gomes et al. 2006), presumably due to reduced copulatory behavior. While neonatally handled rats exhibited regular estrous cycles, they showed a drastic reduction in the number of oocytes on the estrus morning (Gomes et al. 1999; Raineki et al. 2008). Ovulation is a result of a dynamic and complex neuroendocrine process that involves a sharp surge of steroid hormones, gonadotropins, and gonadotropin-releasing hormone (GnRH) in the preovulatory period (Freeman 2006; Schwartz 2000). Notably, many of these neuroendocrine processes are disrupted in rats that were handled during the neonatal period. Specifically, on the afternoon of proestrus, neonatally handled females show increased content of GnRH in the medial preoptic area (MPOA), followed by a reduction in plasma concentrations of estradiol, luteinizing hormone (LH), follicle-stimulating hormone (FSH), and prolactin (Gomes et al. 2005). Deficit in the noradrenergic stimulation of the MPOA seems to be the cause of reduced LHRH (Raineki et al. 2008). In addition to those hormonal alterations, neonatal handling also reduces the size and number of neurons in the MPOA (Camozzato et al. 2009), a key area controlling the production and release of GnRH. Another crucial structure affected by that early-life intervention is the Locus Coeruleus (LC), in which the number of cells is reduced in both male and female rats (Lucion et al. 2003). These neural and hormonal changes may, at least in part, explain the changes in sexual behavior and ovulation of female rats that were handled during infancy.

Although less explored, the effects of maternal separation on the reproductive function of males show similarities to the effects of neonatal handling. Indeed, male rats separated from the mother during the neonatal period (6 h of separation from postnatal day 2 to 10) show deficits in sexual behavior (Rhees et al. 2001). Complete maternal deprivation by artificial rearing also produces deficits in sexual behavior (Akbari et al. 2008), especially in penile reflexes (Lenz et al. 2008). Maternal separation also reduced the testicular weight. However, this reduction did not reflect a reduction of daily sperm production (Lau 1996). In females, the effects of maternal separation on the reproductive function are even less explored, and the results indicate that maternal separation has little or no effects on female reproduction. Indeed, no effect was observed in puberty onset (Lau 1996; Rhees et al. 2001) or regularity of the estrous cycle (Rhees et al. 2001).

This reduction in the reproductive ability of female and males could be considered as a negative effect of the neonatal handling intervention, which is in opposition to the stress response. That is, the neonatal handling intervention would reduce the activity of the HPA axis, considered as a positive effect by the increased animal resilience to stress. On the other hand, the same early-life intervention would also reduce reproductive parameters, considered intuitively as a negative consequence by reducing the ability to pass genes (to preserve the specie).

These effects on the HPA and HPG axes show the complexity of an early-life environmental intervention. The environmental intervention on a developing animal may affect several susceptible neuroendocrine systems. In fact, a balance between stress and reproduction seems crucial. In a recent study, lepers show a very precise interaction between stress responsiveness and reproduction in different natural environments, considering reduced environment temperature and predators (Sheriff et al. 2010). The increased number in predators and reduced temperature reduce reproduction in the lepers and this pattern is transmitted to the offspring. Thus, it seems that the balance between these two systems can be disrupted by early-life environment. When the environmental conditions are stressful, the pups become more prepared to face stressful environments later on in life.

Handling is not an event that naturally occurs during the rat's life. In fact, rats avoid human contact. Considering the scarce human interaction with rats, moreover in a systematic way such as an everyday manipulation, it could be argued that this intervention has no biological significance. Indeed that would be a reasonable conclusion. However, in spite of not being part of the natural repertoire of stimuli, the handling procedure induces vast and profound behavioral, neuroendocrine, and structural effects. Moreover, these effects seem to be induced by the intervention early in life and they persist into adulthood. Thus, it is possible to conclude that rat pups are susceptible to the mentioned intervention. Moreover, the early-life environmental intervention may induce stable changes. Therefore, it is possible to infer that there are "tuned pathways," through which the stimulation of the handling procedure, whatever that stimulation specifically would be, may impact some neuroendocrine systems. Furthermore, we may reason that there must be systems that respond to this intervention and those "tuned pathways" would be susceptible to biological stimuli, that in fact we do not know what would be. There must be natural stimuli that would trigger the same systems as the handling does. The disruption of the mother–infant relationship would be a very likely cause event that could trigger the changes observed in the handled animals. Systems have evolved to be very sensitive to environmental interventions that could interfere with the mother–infant relationship.

7.2.6 Early-Life Stress and Social Behavior

During early life, the infant's social world revolves around the caregiver and the quality of the social experiences during this period lays the foundation for later life social behavior and mental health (Bowlby 1969; Hofer and Sullivan 2008; Veenema 2012). As the individual matures, the spectrum of social behaviors must expand to meet the ever-increasing complexity of social relationships. Indeed, the organism must transition from infancy, which is dominated by the infant–caregiver relationship, to adolescence where the focus is on peers. Finally, the organism must transition into adulthood, which requires the ability of executing a complex repertoire of social behaviors, including reproductive behaviors (discussed below), agonistic actions (e.g., aggression), and affiliative behaviors (Insel 2010). Importantly, the literature has consistently demonstrated that adverse and/or stressful experiences during infancy can have pervasive and negative effects on social behavior that start during the infancy and persist into adulthood.

The effects of neonatal handling on social behavior can be observed even during infancy, where neonatal handling alters the infant's social interaction with the mother. For example, neonatally handled female rat pups showed a reduction in the preference for the maternal odor, indicating a possible deficit in the mother–pup social relationship (Raineki et al. 2009). After infancy, when the animal is approaching the periweaning period, play behavior—also called rough-and-tumble play—is the most common form of social behavior (Pellis and Pellis 1987). When tested as juveniles, neonatally handled rats exhibited more playful behavior than non-handled controls (Aguilar et al. 2009; Siviy and Harrison 2008). Despite this abnormal increase in play behavior, though, neonatal handled rats exhibited a typical reduction in play behavior and more risk assessment (increased environmental investigation) following the introduction of a predator odor. When returned to the same environment where the predator odor had been experienced, however, neonatal handled rats were less likely to exhibit a conditioned suppression of play (Siviy and Harrison 2008). In contrast to neonatal handling, maternal separation (3 h daily) resulted in increased offensive play-fighting in juvenile rats, which is a behavioral precursor to adult aggressive behavior (Veenema and Neumann 2009; Zimmerberg and Sageser 2011). Specifically, male juvenile rats that were separated from the mother during infancy showed an increase in the number of nape attacks, pulling and biting behaviors, and a decrease in submissive play (supine, evading) as compared to controls (Veenema and Neumann 2009).

During adulthood, male rats that were handled in the neonatal period showed a reduction in affiliative social behavior as well as an increase in non-affiliative behaviors such as aggression in the social interaction test (Todeschin et al. 2009). Studies assessing the effects of neonatal handling on aggressive behavior have produced varying results that can be attributed to differences in the neonatal handling protocols. For example, lactating rats handled 1 min per day from postnatal day 1 to 10 showed an enhancement in aggressive behaviors such as biting and lateral attacks against a male intruder (Giovenardi et al. 2005; Padoin et al. 2001). Conversely, a different protocol of neonatal handling where rats were handled 15 min per day from postnatal day 2 to 14 did not induce changes in aggressive behaviors of lactating rats relative to non-handled controls (Boccia and Pedersen 2001). When assessing the effects of maternal separation, though, lactating rats separated (3 h per day) from the mother during the neonatal period attacked the male intruder less quickly and less frequently than control rats, indicating that maternal separation may reduce the aggressive behavior of lactating rats (Boccia and Pedersen 2001).

Another domain of social behavior affected by early-life stress is social learning. Adult rats reared in complete isolation (maternal separation from postnatal day 4 to 21) showed consistent deficits in learning to recognize social stimuli (Levy et al. 2003). Specifically, maternally deprived rats were not able to differentiate between a new and a previously presented (familiar) juvenile conspecific in a social recognition task, nor were they able to develop a preference for a food previously eaten by a familiar conspecific (Levy et al. 2003). In a social learning test, neonatal handled rats showed reduced social investigation during the learning phase, however their ability to differentiate between a novel and familiar social stimulus (social memory) was not affected (Todeschin et al. 2009).

Nonhuman primate studies that use complete maternal separation also reveal an enduring dysregulation of infant social development as demonstrated by inappropriate expression of agonistic interactions, decreased play, and social inhibition (Champoux et al. 1989; Harlow et al. 1965; Kraemer 2003; Suomi et al. 1971). However, even less severe maternal separation (6 days) induced robust changes in social behavior including decreases in exploring, playing, and time away from their mothers after being reunited (Hinde et al. 1966; Hinde and Spencer-Booth 1971), behaviors that suggest features of insecure attachment (Ainsworth et al. 1978).

The underlying neurobiology of social behavior involves the complex interplay of many neural structures and neuroendocrine systems (Adolphs 2001; Ross and Young 2009). Among these structures, the amygdala (Amaral 2003) and PVN (Neumann 2008; Veenema and Neumann 2008) have been identified as two key areas that modulate fundamental aspects of socio-emotional processing across a wide range of species, including humans and nonhuman primates, and rats (Amaral 2003; Thomas et al. 2001). At the neurotransmitter level, the peptides oxytocin and vasopressin appear to support characteristics of social behavior, including social motivation (Lim and Young 2006), social recognition (Bielsky and Young 2004; Engelmann et al. 1994; Ferguson et al. 2002), and aggression (Ferris 1992; Gobrogge et al. 2009). Both neonatal handling and maternal separation have been found to induce changes in the oxytocinergic and vasopressinergic systems (Veenema 2012). In particular, neonatal handling decreased the number of oxytocin-positive parvocellular neurons in the PVN of male and female rats (Todeschin et al. 2009; Winkelmann-Duarte et al. 2007), while maternal separation in mice increased (Tsuda et al. 2011) or had no effect (Veenema et al. 2007) on PVN neurons. In the amygdala, both neonatal handling and maternal separation induced a reduction in oxytocin-positive neurons (Oreland et al. 2010). Regarding the vasopressinergic system, neonatal handling and maternal separation induced an increase in vasopressin-positive neurons in the PVN (Todeschin et al. 2009; Veenema et al. 2006; Veenema and Neumann 2009). However, the effects of maternal separation on vasopressin are not as straightforward, as other studies have also found decreased vasopressin expression in the PVN of mice and rats (Desbonnet et al. 2008; Tsuda et al. 2011). Together, these results suggest that the oxytocinergic and/or vasopressinergic systems may play differential roles in modulating deficits in social behavior induced by neonatal handling or maternal separation paradigms.

7.3 Stress During the Prepubertal Period

The prepubertal period in rodents is chronologically marked by weaning, around 21 days of age and ends around postnatal day 30. The onset of puberty begins around postnatal day 30 to 35, signaling the beginning of adolescent development. During this period, the animals are in constant behavioral and neurobiological transformations such as increased strength, immune function, and cognitive skills (Dahl 2004), maturation of social and cognitive behavior, and increase in exploratory behavior, in social play (Sisk and Foster 2004; Klein et al. 2010).

The pubertal period is critical to complete brain maturation, including circuits controlling energy homeostasis and stress response (Mccormick and Mathews 2007). The maturation rates and plasticity during development are highly variable in different brain regions, including those involved in the processing and regulation of stress and emotion. For example, the prefrontal cortex exhibits late maturation (Gogtay et al. 2004), while the hippocampus presents increased neurogenesis and dendritic spine density just before puberty (< postnatal day 35) that decreased in the adulthood (He and Crews 2007; Yildirim et al. 2008). Besides, the composition of hippocampal signaling proteins also changes between weaning and adulthood (Weitzdorfer et al. 2008). Most neurotransmitter systems, including dopaminergic and glutamatergic systems, mature during this stage of life (Spear 2000). The mesolimbic dopaminergic system is not fully developed until as late as postnatal day 35 to 40 (Spear 2000; Buck Louis et al. 2008; Schulz et al. 2009).

7.3.1 The HPA Axis in the Prepubertal and Pubertal Period

The transitions from neonate to adolescence to adulthood involve significant functional changes in HPA axis. Interestingly, areas of the brain that modulate HPA responsiveness, such as the hippocampus and the prefrontal cortex (Herman et al. 2003), show maturation during the prepubertal period in rodents. Considering that many brain structures that participate in the HPA axis responsiveness mature later, it is not surprising that stressors during this period may affect HPA axis function and lead to long-term changes in HPA reactivity (Matthews 2002; Pryce et al. 2005b; Romeo et al. 2009). Furthermore, HPA hormones are important mediators of physical and physiological transitions in development (Wada 2008).

Rats exposed to acute stressors in the prepubertal period have higher levels of corticosterone than adult animals in the same conditions (Romeo 2010). In addition, adult animals repeatedly exposed to the same stressor showed habituated ACTH and corticosterone responses, while the animals in the prepubertal period exhibit a sensitized response (Romeo et al. 2006; Doremus-Fitzwater et al. 2009; Foilb et al. 2011). Some studies have suggested a possible change in the HPA axis feedback or expression of GR in brain structures involved in the stress response in prepubertal rats (Goldman et al. 1973; Schmidt et al. 2007; Uys et al. 2006a, b).

It is important to consider that both male and female prepubertal rats show a protracted stress-induced ACTH and corticosterone response in contrast to adults. However, females, at both ages, demonstrate a higher peak of corticosterone response, but return faster to baseline, compared to males at both ages (Romeo et al. 2004a, b).

7.3.2 Stressors in the Prepubertal Period

The effects of stress appear to be exacerbated during prepubertal period (Charil et al. 2010). Corticosterone levels increase more and take longer to return to baseline in prepubertal male rats than adults after stressors such as restraint, intermittent foot shock, or ether vapors (Romeo et al. 2006). However, juvenile and adult female rats are able to habituate to chronic stress (Doremus-Fitzwater et al. 2009).

Hyperactivity of HPA axis, indicated by high levels of GCs, has been observed in depression-like disorders (Plotsky et al. 2008; Van Eck et al. 1996), and stressful experiences during childhood and adolescence have been associated with the development of psychiatric disorders later in life (Kessler and Magee 1993; Penza et al. 2003; Heim et al. 2004). For example, exposure to predator scent or chronic variable stress (including forced swim, elevated platform, and foot shock) during postnatal days 27–29 induces a predisposition to either anxiety or depressive-like symptoms following stress in adulthood (Bazak et al. 2009; Tsoory et al. 2007). These findings suggest that stressful experiences occurring in early life leave a trace that makes the brain vulnerable to the development of mood and anxiety disorders, and stress-related disorders (Breslau et al. 2004; Ford and Kidd 1998; Heim and Nemeroff 2001; Meinlschmidt and Heim 2005; Shea et al. 2005).

There is evidence that stressors early in life impact on the cognitive performance of rats (Hodes and Shors 2005). The hippocampus is an important structure for learning and memory, and its physiology and morphology is sensitive to acute and chronic stressors (Howland and Wang 2008). Exposure to stress during postnatal days 27–29 leads to long-term deleterious effect on learning under stressful conditions in adulthood (Tsoory and Richter-Levin 2005).

The stress-induced changes in behavior during prepuberty are accompanied by neurochemical changes. These changes occur mainly in hippocampus, prefrontal cortex, and amygdala, brain structures that continue to mature during this stage of life. These structures show significant morphological stress-induced alterations. Significantly reduced dendritic complexity of pyramidal neurons in hippocampus and prefrontal cortex were observed in males and females rats after chronic variable stress 6 h per day during postnatal days 20–41 (Eiland et al. 2012), while neurons in the basolateral amygdala displayed increased complexity. It is important to mention that other studies showed sex-dependent responses to stress during the prepubertal period: For example, juvenile females appear to be more prone to the effects of stress with palatable food intake, while males are more susceptible to long-term effects of early stress on biochemical parameters, such as increased plasma triacylglycerols (Krolow et al. 2013 in press).

Early in life, when social interactions are necessary for normal emotional development, perturbations of the social environment have significant effects on brain structure and function (Einon and Morgan 1977); stressors can profoundly alter social interactions and play behavior (Klein et al. 2010). In addition, the social environment is a source of stress, both for humans and rodents, especially during the prepubertal period. In the natural environment, rodents live in groups and exhibit high levels of social behavior, both with younger and older animals (McCormick and Mathews 2007; Panksepp and Lahvis 2007). Social interactions are rewarding (Panksepp et al. 2007), while social isolation is an aversive event and increases the activity of the HPA axis (Douglas et al. 2004; McCormick and Mathews 2007). Therefore, social stress is one of the most potent stressors during the development (Arakawa 2005), which at long term induces a variety of behavioral abnormalities, including increased aggressiveness (Koike et al. 2009; Pinna et al. 2003; Pinna et al. 2009), anxiety-related behaviors (Pinna et al. 2006; Wei et al. 2007), cognitive deficits (Pibiri et al. 2008), hypoalgesia, and hyperlocomotion (McEwen 2007). Additionally, this type of stress changes fat distribution and contributes to an increased risk for metabolic disease (Schmidt et al. 2009). Social play-behavior deprivation during juvenility is associated with deregulation of the endogenous opioid systems (Van den Berg et al. 1999, 2000), impairment in the Morris water maze test (Frisone et al. 2002), and in the exploration of novel areas and objects (Einon and Morgan 1977).

Besides the many behavior changes, social stress during development induces long-lasting morphological alterations (Leussis et al. 2008; Radley et al. 2005). The changes include structural aspects of cortical areas, with reduced synaptic density in the infralimbic cortex and cingulated gyrus (Leussis et al. 2008), and may be sex-specific, since in male but not in female rats a reduction in the myelination in prefrontal cortex was observed (Leussis and Andersen 2008).

Isolation stress applied in male rats during postnatal days 21–28 induces neurochemical changes, such as increase in superoxide dismutase and mitochondrial Complex IV activities in the prefrontal cortex that persist into adulthood (Krolow et al. 2012). Juvenile male rats exposed to isolation stress exhibited oxidative imbalance, increased DNA fragmentation, increased mitochondrial potential and early apoptosis, and decreased number of live and necrotic cells in hippocampus (Krolow et al. 2013). These findings suggest that isolation stress in the prepubertal period leads to long-lasting changes in antioxidant enzymes and energetic metabolism in prefrontal cortex and hippocampus.

Therefore, the prepubertal period in rodents is marked by neurochemical, metabolic, and behavioral changes that can be implicated in the regulation of emotion and stress responses. The continued maturation of stress-responsive brain regions and the quantity and type of stressors during this period may contribute to neurochemical and behavioral alterations, and to the development of psychiatric disorders later in life. Further studies are necessary to clarify the possible mechanisms involved in the observed differences in sex and age responses to stress.

7.4 Conclusion

Interventions in early postnatal days lead to complex and stable changes throughout the life of the animal, and even apparently mild interventions may lead to important changes. The relation between the dam and the pups is also changed by these interventions and may be the cause of several changes observed later in life. Different models have been used to study these aspects of behavioral programming and results demonstrate changes in the stress response, reproductive functions, memory, anxiety, feeding, and social behavior. The observed changes are apparently adaptive to environment modifications, possibly preparing the animal to face scarce resources later in life. However, some of these changes may be maladaptive, and mismatches between type of behavioral programming and environment in which the animal will live may occur. It is important to consider that, although stable, the plasticity is an essential feature of the nervous system, and modulation of programming is also possible. The challenge researchers are facing now is to understand the mechanisms underlying the changes induced by early-life interventions and to determine how to reverse the deleterious effects that may result from early-life stress exposure.

Conflict of Interest The authors declare no conflicts of interest.

References

Adler NT (1968) Effects of male's copulatory behavior in the initiation of pregnancy in the female rat. Anatomic Rec 160:305

Adler NT, Resko JA, Goy RW (1970) The effect of copulatory behavior on hormonal change in the female rat prior to implantation. Physiol Behav 5(9):1003–1007

Adolphs R (2001) The neurobiology of social cognition. Curr Opin Neurobiol 11:231–239

Aguilar R, Carames JM, Espinet A (2009) Effects of neonatal handling on playfulness by means of reversal of the desire to play in rats (*Rattus norvegicus*). J Comp Psychol 123:347–356

Aguilera G (1994) Regulation of pituitary ACTH secretion during chronic stress. Front Neuroendocrinol 15:321–350

Ainsworth MDS, Blehar MC, Waters E, Wall S (1978) Patterns of attachment: a psychological study of the strange situation. Lawrence Erlbaum Associates, Hillsdale

Aisa B, Tordera R, Lasheras B, Del Río J, Ramírez MJ (2007) Cognitive impairment associated to HPA axis hyperactivity after maternal separation in rats. Psychoneuroendocrinology 32(3):256–266

Aisa B, Tordera R, Lasheras B, Del Río J, Ramírez MJ (2008) Effects of maternal separation on hypothalamic-pituitary-adrenal responses, cognition and vulnerability to stress in adult female rats. Neuroscience 154(4):1218–1226

Akbari EM, Budin R, Parada M, Fleming AS (2008) The effects of early isolation on sexual behavior and c-fos expression in naïve male long-evans rats. Dev Psychobiol 50:298–306

Altman J, Bayer SA (1990) Migration and distribution of two populations of hippocampal granule cell precursors during the perinatal and postnatal periods. J Comp Neurol 301(3):365–381

Amaral DG (2003) The amygdala, social behavior, and danger detection. Ann N Y Acad Sci 1000:337–347

Arakawa H (2005) Interaction between isolation rearing and social development on exploratory behavior in male rats. Behav Process 70:223–234

Avishai-Eliner S, Eghbal-Ahmadi M, Tabachnik E, Brunson KL, Baram TZ (2001) Down-regulation of hypothalamic corticotropin releasing hormone messenger ribonucleic acid (mRNA) precedes early-life experience-induced changes in hippocampal glucocorticoid receptor mRNA. Endocrinology 142:89–97

Banihashemi L, O'Neill EJ, Rinaman L (2011) Central neural responses to restraint stress are altered in rats with an early life history of repeated brief maternal separation. Neuroscience 192:413–428

Barbosa Neto JB, Tiba PA, Faturi CB, de Castro-Neto EF, da Graça Naffah-Mazacoratti M, de Jesus Mari J, de Mello MF, Suchecki D (2012) Stress during development alters anxiety-like behavior and hippocampal neurotransmission in male and female rats. Neuropharmacology 62(1):518–526

Bartova A (1968) Functioning of the hypothalamo-pituitary-adrenal system during postnatal development in rats. Gen Comp Endocrinol 10:235–239

Bateson P, Barker D, Clutton-Brock T, Deb D, D'Udine B, Foley RA, Gluckman P Godfrey K, Kirkwood T, Lahr MM, McNamara J, Metcalfe NB, Monaghan P, Spencer HG, Sultan SE (2004) Developmental plasticity and human health. Nature 430: 419–421

Bazak N, Kozlovsky N, Kaplan Z, Matar M, Golan H, Zohar J et al (2009) Pre-pubertal stress exposure affects adult behavioral response in association with changes in circulating corticosterone and brain-derived neurotrophic factor. Psychoneuroendocrinology 34:844–858

Benekareddy M, Vadodaria KC, Nair AR, Vaidya VA (2011) Postnatal serotonin type 2 receptor blockade prevents the emergence of anxiety behavior, dysregulated stress-induced immediate early gene responses, and specific transcriptional changes that arise following early life stress. Biol Psychiatry 70(11):1024–1032

Bielsky IF, Young LJ (2004) Oxytocin, vasopressin, and social recognition in mammals. Peptides 25:1565–1574

Boccia ML, Pedersen CA (2001) Brief vs. long maternal separations in infancy: contrasting relationships with adult maternal behavior and lactation levels of aggression and anxiety. Psychoneuroendocrinology 26:657–672

Boufleur N, Antoniazzi CT, Pase CS, Benvegnú DM, Dias VT, Segat HJ, Roversi K, Roversi K, Nora MD, Koakoskia G, Rosa JG, Barcellos LJ, Bürger ME (2012) Neonatal handling prevents anxiety-like symptoms in rats exposed to chronic mild stress: behavioral and oxidative parameters. Stress. [Epub ahead of print]

Bowlby J (1969) Attachment and loss, vol 1. Basic Books, New York

Breslau N, Wilcox H, Storr C, Lucia V, Anthony J (2004) Trauma exposure and posttraumatic stress disorder: a study of youths in urban America. J Urban Health 81:530–544

Buck Louis GM, Gray LE, Marcus M, Ojeda SR, Pescovitz OH, Witchel SF et al (2008) Environmental factors and puberty timing: expert panel research needs. Pediatrics 121:192–207

Caldji C, Tannenbaum B, Sharma S, Francis D, Plotsky P M., Meaney MJ (1998) Maternal care during infancy regulates the development of neural systems mediating the expression of fearfulness in the rat. Proc Natl Acad Sci U S A 95(9):5335–5340

Caldji C, Francis D, Sharma S, Plotsky PM, Meaney MJ (2000) The effects of early rearing environment on the development of GABAA and central benzodiazepine receptor levels and novelty-induced fearfulness in the rat. Neuropsychopharmacology 22(3):219–229

Cameron NM, Champagne FA, Parent C, Fish EW, Ozaki-Kuroda K, Meaney MJ (2005) The programming of individual differences in defensive responses and reproductive strategies in the rat though variations in maternal care. Neurosci Biobehav Rev 29:843–865

Camozzato TSC, Winkekmann-Duarte EC, Padilha CB, Miguel SPR, Bonzanini L, Anselmo-Franci JA, Fernandes MC, Lucion AB (2009) Neonatal handling reduces the number of cells in the medial preoptic area of female rats. Brain Res 1247:92–99

Carlson M, Earls F (1997) Psychological and neuroendocrinological sequelae of early social deprivation in institutionalized children in Romania. Ann N Y Acad Sci 807:419–428

Champoux M, Coe CL, Shanberg S, Kuhn C, Suomi SJ (1989) Hormonal effects of early rearing conditions in the infant rhesus monkey. Am J Primatol 19 111–117

Charil A, Laplante DP, Vaillancourt C, King S (2010) Prenatal stress and brain development. Brain Res Rev 65:56–79

Chrousos GP, Gold PW (1992) The concepts of stress and stress system disorders. Overview of physical and behavioral homeostasis. JAMA 267:1244–1252

Crews F, He J, Hodge C (2007) Adolescent cortical development: a critical period of vulnerability for addiction. Pharmacol Biochem Behav 86:189–199

Dahl RE (2004) Adolescent brain development: a period of vulnerabilities and opportunities. Ann N Y Acad Sci 1021:1–22

Dalle Molle R, Portella AK, Goldani MZ, Kapczinski FP, Leistner-Segala S, Salum GA, Manfro GG, Silveira PP (2012) Associations between parenting behavior and anxiety in a rodent model and a clinical sample: relationship to peripheral BDNF levels. Transl Psychiatry 2(11):2195
De Kloet ER, Wallach G, McEwen BS (1975) Differences in corticosterone and dexamethasone binding to rat brain and pituitary. Endocrinology 96:598–609
Desbonnet L, Garrett L, Daly E, McDermott KW, Dinan TG (2008) Sexually dimorphic effects of maternal separation stress on corticotrophin-releasing factor and vasopressin systems in the adult rat brain. Int J Dev Neurosci 26:259–268
Diehl LA, Silveira PP, Leite MC, Crema LM, Portella AK, Billodre MN, Nunes E, Henriques TP, Fidelix-da-Silva LB, Heis MD, Gonçalves CA, Quillfeldt JA, Dalmaz C (2007) Long lasting sex-specific effects upon behavior and S100b levels after maternal separation and exposure to a model of post-traumatic stress disorder in rats. Brain Res 1144:107–116
Doremus-Fitzwater TL, Varlinskaya EI, Spear LP (2009) Social and non-social anxiety in adolescent and adult rats after repeated restraint. Physiol Behav 97:484–494
Douglas L, Varlinskaya E, Spear LP (2004) Rewarding properties of social interactions in adolescent and adult male and female rates: impact of social versus isolate housing of subjects and partners. Dev Psychobiol 45:153–162
Eiland L, Ramroop J, Hill MN, Manely J, McEwen BS (2012) Chronic juvenile stress produces corticolimbic dendritic architectural remodeling and modulates emotional behavior in male and female rats. Psychoneuroendocrinology 37:39–47
Einon DF, Morgan MJ (1977) A critical period for social isolation in the rat. Dev Psychobiol 10:123–132
Engelmann M, Ludwig M, Landgraft R (1994) Simultaneous monitoring of intracerebral release and behavior: endogenous vasopressin improves social recognition. J Neuroendocrinol 6:391–395
Fenoglio KA, Brunson KL, Avishai-Eliner S, Stone BA, Kapadia BJ, Baram TZ (2005) Enduring, handling-evoked enhancement of hippocampal memory function and glucocorticoid receptor expression involves activation of the corticotropin-releasing factor type 1 receptor. Endocrinology 146(9):4090–4096
Fenoglio KA, Chen Y, Baram TZ (2006) Neuroplasticity of the hypothalamic-pituitary-adrenal axis early in life requires recurrent recruitment of stress-regulating brain regions. J Neurosci 26(9):2434–2442
Ferguson JN, Young LJ, Insel TR (2002) The neuroendocrine basis of social recognition. Front Neuroendocrinol 23:200–224
Ferris C (1992) Role of vasopressin in aggressive and dominant/subordinate behaviors. Ann N Y Acad Sci 652:212–226
Foilb AR, Lui P, Romeo RD (2011) The transformation of hormonal stress responses throughout puberty and adolescence. J Endocrinol 210:391–398
Ford JD, Kidd P (1998) Early childhood trauma and disorders of extreme stress as predictors of treatment outcome with chronic posttraumatic stress disorder. J Trauma Stress 11: 743–761
Francis DD, Meaney MJ (1999) Maternal care and the development of stress responses. Curr Opin Neurobiol 9:128–134
Freeman ME (2006) Neuroendocrine control of the ovarian cycle of the rat. In: Neill JD (ed) Knobil and Neill's physiology of reproduction. Raven Press, New York, pp 2327–2388
Frisone DF, Frye CA, Zimmerberg B (2002) Social isolation stress during the third week of life has age-dependent effects on spatial learning in rats. Behav Brain Res 128:153–160
Garoflos E, Panagiotaropoulos T, Pondiki S, Stamatakis A, Philippidis E, Stylianopoulou F (2005) Cellular mechanisms underlying the effects of an early experience on cognitive abilities and affective states. Ann Gen Psychiatry 4(1):8
Giovenardi M, de Azevedo MS, da Silva SP, Hermel EES, Gomes CM, Lucion AB (2005) Neonatal handling increases fear and aggression in lactating rats. Physiol Behav 86 209–217
Gobrogge KL, Liu Y, Young LJ, Wang Z (2009) Anterior hypothalamic vasopressin regulates pair-bonding and drug-induced aggression in a monogamous rodent. Proc Natl Acad Sci U S A 106:19144–19149

Gogtay N, Giedd JN, Lusk L, Hayashi KM, Greenstein D, Vaituzis AC, Nugent TF, Herman DH, Clasen LS, Toga AW, Rappoport JL, Thompson PM (2004) Dynamic mapping of human cortical development during childhood through early adulthood. Proc Natl Acad Sci U S A 101:8174–8179

Goldman L, Winget C, Hollingshead GW, Levine S (1973) Postweaning development of negative feedback in the pituitary- adrenal system of the rat. Neuroendocrinology 12:199–211

Gomes CM, Frantz PJ, Sanvitto GL, Anselmo-Franci JA, Lucion AB (1999) Neonatal handling induces anovulatory estrous cycles in rats. Braz J of Med Biol Res 32:1239–1242

Gomes CM, Raineki C, Ramos de Paula P, Severino GS, Helena CVV, Anselmo-Franci JA, Franci CR, Sanvitto GL, Lucion AB (2005) Neonatal handling and reproductive function in female rats. J of Endocrinol 184:435–445

Gomes CM, Donadio MVF, Anselmo-Franci JA, Franci CR, Lucion AB, Sanvitto GL (2006) Neonatal handling induces alteration in progesterone secretion after sexual behavior but not in angiotensin II receptor density in the medial amygdala: implication for reproductive success. Life Sci 78:2867–2871

Gross C, Hen R (2004) The developmental origins of anxiety. Nat Rev Neurosci 5(7):545–552

Haltmeyer GC, Denenberg VH, Thatcher J, Zarrow MX (1966) Response of the adrenal cortex of the neonatal rat after subjection to stress. Nature.212:1371–1373

Harlow HF, Dodsworth RO, Harlow MK (1965) Total social isolation in monkeys. Proc Natl Acad Sci U S A 54(1):90–97

He J, Crews FT (2007) Neurogenesis decreases during brain maturation from adolescence to adulthood. Pharmacol Biochem Behav 86:327–333

Heim C, Nemeroff CB (2001) The role of childhood trauma in the neurobiology of mood and anxiety disorders: preclinical and clinical studies. Biol Psychiatry 49:1023–1039

Heim C, Plotsky PM, Nemeroff CB (2004) Importance of studying the contributions of early adverse experience to neurobiological findings in depression. Neuropsychopharmacology 29:641–648

Herman JP, Figueiredo H, Mueller NK, Ulrich-Lai Y, Ostander MM, Choi DC, Cullinan WE (2003) Central mechanisms of stress integration: hierarchical circuitry controlling hypothalamic-pituitary-adrenocortical responsiveness. Front Neuroendocrinol 24:151–180

Hinde RA, Spencer-Booth Y (1971) Towards understanding individual differences in rhesus mother-infant interaction. Anim Behav 19:165–173

Hinde RA, Spencer-Booth Y, Bruce M (1966) Effects of 6-day maternal deprivation on rhesus monkey infants. Nature 210:1021–1023

Hodes GE, Shors TJ (2005) Distinctive stress effects on learning during puberty. Horm Behav 48:163–171

Hofer MA, Sullivan RM (2008) Towards a neurobiology of attachment. In: Nelson CA, Luciana M (eds) Handbook of developmental cognitive neuroscience. MIT Press, Cambridge, pp 787–806

Holmes A, le Guisquet AM, Vogel E, Millstein RA, Leman S, Belzung C (2005) Early life genetic, epigenetic and environmental factors shaping emotionality in rodents. Neurosci Biobehav Rev 29(8):1335–1346

Howland JG, Wang YT (2008) Synaptic plasticity in learning and memory: stress effects in the hippocampus. Prog Brain Res 169:145–158

Hubel D, Wiesel T (2012) David Hubel and Torsten Wiesel. Neuron 75(2):182–184

Huot RL, Thrivikraman KV, Meaney MJ, Plotsky PM (2001) Development of adult ethanol preference and anxiety as a consequence of neonatal maternal separation in Long Evans rats and reversal with antidepressant treatment. Psychopharmacology (Berl) 158(4):366–373

Huot RL, Plotsky PM, Lenox RH, McNamara RK (2002) Neonatal maternal separation reduces hippocampal mossy fiber density in adult Long Evans rats. Brain Res 950(1–2):52–63

Insel TR (2010) The challenge of translation in social neuroscience: a review of oxytocin, vasopressin, and affiliative behavior. Neuron 65:768–779

Ivy AS, Brunson KL, Sandman C, Baram TZ (2008) Dysfunctional nurturing behavior in rat dams with limited access to nesting material: a clinically relevant model for early-life stress. Neuroscience 154(3):1132–1142

Iwasaki S, Inoue K, Kiriike N, Hikiji K (2000) Effect of maternal separation on feeding behavior of rats in later life. Physiol Behav 70(5):551–556
Jahng JW (2011) An animal model of eating disorders associated with stressful experience in early life. Horm Behav 59(2): 213–220
Kember RL, Dempster EL, Lee TH, Schalkwyk LC, Mill J, Fernandes C (2012) Maternal separation is associated with strain-specific responses to stress and epigenetic alterations to Nr3c1, Avp, and Nr4a1 in mouse. Brain Behav 2(4):455–467
Kessler RC, Magee WJ (1993) Childhood adversities and adult depression: basic patterns of association in a US national survey. Psychol Med 23:679–690
Kiosterakis G, Stamatakis A, Diamantopoulou A, Fameli M, Stylianopoulou F (2009) Long-term effects of neonatal handling on mu-opioid receptor levels in the brain of the offspring. Dev Psychobiol 51(5):439–449
Klein ZA, Padow VA, Romeo RD (2010) The effects of stress on play and home cage behaviors in adolescent male rats. Dev Psychobiol 52:62–70
Koike H, Ibi D, Mizoguchi H, Nagai T, Nitta A, Takuma K et al (2009) Behavioral abnormality and pharmacological response in social isolation-reared mice. Behav Brain Res 202:114–121
Kosten TA, Miserendino MJ, Bombace JC, Lee HJ, Kim JJ (2005) Sex-selective effects of neonatal isolation on fear conditioning and foot shock sensitivity. Behav Brain Res 157(2):235–244
Kosten TA, Lee HJ, Kim JJ (2006) Early life stress impairs fear conditioning in adult male and female rats. Brain Res 1087(1):142–150
Kosten TA, Lee HJ, Kim JJ (2007) Neonatal handling alters learning in adult male and female rats in a task-specific manner. Brain Res 1154:144–153
Kosten TA, Kim JJ, Lee HJ (2012) Early life manipulations alter learning and memory in rats. Neurosci Biobehav Rev 36(9):1985–2006
Kraemer GW (2003) Psychobiology of early social attachment in rhesus monkey. Clinical implications. Ann N Y Acad Sci 15:401–408
Krolow R, Noschang C, Weis SN, Pettenuzzo LF, Huffell AP, Arcego DM, Marcolin M, Mota CS, Kolling J, Scherer EB, Wyse AT, Dalmaz C (2012) Isolation stress during the prepubertal period in rats induces long-lasting neurochemical changes in the prefrontal cortex. Neurochem Res 37:1063–1073
Krolow R, Noschang C, Arcego DM, Pettenuzzo LF, Weis SN, Marcolin ML, Huffell AP, Mota CS, Dalmaz C (2013) Isolation stress exposure and consumption of palatable diet during the prepubertal period leads to cellular changes in the hippocampus. Neurochem Res 38:262–272
Ladd CO, Huot RL, Thrivikraman KV, Nemeroff CB, Plotsky PM (2004) Long-term adaptations in glucocorticoid receptor and mineralocorticoid receptor mRNA and negative feedback on the hypothalamo-pituitary-adrenal axis following neonatal maternal separation. Biol Psychiatry 55(4):367–375
Lau C, Klinefelter G, Cameron AM (1996) Reproductive development and functions in the rat after repeated maternal deprivation stress. Fundam Appl Toxicol 30:298–301
Lehmann J, Pryce CR, Bettschen D, Feldon J (1999) The maternal separation paradigm and adult emotionality and cognition in male and female Wistar rats. Pharmacol Biochem Behav 64(4):705–715
Lenz KM, Graham MD, Parada M, Fleming AS, Sengelaub DR, Monks DA (2008) Tactile stimulation during artificial rearing influences adult function and morphology in a sexually dimorphic neuromuscular system. Dev Neurobiol 68:542–557
Leussis MP, Andersen SL (2008) Is adolescence a sensitive period for depression? Behavioral and neuroanatomical findings from a social stress model. Synapse 62:22–30
Leussis MP, Lawson K, Stone K, Andersen SL (2008) The enduring effects of an adolescent social stressor on synaptic density, part II: poststress reversal of synaptic loss in the cortex by adinazolam and MK-801. Synapse 62:185–192
Levine S (1957) Infantile experience and resistance to physiological stress. Science 126:405–406
Levine S, Haltmeyer G, Karas G, Denenberg V (1967) Physiological and behavioral effects of infantile stimulation. Physiol Behav 2:55–59

Levy F, Melo AI, Galef Jr BG, Madden M, Fleming AS (2003) Complete maternal deprivation affects social, but not spatial, learning in adult rats. Dev Psychobiol 43:177–191

Lim MM, Young LJ (2006) Neuropeptidergic regulation of affiliative behavior and social bonding in animals. Horm Behav 50:506–517

Lippmann M, Bress A, Nemeroff CB, Plotsky PM, Monteggia LM (2007) Long-term behavioural and molecular alterations associated with maternal separation in rats. Eur J Neurosci 25(10):3091–3098

Liu D, Diorio J, Tannenbaum B, Caldji C, Francis D, Freedman A, Sharma S, Pearson D, Plotsky PM, Meaney MJ (1997) Maternal care, hippocampal glucocorticoid receptors, and hypothalamic-pituitary-adrenal responses to stress. Science 277(5332):1659–1662

Liu D, Caldji C, Sharma S, Plotsky PM, Meaney MJ (2000) Influence of neonatal rearing conditions on stress-induced adrenocorticotropin responses and norepinepherine release in the hypothalamic paraventricular nucleus. J Neuroendocrinol 12(1):5–12

Lucion AB, Pereira FM, Winkelman EC, Anselmo-Franci JA, Sanvitto GL (2003) Neonatal handling reduces the number of cells in the locus coeruleus of rats. Behav Neurosci 117(5): 894–903

Lummaa V, Clutton-Brock T (2002) Early development, survival and reproduction in humans. Trends Ecol Evol 17:141–147

Lupien SJ, Fiocco A, Wan N, Maheu F, Lord C, Schramek T, Tu MT (2005) Stress hormones and human memory function across the lifespan. Psychoneuroendocrinology 30:225–242

Lutter M, Nestler EJ (2009) Homeostatic and hedonic signals interact in the regulation of food intake. J Nutr 139(3):629–632

Madruga C, Xavier LL, Achaval M, Sanvitto GL, Lucion AB (2006) Early handling, but not maternal separation, decreases emotional responses in two paradigms of fear without changes in mesolimbic dopamine. Behav Brain Res 166(2):241–246

Makena N, Bugarith K, Russell VA (2012) Maternal separation enhances object location memory and prevents exercise-induced MAPK/ERK signalling in adult Sprague-Dawley rats. Metab Brain Dis 27(3):377–385

Matthews SG (2002) Early programming of the hypothalamo-pituitary-adrenal axis. Trend Endocrinol Metabol 13:373–380

Mazaro R, Lamano-Carvalho TL (2006) Prolonged deleterious effects of neonatal handling on reproductive parameters of pubertal male rats. Reprod Fertil Dev 18:497–500

McCormick CM, Mathews IZ (2007) HPA function in adolescence: role of sex hormones in its regulation and the enduring consequences of exposure to stressors. Pharmacol Biochem Behav 86:220–233

McEwen BS (1994) Corticosteroids and hippocampal plasticity. Ann N Y Acad Sci 746:134–142

McEwen BS (2007) Physiology and neurobiology of stress and adaptation: central role of the brain. Physiol Rev 87:873–904

McEwen BS, Weiss JM, Schwartz LS (1968) Selective retention of corticosterone by limbic structure in rat brain. Nature 220:911–912

McIntosh J, Anisman H, Merali Z (1999) Short- and long-periods of neonatal maternal separation differentially affect anxiety and feeding in adult rats: gender-dependent effects. Brain Res Dev Brain Res 113(1–2):97–106

McLeod J, Sinal CJ, Perrot-Sinal TS (2007) Evidence for non-genomic transmission of ecological information via maternal behavior in female rats. Genes Brain Behav 6:19–29

Meaney MJ, Aitken DH (1985) The effects of early postnatal handling on hippocampal glucocorticoid receptor concentrations: temporal parameters. Brain Res 354(2):301–304

Meaney MJ, Aitken DH, Bodnoff SR, Iny LJ, Tatarewicz JE, Sapolsky RM (1985) Early postnatal handling alters glucocorticoid receptor concentrations in selected brain regions. Behav Neurosci 99(4):765–770

Meaney MJ, Aitken DH, Sharma S, Viau V, Sarrieau A (1989) Postnatal handling increases hippocampal type II, glucocorticoid receptors and enhances adrenocortical negative-feedback efficacy in the rat. Neuroendocrinology 51:597–604

Meaney MJ, Mitchell JB, Aitken DH, Bhatnagar S, Bodnoff SR, Iny LJ, Sarrieau A (1991) The effects of neonatal handling on the development of the adrenocortical response to stress: implications for neuropathology and cognitive deficits in later life. Psychoneuroendocrinology 16(1–3):85–103

Meaney MJ, Diorio J, Francis D, Widdowson J, LaPlante P, Caldji C, Sharma S, Seckl JR, Plotsky PM (1996) Early environmental regulation of forebrain glucocorticoid receptor gene expression: implications for adrenocortical responses to stress. Dev Neurosci 18(1–2):49–72

Meerlo P, Horvath KM, Nagy GM, Bohus B, Koolhaas JM (1999) The influence of postnatal handling on adult neuroendocrine and behavioural stress reactivity. J Neuroendocrinol 11:925–933

Meinlschmidt G, Heim C (2005) Decreased cortisol awakening response after early loss experience. Psychoneuroendocrinology 30:568–576

Neumann ID (2008) Brain oxytocin: a key regulator of emotional and social behaviours in both male and females. J Neuroendocrinol 20:858–865

Nishi M, Horii-Hayashi N, Sasagawa T, Matsunaga W (2013) Effects of early life stress on brain activity: implications from maternal separation model in rodents. Gen Comp Endocrinol 181:306–309

Noschang CG, Krolow R, Fontella FU, Arcego DM, Diehl LA, Weis SN, Arteni NS, Dalmaz C (2010) Neonatal handling impairs spatial memory and leads to altered nitric oxide production and DNA breaks in a sex specific manner. Neurochem Res 35(7):1083–1091

Noschang C, Krolow R, Arcego DM, Toniazzo AP, Huffell AP, Dalmaz C (2012) Neonatal handling affects learning, reversal learning and antioxidant enzymes activities in a sex-specific manner in rats. Int J Dev Neurosci 30:285–291

Núñez JF, Ferré P, García E, Escorihuela RM, Fernández-Teruel A, Tobeña A (1995) Postnatal handling reduces emotionality ratings and accelerates two-way active avoidance in female rats. Physiol Behav 57(5):831–835

Ogawa T, Mikuni M, Kuroda Y, Muneoka K, Mori KJ, Takahashi K (1994) Periodic maternal deprivation alters stress response in adult offspring: potentiates the negative feedback regulation of restraint stress-induced adrenocortical response and reduces the frequencies of open field-induced behaviors. Pharmacol Biochem Behav 49(4):961–967

Oomen CA, Soeters H, Audureau N, Vermunt L, van Hasselt FN, Manders EM, Joëls M, Krugers H, Lucassen PJ (2011) Early maternal deprivation affects dentate gyrus structure and emotional learning in adult female rats. Psychopharmacology (Berl) 214(1):249–260

Oreland S, Gustafsson-Ericson L, Nylander I (2010) Short- and long-term consequences of different early environmental conditions on central immunoreactive oxytocin and arginine vasopressin levels in male rats. Neuropeptides 44:391–398

Padoin MJ, Cadore LP, Gomes CM, Barros HMT, Lucion AB (2001) Long-lasting effects of neonatal stimulation on the behavior of rats. Behav Neurosci 115:1332–1340

Panagiotaropoulos T, Pondiki S, Papaioannou A, Alikaridis F, Stamatakis A, Gerozissis K, Stylianopoulou F (2004) Neonatal handling and gender modulate brain monoamines and plasma corticosterone levels following repeated stressors in adulthood. Neuroendocrinology 80:181–191

Panksepp JB, Lahvis GP (2007) Social reward among juvenile mice. Genes Brain Behav 6:661–671

Panksepp JB, Jochman KA, Kim JU, Koy JJ, Wilson ED, Chen Q et al (2007) Affiliative behavior, ultrasonic communication and social reward are influenced by genetic variation in adolescent mice. PLoS ONE 2(4):e351

Pellis SM, Pellis VC (1987) Play-fighting differs from serious fighting in both target of attack and tactics of fighting in the laboratory rat *Rattus norvegicus*. Aggress Behav 13:227–242

Penza KM, Heim C, Nemeroff CB (2003) Neurobiological effects of childhood abuse: implications for the pathophysiology of depression and anxiety. Arch Women Ment Health 6:15–22

Pibiri F, Nelson M, Guidotti A, Costa E, Pinna G (2008) Decreased corticolimbic allopregnanolone expression during social isolation enhances contextual fear: a model relevant for posttraumatic stress disorder. Proc Natl Acad Sci U S A 105:5567–5572

Pinna G, Dong E, Matsumoto K, Costa E, Guidotti A (2003) In socially isolated mice, the reversal of brain allopregnanolone down-regulation mediates the antiaggressive action of fluoxetine. Proc Natl Acad Sci U S A 100:2035–2040

Pinna G, Costa E, Guidotti A (2006) Fluoxetine and norfluoxetine stereospecifically and selectively increase brain neurosteroid content at doses that are inactive on 5-HT reuptake. Psychopharmacology 186:362–372

Pinna G, Costa E, Guidotti A (2009) SSRIs act as selective brain steroidogenic stimulants (SBSSs) at low doses that are inactive on 5-HT reuptake. Curr Opin Pharmacol 9:24–30

Plotsky PM, Meaney MJ (1993) Early, postnatal experience alters hypothalamic corticotropin-releasing factor (CRF) mRNA, median eminence CRF content and stress-induced release in adult rats. Brain Res Mol Brain Res 18(3):195–200

Plotsky PM, Owens MJ, Nemeroff CB (2008) Psychoneuroendocrinology of depression. Hypothalamic-pituitary-adrenal axis. Psychiatr Clin North Am 21:293–307

Pryce CR, Feldon J (2003) The role of corticotropin-releasing factor-norepinephrine systems in mediating the effects of early experience on the development of behavioral and endocrine responses to stress. Neurosci Biobehav Rev 27:57–71

Pryce CR, Ruedi-Bettschen D, Dettling AC, Feldon J (2005a) Early-life environmental manipulations in rodents and primates: potential animal models in depression research. In: Steckler T, Kalin NH, Reul JMHM (eds) Handbook of stress and the brain. Elsevier, Amsterdam, pp 23–50

Pryce CR, Rüedi-Bettschen D, Dettling AC, Weston A, Russig H, Ferger B, Feldon J (2005b) The role of corticotropin-releasing factor-norepinephrine systems in mediating the effects of early experience on the development of behavioral and endocrine responses to stress. Neurosci Biobehav Rev 29(4-5):649–674

Qiang M, Chen YC, Wang R et al (1997) Nitric oxide is involved in the formation of learning and memory in rats: studies using passive avoidance response and Morris water maze task. Behav Pharmacol 8:183–187

Quillfeldt JA (2010) Behavioral methods to study learning and memory in rats. In: Levy-Andersen M, Tufik S (eds) Animal models as tools in ethical biomedical research (pp 227–269). AFIP, São Paulo

Radley JJ, Rocher AB, Janssen WGM, Hof PR, McEwen BS, Morrison JH (2005) Reversibility of apical dendritic retraction in the rat medial prefrontal cortex following repeated stress. Exp Neurol 196:199–203

Raineki C, Szawka RE, Gomes CM, Lucion MK, Barp J, Bello-Klein A, Franci CR, Anselmo-Franci JA, Sanvitto GL, Lucion AB (2008) Effects of neonatal handling on central noradrenergic and nitric oxidergic systems and reproductive parameters in female rats. Neuroendocrinology 87:151–159

Raineki C, Souza MA, Szawka RE, Lutz ML, Vasconcellos LFT, Sanvitto GL Izquierdo I, Bevilaqua LR, Cammarota M, Lucion AB (2009) Neonatal handling and maternal odor preference in rat pups: involvement of monoamines and cyclic AMP response element-binding protein pathway in the olfactory bulb. Neuroscience 159:31–38

Raineki C, Lutz ML, Sebben V, Ribeiro RA, Lucion AB (2013) Neonatal handling induces deficits in infant mother preference and adult partner preference. Dev Psychobiol 55(5):496–507

Rhees RW, Laphart ED, Eliason D (2001) Effects of maternal separation during early postnatal development on male sexual behavior and female reproductive function. Behav Brain Res 123:1–10

Roman E, Gustafsson L, Berg M, Nylander I (2006) Behavioral profiles and stress-induced corticosteroid secretion in male Wistar rats subjected to short and prolonged periods of maternal separation. Horm Behav 50:736–747

Romeo RD (2010) Adolescence: a central event in shaping stress reactivity. Dev Psychobiol 52:244–253

Romeo RD, Lee SJ, Chhua N, McPherson CR, McEwen BS (2004a) Testosterone cannot activate an adult-like stress response in prepubertal male rats. Neuroendocrinology 79:125–132

Romeo RD, Lee SJ, McEwen BS (2004b) Differential stress reactivity in intact and ovariectomized prepubertal and adult female rats. Neuroendocrinology 80:387–393

Romeo RD, Bellani R, Karatsoreos IN, Chhua N, Vernov M, Conrad CD, McEwen BS (2006) Stress history and pubertal development interact to shape hypothalamic pituitary adrenal axis plasticity. Endocrinology 147:1664–1674

Romeo RD, Tang AC, Sullivan RM (2009) Early-life experiences: enduring behavioral, neurological, and endocrinological consequences, In: Pfaff DW, Etgen AM, Fahrbach SE, Rubin RT (eds) Hormones, brain and behavior, Elsevier, New York, 20 pp (1975–2004)

Ross HE, Young LJ (2009) Oxytocin and neural mechanisms regulating social cognition and affiliative behavior. Front Neuroendocrinol 30:534–547

Ryu V, Lee JH, Yoo SB, Gu XF, Moon YW, Jahng JW (2008) Sustained hyperphagia in adolescent rats that experienced neonatal maternal separation. Int J Obes (Lond) 32(9):1355–1362

Ryu V, Yoo SB, Kang DW, Lee JH, Jahng JW (2009) Post-weaning isolation promotes food intake and body weight gain in rats that experienced neonatal maternal separation. Brain Res 1295:127–134

Sánchez MM, Ladd CO, Plotsky PM (2001) Early adverse experience as a developmental risk factor for later psychopathology: evidence from rodent and primate models. Dev Psychopathol 13(3):419–449

Sapolsky RM, Meaney MJ (1986) Maturation of the adrenocortical stress response: neuroendocrine control mechanisms and the stress hyporesponsive period. Brain Res 396(1):64–76

Schmidt MV, Sterlemann V, Ganea K, Liebl C, Alam S, Harbich D et al (2007) Persistent neuroendocrine and behavioral effects of a novel, etiologically relevant mouse paradigm for chronic social stress during adolescence. Psychoneuroendocrinology 32:417–429

Schmidt MV, Czisch M, Sterlemann V, Reinel C, Samann P, Muller MB (2009) Chronic social stress during adolescence in mice alters fat distributionin late life: prevention by antidepressant treatment. Stress 12:89–94

Schulz KM, Molenda-Figueira HA, Sisk CL (2009) Back to the future: the organizational-activational hypothesis adapted to puberty and adolescence. Horm Behav 55:597–604

Schwartz NB (2000) Neuroendocrine regulation of reproductive cyclicity. In: Conn PM, Freeman ME (eds) Neuroendocrinology in physiology and medicine. Humana Press, Clifton, pp 135–146

Severino GS, Fossati IA, Padoin MJ, Gomes CM, Trevizan L, Sanvitto GL, Franci CR, Anselmo-Franci JA, Lucion AB (2004) Effects of neonatal handling on the behavior and prolactin stress response in male and female rats at various ages and estrous cycle phases of females. Physiol Behav 81(3):489–498

Shea A, Walsh C, Macmillan H, Steiner M (2005) Child maltreatment and HPA axis dysregulation: relationship to major depressive disorder and post traumatic stress disorder in females. Psychoneuroendocrinology 30:162–178

Sheriff MJ, Krebs CJ, Boonstra R (2010) The ghosts of predators past: population cycles and the role of maternal programming under fluctuating predation risk. Ecology 91(10):2983–2994

Sieck G, Ramaley JA (1975) Effects of early handling upon puberty: correlations with adrenal stress responsiveness. Physiol Behav 15:487–489

Silveira PP, Portella AK, Clemente Z, Bassani E, Tabajara AS, Gamaro GD, Dantas G, Torres IL, Lucion AB, Dalmaz C (2004) Neonatal handling alters feeding behavior of adult rats. Physiol Behav 80(5):739–745

Silveira PP, Portella AK, Clemente Z, Gamaro GD, Dalmaz C (2005) The effect of neonatal handling on adult feeding behavior is not an anxiety-like behavior. Int J Dev Neurosci 23(1):93–99

Silveira PP, da Silva Benetti C, Ayres C, Pederiva FQ, Portella AK, Lucion AB, Dalmaz C (2006) Satiety assessment in neonatally handled rats. Behav Brain Res 173(2):205–210

Silveira PP, Portella AK, Assis SA, Nieto FB, Diehl LA, Crema LM, Peres W, Costa G, Scorza C, Quillfeldt JA, Lucion AB, Dalmaz C (2010) Early life experience alters behavioral responses to sweet food and accumbal dopamine metabolism. Int J Dev Neurosci 28(1):111–118

Sisk CL, Foster DL (2004) The neural basis of puberty and adolescence. Nat Neurosci 7:1040–1047

Siviy SM, Harrison KA (2008) Effects of neonatal handling on play behavior and fear towards a predator odor in juvenile rats (*Rattus norvegicus*) J Comp Psychol 122:1–8

Slotten HA, Kalinichev M, Hagan JJ, Marsden CA, Fone KC (2006) Long-lasting changes in behavioural and neuroendocrine indices in the rat following neonatal maternal separation: gender-dependent effects. Brain Res 1097(1):123–132

Spear LP (2000) The adolescent brain and age-related behavioral manifestations Neurosci Biobehav Rev 24:417–463

Stevenson CW, Meredith JP, Spicer CH, Mason R, Marsden CA (2009) Early life programming of innate fear and fear learning in adult female rats. Behav Brain Res 198(1):51–57

Suomi SJ, Harlow HF, Domek CJ (1971) Effects of repetitive infant-infant separation of young monkeys. J Abnormal Psychol 76:161–172

Thomas KM, Drevets WC, Dahl RE, Ryan ND, Birmaher B, Eccard CH, Casey BJ (2001) Amygdala response to fearful faces in anxious and depressed children. Arch Gen Psychiatry 58:1057–1063

Todeschin AS, Winkelmann-Duarte EC, Jacob MHV, Aranda BCC, Jacobs S, Fernandes MC, Ribeiro MFM, Sanvitto GL, Lucion AB (2009) Effects of neonatal handling on social memory, social interaction, and number of oxytocin and vasopressin neurons in rats. Horm Behav 56:93–100

Tsoory M, Richter-Levin G (2005) Learning under stress in the adult rat is differentially affected by 'juvenile' or 'adolescent' stress. Int J Neuropsychopharmacol 8:1–16

Tsoory M, Cohen H, Richter-Levin G (2007) Juvenile stress induces a predisposition to either anxiety or depressive-like symptoms following stress in adulthood. Eur Neuropsychopharmacol 17:245–256

Tsuda MC, Yamaguchi N, Ogawa S (2011) Early life stress disrupts peripubertal development of aggression in male mice. Neuroreport 22:259–263

Uys JDK, Marais L, Faure J, Prevoo D, Swart P, Mohammed AH et al (2006a) Developmental trauma is associated with behavioral hyperarousal, altered HPA axis activity, and decreased hippocampal neurotrophin expression in the adult rat. Ann N Y Acad Sci 1071:542–546

Uys JDK, Muller CJF, Marais L, Harvey BH, Stein DJ, Daniels WMU (2006b) Early life trauma decreases glucocorticoid receptors in rat dentate gyrus upon adult re-stress: reversal by escitalopram. Neuroscience 137:619–625

Vallée M, Mayo W, Dellu F, Le Moal M, Simon H, Maccari S (1997) Prenatal stress induces high anxiety and postnatal handling induces low anxiety in adult offspring: correlation with stress-induced corticosterone secretion. J Neurosci 17(7):2626–2636

Van den Berg CL, Pijlman FT, Koning HA, Diergaarde L, Van Ree JM, Spruijt BM (1999) Isolation changes the incentive value of sucrose and social behaviour in juvenile and adult rats. Behav Brain Res 106:133–142

Van den Berg CL, Van Ree JM, Spruijt BM (2000) Morphine attenuates the effects of juvenile isolation in rats. Neuropharmacology 39:969–976

Van Eck M, Berkhof H, Nicolson N, Sulon J (1996) The effects of perceived stress, traits, mood states, and stressful daily events on salivary cortisol. Psychosom Med 58:447–458

Veenema AH (2012) Toward understanding how early-life social experiences alter oxytocin- and vasopressin-regulated social behaviors. Horm Behav 61:304–312

Veenema AH, Neumann ID (2008) Central vasopressin and oxytocin release: regulation of complex social behaviours. Progress Brain Res 170:261–276

Veenema AH, Neumann ID (2009) Maternal separation enhances offensive play-fighting, basal corticosterone and hypothalamic vasopressin mRNA expression in juvenile male rats. Psychoneuroendocrinology 34:463–467

Veenema AH, Blume A, Niederle D, Buwalda B, Neumann ID (2006) Effects of early lide stress on adult male aggression and hypothalamic vasopressin and serotonin. Euro J Neurosci 24:1711–1720

Veenema AH, Bredewold R, Neumann ID (2007) Opposite effects of maternal separation on intermale and maternal aggression in C57BL/6 mice: link to hypothalamic vasopressin and oxytocin immunoreactivity. Psychoneuroendocrinology 32:437–450

Wada H (2008) Glucocorticoids: mediators of vertebrate ontogenetic transitions. Gen Comp Endocrinol 156:441–453

Weaver, IC (2007) Epigenetic programming by maternal behavior and pharmacological intervention. Nature versus nurture: let's call the whole thing off. Epigenetics 2(1):22–28

Wei XY, Yang JY, Dong YX, Wu CF (2007) Anxiolytic-like effects of oleamide in grouphoused and socially isolated mice. Prog Neuropsychopharmacol Biol Psychiatry 31:1189–1195

Weisstaub NV, Zhou M, Lira A, Lambe E, González-Maeso J, Hornung JP, Sibille E, Underwood M, Itohara S, Dauer WT, Ansorge MS (2006) Cortical 5-HT2A receptor signaling modulates anxiety-like behaviors in mice. Science 313(5786):536–540

Weitzdorfer R, Hoger H, Shim KS, Cekici L, Pollak A, Lubec G (2008) Changes of hippocampal signaling protein levels during postnatal brain development in the rat. Hippocampus 18:807–813

Wigger A, Neumann ID (1999) Periodic maternal deprivation induces gender-dependent alterations in behavioral and neuroendocrine responses to emotional stress in adult rats. Physiol Behav 66(2):293–302

Winkelmann-Duarte EC, Todeschin AS, Fernandes MC, Bittencourt LC, Pereira GA, Samios VN, Schuh AF, Achaval ME, Xavier LL, Sanvitto GL, Mandarim-de-Lacerda CA, Lucion AB (2007) Plastic changes induced by neonatal handling in the hypothalamus of female rats. Brain Res 1170:20–30

Yildirim M, Mapp OM, Janssen WGM, Yin WL, Morrison JH, Gore AC (2008) Postpubertal decrease in hippocampal dendritic spines of female rats. Exp Neuro 210:339–348

Yoshimura S, Sakamoto S, Kudo H, Sassa S, Kumai A, Okamoto R (2003) Sex-differences in adrenocortical responsiveness during development in rats. Steroids 68:439–445

Zimmerberg B, Sageser KA (2011) Comparison of two rodent models of maternal separation on juvenile social behavior. Front Psychiatry 2:39

Chapter 8
Perinatal Positive and Negative Influences on the Early Neurobehavioral Reflex and Motor Development

Gabor Horvath, Dora Reglődi, Jozsef Farkas, Gyongyver Vadasz, Barbara Mammel, Timea Kvarik, Greta Bodzai, Blanka Kiss-Illes, Dorottya Farkas, Attila Matkovits, Sridharan Manavalan, Balazs Gaszner, Andrea Tamas and Peter Kiss

Abstract Early life events are critical in the development of the central nervous system. Injuries in this period can cause severe damage with permanent disabilities. The early changes following a perinatal lesion have prognostic significance. The nervous system in young age has a potential for plasticity and regeneration, which can prevent the negative effects of neuronal damage, and the most important objective of rehabilitation is to enhance this inner potential of the developing brain. Experimental examination of the environmental factors affecting this regeneration and remodeling process is very important. Endogenous factors, such as neurotrophic factors, which play a role in neurogenesis, migration, and differentiation of neurons, and development of neuronal circuits, are also in the center of interest. Most studies concerning the effect of positive or negative perinatal treatments focus mainly on long-term effects, and most examinations are carried out on adult animals following perinatal injuries. Less data are available on short-term effects and early neurobehavioral changes. In the past several years, we have shown how different (positive or negative) perinatal events affect the early neuronal development. Applying different tests widely used for behavioral testing, we have established a standardized testing method. This includes measuring parameters of somatic growth and facial development, appearance of basic neurological reflexes and also reflex performance, more complex motor coordination tests, and open-field and novelty-seeking tests. In the present chapter, we summarize data on early neurobehavioral development of newborn rats subjected to negative (perinatal asphyxia, hypoxia, excitotoxic injury, stress) and positive (enriched environment, neurotrophic factor treatment) stimuli during early postnatal life.

G. Horvath (✉) · D. Reglődi · J. Farkas · G. Vadasz · B. Mammel · T. Kvarik · G. Bodzai · B. Kiss-Illes · D. Farkas · A. Matkovits · S. Manavalan · B. Gaszner · A. Tamas · P. Kiss
Department of Anatomy, PTE-MTA Lendulet PACAP Research Team, University of Pecs, Szigeti u 12, Pecs 7624, Hungary
e-mail: gabor.horvath@aok.pte.hu

S. Manavalan
Department of Basic Sciences, National University of Health Sciences, 7200 66th Street North, Pinellas Park, Saint Petersburg, FL 33709, USA

M. C. Antonelli (ed.), *Perinatal Programming of Neurodevelopment*, Advances in Neurobiology 10, DOI 10.1007/978-1-4939-1372-5_8,

Early life events are critical in the development of the central nervous system. Perinatal injuries affecting the nervous system can lead to delayed development and can cause long-term motor and cognitive disabilities. Furthermore, perinatal insults may weaken the endogenous regenerative capacity of the nervous system and thus cause higher vulnerability to different stressors. Positive influences, on the other hand, increase the inner potential of the developing brain, which can have protective effects even after long periods. Early postnatal period is one of the critical age windows, in addition to prenatal life and adolescence, that is open to plastic changes and to positive or negative environmental influences (Marco et al. 2011).

The early changes following a perinatal lesion have prognostic significance. The nervous system in young age has a potential for plasticity and regeneration, which can prevent the negative effects of neuronal damage, and the most important objective of rehabilitation is to enhance this inner potential of the developing brain. Experimental examination of the environmental factors affecting this regeneration and remodeling process is very important. Endogenous factors, such as neurotrophic factors, which play a role in neurogenesis, migration, and differentiation of neurons and development of neuronal circuits, are also in the center of interest. Most studies concerning the effect of positive or negative perinatal treatments focus mainly on long-term effects, and most examinations are carried out on adult animals following perinatal injuries. Less data are available on short-term effects and early neurobehavioral changes. In the past several years, we have shown how different (positive or negative) perinatal events affect the early neuronal development. In the present chapter, we summarize data on early neurobehavioral development of newborn rats subjected to negative and positive stimuli during early postnatal life. Using the same battery of tests, we are able to make comparative analysis based on experiments carried out in the past 10 years. As negative influences, perinatal asphyxia, hypoxia, excitotoxic injury, stress, and neurotrophic factor antagonism have been tested and are described here. As positive influences, we have chosen two different kinds of approaches: enriched environment and treatment with a neurotrophic factor.

In rodents, the first 2 weeks of age is the critical period of neuronal maturation. Pups reach adult level in basic neurological development at the end of the third postnatal week (Altman and Sudarshan 1975, Hill et al. 1991, Smart and Dobbing 1971a, 1971b). Examinations of neurobehavioral development were started on the first postnatal day and were carried out daily between 12 and 3 PM until postnatal day 21. Animals were always cross-fostered to avoid litter differences. Inspections were made for maturation of physical characteristics such as eye opening, incisor eruption, and ear unfolding. Weight was also recorded. The battery of tests consisted of a series of physical signs and neurological reflexes as well as motor coordination. Pups were tested for the appearance of certain neurological reflexes and/or the time to perform a certain task to test the maturation of the nervous system. The neurological signs and reflexes tested are described below.

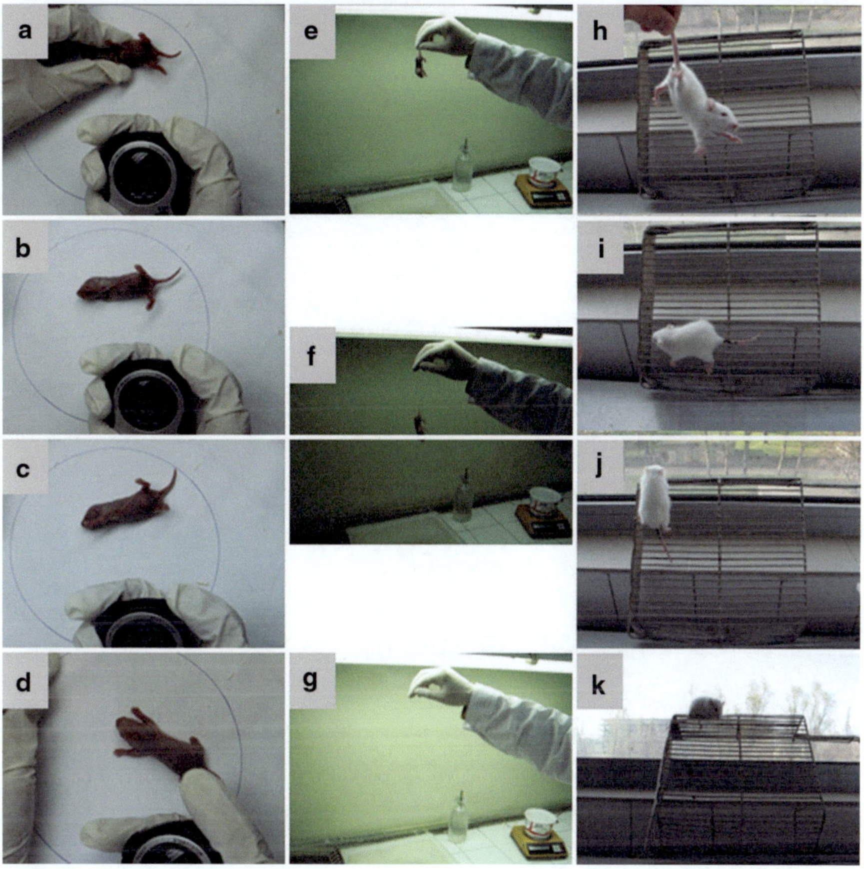

Fig. 8.1 Examination of neurological reflexes 1. **a–d** righting reflex, **e–f** air-righting reflex, **h–k** negative geotaxis

8.1 Testing of Neurological Reflexes

- *Righting reflex:* (a) The time in seconds to turn over to prone position from supine position (Fig. 8.1a, b, c, d) and (b) air righting—the first day of landing on four paws—is recorded when animals are dropped head down onto a soft surface (Fig. 8.1e, f, g).
- *Negative geotaxis:* Animals are placed head down on an inclined wire grid. The day they begin to turn around and climb up the wire grid with their forelimbs reaching the upper rim is observed and recorded (Fig. 8.1h, i, j, k). Time in seconds to reach the upper end of the wire grid is also recorded daily.
- *Sensory reflexes:* The ear and the eyelid are gently touched with a cotton swab and the 1st day of the ear twitch reflex (Fig. 8.2a) and the contraction of the eyelid (Figure. 8.2b) is recorded.

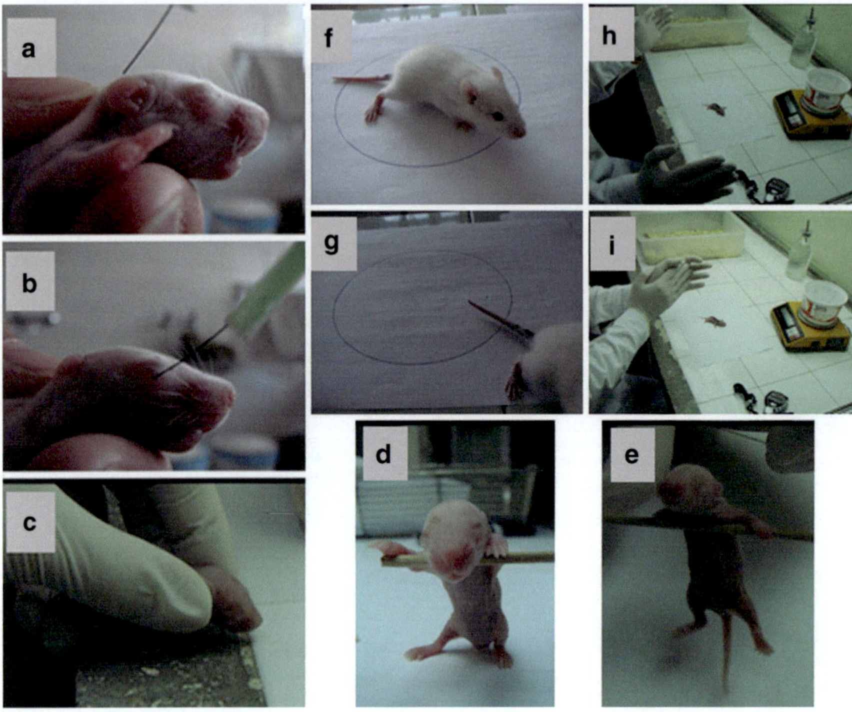

Fig. 8.2 Examination of neurological reflexes. **a** ear twitch reflex, **b** eyelid reflex, **c** forelimb-placing reflex, **d** forelimb grasp reflex negative and **e** positive, **f–g** gait reflex, **h–i** auditory startle reflex

- *Limb placing:* The back of the forepaw (Fig. 8.2c) and hindpaw is touched with the edge of the bench while the animal is suspended and the first day of lifting and placing the paws on the table is noted.
- *Limb grasp:* The fore- (Fig. 8.2d and e) and hindlimbs are touched with a thin rod and the first day of grasping onto the rod is recorded.
- *Gait*: The animals are placed in the center of a circle of 13 cm diameter and the day they began to move off the circle with both forelimbs is recorded (Fig. 8.2f, g). From the day of the appearance, the time in seconds to move off the circle is recorded daily.
- *Auditory* startle: The first day of the startle response to a clapping sound is observed (Fig. 8.2h, i).

8.2 Motor Coordination Tests

Pups were tested for motor coordination twice a week between 2 and 5 weeks of age.

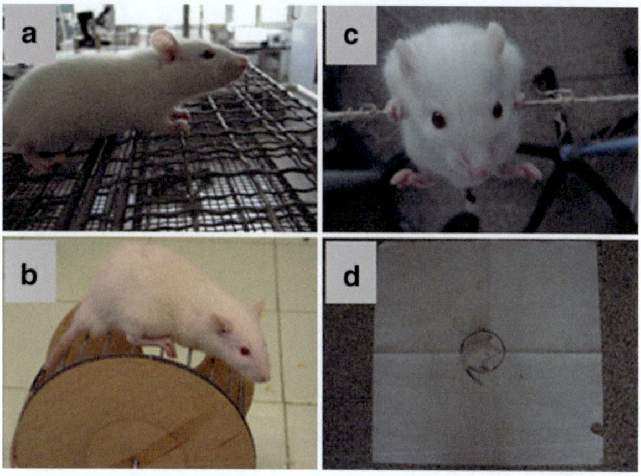

Fig. 8.3 Motor coordination tests. **a** grid-walking and foot-fault test, **b** rota-rod test, **c** rope suspension test, **d** walk initiation test

- *Grid-walking and foot-fault test:* Animals are placed on a stainless steel grid floor (20 × 40 cm with a mesh size of 4 cm^2) elevated 1 m above the floor. For a 1-min observation period, the total number of steps is counted. The number of foot-fault errors, when the animals misplace a forelimb or hindlimb that falls through the grid, is also recorded (Fig. 8.3a).
- *Rota-rod test*: Animals are tested on a commercially available treadmill for small animals with a diameter of 14 cm, attached to a rotating motor. The test is performed at a speed of 13 rpm. The pups are placed on the rotating drum and the time the animal can stay on the rota-rod was measured (max. 2 min; Fig. 8.3b).
- *Rope suspension:* Rats are placed with both forepaws on a rope suspended 1 m above the floor for a maximum of 30 s. The time they stay on the rope is recorded.
- *Walking initiation:* Pups are placed on a horizontal surface in the center of two concentric circles with diameters of 10 and 45 cm. The time taken to move off the circles is recorded (Fig. 8.3d).

The present chapter summarizes findings with different positive and negative influences using the same battery of tests. Below, we describe different lesions or treatments and how they affected neurobehavioral development. Both male and female pups were tested, but they showed differences only in the maternal separation paradigm. Therefore, for the other treatments/influences, we describe the findings for both male and female pups as one group (Table 8.1).

8.3 Excitotoxic Injury: Treatment with Monosodium Glutamate

Several toxic agents have been shown to cause altered neurobehavioral development (Archer et al. 2003; Cheng et al. 2013; Gilbert and Llorens 1993; Rousset et al. 2013; Schuck et al. 2009). One of the mechanisms of neurotoxicity is excitotoxicity

Table 8.1 Changes in neurobehavioral development and motor coordination. Data of physical and reflex development were recorded in day of appearance, reflex times were measured in seconds to perform a certain task, and motor coordination tests were measured in time except for the grid walking, where the number of total steps and the number of foot faults were recorded. Arrows indicate alteration to control groups

Perinatal exposure:		MSG treatment	Hypoxia/ ischemia	Asphyxia	Maternal deprivation		PACAP treatment		Enriched environment
					Male	Female	PACAP38	PACAP6-38	
Physical development	Weight	↓↓↓	↓-↓↓	↓↓-↓↓↓	↑	—	nd	nd	—
	Eye opening	—	↑	—	—	—	↓↓↓	↑↑↑	—
	Incisor eruption	—	—	↑↑	—	—	↓↓↓	—	—
	Ear unfolding	—	—	↑↑↑	—	—	↓↓↓	↑	—
Reflex development	Negative geotaxis	—	↑↑	↑↑↑	—	—	↓	—	↑
	Ear twitch	—	↑	—	—	—	↓↓↓	—	—
	Eyelid reflex	—	—	↑↑↑	—	—	↓↓↓	—	—
	Forelimb placing	↑↑	—	↑↑	—	—	↓	—	—
	Hindlimb placing	—	—	↑↑	—	—	—	↑	—
	Forelimb grasp	↑	↑↑↑	↑↑↑	—	↑	—	—	—
	Hindlimb grasp	—	↑↑↑	↑↑	↓	—	↓↓	—	—
	Gait	—	↑	↑↑	↓	—	↓	—	—
	Auditory startle	—	—	↑↑↑	—	—	↓↓↓	—	—
	Air righting	↑↑	—	↑↑↑	—	↑	—	—	—
Reflex times	Righting reflex	↑	↑↑↑	↑↑	—	↑	—	↑↑	—
	Negative geotaxis	—	↑↑	↑↑	—	—	↓↓↓	—	—
	Gait	—	↑	↑	—	—	↓↓↓	↑	↑

Perinatal exposure:		MSG treatment			Hypoxia/ ischemia			Asphyxia			Maternal deprivation Male			Maternal deprivation Female			PACAP38			PACAP6-38			Enriched environment		
Motor coordination	Age: weeks	w 3	w 4	w 5	w 3	w 4	w 5	w 3	w 4	w 5	w 3	w 4	w 5	w 3	w 4	w 5	w 3	w 4	w 5	w 3	w 4	w 5	w 3	w 4	w 5
	Walk initiation	—	—	nd	nd	nd	nd	↑	↑↑↑	↑	—	—	—	—	—	—	nd	nd	nd	nd	nd	nd	—	—	—
	Rope suspension	—	—	nd	↓	↓	—	nd	nd	nd	nd	nd	nd	nd	nd	nd	nd	nd	nd	nd	nd	nd	—	—	—
	Rotarod	↓	↓	—	nd	nd	nd	—	—	—	—	—	—	—	—	—	nd	nd	nd	nd	nd	nd	—	—	—
	Total steps/grid walking	—	—	—	—	↓	—	↓↓↓	↓	↓↓	—	—	—	—	—	—	nd	nd	nd	nd	nd	nd	—	—	—
	Foot fault	↑↑	↑↑↑	—	↑↑	↑	↑↑↑	↑↑↑	↑	↑↑	↓	—	—	—	—	—	nd	nd	nd	nd	nd	nd	—	—	—

nd no data; — no difference; ↑ higher/more/later; ↓ lower/less/earlier caused by excessive levels of glutamate. Monosodium glutamate (MSG) is a food additive widely used as a flavoring substance (abbreviation: E621). In adult animals and humans, MSG does not cross the blood–brain barrier in a considerable ratio. Therefore, MSG toxicity from exogenous overdosing does not exist in adults. There are some reports on MSG causing minor symptoms. The so called "Chinese restaurant syndrome" is attributed to MSG overdosing, but recent reports question the correlation between MSG and acute or chronic diseases (Williams and Woessner 2009). In contrast to adults, the immature blood–brain barrier in newborn rodent pups allows significant transportation of MSG into the nervous system leading to different lesions. Highly sensitive are the retina and arcuate nucleus of the hypothalamus, but other brain structures are also affected (Atlasz et al. 2009; Babai et al. 2005; Szabadfi et al. 2009). Among others, lesions in the hippocampus, prefrontal cortex, and other cortical areas have been reported after MSG toxicity (Beas-Zarate et al. 2002; Chaparro-Huerta et al. 2002; Gonzalez-Burgos et al. 2001). Even altered autonomic responses and other peripheral nervous system effects have been described (Andreazzi et al. 2011; Karlen-Amarante et al. 2012).

MSG treatment was performed with 4 mg/g body weight MSG on postnatal days 1, 3, 5, 7, and 9 based on earlier descriptions (Dunn and Webster 1985; Klingberg et al. 1987; Kubo et al. 1993). MSG was dissolved in 100 μl saline, and animals were treated subcutaneously. Among the MSG-treated pups, mortality was markedly higher than in the control group (Kiss et al. 2005). Among the physical parameters, only the body weight and length were significantly lower in MSG-treated pups, but no delays were observed in the day of eye opening, ear unfolding, or incisor eruption. Studying the neurobehavioral development of newborn rats, we found minor delays in the appearance of certain neurological reflexes (forelimb placing, forelimb grasp, and air righting), but the appearance of most signs remained unaffected by neonatal MSG treatment. Treated pups performed surface righting in significantly longer times and significantly worse performance was observed in motor coordination tests: MSG-treated pups made more foot faults and could stay on the rotating wheel for shorter times. In the motor coordination tests, however, MSG-treated pups reached control levels by the end of the observation period (fifth week). These results show that MSG treatment does not lead to permanent motor coordination or reflex defects; it leads to temporary delay only.

8.4 Perinatal Stress

Perinatal stress is related to long-term disturbances of several cognitive, behavioral, and emotional functions. Both pre- and postnatal stress models are widely used in neuroendocrinological and psychiatric research (Baier et al. 2012; Gaszner et al. 2012). These stress models may provide important correlation with human psychiatric diseases. For example, major depression and other stress-related mood disorders develop based on genetic predisposition, significant early life events inducing epigenetic alterations, and additional stress in later life. The coincidence and inter-

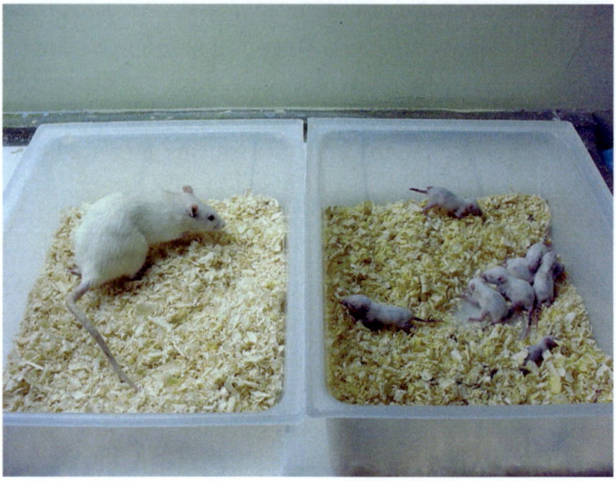

Fig. 8.4 Maternal deprivation

action of these factors precipitate the symptoms of a definitive stress-related mood disorder (de Kloet 2008). Maternal separation of rodent pups from their dam is a well-known model for early life events as it induces long-lasting changes in several stress-related systems (Gaszner et al. 2012; Marco et al. 2011). Maternal separation has been shown to increase hypothalamo–pituitary–adrenal response to stress. Several behavioral alterations can be observed in rats after maternal separation during their postnatal period of life (Li et al. 2013). Among others, fearfulness, changed reactivity to novel environments, morphine-induced sensitization and tolerance, along with several neurochemical changes in the reward circuitry, have been described (Gaszner et al. 2012; Marco et al. 2011).

Maternal deprivation was carried out by daily removal of the dam from the nest (Fig. 8.4) in a randomized manner, from postnatal days 1 to 14 for 3 h. In our experiments, the dam was removed in a semi-randomized order from their home cages and was placed into a separate cage; pups were put together into a container lined with home cage nesting material, in an incubator at 32 °C from days 2 to 8 and at 30 °C from days 8 to 14. After 180 min of maternal separation, pups are returned to their dams. Control rats are only shortly handled, for the duration of the neurobehavioral testing (Farkas et al. 2009). A short separation is considered to be physiologically resembling the mother rat looking for nutrition and being separated from her pups for a few minutes in the wild.

Effects of several agents on neurobehavioral development have been shown to be gender dependent (Nasiraei-Moghadam et al. 2013). The effects of maternal deprivation were also found to be gender dependent. We found that only male pups exposed to maternal separation reacted with faster weight gain than controls. No such difference was observed in female rats. Interestingly, also reflex development was faster in male pups exposed to perinatal stress: most reflexes developed one day earlier, reaching significant level in cases of hindlimb grasping and gait reflexes. However, females had significant delay in forelimb grasp and righting reflexes. The performance of reflexes measured in time was not markedly different after exposing rats to maternal deprivation. The signs of motor coordination showed a difference in foot-fault test: male rats had an enhanced motor coordination development, indicated by the significantly less number of faults made on the elevated wire meshwork on the third week. Taken together, the data indicated that maternal separation did not lead to marked changes in the neurobehavioral development, only minor alterations, which were, however, different between male and female pups. While a subtle enhancement was observed in male rats, a slight delay was detected in females. Other studies have also indicated gender-dependent effects of maternal separation. These include differences in light-induced startle reflex, which only affected female rats (De Jongh et al. 2005). Further differences have been described in plus maze test, depressive behavior, adrenocorticotropic hormone (ACTH) response, learning strategy, and dietary and hormonal effects later in life (Bobrovskaya et al. 2013; Grissom et al. 2012; Leussis et al. 2012; Slotten et al. 2006; Wigger and Neumann 1999). Our results indicate that the maternal separation paradigm most commonly used to test alterations in the stress axis and behavioral consequences later in life does not influence significantly the early neurobehavioral development, and thus, the major signs in the neurological and physical development do not have a predictive value for future alterations. More drastic perinatal stress paradigms have been shown to cause significant delays in the development, such as a 24-h maternal separation (Ellenbroek et al. 2005). In spite of the little predictive value, the results clearly indicate that there are gender-dependent effects of maternal separation already observable in early life.

8.5 Perinatal Asphyxia and Hypoxia

Hypoxic injuries in the perinatal period are still among the most feared perinatal events for their long-term consequences. In spite of intensive neonatal care, perinatal asphyxia and hypoxia during delivery represent a major clinical problem in neonatology. In spite of modern neonatal care, long-term consequences of perinatal hypoxic injuries cause a huge burden on society. There are several methods to induce hypoxia in rat pups. Rat pups resemble premature human babies at the time of their birth and reach the level of human newborn maturity on postnatal day 7. Perinatal asphyxia is usually induced by delivering pups from ready-to-deliver mothers after a period of hypoxia exposure. This model mimics the pathophysiological processes at the time of delivery, more closely resembling asphyxia in premature births. In our experiments, pups were delivered by Cesarean section from the uterine horns (Fig. 8.5.) and were stimulated to breath (Morales et al. 2008; Simola et al. 2008). Control Cesarean-delivered pups were delivered immediately after the sacrifice of their mothers, while other pups were delivered following a 15-min asphyctic period, kept at constant temperature (37 °C). Surviving rats were given to surrogate mothers after a survival period of 40 min at 37 °C. Another model of neonatal hypoxia involves the unilateral ligation of the common carotid artery followed by a period of hypoxic exposure (8 % oxygen and 92 % nitrogen, for 2 h) on postnatal day 7 (Vannucci et al. 1997).

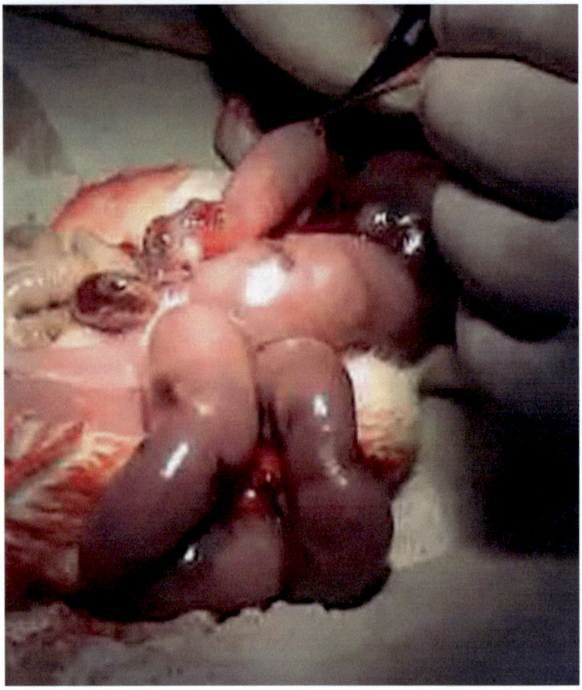

Fig. 8.5 Cesarean section for asphyctic pups

Both models of hypoxia have been associated with numerous deficits later in life (Allende-Castro et al. 2012). Behavioral studies have revealed that perinatal hypoxia leads to cognitive, motor, and social behavioral impairments (Galeano et al. 2011; van de Berg et al. 2003). Several mechanisms have been described in the background, such as different changes in synaptic structures (Grimaldi et al. 2012; van de Berg et al. 2000), neurotransmitter systems (van de Berg et al. 2003; Wixey et al. 2011), metabolic parameters (Souza et al. 2013), cytoskeletal structure (Saraceno et al. 2012), neuronal density (Blutstein et al. 2013), apoptotic pathways (Zhang et al. 2012), inflammation (Wixey et al. 2011), and overexpression of sentinel proteins (Herrera-Marschitz et al. 2012).

In our hands, perinatal asphyxia resulted in markedly increased mortality and an initial weight drop compared to control rats (Kiss et al. 2009). The high mortality observed is in accordance with other observations (Bustamante et al. 2007). Significant delays were also observed in the day of appearance of some physical maturity signs: incisor eruption and ear unfolding. The most significant delays were observed in the neurological reflexes: Almost all reflexes were delayed, some reflexes showed a 3–4-day delay. Among the lesions studied, we observed the most marked delays in this group. The performance of reflexes, such as surface righting, negative geotaxis, and gait reflexes, was also slower in the beginning of development. Among the motor coordination tests, the most pronounced difference was observed in the grid-walking test. Asphyctic pups took fewer steps in total but made more mistakes on the grid, indicating an impairment in walking on a field with obstacles.

Fig. 8.6 50% atrophy after neonatal hypoxia-ischemia

Neonatal hypoxia–ischemia caused an approximately 50% atrophy (Fig. 8.6) of the hypoxic hemisphere (Lubics et al. 2005). It resulted in retarded neurobehavioral development as shown by delayed appearance and worse performance of some neurological reflexes, and retarded development of motor coordination. However, in spite of the permanent cerebral atrophy, most animals reached control levels by 6 weeks of age in most tests, except for the contralateral foot-fault test (Lubics et al. 2005).

Based on our results, most severe deficits were observed in rat pups undergoing hypoxic ischemic injuries. The appearance of most neurological reflexes was delayed, in some cases for even 3–4 days. Reflex performance was also impaired. In most signs, hypoxic/asphyctic pups could catch up in their growth. However, fine motor skills remained impaired even 5 weeks later. Walking on the elevated wire meshwork without foot faults requires well-developed motor skills and has proven to be a good indicator of motor injury. We found that rats suffering hypoxic insults have long-term deficit in this fine motor test.

8.6 Treatment with a Neurotrophic Peptide and its Antagonist

Pituitary adenylate cyclase-activating polypeptide (PACAP) is a neuropeptide first isolated from the hypothalamus but later identified in various parts of the central and peripheral nervous system. PACAP belongs to the VIP–secretin–glucagon peptide family and exists in two biologically active forms, PACAP27 and PACAP38. Shorter fragments are usually described as receptor antagonists, although agonist effects have also been described (Harmar et al. 2012; Vaudry et al. 2009). Soon after its discovery in 1989, it became evident that PACAP promotes neurite growth and neuronal differentiation and plays various roles during the development of the nervous system (Basille-Dugay et al. 2013; Holighaus et al. 2012; Kambe and Miyata 2012; Nakamachi et al. 2011; Yan et al. 2013). Mice deficient in PACAP or its specific PAC1 receptor show various developmental deficits (Allais et al. 2007; Falluel-Morel et al. 2008). The expression of PACAP decreases after maturation of the brain, but the neuropeptide still plays important roles in the adult nervous system. It is widely accepted that in cases of injury or exposure to harmful stimuli, the nervous system reemploys mechanisms used during its development. PACAP is among those substances which are strongly upregulated upon various types of injury. Not surprisingly, PACAP has strong neuroprotective effects in traumatic, ischemic, and toxic nervous system injuries and different models of neurodegenerative diseases (Bourgault et al. 2009; Nakamachi et al. 2012; Reglodi et al. 2011; Tsuchikawa et al. 2012; Waschek 2013). Furthermore, PACAP-deficient mice have been shown to be highly vulnerable to different toxic or hypoxic injuries (Hori et al. 2013; Reglodi et al. 2012). The reemployment of embryonic mechanisms requires a well-organized network coordinating environmental circumstances, endogenous mechanisms, and exogenous stimuli. Studies performed in our group have mainly focused on the protective effects of PACAP. In light of the aforementioned relation between embryonic and adult protective mechanisms, we decided to examine what the effects of neonatal administration of PACAP38 and its antagonist PACAP6–38 might be.

Rat pups were treated subcutaneously with PACAP1–38 or its antagonist, PACAP6–38, for 2 weeks after birth. We found that neonatal rats treated with PACAP1–38 showed enhanced neurobehavioral development, as all physical maturation signs and most neurological reflexes appeared earlier. PACAP1–38 treatment led to significant advance not only on the day of appearance of most reflexes, but animals performed significantly better in tests, in which we measured the time to execute the task. PACAP1–38-treated pups completed the gait and negative geotaxis tasks in a significantly shorter time throughout the examination period. In contrast, pups treated with the PACAP antagonist PACAP6–38 showed delay in the day of eye opening, ear unfolding, and hindlimb placing. Also, these animals performed worse in the righting reflex.

These results let us conclude that PACAP can enhance facial development and neurobehavioral maturation, while its antagonist delays it, pointing to the neurotrophic role of endogenous PACAP. Our results are in accordance with those of Hill

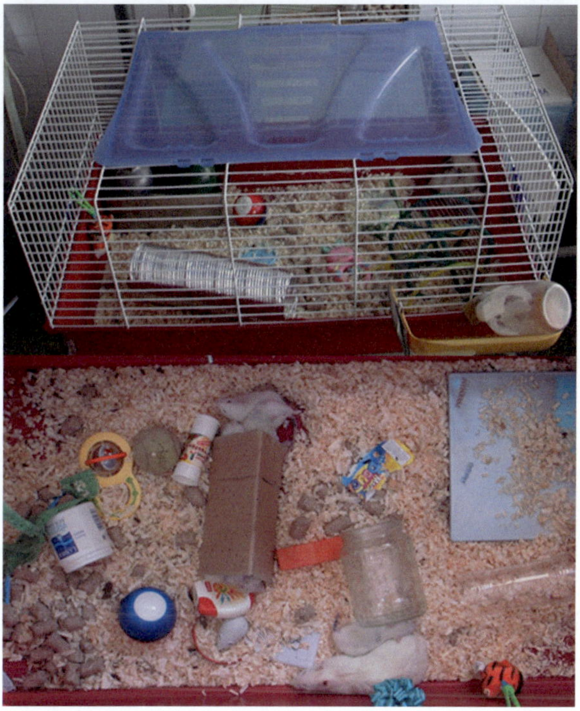

Fig. 8.7 Enriched environment cages

et al. (1991) who found that VIP antagonists delay neurobehavioral development. Taken together, these observations show that the peptides of the VIP/PACAP family and their receptors probably play a significant role in the neurobehavioral development of newborn rats.

8.7 Enriched Environment

The first description of enriched environment is dated back to 1947, when Hebb described that rats raised as pets performed better in problem-solving tasks (Hebb 1947). Since then, hundreds of studies have described the positive effects of environmental enrichment. Rats raised in enriched conditions are less vulnerable to various toxic, ischemic, and traumatic injuries (Jain et al. 2013; Johnson et al. 2013; Szabadfi et al. 2009). Various morphological and biochemical changes account for the improved functional performance observed in these animals (Lee et al. 2013).

In our study, we aimed at investigating whether raising rats in an enriched environment alters neurobehavioral development. In order to investigate this, animals of both sexes were cross-fostered immediately after birth, to minimize litter differences. Pups were placed in either regular (control) cages with 43 × 30 × 20-cm dimensions or in large cages with 88 × 50 × 44-cm dimensions (Fig. 8.7) supplemented with a

Fig. 8.8 Life in enriched environment

complex environmental enrichment (Fig. 8.8). Rats were continuously exposed to intensive multisensory stimulation. The cage contained different toys, objects, running tunnels, and rotating rods with various shapes, materials, and colors. Half of the objects were changed daily, while the other half were left unchanged. We found that there were no major alterations in the appearance of physical signs or neurological reflexes. Physical maturation signs were not altered by enriched environment, and among the neurological reflexes, only negative geotaxis showed a slight delay. Gait performance was also slightly slower. These observations were a little surprising, as it was expected that an enriched environment would enhance development. The slight delay observed in these groups could be explained by the previous experience of exposure to different objects. As with most neonatal treatments, the nervous system forms a "reservoir" of protective strategies that can be counted on in later challenges.

8.8 Combination of Different Treatments

It is well known that the effect of perinatal events can accumulate or can counteract each other. For example, it has been shown that the lack of PACAP can be masked by an enriched environment: several abnormal behavioral symptoms of PACAP deficient mice were not observed when mice were kept in enriched conditions (Ishihama et al. 2010). When combining PACAP deficiency, however, with maternal deprivation, we found that PACAP-deficient mice that have undergone maternal deprivation showed a depression phenotype and could serve as an animal model of stress-related mood disorders (Gaszner et al. 2012). Similar findings have been described with maternal deprivation: early favorable environment can compensate the effects of maternal separation (Vivinetto et al. 2013). It is expected that combining positive influences could help reaching a more significant protection. However, we have shown that in spite of PACAP treatment as well as enriched environment counteracting the deleterious effects of MSG-induced excitotoxic lesion, the combination of the two treatments did not result in an additive effect (Kiss et al. 2006, 2011).

8.9 Summary

In summary, we were able to compare the effects of different positive and negative stimuli on the neurobehavioral development of newborn rats. We found that neonatal hypoxia and asphyxia led to the most severe retardations in our developmental model. However, maternal deprivation or excitotoxic injury caused by MSG treatment led to only transient delay in the development of physical signs, neurological reflexes, and motor coordination. Recently, similar results have been found in neonatal inflammation (Rousset et al. 2013): maternal exposure to lipopolysaccharide leads to transient motor dysfunction in rats. As for the positive influences, we found that an enriched environment did not lead to profound changes, while treatment with a neurotrophic factor, PACAP, led to significant enhancement of neurobehavioral development. Our results also show that the battery of tests used are most suitable for detecting drastic changes induced by hypoxic injuries and to test potential neuroprotective agents in these insults. Furthermore, the transient delays can indicate a future vulnerability to other insults, as has been shown in rats undergoing maternal separation or in animals treated with MSG (Beas-Zarate et al. 1998; Skultetyova et al. 1998; Swinny et al. 2010).

Acknowledgments This study was supported by OTKA K104984, PD 109644, TAMOP (4.2.2.A-11/1/KONV-2012–0024 and 4.2.4.A/2-11-1-2012-0001 “National Excellence Program”), Arimura Foundation, PTE-MTA “Lendület” Program, Bolyai Scholarship, Hungarian Brain Research Program-Grant No. KTIA_13_NAP-A-III/5.

Conflict of Interest The authors declare no conflicts of interest.

References

Allais A, Burel D, Isaac ER, Gray SL, Basille M, Ravni A, Sherwood NM, Vaudry H, Gonzalez BJ (2007) Altered cerebellar development in mice lacking pituitary adenylate cyclase-activating polypeptide. Eur J Neurosci 25(9):2604–2618

Allende-Castro C, Espina-Marchant P, Bustamante D, Rojas-Mancilla E, Neira T, Gutierrez-Hernandez MA, Esmar D, Valdes JL, Morales P, Gebicke-Haerter PJ, Herrera-Marschitz M (2012) Further studies on the hypothesis of PARP-1 inhibition as a strategy for lessening the long-term effects produced by perinatal asphyxia: effects of nicotinamide and theophylline on PARP-1 activity in brain and peripheral tissue: nicotinamide and theophylline on PARP-1 activity. Neurotox Res 22(1):79–90

Altman J, Sudarshan K (1975) Postnatal development of locomotion in the laboratory rat. Anim Behav 23:896–920
Andreazzi AE, Grassiolli S, Marangon PB, Martins AG, de Oliveira JC, Torrezan R, Gravena C, Garcia RM, Mathias PC (2011) Impaired sympathoadrenal axis function contributes to enhanced insulin secretion in prediabetic obese rats. Exp Diabetes Res 2011:947917
Atlasz T, Szabadfi K, Reglodi D, Kiss P, Tamas A, Toth G, Molnar A, Szabo K, Gabriel R (2009) Effects of pituitary adenylate cyclase activating polypeptide (PACAP1-38) and its fragments on retinal degeneration induced by neonatal MSG treatment. Ann NY Acad Sci 1163:348–352
Archer T, Schroder N, Fredriksson A (2003) Neurobehavioural deficits following postnatal iron overload: II Instrumental learning performance. Neurotox Res 5:77–94
Babai N, Atlasz T, Tamás A, Reglődi D, Kiss P, Gábriel R (2005) Degree of damage compensation by various PACAP treatments in monosodium glutamate-induced retina degeneration. Neurotox Res 8:227–233
Baier CJ, Katunar MR, Adrover E, Pallarés ME, Antonelli MC (2012) Gestational restraint stress and the developing dopaminergic system: an overview. Neurotox Res 22(1):16–32
Basille-Dugay M, Vaudry H, Fournier A, Gonzalez B, Vaudry D (2013) Activation of pac1 receptors in rat cerebellar granule cells stimulates both calcium mobilization from intracellular stores and calcium influx through n-type calcium channels. Front Endocrinol (Lausanne) 4:56
Beas-Zárate C, Sánchez-Ruíz MY, Ureña-Guerrero ME, Feria-Velasco A (1998) Effect of neonatal exposure to monosodium L-glutamate on regional GABA release during postnatal development. Neurochem Int 33(3):217–232
Beas-Zarate C, Perez-Vega MI, Gonzalez-Burgos I (2002) Neonatal exposure to monosodium L-glutamate induces loss of neurons and cytoarchitectural alterations in hippocampal CA1 pyramidal neurons of adult rats. Brain Res 952:275–281
Blutstein T, Castello MA, Viechweg SS, Hadjimarkou MM, McQuail JA, Holder M, Thompson LP, Mong JA (2013) Differential responses of hippocampal neurons and astrocytes to nicotine and hypoxia in the fetal Guinea pig. Neurotox Res 24(1):80–93
Bourgault S, Vaudry D, Dejda A, Doan ND, Vaudry H, Fournier A (2009) Pituitary adenylate cyclase-activating polypeptide: focus on structure-activity relationships of a neuroprotective peptide. Curr Med Chem 16(33):4462–4480
Bobrovskaya L, Maniam J, Ong LK, Dunkley PR, Morris MJ (2013) Early life stress and post-weaning high fat diet alter tyrosine hydroxylase regulation and AT1 receptor expression in the adrenal gland in a sex dependent manner. Neurochem Res 38(4):826–833
Bustamante D, Morales P, Pereyra JT, Goiny M, Herrera-Marschitz M (2007) Nicotinamide prevents the effect of perinatal asphyxia on dopamine release evaluated with in vivo microdialysis 3 months after birth. Exp Brain Res 177:358–369
Chaparro-Huerta V, Rivera-Cervantes MC, Torres-Mendoza BM, Beas-Zarate C (2002) Neuronal death and tumor necrosis factor-α response to glutamate-induced excitotoxicity in the cerebral cortex of neonatal rats. Neurosci Lett 333:95–98
Cheng J, Fujimura M, Zhao W, Wang W (2013) Neurobehavioral effects, c-Fos/Jun expression and tissue distribution in rat offspring prenatally co-exposed to mehg and PFOA: PFOA impairs Hg retention. Chemosphere 91(6):758–764
De Jongh R, Geyer MA, Olivier B, Groenink L (2005) The effects of sex and neonatal maternal separation on fear-potentiated and light-enhanced startle. Behav Brain Res 161(2):190–196
de Kloet ER (2008) About stress hormones and resilience to psychopathology. J Neuroendocrinol 20(6):885–892
Dunn AJ, Webster EL (1985) Neonatal treatment with monosodium glutamate does not alter grooming behavior induced by novelty or adrenocorticotropic hormone. Behav Neur Biol 44:80–89
Ellenbroek BA, Derks N, Park HJ (2005) Early maternal deprivation retards neurodevelopment in Wistar rats. Stress 8(4):247–257
Falluel-Morel A, Tascau LI, Sokolowski K, Brabet P, DiCicco-Bloom E (2008) Granule cell survival is deficient in PAC1-/- mutant cerebellum. J Mol Neurosci 36(1–3):38–44

Farkas J, Reglodi D, Gaszner B, Szogyi D, Horvath G, Lubics A, Tamas A, Falko F, Besirevic D, Kiss P (2009) Effects of maternal separation on the neurobehavioral development of newborn Wistar rats. Brain Res Bull 79:208–214

Galeano P, Blanco Calvo E, Madureira de OD, Cuenya L, Kamenetzky GV, Mustaca AE, Barreto GE, Giraldez-Alvarez LD, Milei J, Capani F (2011) Long-lasting effects of perinatal asphyxia on exploration, memory and incentive downshift. Int J Dev Neurosci 29(6):609–619

Gaszner B, Kormos V, Kozicz T, Hashimoto H, Reglodi D, Helyes Z (2012) The behavioral phenotype of pituitary adenylate cyclase activating polypeptide deficient mice in anxiety and depression tests is accompanied by blunted c-Fos expression in the bed nucleus of the stria terminalis, central projecting Edinger Westphal nucleus, ventral lateral septum and dorsal raphe nucleus. Neuroscience 202:283–299

Gonzalez-Burgos I, Perez-Vega MI, Beas-Zarate C (2001) Neonatal exposure to monosodium glutamate induces cell death and dendritic hypotrophy in rat prefrontocortical pyramidal neurons. Neurosci Lett 297:69–72

Gilbert ME, Llorens J (1993) Delay in the development of amygdala kindling following treatment with 3,3'-iminodipropionitrile. Neurotoxicol Teratol 15(4):243–250

Grimaldi M, Romer I, de Apodaca MT, Iturbe L, Catania ID, González J, Kolliker-Fres R, Barreto G, Capani F (2012) Early changes in the synapses of the neostriatum induced by perinatal asphyxia. Nutr Neurosci 15(3):103–110

Grissom EM, Hawley WR, Bromley-Dulfano SS, Marino SE, Stathopoulos NG, Dohanich GP (2012) Learning strategy is influenced by trait anxiety and early rearing conditions in prepubertal male, but not prepubertal female rats. Neurobiol Learn Mem 98(2):174–181

Harmar AJ, Fahrenkrug J, Gozes I, Laburthe M, May V, Pisegna JR, Vaudry D, Vaudry H, Waschek JA, Said SI (2012) Pharmacology and functions of receptors for vasoactive intestinal peptide and pituitary adenylate cyclase-activating polypeptide: IUPHAR review 1. Br J Pharmacol 166(1):4–17

Hebb DO (1947) The effects of early experience on problem solving at maturity. Am Psychol 2:306–307

Hill JM, Gozes I, Hill JL, Fridkin M, Brenneman DE (1991) Vasoactive intestinal peptide antagonist retards the development of neonatal behaviors in the rat. Peptides 12:187–192

Holighaus Y, Weihe E, Eiden LE (2012) STC1 induction by PACAP is mediated through cAMP and ERK1/2 but not PKA in cultured cortical neurons. J Mol Neurosci 46(1):75–87

Hori M, Nakamachi T, Rakwal R, Shibato J, Ogawa T, Aiuchi T, Tsuruyama T, Tamaki K, Shioda S (2012) Transcriptomics and proteomics analyses of the PACAP38 influenced ischemic brain in permanent middle cerebral artery occlusion model mice. J Neuroinflammation 9:256

Ishihama T, Ago Y, Shintani N, Hashimoto H, Baba A, Takuma K, Matsuda T (2010) Environmental factors during early developmental period influence psychobehavioral abnormalities in adult PACAP-deficient mice. Behav Brain Res 209(2):274–280

Jain V, Baitharu I, Prasad D, Ilavazhagan G (2013) Enriched environment prevents hypobaric hypoxia induced memory impairment and neurodegeneration: role of BDNF/PI3K/GSK3β pathway coupled with CREB activation. PLoS One 8(5):e62235

Johnson EM, Traver KL, Hoffman SW, Harrison CR, Herman JP (2013) Environmental enrichment protects against functional deficits caused by traumatic brain injury. Front Behav Neurosci 7:44

Kambe Y, Miyata A (2012) Role of mitochondrial activation in PACAP dependent neurite outgrowth. J Mol Neurosci 48(3):550–557

Karlen-Amarante M, da Cunha NV, de Andrade O, de Souza HC, Martins-Pinge MC (2012) Altered baroreflex and autonomic modulation in monosodium glutamate-induced hyperadipose rats. Metabolism 61(10):1435–1442

Kiss P, Tamás A, Lubics A, Szalai M, Szalontay L, Lengvári I, Reglődi D (2005) Development of neurological reflexes and motor coordination in rats neonatally treated with monosodium glutamate. Neurotox Res 8:235–244

Kiss P, Tamás A, Lubics A, Lengvári I, Szalai M, Hauser D, Horváth Z, Rácz B, Gábriel R, Babai N, Tóth G, Reglődi D (2006) Effects of systemic PACAP treatment in monosodium glutamate-induced behavioral changes and retinal degeneration. Ann NY Acad Sci 1070:365–370

Kiss P, Szogyi D, Reglodi D, Horvath G, Farkas J, Lubics A, Tamas A, Atlasz T, Szabadfi K, Babai N, Gabriel R, Koppan M (2009) Effects of perinatal asphyxia on the neurobehavioral and retinal development of newborn rats. Brain Res 1255:42–50

Kiss P, Atlasz T, Szabadfi K, Horvath G, Griecs M, Farkas J, Matkovits A, Toth G, Lubics A, Tamas A, Gabriel R, Reglodi D (2011) Comparison between PACAP- and enriched environment-induced retinal protection in MSG-treated newborn rats. Neurosci Lett 487:400–405

Klingberg H, Brankack J, Klingberg F (1987) Long-term effects on behavior after postnatal treatment with monosodium-L-glutamate. Biomed Biochim Acta 46:705–711

Kubo T, Kohira R, Okano T, Ishikawa K (1993) Neonatal glutamate can destroy the hippocampal CA1 structure and impair discrimination learning in rats. Brain Res 616:311–314

Lee MY, Yu JH, Kim JY, Seo JH, Park ES, Kim CH, Kim H, Cho SR (2013) Alteration of synaptic activity-regulating genes underlying functional improvement by long-term exposure to an enriched environment in the adult brain. Neurorehabil Neural Repair 27(6):561–574

Leussis MP, Freund N, Brenhouse HC, Thompson BS, Andersen SL (2012) Depressive-like behavior in adolescents after maternal separation: sex differences, controllability, and GABA. Dev Neurosci 34(2–3):210–217

Li M, Xue X, Shao S, Shao F, Wang W (2013) Cognitive, emotional and neurochemical effects of repeated maternal separation in adolescent rats. Brain Res 1518:82–90.

Lubics A, Reglődi D, Tamás A, Kiss P, Szalai M, Szalontay L, Lengvári I (2005) Neurological reflexes and early motor behavior in rats subjected to neonatal hypoxic/ischemic injury. Behav Brain Res 157:157–165

Marco EM, Macrì S, Laviola G (2011) Critical age windows for neurodevelopmental psychiatric disorders: evidence from animal models. Neurotox Res 19(2):286–307

Morales P, Fiedler JL, Andres S, Berrios C, Huaiquin P, Bustamante D, Cardenas S, Parra E, Herrera-Marschitz M (2008) Plasticity of hippocampus following perinatal asphyxia: effects on postnatal apoptosis and neurogenesis. J Neurosci Res 86:2650–2662

Nasiraei-Moghadam S, Sherafat MA, Safari MS, Moradi F, Ahmadiani A, Dargahi L (2013) Reversal of prenatal morphine exposure-induced memory deficit in male but not female rats. J Mol Neurosci 50(1):58–69

Nakamachi T, Farkas J, Watanabe J, Ohtaki H, Dohi K, Arata S, Shioda S (2011) Role of PACAP in neural stem/progenitor cell and astrocyte-from neural development to neural repair. Curr Pharm Des 17(10):973–984

Nakamachi T, Tsuchida M, Kagami N, Yofu S, Wada Y, Hori M, Tsuchikawa D, Yoshikawa A, Imai N, Nakamura K, Arata S, Shioda S (2012) IL-6 and PACAP receptor expression and localization after global brain ischemia in mice. J Mol Neurosci 48(3):518–525

Reglodi D, Kiss P, Lubics A, Tamas A (2011) Review of the protective effects of PACAP in models of neurodegenerative diseases in vitro and in vivo. Curr Pharm Des 17:962–972

Reglodi D, Kiss P, Szabadfi K, Atlasz T, Gabriel R, Horvath G, Szakaly P, Sandor B, Lubics A, Laszlo E, Farkas J, Matkovits A, Brubel R, Hashimoto H, Ferencz A, Vincze A, Helyes Z, Welke L, Lakatos A, Tamas A (2012) PACAP is an endogenous protective factor - insights from PACAP deficient mice. J Mol Neurosci 48:482–492

Rousset CI, Kassem J, Aubert A, Planchenault D, Gressens P, Chalon S, Belzung C, Saliba E (2013) Maternal exposure to lipopolysaccharide leads to transient motor dysfunction in neonatal rats. Dev Neurosci 35(2–3):172–181

Saraceno GE, Castilla R, Barreto GE, Gonzalez J, Kölliker-Frers RA, Capani F (2012) Hippocampal dendritic spines modifications induced by perinatal asphyxia. Neural Plast 2012:873532

Schuck PF, Ferreira Gda C, Viegas CM, Tonin AM, Busanello EN, Pettenuzzo LF, Netto CA, Wajner M (2009) Chronic early postnatal administration of ethylmalonic acid to rats causes behavioral deficit. Behav Brain Res 197(2):364–370

Slotten HA, Kalinichev M, Hagan JJ, Marsden CA, Fone KC (2006) Long-lasting changes in behavioural and neuroendocrine indices in the rat following neonatal maternal separation: gender-dependent effects. Brain Res 1097(1):123–132

Simola N, Bustamante D, Pinna A, Pontis S, Morales P, Morelli M, Herrera-Marschitz M (2008) Acute perinatal asphyxia impairs non-spatial memory and alters motor coordination in adult male rats. Exp Brain Res 185:595–601

Skultétyová I, Tokarev D, Jezová D (1998) Stress-induced increase in blood-brain barrier permeability in control and monosodium glutamate-treated rats. Brain Res Bull 45(2):175–178

Smart JL, Dobbing J (1971a) Vulnerability of developing brain. II. Effects of early nutritional deprivation on reflex ontogeny and development on behavior in the rat. Brain Res 28:85–95

Smart JL, Dobbing J (1971b) Vulnerability of developing brain. VI. Relative effects of foetal and early postnatal undernutrition on reflex ontogeny and development of behavior in the rat. Brain Res 33:303–314

Souza SK, Martins TL, Ferreira GD, Vinagre AS, Silva RS, Frizzo ME (2013) Metabolic effects of perinatal asphyxia in the rat cerebral cortex. Metab Brain Dis 28(1):25–32

Swinny JD, O'Farrell E, Bingham BC, Piel DA, Valentino RJ, Beck SG (2010) Neonatal rearing conditions distinctly shape locus coeruleus neuronal activity, dendritic arborization, and sensitivity to corticotrophin-releasing factor. Int J Neuropsychopharmacol 13(4):515–525

Szabadfi K, Atlasz T, Horvath G, Kiss P, Hamza L, Farkas J, Tamas A, Lubics A, Gabriel R, Reglodi D (2009) Early postnatal enriched environment decreases retinal degeneration induced by monosodium glutamate treatment. Brain Res 1259:107–112

Tsuchikawa D, Nakamachi T, Tsuchida M, Wada Y, Hori M, Farkas J, Yoshikawa A, Kagami N, Imai N, Shintani N, Hashimoto H, Atsumi T, Shioda S (2012) Neuroprotective effect of endogenous pituitary adenylate cyclase-activating polypeptide on spinal cord injury. J Mol Neurosci 48(3):508–517

Van de Berg WD, Blockland A, Cuello AC, Schmitz C, Vreuls W, Steinbusch HW, Blanco CE (2000) Perinatal asphyxia results in changes in presynaptic bouton number in striatum and cerebral cortex - a stereological and behavioral analysis. J Chem Neuroanat 20:71–82

Van de Berg WD, Kwaijtaal M, de Louw AJ, Lissone NP, Schmitz C, Faull RL, Blokland A, Blanco CE, Steinbusch HW (2003) Impact of perinatal asphyxia on the GABAergic and locomotor system. Neuroscience 117:83–96

Vannucci RC, Rossini A, Towfight J, Vannucci SJ (1997) Measuring the accentuation of the brain damage that arises from perinatal cerebral hypoxia-ischemia. Biol Neonate 72:187–191

Vaudry D, Falluel-Morel A, Bourgault S, Basille M, Burel D, Wurtz O, Fournier A, Chow BK, Hashimoto H, Galas L, Vaudry H (2009) Pituitary adenylate cyclase-activating polypeptide and its receptors: 20 years after the discovery. Pharmacol Rev 61:283–357

Vivinetto AL, Suarez MM, Rivarola MA (2013) Neurobiological effects of neonatal maternal separation and post-weaning environmental enrichment. Behav Brain Res 240:110–118

Waschek J (2013) VIP and PACAP: neuropeptide modulators of CNS inflammation, injury, and repair. Br J Pharmacol 169(3):512–523

Wigger A, Neumann ID (1999) Periodic maternal deprivation induces gender-dependent alterations in behavioral and neuroendocrine responses to emotional stress in adult rats. Physiol Behav 66(2):293–302

Williams AN, Woessner KM (2009) Monosodium glutamate 'allergy': menace or myth? Clin Exp Allergy 39(5):640–646

Wixey JA, Reinebrant HE, Buller KM (2011) Inhibition of neuroinflammation prevents injury to the serotonergic network after hypoxia-ischemia in the immature rat brain. J Neuropathol Exp Neurol 70(1):23–35

Yan Y, Zhou X, Pan Z, Ma J, Waschek JA, DiCicco-Bloom E (2013) Pro- and anti-mitogenic actions of pituitary adenylate cyclase-activating polypeptide in developing cerebral cortex: potential mediation by developmental switch of PAC1 receptor mRNA isoforms. J Neurosci 33(9):3865–3878

Zhang L, Qu Y, Tang B, Zhao F, Xiong T, Ferriero D, Mu D (2012) Integrin β8 signaling in neonatal hypoxic-ischemic brain injury. Neurotox Res 22(4):280–291

Chapter 9
Short- and Long-Term Consequences of Perinatal Asphyxia: Looking for Neuroprotective Strategies

M. Herrera-Marschitz, T. Neira-Peña, E. Rojas-Mancilla, P. Morales, D. Bustamante, L. Leyton and P. Gebicke-Haerter

Abstract Perinatal asphyxia constitutes a prototype of obstetric complications occurring when pulmonary oxygenation is delayed or interrupted. A primary insult is first produced by the length of the time without oxygenation, leading to hypoxia/ischemia and death if oxygenation is not promptly established. A second insult is produced by re-oxygenation, eliciting a cascade of biochemical events for restoring function, implying, however, improper homeostasis. The effects observed long after perinatal asphyxia can be explained by over-expression of sentinel proteins, such as poly(ADP-ribose) polymerase-1 (PARP-1), competing for oxidised nicotinamide adenine dinucleotide (NAD^+) during re-oxygenation. Asphyxia also induces transcriptional activation of pro-inflammatory factors, including nuclear factor κB (NFκB) and its subunit p65, whose translocation to the nucleus is significantly increased in brain tissue from asphyxia-exposed animals, in tandem with PARP-1 overactivation, leading to the idea that sentinel protein inhibition constitutes a suitable therapeutic strategy. It is proposed that PARP-1 inhibition also down-regulates the expression of pro-inflammatory cytokines.

Nicotinamide is a suitable PARP-1 inhibitor, whose effects have been studied in an experimental model of global perinatal asphyxia in rats, inducing the insult by immersing rat foetuses into a water bath for various periods of time. Following asphyxia, the pups are delivered, immediately treated, or given to surrogate dams for nursing, pending further experiments. Systemic administration of nicotinamide 1 h after the insult inhibited PARP-1 overactivity in peripheral and brain tissue, preventing several of the long-term consequences elicited by perinatal asphyxia, supporting the idea that it constitutes a lead for exploring compounds with similar or better pharmacological profiles.

M. Herrera-Marschitz (✉) · T. Neira-Peña · E. Rojas-Mancilla · P. Morales · D. Bustamante · L. Leyton · P. Gebicke-Haerter
Millenium Institute BNI-Chile, Programme of Molecular & Clinical Pharmacology, ICBM, BNI, Medical Faculty, University of Chile, Santiago 7, P. O. Box 70.000, Chile
e-mail: mh-marschitz@med.uchile.cl

P. Gebicke-Haerter
Department of Psychopharmacology, Central Institute of Mental Health J5, Mannheim, Germany

M. C. Antonelli (ed.), *Perinatal Programming of Neurodevelopment*,
Advances in Neurobiology 10, DOI 10.1007/978-1-4939-1372-5_9,

Abbreviations

AIF	Apoptosis-inducing factor
ADP	Adenosine diphosphate
ATP	Adenosine triphosphate
ATPase	ATP polymerase
BAD	Bcl-2-associated death promoter
BAX	Bcl-2-associated X protein
BBB	Blood–brain barrier
bFGF	Basic fibroblast growth factor
BCL-2	B-cell lymphoma-2
CA	*Cornus Ammonis* (CA1, CA2, CA3)
CAM	Cell adhesion molecule
CNS	Central nervous system
CS	Caesarean-delivered saline-treated animal
COX-2	Cyclooxygenase-2
DG	Dentate gyrus of the hippocampus
DNA	Deoxyribonucleic acid
EAAC1	Excitatory amino acid carrier 1
Elk1	ETS domain-containing protein1
ERCC2	Excision repair cross-complementing rodent repair group 2
ERK	Extracellular signal-regulated kinases
FGFR	bFGF receptors
FLRT3	Leucine-rich repeat transmembrane protein
HIF	Hypoxia-inducible factor
IκB	Inhibitor of kappa B protein
iNOS	Inducible NOS
IL-1β/-6	Interleukin-1β/-6
IFG-1	Insulin-like growth factor 1
ICAM-1	TNFα adhesion molecule-1
L1	L1CAM
LPS	Lipopolysaccharides
MAP-2	Microtubule-associated protein-2
MAPK	Mitogen-activated protein kinase
NAD^+	Oxidised nicotinamide adenine dinucleotide
NADH	Reduced nicotinamide adenine dinucleotide
NFκB	Nuclear factor κB
Ngn2	Neurogenin-2
NgR	Nogo receptor
NMDA	N-methyl-D-aspartate
NOS	Nitric oxide synthase
nNOS	Neuronal NOS
p65/p50	Protein subunits of 65/50k Dalton MW
P1	Postnatal day 1
PARP-1	Poly(ADP-ribose) polymerase-1

PIP2	Phosphatidylinositol-4,5-bisphosphate
PKC	Protein kinase C
PSD95	Postsynaptic density protein 95
RhoA	Ras homolog gene family, member A, small GTPase protein
ROS	Reactive oxygen species
Sef	Similar expression fgf gene
SIRT	Sirtuin
SEM	Standard error of the means
Spry	Sprouty
SRY	Sex-determining region Y
SVZ	Subventricular zone
TH	Tyrosine hydroxylase
Thy-1	Thymocyte differentiation antigen 1
TNF-α	Tumour necrosis factor-alpha cytokine
TUNEL	TdT-mediated dUTP nick-end labelling
VTA	Ventral tegmental area
XRCC1	X-ray cross-complementing factor 1

9.1 Obstetric Complications are Associated with Psychiatric and Neurological Disorders Characterised by a Delayed Clinical Onset

Obstetric complications have been associated with psychiatric and neurological disorders characterised by a delayed clinical onset (Cannon et al. 2002). Among many different obstetric complications, hypoxia appears as a main factor, known to prime brain development by mechanisms not yet established (Low 2004; Basovich 2010; Herrera-Marschitz et al. 2011).

Delay in starting pulmonary ventilation at birth implies decrease of oxygen saturation in blood and oxygen supply to the brain, which depends on aerobic metabolism for maintaining the respiratory chain and mitochondrial ATP polymerase (ATPase) activity. Whenever hypoxia is sustained, there is a switch to glycolysis, which for neurons is a poor metabolic alternative, because of low stores of glucose in brain tissue and deficient ATP output by the glycolysis pathway. Glycolysis implies production of lactate, which is accumulated in extracellular compartments, causing acidosis (see Wyss et al. 2011). Prolonged hypoxia not only implies decreased gene expression and translation but also activation of genes, such as hypoxia-inducible factor (HIF) and its target molecules (Iyer et al. 1998).

Re-oxygenation is a requisite for survival, but it implies uneven metabolism, with metabolically privileged (e.g. heart, brain and adrenal medulla) and less privileged (e.g. muscles, kidneys and carcass) organs, including differences among brain regions. During the re-oxygenation period, extracellular levels of glutamate are increased, enhancing the activation of Na^+/K^+ ATPase, increasing further ATP consumption. Extracellular glutamate levels overpass the buffer capacity of astrocytes, resulting in overactivation of glutamate receptors, mainly the *N*-methyl-D-aspartate (NMDA) subtype, increasing Ca^{2+} conductance and further improper homeostasis. Extracellular lactate levels are increased in brain tissue even 8 days after perinatal asphyxia (Chen et al. 1997a). All these changes imply partial recovery, and sustained over-expression of alternative metabolic pathways, prolonging the energy deficit and oxidative stress. Oxidative stress is inherent to re-oxygenation, resulting in overactivation and inactivation of a number of buffering enzymes, including those modulating mitochondrial enzymes (see Gitto et al. 2002). In the clinical scenario, after resuscitation, the emphasis is on supportive therapy, with few interventions for preventing the long-term consequences of the insult, apart from hypothermia, which still is a controversial issue (van den Broek et al. 2010; Robertson et al. 2012).

The incidence of perinatal asphyxia is still high (2–6/1000 term births; see De Hann et al. 2006; Kurinczuk et al. 2010), occurring with higher prevalence in developing countries (Lawn et al. 2010). After asphyxia, infants have been observed to suffer from long-term neurological sequelae, the severity of which depends upon the extent and the recovery from the insult. Severe asphyxia has been linked to cerebral palsy, mental retardation and epilepsy, while mild-severe asphyxia has been associated with attention deficits, hyperactivity and schizophrenia (Robertson and Perlman 2006).

9.2 Early and Delayed Cell Death

Cell death can occur by different mechanisms, in a continuum from apoptosis to necrosis. Apoptosis occurs when the organism has time to organise a programmed death, while necrosis is the leading mechanism when all adaptive responses collapse following fulminate insults (Bonfocco et al. 1995). No gross morphology indicating necrotic cell death has been observed in rat pups surviving short periods of perinatal asphyxia (Dell'Anna et al. 1997), although long-lasting signs of neurodegeneration have been reported (Kohlhauser et al. 1999a). Indeed, several neurochemical and immunocytochemical studies have shown specific neuronal changes long after the insult (Andersson et al. 1995; Dell'Anna et al. 1997; Chen et al. 1997a, b, c), which may explain behavioural and cognitive deficits recurrently reported following perinatal asphyxia (Chen et al. 1995; Boksa et al. 1995; Iuvone et al. 1996; Hoeger et al. 2000; Simola et al. 2008; Morales et al. 2010).

Delayed cell death has been described as an important end point for perinatal asphyxia, associated to caspase-dependent and caspase-independent mechanisms (Northington et al. 2001). Indeed, pro-apoptotic proteins have been observed to be increased following perinatal asphyxia, including B-cell lymphoma-2 (Bcl-2)-associated X (BAX) and Bcl-2 associated death (BAD) factors, and anti-apoptotic proteins, including Bcl-2, extracellular signal-regulated kinase 2 (ERK2) and basic fibroblast growth factor (bFGF), suggesting the activation of neuroprotective and repair pathways (Morales et al. 2008). Extensive and regionally selective nuclear fragmentation has been observed in control and asphyctic rat pups, depending upon the stage of development and the analysed brain region (Dell'Anna et al. 1997). Signs of apoptosis have been observed in para- and presubiculum of both control and asphyxia-exposed animals, independently upon the severity of the insult, with the highest number of cells showing nuclear fragmentation observed at postnatal day 1–2 (P1–2), decreasing thereafter, with no cells showing nuclear fragmentation at P8. In contrast, no significant nuclear fragmentation is observed in neostriatum and neocortex of control pups, but a significant increase is observed in asphyxia-exposed animals at P8 (Fig. 9.1).

When assayed with an ApopTag® TUNEL kit (Millipore, Temecula, CA), the number of apoptotic cells was found to be increased in *cornus ammonis* (CA1 and CA3) regions of the hippocampus of asphyxia-exposed pups, compared to that from control pups (Fig. 9.2a). A large increase was also observed in the retrosplenial granular cortex (Fig. 9.2b), a region connecting the hippocampus with the entorhinal cortex (Wyss and Van Groen 1992), and receiving dopamine nerve terminals from mesencephalon (Hökfelt et al. 1974), where apoptosis is increased in substantia nigra and ventral tegmental area (VTA) of animals exposed to perinatal asphyxia (Fig. 9.2c; Neira-Peña et al. 2013).

9.3 Regional Vulnerability

Neurocircuitries of the basal ganglia, but also of hippocampus, are vulnerable to global anoxia/ischemia occurring at neonatal (Pasternak et al. 1991; Pastuzko 1994; Cowan et al. 2003; Miller et al. 2005; Barkovich 2006) and adult (Pulsinelli et al. 1982; see Haddad and Jiang 1993; Calabresi et al. 2000; Venkatesan and Frucht 2006) stages. Our previous work confirmed that vulnerability, assayed with immunochemistry (Dell'Anna et al. 1997; Chen et al. 1997a, b; Morales et al. 2003, 2005; Klawitter et al. 2005), molecular biology (Andersson et al. 1995; Gross et al. 2000, 2005), in vivo (Dell'Anna et al. 1995, 1997; Chen et al. 1997c), in vitro (Morales et al. 2003; Klawitter et al. 2005, 2006, 2007) and/or ex vivo (Ungethüm et al. 1996; Chen et al. 1997c; Bustamante et al. 2003) biochemistry, supporting the idea that the regional impact of the insult is related to (1) the severity of the insult, (2) the metabolic imbalance during the re-oxygenation period and (3) the developmental stage of the affected region.

The immaturity of the brain, or of particular brain regions, provides a factor of vulnerability to metabolic insults, including oxidative stress. In mammals, neuronal and glial growth and differentiation are predominantly postnatal events, with different time courses for neuronal migratory pathways. Thus, among monoamine pathways, dopamine-containing neuronal pathways have an earlier

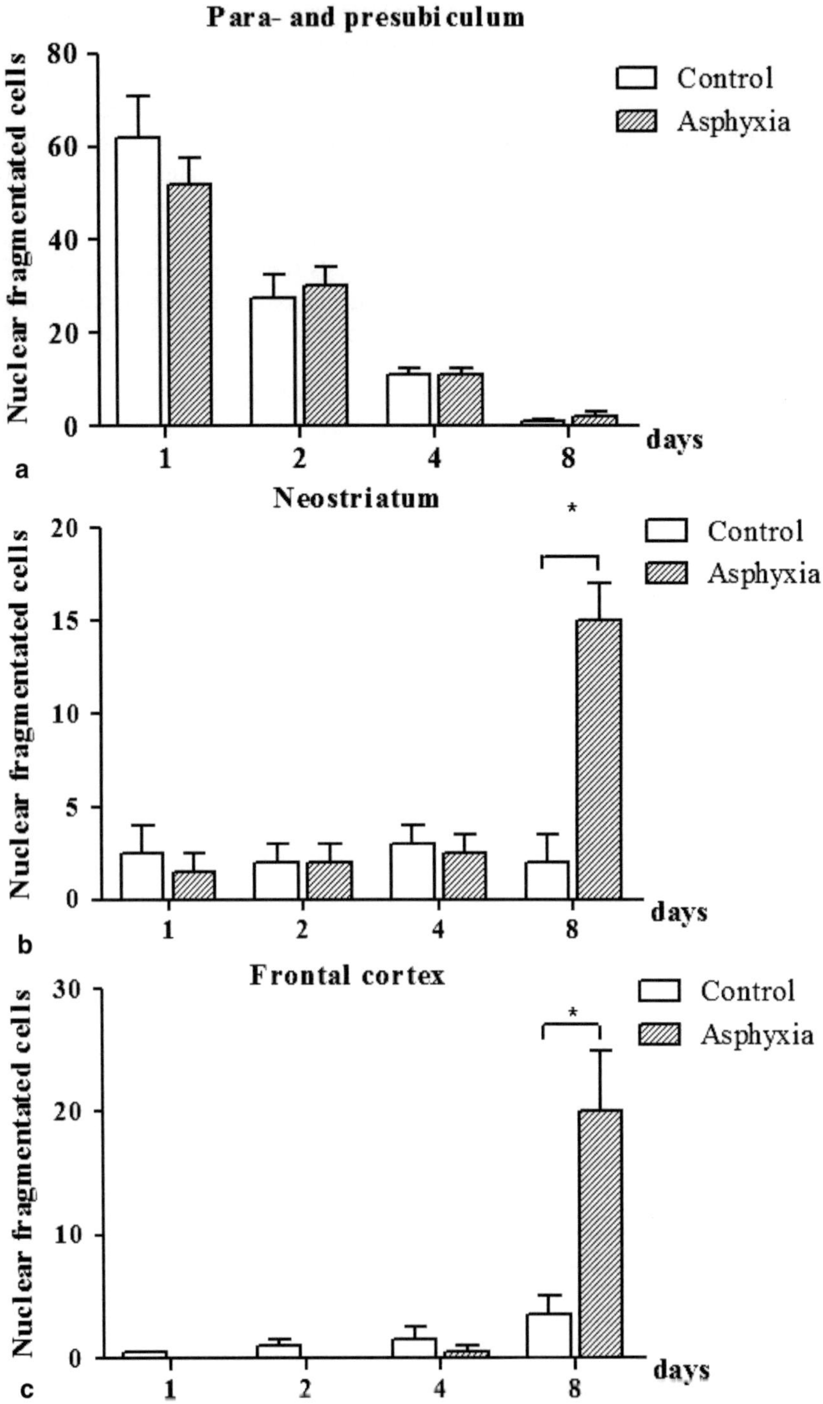

Fig. 9.1 Number of neurons showing nuclear fragmentation in para- and presubiculum (**a**), neostriatum (**b**) and frontal cortex (**c**) of asphyxia-exposed and control rat pups 1–8 days after birth.

and faster development than noradrenaline and 5-hydroxytryptamine pathways (Loizou 1972). Similarly, mesencephalic and telencephalic structures are differently developed at birth, and the functional development of a pivotal region such as neostriatum depends upon mesencephalic and neocortical inputs. At P1, the brain of rats possesses the same number of dopamine cell bodies as in adulthood, but while dopamine fibres start to invade the neostriatum before birth (Seiger and Olson 1972), dopamine-containing axon terminals reach a peak at the 4th postnatal week and a mature targeting only after several postnatal weeks, when patches are replaced by a diffuse dopamine innervation pattern (Olson and Seiger 1972; Loizou 1972; Seiger and Olson 1972; Voorn et al. 1988; Antonopoulos et al. 2002). Dopaminergic axons continue to grow at a slow rate during adulthood (Loizou 1972; Voorn et al. 1988), with naturally occurring waves of dopamine cell death, increasing the susceptibility of surviving neurons to energy failure (Oo and Burke 1997; Antonopoulos et al. 2002). The neocortex and hippocampus also mature at postnatal stages. In the rat, neocortical pyramidal projections become physiologically viable only 1 week after birth (Li and Martin 2000; Meng and Martin 2003; Meng et al. 2004), and in humans, the prefrontal cortex achieves a full mature stage long after post-adolescent stages (Sowell et al. 1999; Segalowitz and Davies 2004).

It is not surprising then that monoamine pathways have been found to be vulnerable to perinatal asphyxia, in particular nigrostriatal and mesolimbic dopamine pathways. Long-lasting decreases of tyrosine hydroxylase (TH)-positive cell bodies have been observed in substantia nigra and VTA of rats exposed to severe perinatal asphyxia (Chen et al. 1995, 1997b), together with decreased dopamine utilisation in neostriatum, accumbens and olfactory tubercle (Ungethüm et al. 1996; Bustamante et al. 2003) 1 month after birth. Also, when asphyxia-exposed animals were evaluated with in vivo microdialysis 3 months after birth (Bustamante et al. 2007), it was found that dopamine release was significantly decreased under basal and D-amphetamine-stimulated conditions. Decreases in TH immunohistochemistry have been observed in neostriatum, and also in hippocampus, thalamus, frontal cortex and cerebellum of asphyxia-exposed rats evaluated 1–3 months after birth (Kohlhauser et al. 1999a, b). In the same animals, the excitatory amino acid carrier 1 (EAAC1), a marker for glutamate neuronal phenotype, was, however, increased in the frontal cortex (Kohlhauser et al. 1999b). The regional selectivity of the insult has been further investigated with triple organotypic cultures, finding a selective decrease in the number of dopamine neurons in cultures from asphyxia-exposed animals. In the same cultures, nitric oxide synthase (NOS)-positive neurons were increased in substantia nigra, decreased in neostriatum and not changed in neocortex (Klawitter et al. 2007).

Brain sections (sagittal 20 μm serial formalin-fixed sections) were obtained from asphyxia-exposed (20–21 min of asphyxia) (n=5–7 animals per groups) (*dashed columns*) and the corresponding control (n=9–16) (*open columns*) animals, 1–8 days after delivery. The sections were stained for haematoxylin-eosin and in situ DNA double-strand breaks and analysed under light microscopy at 40× magnification to detect cellular alterations using Foster atlas (1998) for proper brain region identification (see Dell'Anna et al. 1997). Values are expressed as mean ± SEM. (*$p<0.05$; Student's t test)

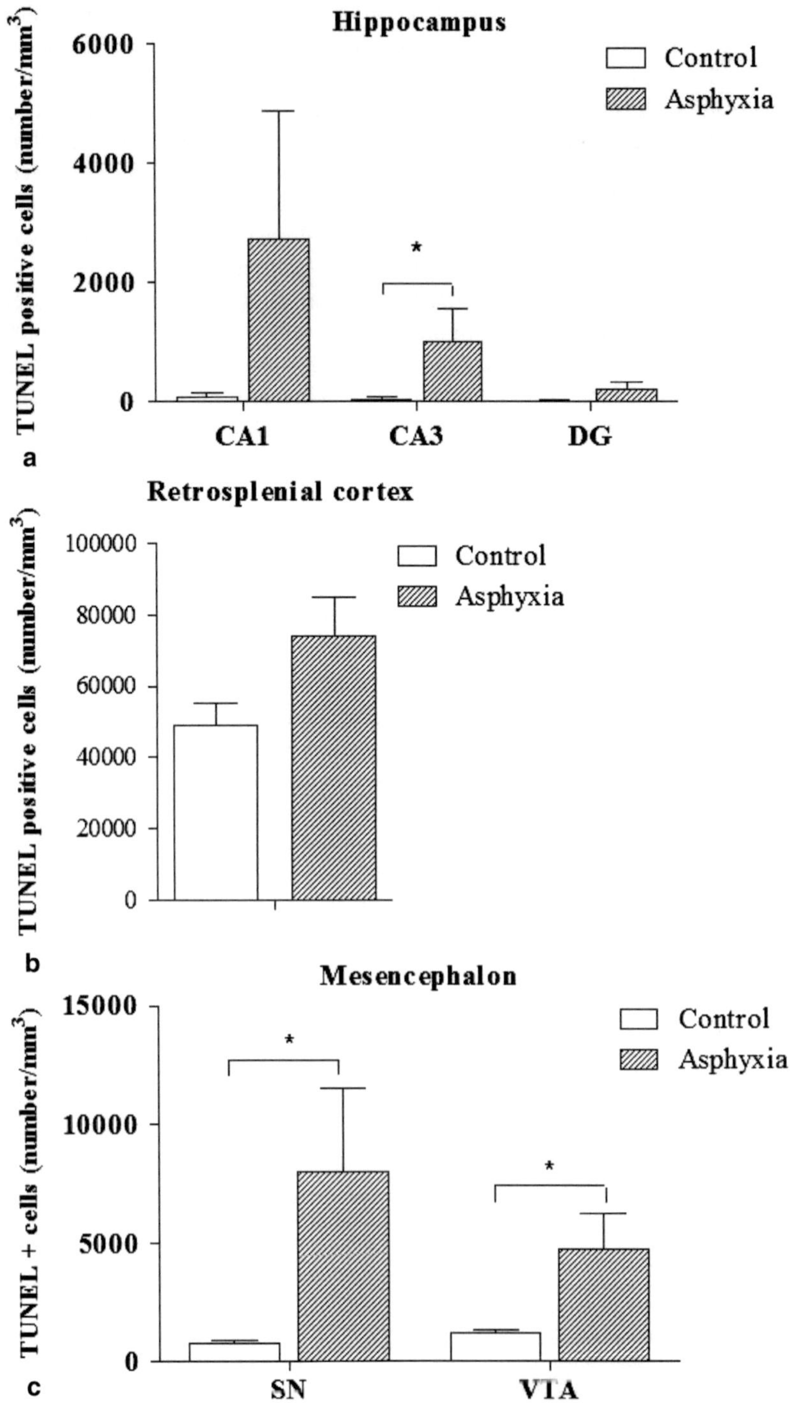

Fig. 9.2 Delayed cell death following perinatal asphyxia. Twenty-four hours after delivery, asphyxia-exposed and control animals were sacrificed, the brains rapidly removed and fixed in

9.4 Postnatal Neurogenesis, Neurite Growth, Branching and Synaptogenesis

In mammals, postnatal neurogenesis occurs in dentate gyrus (DG) of the hippocampus, as well as in other brain regions such as the subventricular zone (SVZ) (Alvarez-Buylla and Lim 2004; Ming and Song 2005; Zhao et al. 2008), probably for neuronal replacement (Arvidsson et al. 2002). Increased neurogenesis has been observed in the SVZ following anoxic/ischemic insults, producing neuroblasts that migrate, differentiate and develop to functionally competent neurons, being integrated to injured areas (Collin et al. 2005; Bédard et al. 2006).

The neuronal pathways of the hippocampus are largely established at postnatal stages, and the fine-tuning of their functioning probably takes place along the life. Neuroblasts from the subgranular zone of the DG migrate to the inner granular cell layer, projecting axons to the CA3 and dendrites to the molecular layer (Hastings and Gould 1999). Several reports have shown that neurogenesis in the DG is increased following anoxic/ischemic insults (Scheepens et al. 2003; Bartley et al. 2005; Morales et al. 2005, 2008) by mechanisms not well established.

bFGF promotes cell survival and neurogenesis (Takami et al. 1992; Cheng et al. 2002; Ganat et al. 2002) through activation of the mitogen-activated protein kinase (MAPK)/extracellular signal-regulated kinase (ERK) pathway (Morales et al. 2008; Mudò et al. 2009). Along these lines, the expression of bFGF has been observed to be up-regulated in DG and SVZ following perinatal asphyxia (Plane et al. 2004; Morales et al. 2008; Suh et al. 2009), possibly to prevent cell death (Han and Holtzman 2000). Several proteins have been identified as modulators of the transduction cascade elicited by bFGF receptors (FGFR) during embryogenesis, including Spry, Sef and leucine-rich repeat transmembrane protein 3 (FLRT3) (Tsang and Dawid 2004). Spry and Sef provide inhibitory regulation, while FLRT3 stimulates the activation of FGFR and ERK (Tsuji et al. 2004). Interestingly, ERK2 phosphorylation is modulated by PARP-1, promoting the expression of c-Fos (Cohen-Armon et al. 2007). bFGF requires proteoglycans to achieve full activation of FGFR, involving the action of phosphatases and the formation of a ternary complex, comprising the proteoglycan syndecan-4, phosphatidylinositol-4,5-bisphosphate (PIP2) and protein kinase Cα (PKCα) (Horowitz et al. 2002), for modulating the cellular responses required to be activated after postnatal injuries.

0.1 M pH 7.4 phosphate-buffered saline (*PBS*) containing 4 % paraformaldehyde, kept in 20 % sucrose, embedded in cryomatrix (Thermo Electron Corp, Pittsburgh, PA, USA) and stored at −80 °C. DNA fragmentation was evaluated in coronal brain sections (20 μm thick) with the TUNEL assay (ApopTag®; Millipore, Temecula, CA). TUNEL positive cells were manually counted in a Nikon TS100 microscope (magnification 100 ×). The number of TUNEL positive cells per mm^3 was determined in sections showing hippocampus (**a**), retrosplenial granular zone (**b**) and mesencephalon (**c**) (Foster 1998) from asphyxia-exposed (*dashed bars*) and the corresponding controls (*open bars*). Comparisons were tested with a Student's *t* test (*$p < 0.05$) (Neira-Peña et al. 2013). *CA1, CA3* Cornus Ammonis, *DG* dentate gyrus of the hippocampus, *VTA* ventral tegmental area

Several cell adhesion molecules of the immunoglobulin superfamily are involved in cell migration, neurite outgrowth and synaptic targeting. L1 cell adhesion molecule (L1) and β1 integrins converge into growth factor signalling, activating the MAPK/ERK pathways. L1 also interacts with Ephrin/EphB proteins, promoting axon branching (see Schmid and Maness 2008). Additionally, L1 promotes homophilic interactions with L1 itself, as well as a number of different heterophilic interactions with molecules, including αvβ3 integrin, axonin-1 and contactin/F3, to promote neurite outgrowth (Zhao et al. 1998; De Angelis et al. 1999; Ruppert et al. 1995; Montgomery et al. 1996). More than 100 mutations in L1 have been found in humans, associated with severe neurological dysfunctions, including agenesis of the corticospinal tract and the corpus callosum, spastic paraplegia and mental retardation (Kenwrick et al. 2000), supporting its role in central nervous system (CNS) development. L1 localises in the same compartment as microtubule-associated protein-2 (MAP-2), a somatodendritic marker (Demyanenko et al. 2001). Furthermore, it has been reported that the hippocampus of L1 mutant mice is smaller than in normal animals, suggesting that L1 is relevant for the regulation of hippocampal development (Demyanenko et al. 2001). L1 also plays a function in the organisation of dopaminergic neurons in mesencephalon and diencephalon, modulating the extension of growth cones to different synaptic targets (Demyanenko et al. 2001).

Neuritogenesis and synaptogenesis require specificity, which has to be preserved when building functional neurocircuitries, hence, the relevance of inhibitory proteins, such as thymocyte differentiation antigen 1 (Thy-1) and Nogo receptor (NgR), counteracting the effect of stimulatory proteins, such as L1. Thy-1 is abundantly expressed on the surface of most neurons in the CNS, but its expression is regulated during development, appearing only in the early postnatal period (Morris et al. 1985). Thy-1 plays an inhibitory role for axonal growth (Morris et al. 1992; Chen et al. 2005), stabilising neuronal connections during postnatal development (Morris et al. 1992; Barlow and Huntley 2000).

The inhibitory function of Thy-1 occurs by binding to αvβ3 integrin and syndecan-4 proteins expressed by astrocytes (Tiveron et al. 1992; Dreyer et al. 1995). There is evidence indicating that upon binding to both αvβ3 integrin and syndecan-4, Thy-1 triggers tyrosine phosphorylation of focal adhesion proteins, increasing Ras homolog member A protein (RhoA) activity, promoting the attachment and spreading of astrocytes (Leyton et al. 2001; Avalos et al. 2002, 2004, 2009; Hermosilla et al. 2008). Indeed, there is evidence indicating that αvβ3 integrin is a ligand for Thy-1 modulating neurite and axon growth following neuronal damage (Herrera-Molina et al. 2012).

NgR is also a neuronal protein inhibiting neurite outgrowth. The expression of NgR and its ligand Nogo-A are up-regulated in rats subjected to hypoxia–ischemia at P7, indicating their relevance for metabolic insults affecting neonates (Wang et al. 2006). Nogo-A is expressed in myelin-forming oligodendrocytes, and inhibition of either the receptor or the ligand leads to axonal sprouting and functional recovery of damaged neurons (Wiessner et al. 2003; Li et al. 2005), supporting the idea of NgR as an inhibitory protein regulating neuronal processes. Interestingly, myelination deficits have been observed in the brain of rats assayed 3 months after perinatal asphyxia (Kohlhauser et al. 2000). Astrocyte αvβ3 integrin and other oligodendrocyte, and also neuronal proteins, provide targets for modulating brain regeneration, opening therapeutic avenues for individuals affected by hypoxia. Indeed, perinatal asphyxia has been associated with increased sprouting in stratum oriens of dorsal hippocampus, promoting the formation of new recurrent excitatory branches from the mossy fibre projection to the CA3 region (Morales et al. 2010), a feature leading to epileptic seizures (Gill et al. 2012).

We have reported (Morales et al. 2003; Klawitter et al. 2005, 2007) that there is a decrease of neurite length and branching of neurons with dopamine and NOS phenotypes in brain tissue from rats exposed to perinatal asphyxia. A similar effect has been observed in hippocampus, affecting neurite length and branching, as well as synaptophysin and postsynaptic density protein 95 (PSD95) expression found to be decreased at P30 in tissue from asphyxia-exposed animals (Rojas-Mancilla et al. 2013).

9.5 Inflammatory and Immunosignalling

The immature CNS is capable of mounting innate and adaptive immune responses through microglia and astrocytes. It has been argued that the immune responses of the developing CNS differ from that of the adult, in part due to immaturity of blood–brain barrier (BBB). Tight junctions are, however, present early during embryonic development (Kniesel et al. 1996), controlling the interchange of proteins, including leukocytes (Engelhardt 2003; Vexler and Yenari 2009). The developing brain is, however, vulnerable to increase in inflammatory cytokines, leading to a substantial crosstalk between peripheral and local brain immune components (see Ransohoff et al. 2003; Vexler and Yenari 2009). Indeed, Qiao et al. (2001) have demonstrated that in neonatal brain, disruption of the BBB to proteins occurs earlier after hypoxic–ischemic insult than in the mature brain.

Asphyxia induces transcriptional activation of pro-inflammatory factors, including NFκB that is normally located in the cytoplasm as a heterodimer, which is composed of a protein of 65 kDa (p65) and p50 subunits, coupled to an inhibitor kappa B protein (IκB). IκB dissociates from the complex when phosphorylated, liberating p65/p50 subunits, translocated to the nucleus for inducing the transcription of tumour necrosis factor-alpha cytokine (TNF-α), intercellular adhesion molecule-1 (ICAM-1), cyclooxygenase-2 (COX-2), inducible NOS (iNOS) and interleukin-6 (IL-6), all genes involved in inflammation (see Hagberg et al. 1996; Koh et al. 2004; Girard et al. 2009).

Microglia provide immunosurveillance to the brain by stimulus-dependent activation (see Vexler and Yenari 2009). Microglia populate the developing brain at birth, and activated microglia release a number of cytokines, including IL-1β, insulin-like growth factor 1 (IFG-1) and TNF-α. It is not clear yet whether microglia activation contributes or protects against the long-term deficits induced by perinatal asphyxia, but it has been reported that minocycline, a tetracycline derivative with anti-inflammatory properties, protects the neonatal brain against ischemia, partly by inhibiting microglia activation and monocyte infiltration (Arvin et al. 2002; Dommergues et al. 2003; Denker et al. 2007; also Alano et al. 2006).

Translocation of NFκB has been investigated in tissue samples from mesencephalon, hippocampus and telencephalon of asphyxia-exposed and control rat pups, 1–24 h after birth, with an antibody specific to p65. It was observed (Neira-Peña et al. 2013) that translocation of p65 to the nucleus was significantly increased in mesencephalon of both control and asphyxia-exposed pups 24 h after birth, but that increase was remarkable already 8 h after birth in asphyxia-exposed animals (Fig. 9.3a). In hippocampus, translocation of p65 was enhanced 1 h after birth in control animals, decreasing at 8–24 h, while in asphyxia-exposed animals, translocation of p65 was remarkably enhanced in hippocampus 8 h after birth, decreasing thereafter to almost negligible levels at 24 h (Fig. 9.3b; Neira-Peña et al. 2013).

DNA damage and cell death have the potential to activate PARP-1 and NFκB (see Skaper 2003; Gagne et al. 2008). Ullrich et al. (2001) showed that microglial migration towards the site of neuronal injury is controlled by PARP-1 overactivation, correlating with NFκB translocation. To evaluate the role of PARP-1 in inflammatory processes, PARP-1 $^{-/-}$ mice were challenged with lipopolysaccharides (LPS), finding that the knockout mice were resistant to the endotoxin shock compared to that observed in wild-type animals. The knockout mice also showed low levels of TNF-α due to reduced transcriptional activity of NFκB (Pétrilli et al. 2004).

9.6 Sentinel Proteins: PARP-1 as a Target for Neuroprotection

Suppression and/or overactivation of gene expression occur immediately or during the re-oxygenation period following perinatal asphyxia (Labudova et al. 1999; Seidl et al. 2000; Mosgoeller et al. 2000; Lubec et al. 2002). When the DNA integrity is compromised, a number of sentinel proteins are activated, including PARPs (Amé et al. 2004), X-ray cross-complementing factor 1 (XRCC1; Green et al. 1992), DNA ligase IIIα (Leppard et al. 2003), DNA polymerase β (Wilson 1998; Mishra et al. 2003), excision repair cross-complementing rodent repair group 2 (ERCC2; Sung et al. 1993; Chiappe-Gutierrez et al. 1998; Lubec et al. 2002) and DNA-dependent protein kinases (de Murcia and Menissier de Murcia 1994).

PARP proteins act as ADP-ribose transferases, transferring adenosine diphosphate (ADP)-riboses from nicotinamide dinucleotide (NAD^+) to glutamic and aspartic residues of the PARPs and their substrates. PARPs also catalyse the polymerisation of ADP-riboses via glycosidic bonds, creating long and branched ADP-ribose polymers. PARP-1 is the most abundant and conserved member of a large superfamily comprising at least 18 PARP proteins, encoded by different genes, but displaying a conserved catalytic domain (see Amé et al. 2004). PARP-1 is involved in DNA repair, but it also might promote cell death (see De Murcia and Menissier de Murcia 1994; Kauppinen and Swanson 2007; Cohen-Armon 2008). When DNA damage is mild, PARP-1 is involved in the maintenance of chromatin integ-

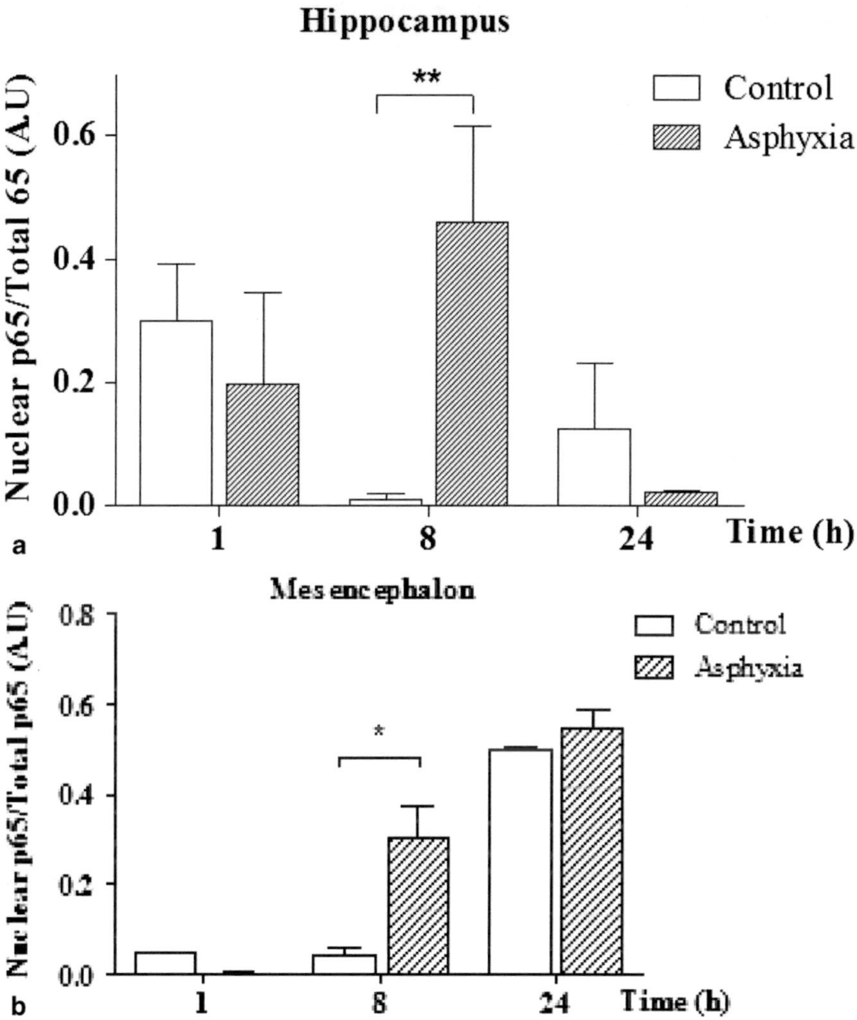

Fig. 9.3 Effect of re-oxygenation on p65 activation. Asphyxia and control rats were sacrificed 1, 8 and 24 h post birth; the brain was rapidly removed, taking up tissue samples from hippocampus (**a**) and mesencephalon (**b**), placed in Eppendorf tubes, frozen in liquid nitrogen and stored at −80 °C, pending further analysis. Cytosolic and nuclear protein were extracted with a ProteoJET™ kit (Fermentas). P65 levels were determined in cytosolic and nuclear protein fractions of tissue samples from asphyxia-exposed (*dashed bars*) and control (*open bars*) rat pups. Western blot was performed using an anti-NFκB p65 antibody (65 KDa) (Santa Cruz Biotechnology, Inc, Santa Cruz, CA). Tubulin and histone H4 were used as controls for the respective fractions. A densitygraphic analysis of the bands was performed. Data are represented as nuclear p65/total p65 ratio (comparisons were analysed with a Student's *t* test; $^*p<0.05$). (Neira-Peña et al. 2013)

rity by signalling cell-cycle arrest or activating DNA repairing molecular cascades. PARP-1 is also involved in the regulation of cell proliferation and differentiation, modulating the transcription of several inflammatory signals, including NFκB

(Hassa and Hottinger 1999). Excessive PARP-1 activation leads to NAD^+ exhaustion and energy crisis (Berger 1985), and to caspase-independent apoptosis, via translocation of the mitochondrial pro-apoptotic protein apoptosis-inducing factor (AIF) to the nucleus, producing nuclear condensation (Jiang et al. 1996; Yu et al. 2002; Hong et al. 2004). PARP-1 is involved in the long-term effects produced by perinatal asphyxia (Martin et al. 2005), interacting with other sentinel proteins (see Herrera-Marschitz et al. 2011).

PARP-1, XRCC1, DNA ligase IIIα and DNA polymerase-β are molecular partners working in tandem to repair single-strand breaks. DNA ligase IIIα has an N-terminal zinc finger interacting with the DNA-binding domain of PARP-1 and DNA strand breaks. DNA ligase IIIα interacts with XRCC1, resulting in the formation of a DNA ligase IIIα-XRCC1 complex (see Ellenberger and Tomkinson 2008). DNA polymerase-β can couple with PARP-1, DNA ligase IIIα and XRCC1, contributing to the overall stability of the repair complex, promoting catalysis and fidelity (Sawaya et al. 1997). PARP-1, DNA polymerase-β and XRCC1 expression are increased by hypoxia induced in newborn piglets (Mishra et al. 2003) and rats (Chiappe-Gutierrez et al. 1998). With an ischemic preconditioning model, Li et al. (2007) demonstrated a fivefold increase of XRCC1 levels 30 min after ischemia, reaching a maximal expression after 4h. DNA polymerase-β and DNA ligase III levels were also increased and co-expressed in neurons and in glial cells (Li et al. 2007). PARP-1 can be phosphorylated by ERK, probably via the isoform 2 (ERK2), as a requirement for maximal PARP-1 activation after DNA damage (Kauppinen et al. 2006). Cohen-Armon et al. (2007) have reported evidence for PARP-1 activation by phosphorylated ERK2, in the absence of DNA damage, via the signalling cascade of the transcription factor ETS domain-containing protein1 (Elk1), increasing the expression of the immediate early gene c-Fos, stimulating cell growth and differentiation. Cell-type specificity and regional distribution of sentinel proteins may constitute regulatory mechanisms by which the long-term effects of metabolic insults occurring at birth are heterogeneous, targeting some, but leaving other brain regions apparently untouched.

PARP-1 inhibition is a target for neuroprotection following hypoxia/ischemia. Several PARP inhibitors, with increasing degrees of potency, have been shown to decrease brain damage, improving the neurological outcome of perinatal brain injury (Zhang et al. 1995; Ducrocq et al. 2000; Sakakibara et al. 2000; see Virag and Szabo 2002; Jagtap and Szabo 2005; Kauppinen et al. 2009).

The idea that PARP-1 activation is beneficial has also been explored, depending upon the actual levels of cellular NAD^+. While PARP inhibitors offer remarkable protection under conditions of NAD^+ and ATP depletion, inhibition of PARP-1 in the presence of NAD^+ sensitises cells to DNA damage and subsequent increase of cell death (Nagayama et al. 2000). It has also been reported that inhibition of PARP-1 induces apoptosis in rapidly dividing cells (Saldeen and Welsh 1998), probably by blocking the access of repairing enzymes. Thus, PARP-1 acts as both a cell survival- and cell death-inducing factor by regulation of DNA repair, chromatin remodelling and regulation of transcription.

There is concern about applying ultrapotent PARP inhibitors during development, since PARP-1 is required for efficient repair of damaged DNA (Trucco et al. 1998; Schultz et al. 2003). It has been suggested that moderate PARP-1 inhibitors are the choice for neuronal protection when administered to paediatric patients (Moonen et al. 2005; Geraets et al. 2006). Several naturally occurring compounds have been suggested to inhibit PARP-1 overactivation. Caffeine (1,3,7-trimethylxanthine) metabolites have been shown to inhibit PARP-1 activity at physiological concentrations (Geraets et al. 2006), including 1,3-dimethylxanthine (theophylline) (Moonen et al. 2005) and 1,7-dimethylxanthine (paraxanthine) (Geraets et al. 2006).

Nicotinamide is a reference compound regarding PARP-1 inhibition (Virag and Szabo 2002). Because of its relative low potency, it has advantage when used with developing animals, antagonising the effects elicited by PARP-1 overactivation without impairing DNA repair or cell proliferation. Nicotinamide has also been proposed as an antioxidant compound (Yan et al. 1999; Wan et al. 1999), and its pharmacodynamic properties can provide advantages over more selective compounds. Importantly, nicotinamide has already been tested in human clinical trials (Macleod et al. 2004).

We have reported that nicotinamide prevents several of neuronal (Klawitter et al. 2007; Morales et al. 2010), neurochemical (Bustamante et al. 2003, 2007) and behavioural (Simola et al. 2008; Morales et al. 2010) effects produced by perinatal asphyxia. At therapeutic doses (0.8 mmol/kg, i.p.), nicotinamide rapidly distributes into the brain, achieving concentrations $>30\ \mu M$ for longer than 30 min, producing a long-lasting inhibition of PARP-1 activity in asphyxia-exposed and control animals (Allende-Castro et al. 2012).

9.7 An Experimental Model of Perinatal Asphyxia

While the short- and long-term clinical outcome of perinatal asphyxia is well established, pre-clinical research is still at an exploratory phase, mainly because of a lack of consensus on a reliable and predictable experimental model. A model for investigating the issue was proposed at the Karolinska Institutet, Stockholm, Sweden, at the beginning of the 1990s (Bjelke et al. 1991; Andersson et al. 1992; Herrera-Marschitz et al. 1993). The model is run by several laboratories around the world, although its acceptance has been marshalled, because the model works with on term pups and not with neonates at P7. The main argument for criticising the model has been that the brain of neonate rats is premature when compared to the neonatal human brain, a statement mainly referring to the neocortex (see Romijn et al. 1991) and to the pattern of oligodendrocyte lineage progression required for cerebral myelination (Craig et al. 2003). The degree of maturity depends upon the tissue and functions selected for the comparisons. Vulnerability is probably related to both the timing and the location of the insult (Craig et al. 2003; see also de Louw et al. 2002).

In the present model, asphyxia is induced at the time when rats deliver, which is controlled by a strictly planned mating following evaluation of the oestral cycle of young female rats (~2 months of age). The female is exposed to a male at the time of the pro-oestrous cycle for one night and thereafter the time of delivery is calculated, supported by ethological and clinical observations. At the time of delivery, a first spontaneous delivery can be observed before subjecting the dams to neck dislocation, caesarean section and hysterectomy. The uterine horns containing the foetuses are then immediately immersed into a water bath at 37 °C for various periods of time (0–22 min). Following asphyxia, the pups are removed from the uterine horns and resuscitated by cleaning the faces of the animals from fluid and amniotic tissue, freeing the mouth and the nose. Further additional care is taken to induce pulmonary breathing by stimulating the tip of the nose and the mouth, as well as pressing the thorax. Pups exposed to caesarean-delivery only (CS, 0 asphyxia), or to mild asphyxia (2–10 min), are rapidly resuscitated, without requiring anything else but removing fluid and amniotic tissue from the head. Pups exposed to zero or mild asphyxia start breathing with a gluttonous gasp, which is rapidly replaced by regular and synchronised breathing. For pups exposed to longer periods of asphyxia (19–21 min), resuscitation implies expert and skilful handling, and it takes a long time (4–6 min) for a first gasping, and even longer for establishing a more or less regular breathing, always supported by gasping. Approximately 1 h after delivery, the pups are given to surrogate dams for nursing, pending further experiments (Fig. 9.4). An Apgar scale is applied during the recovery period, similar to that observed in neonatal units, recording body weight, sex, colour of the skin, respiratory frequency and presence of gasping, vocalisation, muscular rigidity and spontaneous movements (Dell'Anna et al. 1997; Morales et al. 2010; see Herrera-Marschitz et al. 2011) (Table 9.1). The Apgar evaluation is a critical parameter for this experimental model, because it assesses whether the pups are subjected to mild or to severe asphyxia, which is directly determined by the percentage of survival and recovery (Herrera-Marschitz et al. 1993, 1994; see Herrera-Marschitz et al. 2011). Survival is a straightforward parameter. A 100 % survival is observed whenever foetuses-containing uterine horns are immersed for up to 15 min in a water bath at 37 °C. Thereafter, the rate of survival drops rapidly, until no survival is observed following 22 min of asphyxia (Fig. 9.5). The temperature of the water bath is a critical parameter, because each degree below 37 °C is associated with longer survival (Herrera-Marschitz et al. 1993, 1994; Engidawork et al. 2001). The recovery period can also add or prolong a hypoxic/ischemic condition, as the surviving pups may show a decreased breathing rate, decreased cardiovascular function and low peripheral and/or central blood perfusion. The Apgar evaluation also provides information about the condition shown by the caesarean-delivered control pups, which has to be similar to that shown by vaginally delivered pups. Thus, the Apgar evaluation is a requirement when using the present model of perinatal asphyxia, because it permits to compare results obtained by different laboratories and/or different treatments. The quality of the handling of the pups and the experience of the surrogate dam are important factors for the acceptance and nursing of both asphyxia-exposed and control pups.

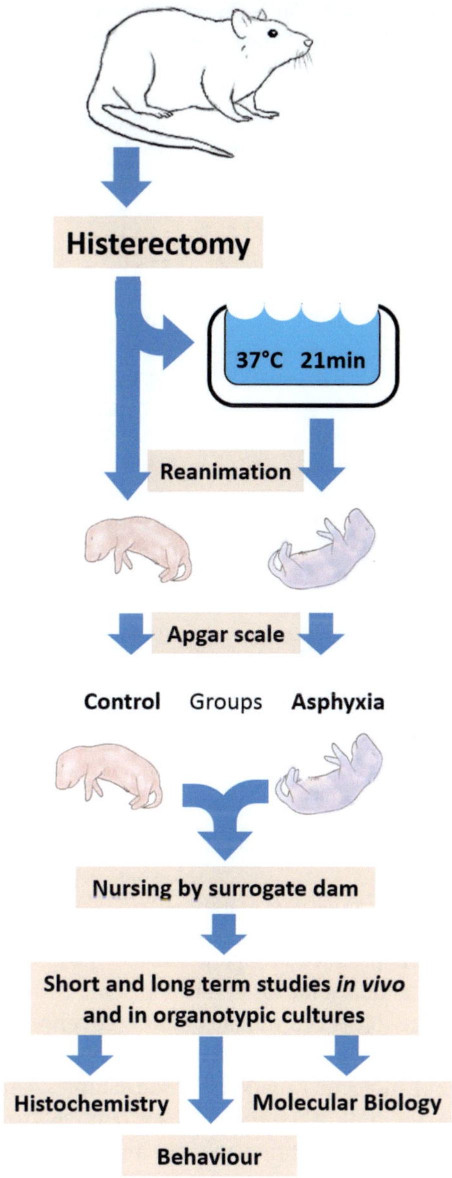

Fig. 9.4 The Karolinska Institutet experimental model of perinatal asphyxia. The experiment starts by an evaluation of the oestral cycle of young female Wistar rats (~2 months of age), in order to plan for a programmed mating. The female is exposed to a male at the time of the pro-oestrous for one night and thereafter the presence of a vaginal clot is evaluated. Pregnancy is continuously monitored until delivery of a first pup is observed (representing a vaginally delivered control), or when the maturity of the foetuses is assessed by clinical abdominal palpation, indicating that the dam is ready for delivery (~22 days after the identification of a vaginal clot). The animal is anaesthetised, killed by neck dislocation and subjected to a caesarean section and hysterectomised. The uterine horns are then immersed into a temperature-controlled water bath (37 °C) for various

Table 9.1 An Apgar scale for rodents. An Apgar scale for evaluating the consequences of perinatal asphyxia 40–80 min after birth. Data expressed as the mean ± SEM (*n* number of pups, *m* number of dams) (experimental cohorts from period 2012–2013). (*n.d.* no data)

Parameters	Spontaneous delivered pups ($n=81$; $m=49$)	Caesarean-delivered pups (0 min asphyxia) ($n=451$; $m=128$)	20-min asphyxia ($n=489$; m=130)
Body weight (g)	5.82±0.10	5.52±0.03	5.52±0.11
Sex (% of males)	52.5±6.2	49.2±2.5	51.2±1.8
Rate of survival (%)	100±0.0	99.2±0.6	65±2.6
Respiratory frequency (events × min^{-1})	78±2	77±1	28±1
Presence of gasping (yes; %)	0±0	0.4±0.4	55.5±3.9
Skin colour (%)			
Pink	100.0±0.0	99.3±0.5	0.8±0.5
Pink-blue	0±0	0.3±0.3	56.8±3.4
Blue-pink	0±0	0.4±0.4	38.5±3.3
Blue	0±0	0±0	3.8±1.4
Presence of vocalisations (yes, %)	100±0	98.0±1.2	10.4±2.4
Spontaneous movements			
No movements, akinesia, rigidity (0) (%)	0±0	0.4±0.4	79.5±3.0
Single movements of front legs, or head alone (1) (%)	0±0	0 ± 0	13.0±2.3
Movements of two body structures (2) (%)	0±0	1.5±1.0	6.5±1.8
Movements of all body structures (3) (%)	0±0	3.9±1.3	0.1±0.1
Intensive movements shown by wriggling (4) (%)	100±0	84.2±1.7	0.9±0.9
Lack of reception by surrogate dams (at 24 h) (5)	n.d.	1.6 (from the total cohorts)	1.5 (from the total cohorts)

Apart from the effects produced by perinatal asphyxia on the survival rate, the model has been shown to be useful for describing some early molecular, metabolic and physiological effects observed minutes after recovering from a caesarean delivery, without any asphyxia, or from mild or severe insults. Tissue sampling starts soon after delivery. For molecular markers, tissue samples can be collected immediately after delivery, the time when the pups are removed from the uterine horns (0 min, with or without previous immersion into a water bath).

periods of time. One or two pups are immediately delivered after hysterectomy, representing a caesarean-delivered control. After asphyxia, the pups are removed from the uterine horns and resuscitated by cleaning the faces of the animals from fluid and amniotic tissue, freeing the mouth and the nose, and stimulated to pulmonary breathing. Pulmonary breathing is further monitored and the surviving pups are evaluated by an Apgar scale 40–60 min after delivery. Thereafter, the pups are given to surrogate dams for nursing (delivering immediately before the hysterectomised dam), pending further experiments. (See Dell'Anna et al. 1997; Klawitter et al. 2007; Herrera-Marschitz et al. 2011)

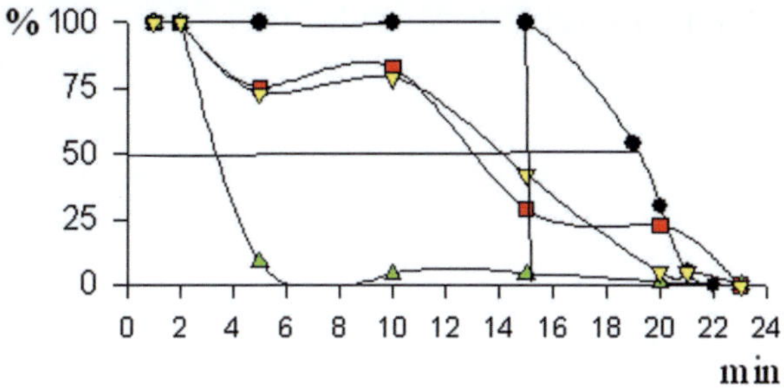

Fig. 9.5 Survival and ATP levels in peripheral organs and brain. Survival rate and ATP levels (%) found 10 min after delivery of asphyxia-exposed and control (caesarean delivery only, zero asphyxia) rat pups. Perinatal asphyxia was performed by immersion of pup containing uterine horns into a water bath at 37 °C, removed by a caesarean section from ready-to-deliver rats. Two minutes after delivery, asphyxia-exposed and control animals were decapitated for removing brain, heart and kidneys, to be treated according to Engidawork et al. (2001), for measuring energy-rich phosphates by high-performance liquid chromatography (*HPLC*) coupled to a ultraviolet (*UV*) detector according to Ingebretsen et al. (1982). In control animals, ATP levels were 1.2±0.2, 4.7±0.1 and 1.6±0.3 μmol/g wet weight in brain ($n=7$), heart ($n=16$) and kidneys ($n=16$), respectively, 10 min after caesarean delivery ($n=7$). Filled circles, survival rate; *filled inverted triangles,* ATP in brain; *filled squares,* ATP in heart; *filled triangle,* ATP in kidneys. Vertical line marks the time (15 min) when survival rate starts to drop; *horizontal line* marks 50 %. Abscissa, time (min) of immersion into a water bath at 37 °C (asphyxia); ordinate, survival in percentage (%). Compared to zero asphyxia.

Energy metabolism has been evaluated in peripheral organs (heart, kidneys) and brain, removed from asphyxia-exposed and control animals, short after delivery (Lubec et al. 1997a, b; Seidl et al. 2000; Engidawork et al. 2001). Among the monitored energy-rich phosphates, ATP levels showed the most evident changes, following or anticipating the final outcome of perinatal asphyxia. In Fig. 9.5, the rate of survival is plotted together with the changes (in percentage) in ATP levels in brain, heart and kidneys. Kidneys were the first organs reacting to hypoxia, already 2 min after the mother circulation was interrupted, decreasing by >90 % after 5 min of perinatal asphyxia, while ATP levels in heart and brain were only decreased by ~20 % compared with control (no asphyxia) levels. After 15 min of asphyxia, ATP levels in heart and brain were decreased by ~70 % compared with the controls, together with a simultaneous drop in the rate of survival. After longer periods of asphyxia, ATP levels were decreased by >90 % in the brain, but only by approximately 80 % in the heart after 20 min of asphyxia, when the survival rate dropped by >70 %. These changes in ATP levels agree with the idea of re-compartmentalisation regarding energy pools. The kidneys are bypassed in favour of heart and brain, but brain bypassed in favour of the heart. Significant changes in pH were already observed after 5 min of asphyxia, decreasing in parallel in peripheral and brain tissue (Engidawork et al. 2001).

9.8 Cognitive Deficits

Motor and cognitive alterations of variable severity, including cerebral palsy, seizures, spasticity, attention deficit, hyperactivity, mental retardation and/or neuropsychiatric syndromes with delayed clinical onset, have been associated with perinatal asphyxia (du Plessis and Volpe 2002; Van Erp et al. 2002; Kaufman et al. 2003; Vannuci and Hagberg 2004; de Hann et al. 2006; Odd et al. 2009). In rats, several studies have investigated the behavioural effects associated with perinatal asphyxia, addressing motor function (Bjelke et al. 1991; Chen et al. 1995), emotional behaviour (Dell'Anna et al. 1991; Hoeger et al. 2000; Venerosi et al. 2004, 2006; Simola et al. 2008; Morales et al. 2010) and spatial memory (Boksa et al. 1995; Iuvone et al. 1996; Hoeger et al. 2000, 2006; Loidl et al. 2000; Van de Berg et al. 2003; Venerosi et al. 2004).

We have investigated whether perinatal asphyxia may produce long-term effects on cognitive performance using the object recognition test (Ennaceur and Delacour 1988). The test consists of discriminating between objects differing in shape and colour, without any genuine significance to the rat, or associated with any rewarding or aversive stimuli. During a first session, two copies of the same object are presented to the rat for 4 min, and then again, during a second session, when one of the previously presented objects is replaced by a novel one, similar in size but different in shape and/or colour. The idea is that the rat has to recognise the novel object, spending longer time exploring the novel than the previously presented object. The first and the second sessions can be separated by different time intervals for evaluating the consolidation of learning. A good memory would be able to recognise a previously presented object after a long time elapsing between a first and second session, meaning that the animal would concentrate on exploring the novel object. In our studies, we used a 15- or 60-min interval, and the animals were studied at 3 months of age (Simola et al. 2008). No differences were observed between asphyxia-exposed (20-min asphyxia) and control animals when a 15-min interval elapsed between the first and the second session. Both asphyxia-exposed and control rats recognised the novel stimulus similarly well, spending longer time exploring the novel object. However, when 60 min elapsed between the first and second session, asphyxia-exposed animals spent less time exploring the novel object, indicating that asphyxia-exposed rats could not recognise its novelty (Simola et al. 2008). This is a straightforward experiment showing a subtle consequence of a metabolic insult (anoxia) occurring at birth, impairing a cognitive function that will show up only after a proper challenge. It is very much reminiscent to the clinical experience revealing effects only when the child starts primary school (see Odd et al. 2009; Strackx et al. 2010).

9.9 Conclusions

Perinatal asphyxia is still a health concern worldwide, a risk factor for several mental and neurological disorders with a delayed clinical onset. Hypoxia implies a severe energetic crisis, leading to death if re-oxygenation is not promptly restored. The functional constraints produced by the lack of oxygen can be exacerbated by and during the re-oxygenation period, implying oxidative stress, synthesis and release of metabolic by-products delaying the onset of proper homeostasis and recovery. A number of sentinel proteins are rapidly activated whenever there is a risk of genome damage, stimulating base excision repair. PARP-1 has been shown to play a pivotal role for repairing damaged DNA and for eliciting caspase-independent cell death when repair is not viable, modulating pro- and anti-inflammatory signalling. PARP-1 overactivation leads to NAD^+ exhaustion, worsening the energy crisis, which is the basis for the hypothesis that PARP-1 is a suitable target for therapeutic interventions preventing the long-term effects of perinatal asphyxia, nicotinamide being a prototype for counteracting PARP-1 overactivation.

Acknowledgments Contract grant sponsors: FONDECYT-Chile (contracts:1120079;1110263;11 10149); Millenium Institute Initiative (BNI P09-015-F); BMBF (NGFN+ TP9) and DAAD (415/ alechile)-Germany. TN-P is a BNI fellow; ER-M is CONICYT (#21090557), an ICBM fellow.

Conflict of Interest The authors declare no conflicts of interest.

References

Alano CC, Kauppinen TM, Valls AV, Swanson RA (2006) Minocycline inhibits poly(ADP-ribose) polymerase-1 at nanomolar concentrations. Proc Natl Acad Sci U S A 103:9685–9690

Allende-Castro C, Espina-Marchant P, Bustamante D, Rojas-Mancilla E, Neira T, Gutierrez-Hernandez MA, Esmar D, Valdes JL, Morales P, Gebicke-Haerter PJ, Herrera-Marschitz M (2012) Further studies on the hypothesis of PARP-1 inhibition as a strategy for lessening the long-term effects produced by perinatal asphyxia: effects of nicotinamide and theophylline on PARP-1 activity in brain and peripheral tissue. Neurotox Res 22:79–90

Alvarez-Buylla A, Lim DA (2004) For the long run: maintaining germinal niches in the adult brain. Neuron 41:683–686

Amé J-C, Spenlehauer C, de Murcia G (2004) The PARP superfamily. Bioessays 26:882–893

Andersson K, Bjelke B, Bolme P, Ögren SO (1992) Asphyxia-induced lesion of the rat hippocampus (CA1, CA3) and the nigro-striatal dopamine system. In: J Gross (ed) Hypoxia and ischemia. CNS, vol 41. Wissenschafliche Publikationen der Humboldt-Universität zu Berlin, B. Medizin, Berlin, pp 71–76

Andersson K, Blum M, Chen Y, Eneroth P, Gross J, Herrera-Marschitz M, Bjelke B, Bolme P, Diaz R, Jamison L, Loidl F, Ungethüm U, Åström G, Ögren SO (1995) Perinatal asphyxia increases bFGF mRNA levels and DA cell body number in mesencephalon of rats. Neuroreport 6:375–378

Antonopoulos J, Dori I, Dinopoulos A, Chiotelli M, Parnavelas JG (2002) Postnatal development of the dopaminergic system of the striatum of the rat. Neuroscience 110:245–256

Arvidsson A, Collin T, Kirik D, Kokaia Z, Lindvall O (2002) Neuronal replacement from endogenous precursors in the adult brain after stroke. Nat Med 9:963–970

Arvin KL, Han BH, Du Y, Lin SZ, Paul SM, Holtzman DM (2002) Minocycline markedly protects the neonatal brain against hypoxic-ischemic injury. Ann Neurol 52:54–61
Avalos AM, Labra CV, Quest AF, Leyton L (2002) Signaling triggered by Thy-1 interaction with b3 integrin on astrocytes is an essential step towards unravelling neuronal Thy-1 function. Biol Res 35:231–238
Avalos AM, Arthur WT, Schneider P, Quest A F, Burridge K, Leyton L (2004) Aggregation of integrins and RhoA activation are required for Thy-1-induced morphological changes in astrocytes. J Biol Chem 279:39139–39145
Avalos AM, Valdivia A, Tapia JC, Muñoz N, Lavandero S, Chiong M, Burridge K, Schneider P, Quest AFG, Leyton L (2009) Neuronal Thy-1 induces adhesion by aVb3 integrin and Syndecan-4-dependent activation of PKCa and RhoA. J Cell Sci 122:3462–3471
Barkovich AJ (2006) MR imagining of the neonatal. Neuroimaging Clin N Am 16:117–135
Barlow JZ, Huntley GW (2000) Developmentally regulated expression of Thy-1 in structures of the mouse sensory-motor system. J Comp Neurol 421:215–233
Bartley J, Soltau T, Wimborne H, Kim S, Martin-Sttudard A, Hess D, Hill W, Waller J, Carrol J (2005) BrdU-positive cells in the neonatal mouse hippocampus following hypoxic-ischemic brain injury. BMC Neurosci 6:15
Basovich SN (2010) The role of hypoxia in mental development and in the treatment of mental disorders. Biosci Trends 4:288–296
Bédard A, Gravel C, Parent A (2006) Chemical characterization of newly generated neurons in the striatum of adult primates. Exp Brain Res 170:501–512
Berger NA (1985) Poly (ADP-ribose) in the cellular response to DNA damage. Radiat Res 1001:4–15
Bjelke B, Andersson K, Ögren SO, Bolme P (1991) Asphyctic lesion: proliferation of tyrosine hydroxylase immunoreactive nerve cell bodies in the rat substantia nigra and functional changes in dopamine transmission. Brain Res 543:1–9
Boksa P, Krishnamurthy A, Brooks W (1995) Effects of a period of asphyxia during birth on spatial learning in the rat. Pediatr Res 37:489–496
Bonfocco E, Krainc D, Ankarcrona M, Nicotera P, Lipton SA (1995) Apotosis and necrosis: two distinct events induced, respectively, by mild and intense insults with N-methyl-D-aspartate or nitric oxide/superoxide in cortical cell cultures. Proc Natl Acad Sci U S A 92:7162–7166
Bustamante D, Goiny M, Åström G, Gross J, Andersson K, Herrera-Marschitz M (2003) Nicotinamide prevents the long-term effects of perinatal asphyxia on basal ganglia monoamine systems in the rat. Exp Brain Res 148:227–232
Bustamante D, Morales P, Torres-Pereyra J, Goiny M, Herrera-Marschitz M (2007) Nicotinamide prevents the effect of perinatal asphyxia on dopamine release evaluated with in vivo microdialysis three months after birth. Exp Brain Res 177:358–369
Calabresi P, Centonze D, Bernardi G (2000) Cellular factors controlling neuronal vulnerability in the brain: a lesson from the striatum. Neurology 55:1249–1255
Cannon M, Jones PB, Murray RM (2002) Obstetric complications and schizophrenia: historical and meta-analytic review. Am J Psychiatry 159:1080–1092
Chen Y, Ögren SO, Bjelke B, Bolme P, Eneroth P, Gross J, Loidl F, Herrera-Marschitz M, Andersson K (1995) Nicotine treatment counteracts perinatal asphyxia-induced changes in the mesostriatal/limbic dopamine systems and in motor behaviour in the four-week old male rat. Neuroscience 68:531–538
Chen Y, Engidawork E, Loidl F, Dell'Anna E, Gony M, Lubec G, Andersson K, Herrera-Marschitz M (1997a) Short- and long-term effects of perinatal asphyxia on monoamine, amino acids and glycolysis product levels measured in the basal ganglia of the rat. Dev Brain Res 104: 19–30
Chen Y, Herrera-Marschitz M, Bjelke B, Blum M, Gross J, Andersson K (1997b) Perinatal asphyxia-induced changes in rat brain tyrosine-hydroxylase-immunoreactive cell body number: effects of nicotine treatment. Neurosci Lett 221:77–80
Chen Y, Hillefors-Berglund M, Herrera-Marschitz M, Bjielke B, Gross J, Andersson K, von Euler G (1997c) Perinatal asphyxia induces long-term changes in dopamine D1, D2, D3 receptor binding in the rat brain. Exp Neurol 146:74–80

Cheng Y, Black IB, DiCicco-Bloom E (2002) Hippocampal granule neuron production and population size are regulated by levels of bFGF. Eur J Neurosci 15:3–12
Chen CH, Wang SM, Yang SH, Jeng CJ (2005) Role of Thy-1 in in vivo and in vitro neural development and regeneration of dorsal root ganglionic neurons. J Cell Biochem 94:684–694
Chiappe-Gutierrez M, Kitzmueller E, Labudova O, Fuerst G, Hoeger H, Hardmeier R, Nohl H, Gille L, Lubec B (1998) mRNA levels of the hypoxia inducible factor (HIF-1) and DNA repair genes in perinatal asphyxia of the rat. Life Sci 63:1157–1167
Cohen-Armon M (2008) PARP-1 activation in the ERK signalling pathway. Trends Pharmacol Sci 28:556–560
Cohen-Armon M, Visochek L, Rozendal D, Kalal A, Geistrikh I, Klein R, Bendetz-Nezer S, Yao Z, Seger R (2007) DNA-independent PARP-1 activation by phosphorylated ERK2 increases Elk1 activity: a link to histone acetylation. Mol Cell 25:297–308
Collin T, Arvidsson A, Kokaia Z, Lindvall O (2005) Quantitative analysis of the generation of different striatal neuronal subtypes in the adult brain following excitotoxic injury. Exp Neurol 195:71–80
Cowan F, Rutherford M, Groenendaal F, Eken P, Mercuri E, Bydder GM, Meiners LC, Dubowitz LMS, de Vries LS (2003) Origin and timing of brain lesions in term infants with neonatal encephalopathy. Lancet 361:736–742
Craig A, Luo NL, Beardsley DJ, Wingate-Pearse N, Walker D, Hohimer AR, Back SA (2003) Quantitative analysis of perinatal rodent oligodendrocyte lineage progression and its correlation with human. Exp Neurol 181:231–240
De Angelis E, MacFarlane J, Du JS, Yeo G, Hicks R, Rathjen FG, Kenwrick S, Brummendorf T (1999) Pathological missense mutations of neural cell adhesion molecule L1 affect homophilic and heterophilic binding activities. EMBO J 18:4744–4753
De Hann M, Wyatt JS, Vargha-Khadem F, Gadian D, Mishki M (2006) Brain and cognitive behavioural development after asphyxia at term birth. Dev Sci 9:350–358
De Louw AJA, De Vente J, Steinbusch HPJ, Gavilanes AWD, Steinbusch HWM, Blanco CE, Troost J, Vles JSH (2002) Apoptosis in the rat spinal cord during postnatal development: the effect of perinatal asphyxia on programmed cell death. Neuroscience 112:751–758
De Murcia G, Menissier de Murcia J (1994) Poly(ADP-ribose) polymerase: a molecular nick-sensor. Trends Biochem Sci 19:172–176
Dell'Anna E, Calzolari S, Milinari M, Luvone L, Calimici R (1991) Neonatal anoxia induces transitory hyperactivity, permanent spatial memory deficits and CA1 cell density reduction in the developing rats. Behav Brain Res 45:125–134
Dell'Anna E, Chen Y, Engidawork E, Andersson K, Lubec G, Luthman J, Herrera-Marschitz M (1995) Short term effects of perinatal asphyxia studied with Fos-immunocytochemistry and in vivo microdialysis in the rat. Exp Neurol 131:279–287
Dell'Anna E, Chen Y, Engidawork E, Andersson K, Lubec G, Luthman J, Herrera-Marschitz M (1997) Delayed neuronal death following perinatal asphyxia in rat. Exp Brain Res 115:105–115
Demyanenko GP, Shibata Y, Maness PF (2001) Altered distribution of dopaminergic neurons in the brain of L1 null mice. Brain Res Dev Brain Res 126:21–30
Denker S, Ji S, Lee SY, Dingman A, Derugin N, Wendland M, Vexler ZS (2007) Macrophages are comprised of resident brain microglia not infiltrating peripheral monocytes acutely after neonatal stroke. J Neurochem 100:893–904
Dommergues MA, Plaisant F, Verney C, Gressens P (2003) Early microglia activation following neonatal excitotoxic brain damage in mice: a potential target for neuroprotection. Neuroscience 121:619–628
Dreyer EB, Leifer D, Heng JE, McConnell JE, Gorla M, Levin LA, Barnstable CJ, Lipton, SA (1995) An astrocytic binding site for neuronal Thy-1 and its effects on neurite outgrowth. Proc Natl Acad Sci U S A 92:11195–11199
Ducrocq S, Benjelloun N, Plotkine M, Ben-Ari Y, Charriaut-Marlangue C (2000) Poly(ADP-ribose) synthase inhibition reduces ischemic injury and inflammation in neonatal rat brain. J Neurochem 74:2504–2511

du Plessis AJ, Volpe JJ (2002) Perinatal brain injury in the preterm and term newborn. Curr Opin Neurol 15:151–157

Ellenberger T, Tomkinson AE (2008) Eukaryotic DNA ligases: structural and functional insights. Annu Rev Biochem 77:313–338

Engelhardt B (2003) Development of the blood-brain barrier. Cell Tissue Res 314:119–129

Engidawork E, Loidl F, Chen Y, Kohlhauser C, Stoeckler S, Dell'Anna E, Lubec B, Lubec G, Goiny M, Gross J, Andersson K, Herrera-Marschitz M (2001) Comparison between hypothermia and glutamate antagonism treatments on the immediate outcome of perinatal asphyxia. Exp Brain Res 138:375–383

Ennaceur A, Delacour J (1988) A new one-trial test for neurobiological studies of memory in rats. I: behavioral data. Behav Brain Res 31:47–59

Foster GA (1998) Chemical neuroanatomy of the prenatal rat brain. Oxford University Press, New York

Gagne J-P, Isabelle M, Lo KS, Bourassa S, Hendzel MJ, Dawson VL, Daeson TM, Poirier GG (2008) Proteome-wide identification of poly(ADP-ribose) binding proteins and poly(ADP-ribose)-associated protein complexes. Nucleic Acids Res 36:6959–6976

Ganat Y, Soni S, Chacon M, Schwartz ML, Vaccarino FM (2002) Chronic hypoxia up-regulates fibroblast growth factor ligands in the perinatal brain and induces fibroblast growth factor-responsive radial glial cells in the sub-ependymal zone. Neuroscience 112:977–991

Geraets L, Moonen HJJ, Wouters EFM, Bast A, Hageman GJ (2006) Caffeine metabolites are inhibitors of the nuclear enzyme poly(ADP-ribose)polymerase-1 at physiological concentrations. Biochem Pharmacol 72:902–910

Gill DA, Perry MA, McGuire EP, Perez-Gomez A, Tasker RA (2012) Low-dose neonatal domoic acid causes persistent changes in behavioural and molecular indicator of stress response in rats. Behav Brain Res 230:409–417

Girard S, Kadhim H, Roy M, Lavoie K, Brochu ME, Larouche A, Sebire G (2009) Role of perinatal inflammation in cerebral palsy. Pediatr Neurol 40:168–174

Gitto E, Reiter RJ, Karbownik M, Tan D-X, Gitto P, Barberi S, Barberi I (2002) Causes of oxidative stress in the pre- and perinatal period. Biol Neonate 81:146–157

Green A, Prager A, Stoudt PM, Murray D (1992) Relationships between DNA damage and the survival of radiosensitive mutant Chinese hamster cell lines exposed to gamma-radiation. Part 1: intrinsic radiosensitivity. Int J Radiat Biol 61:465–472

Gross J, Müller I, Chen Y, Elizalde M, Leclere N, Herrera-Marschitz M, Andersson K (2000) Perinatal asphyxia induces region specific long-term changes in mRNA levels of tyrosine hydroxylase and dopamine D1 and D2 receptors in rat brain. Mol Brain Res 79:110–117

Gross J, Andersson K, Chen Y, Müller I, Andreeva N, Herrera-Marschitz M, (2005) Effect of perinatal asphyxia on tyrosine hydroxylase and D2 and D1 dopamine receptor mRNA levels expressed during early postnatal development in rat brain. Mol Brain Res 134:275–281

Haddad GG, Jiang C (1993) O2 deprivation in the central nervous system: on the mechanisms of neuronal response, differential sensitivity and injury. Prog Neurobiol 40:277–318

Hagberg H, Gilland E, Bona E, Hanson LA, Hahlin-Zoric M, Biennow M, Holst M, McRae A, Söder O (1996) Enhanced expression of interleukin (IL)-1 and IL-6 messenger RNA and bioactive protein after hypoxia-ischemia in neonatal rats. Pediatr Res 40:603–609

Han BH, Holtzman DM (2000) BDNF protects the neonatal brain from hypoxic-ischemic injury in vivo via the ERK pathway. J Neurosci 20:5775–5781

Hassa PO, Hottinger MO (1999) A role of poly(ADP-ribose) polymerase in NF-κB transcriptional activation. Biol Chem 380:953–959

Hastings NB, Gould E (1999) Rapid extension of axons into the CA3 region by adult-generated granule cells. J Comp Neurol 413:146–154 (Erratum: rapid extension of axons into the CA3 region by adult-generated granule cells. J Comp Neurol 415:144)

Hermosilla T, Munoz D, Herrera-Molina R, Valdivia A, Munoz N, Nham SU, Schneider P, Burridge K, Quest AF, Leyton L (2008) Direct Thy-1/alphaVbeta3 integrin interaction mediates neuron to astrocyte communication. Biochem Biophys Acta 1783:1111–1120

Herrera-Marschitz M, Loidl CF, Andersson K, Ungerstedt U (1993) Prevention of mortality induced by perinatal asphyxia: hypothermia or glutamate antagonism? Amino Acids 5:413–419

Herrera-Marschitz M, Loidl CF, You Z-B, Andersson K, Silveira R, O'Connor WT, Goiny M (1994) Neurocircuitry of the basal ganglia studied by monitoring neurotransmitter release. Effects of intracerebral and perinatal asphyctic lesions. Mol Neurobiol 9:171–182

Herrera-Marschitz M, Morales P, Leyton L, Bustamante D, Klawitter V, Espina-Marchant P, Allende C, Lisboa F, Cunich G, Jara-Caviedes A, Neira T, Gutierrez-Hernandez MA, Gonzalez-Lira V, Simola N, Schmitt A, Morelli M, Tasker RA, Gebicke-Haerter PJ (2011) Perinatal asphyxia: current status and approaches towards neuroprotective strategies, focus on sentinel proteins. Neurotox Res 19:603–627

Herrera-Molina R, Frischknecht R, Maldonado H, Seidenbecher CI, Gundelfinger ED, Hetz C, Aylwin ML, Schneider P, Quest AFG, Leyton L (2012) Astrocytic αvβ3 integrin inhibits neurite outgrowth and promotes retraction of neuronal process by clustering Thy-1. PLoS ONE 7(3):e34295

Hoeger H, Engelmann M, Bernet G, Seidl R, Bubna-Littitz H, Mpsgoeller W, Lubec B, Lubec G (2000) Long term neurological and behavioral effects of graded perinatal asphyxia in the rat. Life Sci 66:947–962

Hoeger H, Engidawork E, Stolzlechner D, Bubna-Littitz H, Lubec B (2006) Long-term effect of moderate and profound hyperthermia on morphology, neurological, cognitive and behavioural functions in a rat model of perinatal asphyxia. Amino Acids 31:385–396

Hökfelt T, Ljungdahl Å, Fuxe K, Johansson O (1974) Dopamine nerve terminals in the rat limbic cortex: aspects of the dopamine hypothesis of schizophrenia. Nature 184:177–179

Hong SJ, Dawson TM, Dawson VL (2004) Nuclear and mitochondrial conversations in cell death: PARP-1 and AIF signalling. Trends Pharmacol Sci 25:259–264

Horowitz A, Tkachenko E, Simons M (2002) Fibroblast growth factor-specific modulation of cellular response by syndecan-4. J Cell Biol 157:715–725

Ingebretsen OC, Bakken AM, Segadal L, Farstad M (1982) Uric acid determinations: reversed-phase liquid chromatography with ultraviolet detection compared with kinetic and equilibrium adaptations of the uricase method. J Chromatogr 242:119–126

Iuvone L, Geloso MC, Dell'Anna E (1996) Changes in open field behavior, spatial memory, and hippocampal parvalbumin immunoreactivity following enrichment in rats exposed to neonatal anoxia. Exp Neurol 139:25–33

Iyer NV, Kotch LE, Agani F, Leung SW, Laugner E, Wenger RH, Gassmann M, Gearhart JD, Lawler AM, Yu AY, Semenza GL (1998) Cellular and developmental control of O2 homeostasis by hypoxia-inducible factor 1 alpha. Genes Dev 12:149–162

Jagtap P, Szabo C (2005) Poly(ADP-ribose) polymerase and the therapeutic effects of its inhibitors. Nat Rev 4:421–440

Jiang B-H, Rue E, Wang GL, Roe R, Semenza GL (1996) Dimerization, DNA binding, and transactivation properties of hypoxia-inducible factor 1. J Biol Chem 271:17771–17778

Kaufman SA, Miller SP, Ferreiro DM, Glidden DH, Barkovich AJ, Partridge JC (2003) Encephalopathy as a predictor of magnetic resonance imaging abnormalities in asphyxiated newborns. Pediatr Neurol 28:342–346

Kauppinen TM, Swanson RA (2007) The role of poly(ADP-ribose) polymerase-1 in CNS disease. Neuroscience 147:1267–1272

Kauppinen TM, Chan WY, Suh SW, Wiggins AK, Huang EJ, Swanson RA (2006) Direct phosphorylation and regulation of poly(ADP-ribose) polymerase-1 by extracellular signal-regulated kinases1/2. Proc Natl Acad Sci U S A 103:7136–7141

Kauppinen TM, Suh SW, Berman AE, Hamby AM, Swanson RA (2009) Inhibition of poly(ADP-ribose) polymerase suppresses inflammation and promotes recovery after ischemic injury. J Cereb Blood Flow Metab 29:820–829

Kenwrick S, Watkins A, De Angelis E (2000) Neural cell recognition molecule L1: relating biological complexity to human disease mutations. Hum Mol Genet 9:879–886

Klawitter V, Morales P, Bustamante D, Gomez-Urquijo S, Hökfelt T, Herrera-Marschitz M (2007) Neuronal plasticity of basal ganglia following perinatal asphyxia: neuroprotection by nicotinamide. Exp Brain Res 180:139–152
Klawitter V, Morales P, Johansson S, Bustamante D, Goiny M, Gross J, Luthman J, Herrera-Marschitz M (2005) Effect of perinatal asphyxia on cell survival, neuronal phenotype and neurite growth evaluated with organotypic triple cultures. Amino Acids 28:149–155
Klawitter V, Morales P, Bustamante D, Goiny M, Herrera-Marschitz M (2006) Plasticity of the central nervous system (CNS) following perinatal asphyxia> Does nicotinamide provide neuroprotection? Amino Acids 31:377–384
Kniesel U, Risau W, Wolburgh H (1996) Development of blood-brain barrier tight junctions in the rat cortex. Brain Res Dev Brain Re 96:229–240
Koh S-H, Park Y, Song CW, Kim JG, Kim K, Kim J, Kim M-H, Lee SR, Kim DW, Yu H-J, Chang D, Hwang SJ, Kim SH (2004) The effect of PARP inhibitor on ischemic cell death, its related inflammation and survival signals. Eur J Neurosci 20:1461–1472
Kohlhauser C, Kaehler S, Mosgoeller W, Singewald N, Koulevas D, Prast H, Hoeger H, Lubec B (1999a) Histological changes and neurotransmitter levels three months following perinatal asphyxia in the rat. Life Sci 64:2109–2124
Kohlhauser C, Mosgoeller W, Hoeger H, Lubec G, Lubec B (1999b) Cholinergic, monoaminergic and glutamatergic changes following perinatal asphyxia in the rat. Cell Mol Life Sci 55:1491–1501
Kohlhauser C, Mosgoeller W, Hoeger H, Lubec B (2000) Myelination deficits in brain of rats following perinatal asphyxia. Life Sci 67:2355–2368
Kurinczuk JJ, White-Koning M, Badawi N (2010) Epidemiology of neonatal encephalopathy and hypoxic-ischemic encephalopathy. Early Hum Dev 86:329–338
Labudova O, Schuller E, Yeghiazaryan K, Kitzmueller E, Hoeger H, Lubec G, Lubec B (1999) Genes involved in the pathophysiology of perinatal asphyxia. Life Sci 64:1831–1838
Lawn JE, Kerber K, Enweronu-Laryea C, Cousens S (2010) 3.6 million neonatal deaths-what is progressing and what is not? Semin Perinatol 34:371–386
Leppard JB, Dong Z, Mackey ZB, Tomkinson AE (2003) Physical and functional interaction between DNA ligase IIIalpha and poly(ADP-Ribose) polymerase 1 in DNA single-strand break repair. Mol Cell Biol 23:5919–5927
Leyton L, Schneider P, Labra CV, Rüegg C, Hetz CA, Quest AF, Bron C (2001) Thy-1 binds to the integrin b3 on astrocytes and triggers formation of focal contact sites. Curr Biol 11:1028–1038
Li Q, Martin JH (2000) Postdevelopmental of differential projections from the caudal and rostral motor cortex subregions. Exp Brain Res 134:187–198
Li S, Kim JE, Budel S, Hampton TG, Strittmatter SM (2005) Transgenic inhibition of Nogo-66 receptor function allows axonal sprouting and improved locomotion after spinal injury. Mol Cell Neurosci 29:26–39
Li N, Wu H, Yang S, Chen D (2007) Ischemic preconditioning induces XRCC1, DNA polymerase-beta, and DNA ligase III and correlates with enhanced base excision repair. DNA Repair (Amst) 6:1297–1306
Loidl F, Gavilanes AW, Van Dijk EH, Vreuls W, Blokland A, Vies JS, Steinbusch HW, Blanco CE (2000) Effects of hypothermia and gender on survival and behaviour after perinatal asphyxia in rats. Physiol Behav 68:263–269
Loizou LA (1972) The postnatal ontogeny of monoamine-containing neurones in the central nervous system of the albino rat. Brain Res 40:395–418
Low JA (2004) Determining the contribution of asphyxia to brain damage in the neonate. J Obstet Gynaecol Res 30:276–286
Lubec B, Mark M, Herrera-Marschitz M, Labudova O, Hoeger H, Gille L, Nohl H, Mosgoeller W, Lubec G (1997a) Decrease of heart protein kinase C and cyclin-dependent kinase precedes death in perinatal asphyxia of the rat. FASEB J 11:482–492
Lubec B, Dell'Anna E, Fang-Kircher S, Mark M, Herrera-Marschitz M, Lubec G (1997b) Decrease of brain protein kinase C, protein kinase A, and cyclin-dependent kinase correlating with pH precedes neuronal death in neonatal asphyxia of the rat. J Invest Med 45:284–294

Lubec B, Labudova O, Hoeger H, Kirchner L, Lubec G (2002) Expression of transcription factors in the brain of rats with perinatal asphyxia. Biol Neonate 81:266–278
Macleod MR, O'Collins T, Howells DW, Donnan GA (2004) Pooling of animal experimental data reveals influence of study design and publication bias. Stroke 35:1203–1208
Martin SS, Perez-Polo JR, Noppens KM, Grafe MR (2005) Biphasic changes in the levels of poly(ADPribose) polymerase-1 and caspase 3 in the immature brain following hypoxia-ischemia. Int J Dev Neurosci 23:673–686
Meng Z, Martin JH (2003) Postnatal development of corticospinal postsynaptic action. J Neurophysiol 90:683–692
Meng Z, Li Q, Martin JH (2004) The transition from development to motor control function in the corticospinal system. J Neurosci 24:605–614
Miller SP, Ramaswamy V, Michelson D, Barkovich J, Holshouser B, Wycliff N, Glidden DV, Deming D, Partridge JC, Wu YW, Ashwal S, Ferreiro DM (2005) Patters of brain injury in term neonatal encephalopathy. J Pediatr 146:453–460
Ming GL, Song H (2005) Adult neurogenesis in the mammalian central nervous system. Annu Rev Neurosci 28:223–250
Mishra OP, Akhter W, Ashraf QM, Delivoria-Papadopoulus M (2003) Hypoxia-induced modification of poly (ADP-Ribose) Polymerase and DNA polymerase β activity in cerebral cortical nuclei of newborn piglets: role of nitric oxide. Neuroscience 119:1023–1032
Montgomery AM, Becker JC, Siu CH, Lemmon VP, Cheresh DA, Pancook JD, Zhao X, Reisfeld RA (1996) Human neural cell adhesion molecule L1 and rat homologue NILE are ligands for integrin alpha v beta 3. J Cell Biol 132:475–485
Moonen HJJ, Geraets L, Vaarhorst A, Wouters EFM, Bast A, Hageman GJ (2005) Theophylline prevents NAD+ depletion via PARP-1 inhibition in human pulmonary epithelial cells. Biochem Biophys Res Commun 338:1805–1810
Morales P, Klawitter V, Johansson S, Huaiquin P, Barros VG, Avalos AM, Fiedler J, Bustamante D, Gomez-Urquijo S, Goiny M, Herrera-Marschitz M (2003) Perinatal asphyxia impairs connectivity and dopamine branching in organotypic triple culture from rat substantia nigra. Neurosci Lett 348:175–179
Morales P, Reyes P, Klawitter V, Huaiquin P, Bustamante D, Fiedler JL, Herrera-Marschitz M (2005) Effects of perinatal asphyxia on cell proliferation and neuronal phenotype evaluated with organotypic hippocampal cultures. Neuroscience 135:421–431
Morales P, Fiedler JL, Andres S, Berrios C, Huaiquin P, Bustamante D, Cardenas S, Parra E, Herrera-Marschitz M (2008) Plasticity of hippocampus following perinatal asphyxia: effects on postnatal apoptosis and neurogenesis. J Neurosci Res 86:2650–2662
Morales P, Simola N, Bustamante D, Lisboa F, Fiedler J, Gebicke-Haerter P, Morelli M, Tasker RA, Herrera-Marschitz M (2010) Nicotinamide prevents the effect of perinatal asphyxia on apoptosis, non-spatial working memory and anxiety in rats. Exp Brain Res 202:1–14
Morris RJ, Beech JN, Barber PC, Raisman G (1985) Early stages of Purkinje cell maturation demonstrated by Thy-1 immunohistochemistry on postnatal rat cerebellum. J Neurocytol 14:427–452
Morris RJ, Tiveron MC, Xue GP (1992) The relation of the expression and function of the neuronal glycoprotein Thy-1 to axonal growth. Biochem Soc Trans 20:401–405
Mosgoeller W, Kastner P, Fang-Kircher S, Kitzmueller E, Hoeger H, Sether P, Labudova O, Lubec G, Lubec B (2000) Brain RNA polymerase and nucleolar structure in perinatal asphyxia of the rat. Exp Neurol 161:174–182
Mudò G, Bonomo A, Di Liberto V, Frinchi M, Fuxe K, Belluardo N (2009) The FGF-2/FGFRs neurotrophic system promotes neurogenesis in the adult brain. J Neural Transm
Nagayama T, Simon RP, Chen D, Henshall DC, Pei W, Stetler RA, Chen J (2000) Activation of Poly(ADP-Ribose) polymerase in the rat hippocampus may contribute to cellular recovery following sublethal transient global ischemia. J Neurochem 74:1636–1645
Neira-Peña T, Espina-Marchant P, Rojas-Mancilla E, Esmar D, Kraus C, Munoz V, Perez B, Rivera B, Bustamante D, Valdes JL, Hermoso M, Gebicke-Haerter P, Morales P, Herrera-Marschitz M (2013) Molecular, cellular, and behavioural effects produced by perinatal asphyxia: protection

by poly(ADP-ribose) polymerase 1 (PARP-1) inhibition. In: Kostrzewa RM (ed) Handbook of neurotoxicity. Springer, New York. doi:10.1007/978-1-4614-5836-4115
Northington FJ, Ferreiro DM, Graham EM, Traystman RJ, Martin LJ (2001) Early neurodegeneration after hypoxia-ischemia in neonatal rat is necrosis while delayed neuronal death is apoptosis. Neurobiol Dis 8:207–219
Odd DE, Lewis G, Whitelaw A, Gunnell D (2009) Resuscitation at birth and cognition at 8 years of age: a cohort study. Lancet 373(9675):1615–1622
Olson L, Seiger Å (1972) Early prenatal ontogeny of central monoamine neurons in the rat: fluorescence histochemical observations. Z Anat Entwicklungsgesch 137:301–316
Oo TF, Burke RE (1997) The time course of the development cell death in phenotypically defined dopaminergic neurons of the substantia nigra. Dev Brain Res 98:191–196
Pasternak JF, Predey TA, Mikhael MA (1991) Neonatal asphyxia: vulnerability of basal ganglia, thalamus and brain stem. Pediatr Neurol 7:147–149
Pastuzko A (1994) Metabolic responses of the dopaminergic system during hypoxia in newborn brain. Biochem Med Metab Biol 51:1–15
Pétrilli V, Herced Z, Hassa PO, Patel NS, Di Paola R, Cortes U, Dugo L, Filipe HM, Thiemermann C, Hottiger MO, Cuzzocrea S, Wang ZQ (2004) Noncleavable poly(ADP-ribose) polymerase-1 regulates inflammation response in mice. J Clin Invest 114:1072–1081
Plane JM, Liu R, Wang TW, Silverstein FS, Parent JM (2004) Neonatal hypoxic-ischemic injury increases forebrain subventricular zone neurogenesis in the mouse. Neurobiol Dis 16:16585–16595
Pulsinelli WA, Brierley JB, Plum F (1982) Temporal profile of neuronal damage in a model of transient forebrain ischemia. Ann Neurol 11:491–498
Qiao M, Malisza KL, Del Biio MR, Tuor UL (2001) Correlation of cerebral hypoxic-ischemic T2 changes with tissue alterations in water content and protein extravasation. Stroke 32:958–963
Ransohoff RM, Kivisakk P, Kidd G (2003) Three or more routes for leukocyte migration into the central nervous system. Nat Rev Immunol 3:569–581
Robertson C, Perlman M (2006) Follow-up of the term infant after hypoxic-ischemic encephalopathy. Paediatr Child Health 11:278–282
Robertson NJ, Tan S, Groenedaal F, van Bel F, Juul SE, Bennet L, Derrick M, Back SA, Valdez RC, Northington F, Gunn AJ, Mallard C (2012) Which neuroprotective agents are ready for bench to bedside translation in the newborn infant? J Pediatr 160:544–552
Rojas-Mancilla E, Neira-Pena T, Kraus C, Rivera B, Perez R, Alvarez A, Bustamante D, Morales P, Leyton L, Herrera-Marschitz M (2013) Perinatal asphyxia decreases neuronal branching and synaptogenesis in rat hippocampus. Implication of astrocyte reactivity and HIF-1α.XXVIth international symposium on cerebral flow, metabolism and function & XIth international conference on quantification of brain function with PET. Shanghai, China, May 20–23, 2013
Romijn HJ, Hofman MA, Gramsbergen A (1991) At what age is the developing cerebral cortex of the rat comparable to that of the full-term newborn human baby? Early Hum Dev 26:61–67
Ruppert M, Aigner S, Hubbe M, Yagita H, Altevogt P (1995) The L1 adhesion molecule is a cellular ligand for VLA-5. J Cell Biol 131:1881–1891
Sakakibara Y, Mitha AP, Ogilvy CS, Maynard KI (2000) Post-treatment with nicotinamide (vitamin B(3)) reduces the infarct volume following permanent focal cerebral ischemia in female Sprague-Dawley and Wistar rats. Neurosci Lett 281:111–114
Saldeen J, Welsh N (1998) Nicotinamide-induced apoptosis in insulin producing cells is associated with cleavage of poly(ADP-ribose)polymerase. Mol Cell Endocrinol 139:99–107
Sawaya MR, Prasad R, Wilson SH, Kraut J, Pelletier H (1997) Crystal structures of human DNA polymerase beta complexed with gapped and nicked DNA: evidence for an induced fit mechanism. Biochemistry 36:11205–11215
Scheepens A, Wassink G, Piersma MJ, Van de Berg WDJ, Blanco CE (2003) A delayed increase in hippocampus following global asphyxia in the neonatal rat. Brain Res Dev Brain Res 142:67–76
Schmid RS, Maness PF (2008) L1 and NCAM adhesion molecules as signaling coreceptors in neuronal migration and process outgrowth. Curr Opin Neurobiol 18(3):245–250

Schultz N, Lopez E, Saleh-Gohari N, Helleday T (2003) Poly(ADP-ribose) polymerase (PARP-1) has a controlling role in homologous recombination. Nucl Acids Res 31:4959–4964

Segalowitz SJ, Davies PL (2004) Charting the maturation of the frontal lobe: an electrophysiological strategy. Brain Cogn 55:116–133

Seidl R, Stoeckler-Ipsiroglu S, Rolinski B, Kohlhauser C, Herkner KR, Lubec B, Lubec G (2000) Energy metabolism in graded perinatal asphyxia of the rat. Life Sci 67:421–435

Seiger Å, Olson L (1972) Late prenatal ontogeny of central monoamine neurons in the rat: fluorescence histochemical observations. Z Anat Entwicklungsgesch 140:281–318

Simola N, Bustamante D, Pinna A, Pontis S, Morales P, Morelli M, Herrera-Marschitz M (2008) Acute perinatal asphyxia impairs non-spatial memory and alters motor coordination in adult male rats. Exp Brain Res 185:595–601

Skaper SD (2003) Poly(ADP-ribosyl)ation enzyme-1 as a target for neuroprotection in acute central nervous system injury. Curr Drug Targets CNS Neurol Disord 2:279–291

Sowell ER, Thompson PM, Holmes CJ, Jernigan TL, Toga AW (1999) In vivo evidence for post-adolescent brain maturation in frontal and striatal regions. Nat Neurosci 2:859–861

Strackx E, Van den Hove DL, Prickaerts J, Zimmermann L, Steinbusch HW, Blanco CE, Gavilanes AW, Vles JS (2010) Fetal asphyctic preconditioning protects against perinatal asphyxia-induced behavioral consequences in adulthood. Behav Brain Res 208:343–351

Suh H, Deng W, Cage F (2009) Signaling in adult neurogenesis. Annu Rev Cell Dev Biol 25:253–275

Sung P, Bailly V, Weber C, Thompson LH, Prakash L, Prakash S (1993) Human xeroderma pigmentosum group D gene encodes a DNA helicase. Nature 365:852–855

Takami K, Iwane M, Kiyota Y, Miyamoto M, Tsukuda R, Shiosaka S (1992) Increase of basic fibroblast growth factor immunoreactivity and its mRNA level in rat brain following transient forebrain ischemia. Exp Brain Res 90:1–10

Tiveron MC, Barboni E, Pliego Rivero FB, Gormley AM, Seeley PJ, Grosveld F, Morris RJ (1992) Selective inhibition of neurite outgrowth on mature astrocytes by Thy-1 glycoprotein. Nature 355:745–748

Trucco C, Oliver FJ, deMurcia G, Menissier de Murcia J (1998) DNA repair defect in poly(ADP-ribose) polymerase-deficient cell lines. Nucl Acids Res 26:2644–2649

Tsang M, Dawid IB (2004) Promotion and attenuation of FGF signaling through the Ras-MAPK pathway. Sci STKE 228:17

Tsuji L, Yamashita T, Kubo T, Madura T, Tanaka H, Hosokawa K, Tohyama M (2004) FLRT3, a cell surface molecule containing LRR repeats and a FNIII domain, promotes neurite outgrowth. Biochem Biophys Res Commun 313:1086–1091

Ullrich O, Diestel A, Evüpoglu IY, Nitsch R (2001) Regulation of microglial expression of integrins by poly(ADP-ribose) polymerase-1. Nat Cell Biol 3:1035–1042

Ungethüm U, Chen Y, Gross J, Bjelke B, Bolme P, Eneroth P, Heldt J, Loidl CF, Herrera-Marschitz M, Andersson K (1996) Effects of perinatal asphyxia on the mesostriatal/mesolimbic dopamine system of neonatal and 4-week-old male rats. Exp Brain Res 112:403–410

Van de Berg WD, Kwaijtaal M, de Louw AJ, Lissone NP, Schmitz C, Faull RL, Blokland A, Blanco CE, Steinbusch HW (2003) Impact of perinatal asphyxia on the GABAergic and locomotor system. Neuroscience 117:83–96

Van den Broek MPH, Groenendaal F, Egberts ACG, Rademaker CMA (2010) Effects of hypothermia on pharmacokinetics and pharmacodynamics. Clin Pharmacokinet 49:277–294

Van Erp TGM, Saleh PA, Rosso PA, Huttunen M, Lönnqvist J, Pirkola T, Salonen O, Valanne L, Poutanen V-P, Standersköld-Nordenstam C-G, Cannon TD (2002) Contributions of genetic risk and fetal hypoxia to hippocampal volume in patients with schizophrenia or schizoaffective disorder, their unaffected siblings and healthy unrelated volunteers. Am J Psychiatry 159:1514–1520

Vannuci S, Hagberg H (2004) Hypoxia-ischemia in the immature brain. J Exp Biol 207:3149–3154

Venerosi A, Valanzano A, Cirulli F, Alleva E, Calamandrei G (2004) Acute global anoxia during C-section birth affects dopamine-mediated behavioural responses and reactivity to stress. Behav Brain Res 154:155–164

Venerosi A, Cuturi D, Chiarotti F, Calamandrei G (2006) C-section birth per se or followed by global asphyxia altered emotional behaviour in neonate and adult rats. Behav Brain Res 168:56–63
Venkatesan A, Frucht S (2006) Movement disorders after resuscitation from cardiac arrest. Neurol Clin 24:123–132
Vexler ZS, Yenari MA (2009) Does inflammation after stroke affect the developing brain differently than adult brain? Dev Neurosci 31:378–393
Virag L, Szabo C (2002) The Therapeutic potential of poly(ADP-ribose) polymerase inhibitors. Pharmacol Rev 54:375–429
Voorn P, Kalsbeck A, Jorritsma-Byham B, Groenewegen HJ (1988) The pre- and postnatal development of the dopaminergic cell groups in the ventral mesencephalon and the dopaminergic innervation of the striatum of the rat. Neuroscience 25:857–887
Wan FJ, Lin HC, Kang BH, Tseng CJ, Tung CS (1999) D-amphetamine-induced depletion of energy and dopamine in the rat striatum is attenuated by nicotinamide pretreatment. Brain Res Bull 50:167–171
Wang H, Yao Y, Jiang X, Chen D, Xiong Y, Mu D (2006) Expression of Nogo-A and NgR in the developing rat brain after hypoxia-ischemia. Brain Res 1114:212–220
Wiessner C, Bareyre FM, Allegrini PR, Mir AK, Frentzel S, Zurini M, Schnell L, Oertle T, Schwab ME (2003) Anti-Nogo-A antibody infusion 24 hours after experimental stroke improved behavioral outcome and corticospinal plasticity in normotensive and spontaneously hypertensive rats. J Cereb Blood Flow Metab 23:154–165
Wilson SH (1998) Mammalian base excision repair and DNA polymerase β. Mutation Res 407:203–215
Wyss JM, Van Groen T (1992) Connections between the retrosplenial cortex and the hippocampal formation in the rat: a review. Hippocampus 2:1–11
Wyss MT, Jolivet R, Buck A, Magistretti PJ, Weber B (2011) In vivo evidence for lactate as a neuronal energy source. J Neurosci 31:7477–7485
Yan Q, Briehl M, Crowley CL, Payne CM, Bernstein H, Bernstein C (1999) The NAD+ precursors, nicotinic acid and nicotinamide upregulate glyceraldehyde-3-phosphate dehydrogenase and glucose-6-phosphate dehydrogenase mRNA in Jurkat cells. Biochem Biophys Res Commun 255:133–136
Yu SW, Poitras MF, Coombs C, Bowers WJ, Federoff HJ, Poirier GC, Dawson TM, Dawson VL (2002) Mediation of poly(ADP-ribose) polymerase-1 dependent cell death by apoptosis-inducing factor. Science 297:250–263
Zhang J, Pieper A, Snyder SH (1995) Poly(ADP-ribose) synthase activation: an early indicator of neurotoxic DNA damage. J Neurochem 65:1411–1414
Zhao X, Yip PM, Siu CH (1998) Identification of a homophilic binding site in immunoglobulin-like domain 2 of the cell adhesion molecule L1. J Neurochem 71:960–971
Zhao C, Deng W, Gage FH (2008) Mechanisms and functional implications of adult neurogenesis. Cell 132:645–660

Chapter 10
Affective, Cognitive, and Motivational Processes of Maternal Care

Mariana Pereira and Annabel Ferreira

Abstract The present chapter reviews current knowledge of the neurobiology of maternal behavior in mammals. In the first section, we present existing information of the affective, motivational, and cognitive processes that characterize maternal behavior, primarily discussing research findings in rats and humans, because most of the work on the neurobiological basis of this behavior has been done in these species. The second section outlines the maternal neural circuitry, with a special emphasis on the mechanisms that underlie the affective, motivational, and cognitive processes of motherhood. Finally, we summarize some of the main themes raised in the chapter and issues yet to be explored.

10.1 Maternal Behavior: Definition and Generalities

The study of motherhood did not receive direct scientific attention until the mid-twentieth century (Beach 1939; Schneirla and Rosenblatt 1961; Rosenblatt and Lehrman 1963) when, finally, beyond the mythical exaltation of its sublime character, with the load restrictions and impositions that the myth of motherhood implied for women, began a systematic and objective scientific observation and experimentation of the psychobiological processes that regulate this behavior. Much has happened in this field since then, notably advancing our understanding on the physiology of mammalian maternal behavior. However, because of the extent of such research, this chapter does not attempt to be complete and exhaustive. Our major interest in this chapter is to address the proximal psychobiological mechanisms of maternal behavior, with a special emphasis on the affective, motivational, and cognitive processes.

A. Ferreira (✉)
Sección Fisiología y Nutrición, Facultad de Ciencias,
Universidad de la República,
Montevideo, Uruguay
e-mail: anna@fcien.edu.uy

M. Pereira
Center for Molecular and Behavioral Neuroscience, Rutgers,
The State University of New Jersey, Newark, NJ, USA

M. C. Antonelli (ed.), *Perinatal Programming of Neurodevelopment*,
Advances in Neurobiology 10, DOI 10.1007/978-1-4939-1372-5_10,

A number of theories have focused on different aspects of motherhood. Hence, the process of new motherhood has been referred as a period in life, a personal decision, a biological transition, a cultural creation, an evolutive process, a state of mind, and a basic emotion (Mayes et al. 2005). From a psychobiological perspective, it could be proposed that mothering and bonding occur as a result of a particular mental state (Winnicott 1956) and a basic care emotion (Panksepp 1998) evoked by immature and helpless individuals that promote in the caregiver a selective motivation to care for and protect them. Epigenetic processes and learning shape caregiving behaviors and thoughts, allowing mothers to cope with the physical and psychological needs of the young and ultimately promoting a strong affective bond between them.

In mammals, since the female lactates, it is often the mother who is the primary caregiver of her offspring, and hence most of what we know about the neurobiological mechanisms of mothering is derived from research investigating the mother. However, it needs to be emphasized that the father and other individuals can develop caregiving activities toward young as efficiently as the mother, and converging evidence has outlined a core neural network for mothering.

Even though motherhood is widely conserved in the animal kingdom, there is no universal maternal care in nature but considerable variations among species and individuals exists depending on the context, the culture and historical epoch, the social structure they belong to, the degree of development of the young at birth, and individual factors including the decision about motherhood. In addition, the way in which mothers interact with their infants also show significant individual differences, likely reflecting differences in mothers' affect, cognition, and motivation during their process of new motherhood.

10.2 Affective, Motivational and Cognitive Processes Involved in Maternal Behavior

This section aims to describe the affective, motivational, and cognitive processes that underlie maternal behavior. This division is not new. Already in 348–328 BC, Plato argued that the soul is divided into three parts: the spirited, the appetitive, and the logical. Along the same line, but in more recent times, Berridge and Robinson (1998) proposed that there are three dissociable components of reward: "liking" or the emotional, hedonic impact or pleasure of an experience; "wanting" or the appetitive desire, actual motivation or the incentive salience of an experience; and "learning" in which both hedonic impact and incentive salience are modified by learning, experience and "logical reasoning."

10.2.1 What Is Affect?

Affect refers to the observable reactions of emotional states (Panksepp 1998) and is a key process of organisms' interaction with social stimuli. Interestingly, Panksepp (who fostered the field of Affective Neuroscience, 1998) proposed that one of the basic affects or emotions in both human and non-human animals is the maternal care. Without doubt, the subjective feeling corresponding to maternal affect in humans is love (Harlow 1958; Panksepp 1998). Indeed, most mothers report experiencing a strong feeling of love about their babies together with feelings of well-being when interacting with them (Fleming et al. 1997).

These intense feelings of love are accompanied by "primary maternal preoccupations," a state in which the mother becomes totally focused on her baby to the exclusion of everything else (Winnicott 1956). This mental state, "almost an illness," begins at late pregnancy and continues through the first months of the infant's life. It is because of these feelings of love and concerns about the baby that the mother increases her sensitivity to the needs of the child and withstands difficult situations (i.e., lack of sleep, stress), remaining alert and in close proximity to her newborn (Feldman et al. 1999; Barrett and Fleming 2011; Kim et al. 2013).

These intense feelings of love and exclusivity with the infant are reflected in part by the behavioral synchrony between mothers and their babies from the first hours after birth. New mothers speak more slowly and in a higher pitch and often exaggerate the pronunciation of vowels to adjust their speech when interacting with their baby (Ingram 1989; Newport 1972, 1974, 1975). Mothers also interact emotionally with their newborns through exaggerated facial expressions and sustained mutual gaze (extensive face-to-face interactions with their infant; Klaus et al. 1975; Leckman et al. 1999). This subjective experience of love in human mothers can be assessed by self-reports (e.g., Wojnar 2004) and questionnaires, including Parental Bonding Questionnaire (Papousek and Papousek 1983; Brockington et al. 2001) and Parenting Stress Questionnaire (Loyd and Abidin 1985).

It is also possible to measure the affect component in non-human animals, and a variety of methods exist for monitoring aspects of emotional processes by measuring behavioral and physiological affective responses. One such example is the affective facial reactions elicited by sweet and bitter tastes in human neonates, orangutans, chimpanzees, monkeys, and even rats and mice (Steiner et al. 2001; Berridge and Kringelbach 2008). Unfortunately, very little work has been done to date in examining maternal affect and feelings of love in non-human animals, which is probably the most important process of the mother–infant bond.

10.2.2 Other Affective Changes During Motherhood

This positive affective state of mothers while interacting with their young brings about dramatic changes in their behavioral and physiological response to other environmental stimuli that aid in the healthy development and survival of the young, including a generalized reduction in fearfulness, anxiety, and other aspects of emotional reactivity, and aggressiveness toward intruders. For example, postpartum mother rats exhibit a reduced freezing response toward a sudden loud noise (Hård and Hansen 1985; Ferreira et al. 2002), as well as anxiolytic-like behaviors in the open field paradigm (Fleming and Luebke 1981), conflict test (Ferreira et al 1989) and plus maze test (Bitran et al. 1991; Lonstein 2005; Pereira et al. 2005) compared with males and females in other stages of their reproductive cycle. Interestingly, virgin females that develop maternal behavior toward neonates also exhibit blunted emotional reactivity similar to that of postpartum rats (Ferreira et al. 2002; Pereira et al. 2005; Agrati et al. 2008). In both cases, such changes are highly dependent on the presence of the offspring or at least require recent physical contact with them (Hård and Hansen 1985; Lonstein 2005; Pereira et al. 2005), suggesting that the pups may encourage the mother to confront potentially environmental threats. Notably, only postpartum females, but not maternal virgin females, exhibit a blunted hypothalamic–pituitary–adrenal (HPA) response to a variety of physical and emotional threats (Lightman et al. 2001; Agrati et al. 2008).

Maternal emotional changes of decreased fearfulness, anxiety, and stress reactivity have also been reported in women. The physical contact with the baby, including but not limited to breastfeeding behavior, has been associated with lower anxiety scores in human mothers, although only breastfeeding mothers exhibit a reduced adrenocorticotropic hormone and cortisol responses to stressors compared with nonbreastfeeding mothers (Carter et al. 2001; Heinrichs et al. 2001).

Concomitant with the blunted emotional reactivity, mothers engage in aggressive displays in defense of their young or the nest. In the rat, the expression and maintenance of maternal aggression is highly dependent on the presence of the pups and associated cues (Ferreira and Hansen 1986; Ferreira et al. 1987; Mayer and Rosenblatt 1993; Kolunie and Stern 1995; Ferreira et al. 2002). Although maternal aggression has been studied most extensively in rodents, mothers in most mammalian species, including human (Hahn-Holbrook et al. 2011) and non-human primates (Harlow et al. 1963; Jay 1963; Weisbard and Goy 1976; Hrdy 1977; Troisi et al. 1988; Maestripieri 1994) display increased aggression in potentially hostile situations for the young.

These affective changes of the mother toward her newborn and the environment are reflected in a mixture of intense feelings of love and positive affect together with a vigilant protectiveness of her young, and allow for the construction of a selective motivation to take care of the neonate. A common experience for new mothers are symptoms such as emotional lability, tearfulness, anxiety, or irritability associated with postpartum blues, which can begin in the first days after delivery, and usually resolve by 10–14 days postpartum. For most women, these symptoms are mild and do not consistently affect the women's ability to mother and may even enrich their emotional response to their babies in their first and subsequent interactions. However, 10–15 % of women develop postpartum depression, a serious condition that, if not recognized and treated timely, not only has deleterious effects on the mother but also poses a serious risk for the mother–infant relationship. One or two in a thousand women will develop postpartum psychosis—a very serious illness that needs quick intervention, usually including hospitalization (World Health Organization 2009).

10.2.3 What Is Motivation?

Motivation (a reason to move; the base of an action) can be defined as the set of processes through which organisms modulate the probability, proximity, and availability of biologically significant stimuli (Salamone and Correa 2002), as are the young for a mother.

Maternal behavior is a highly motivated behavior. For example, the maternal rat will allocate most of her time and energy caring and protecting her pups, will overcome stressful and risky situations to retrieve pups (Nissen 1930; Fahrbach and Pfaff 1982), will choose pups over food (Fleming et al. 1994; Afonso et al. 2009) or a male during the postpartum estrus (Agrati et al. 2008), will lever press to gain access to pups (Wilsoncroft 1969, Lee et al. 2000), and will choose an environment associated with pups over one associated with food (Fleming et al. 1994) or cocaine (Mattson et al. 2001).

Human mothers also show a strong attraction to infant cues after delivery that allows for the development of a strong maternal motivation. This motivation is reflected in the maternal expenditure of time, and energy, including sustained attention and disrupted sleep–wake cycles, as well as the financial costs associated with raising a baby to adulthood (Leckman et al. 1999). Mothers also experience a loss of motivation in the world outside the dyad, including other people, occupational goals, and hobbies, which are substituted by the exclusive interest in their newborn (Leckman et al. 1999).

10.2.4 What Is Cognition?

The term cognition (Latin: cognoscere, "to know," "to conceptualize," or "to recognize") refers to a faculty for the processing of information, applying knowledge, and changing preferences. Cognition, or cognitive processes, can be natural or artificial, conscious, or unconscious. But, what is the role of cognition in maternal behavior? Obviously, maternal behavior requires complex cognitive skills even in non-humam animals; new mothers must acquire and update, efficiently process and store, evaluate, organize, and act on the constantly changing information from the young to flexibly adjust their caregiving responses. Orientation, expectation, memory, and decision making are complex cognitive processes necessary for the sensitive performance of maternal activities. Moreover, mothers must do so in a complex dynamic environment, where many other incentives are available and compete for her attention.

Mothers of many species, including humans, have been found to perform better on a number of tasks that assess aspects of cognitive functions, including attention, behavioral flexibility, working memory, and spatial learning and memory, when compared to nonmothers (Lovic and Fleming 2004; Lambert et al. 2005; Kinsley et al. 2006; Leuner and Gould 2010). Such improved cognitive skills involve reorganization of brain circuits (Pawluski and Galea 2007; Kinsley and Lambert 2008; Leuner and Gould 2010), and collectively are a fundamental component of mothering and key to maternal sensitivity.

Frank Beach, already in 1939, showed that mother rats who make fewer errors in a maze build better nests, show shorter latencies to retrieve the pups to the nest, and more licking behavior than those making more errors. These initial investigations have been confirmed and expanded by the studies of Alison Fleming and collaborators on attention and mothering. Specifically, a better performance on tests of executive functions, including the attentional set-shifting task and prepulse inhibition of the startle response, was positively associated with levels of pup licking by these females when they later became mothers (Lovic and Fleming 2004). In terms of human mothers, recent work also in Fleming's laboratory revealed an association between sensitive mothering and measurements of executive functions, including cognitive flexibility, working memory, and attentional control (Gonzalez et al. 2012). In particular, less sensitive parenting was associated with a poorer strategy in spatial working memory and less cognitive flexibility (Gonzalez et al. 2012).

Empathy, defined as appropriate perception, experience, and response to another's emotion, is especially relevant to parenting in which infants' needs are great, yet most communication is exclusively nonverbal and correctly identifying and responding to infant signals or intentions is essential for sensitive parenting (Meins et al. 2003; Slade 2005; Fonagy et al. 2007). The application of the concepts of cognitive empathy and theory of mind to mothering is relatively new and exciting. In a functional MRI (fMRI) study, Leibenluft et al. (2004) showed that viewing of one's own child evoked a unique pattern of neural activation in mothers that reflected maternal attachment and was associated with brain regions involved with representing the mental state of others, including amygdala and prefrontal cortical regions. It may be that mothers with good "theory of mind" and empathy are more sensitive in their interactions with their infants (Meins et al. 2003; Slade 2005; Fonagy et al. 2007).

10.3 Neuroendocrine Mechanism Involved in Affective, Motivational and Cognitive Processes of Maternal Behavior

10.3.1 Hormonal Regulation of Maternal Behavior

The biological processes that synchronize the rapid development of the mother's affect, motivation, and the cognitive ability to care for neonates are closely related to the hormonal changes around parturition (Terkel and Rosenblatt 1972; Bridges 1990). In the rat, a sharp fall in progesterone and an increase in estrogens, oxytocin, placental lactogens, cortisol, and prolactin enhance the mothers' sensory perception of infant stimuli, and are initially responsible for the rapid development of maternal responsiveness, underpinning the mother's initial attraction to her young and promoting the mother–infant interaction that starts a positive reinforcing cycle for attachment (Bridges 1990, Numan and Insel 2003).

In human mothers, the success of adoption (Singer et al. 1985; van Ijzendoorn and Juffer 2006) has led to the idea that maternal behavior of female primates is less dependent on reproductive hormones compared to many non-primate mammalian females. Therefore, even though the function of reproductive hormones on caregiving behavior is well established in non-primate females, little investigation has been devoted to understand the function of these hormones in human parental behavior. However, recent studies suggest that hormonal factors do influence the intensity of maternal responsiveness in humans. Thus, mothers who experienced a stronger attachment to their new babies after birth had an increase in the estradiol-to-progesterone ratio throughout pregnancy, whereas those with low attachment experienced a decreased in the estradiol/progesterone ratio over this same pregnancy period (Fleming et al. 1997).

Oxytocin is another key neuroendocrine factor related to caregiving activities and social behaviors in mammals (for reviews, see Uvnäs-Moberg and Eriksson 1996; Carter 1998; Insel and Young 2001; Feldman 2012). In humans, plasma oxytocin levels in pregnancy predict attachment to the baby, whereas postpartum serum and salivary oxytocin levels predict maternal behaviors toward infants (Levine et al. 2007; Feldman et al. 2010). For instance, serum oxytocin levels in mothers following interaction with their infant was significantly higher in those mothers with a secure attachment style, compared with mothers with an insecure-avoidant attachment style, a result that was strongly correlated with activation of hypothalamus/pituitary region and the ventral striatum in response to smiling and crying own-infant face cues in the fMRI scanner (Strathearn et al. 2009).

Cortisol is another hormone implicated in the emotional regulation of infant stimuli in both human and non-human mothers. Cortisol levels peak during parturition and decrease rapidly in the postpartum period (Storey et al. 2000). In humans, high cortisol levels have been strongly associated with affectionate touching of the infant in first-time mothers, as well as with the success at recognizing one's own infant in multiparous mothers (Fleming et al. 1997; Gonzalez et al. 2009). These studies highlight the role of cortisol as one main hormone associated to accuracy in sensory recognition and pleasantness of infant stimuli.

However, the hormonal profile of parturition is not the only mechanism through which maternal behavior can be initiated in mammals. Thus, nonhormonal parental behaviors appear in humans and other animals during adoption, and high degrees of maternal-like behavior occur when group members other than the biological parents help to care for and protect the young (Hrdy 1977). In the laboratory, continuous exposure (6–10 days) to young pups can stimulate maternal behavior in juvenile, adult male, and cycling female rats, a process called pup induction of maternal behavior or sensitization, indicating the existence of a nonhormonal route through which infant stimuli can gain access to the maternal circuitry (Wiesner and Sheard 1933; Rosenblatt 1967).

However, despite the absence of any hormonal changes of gestation, parturition, and lactation in sensitized female and male rats and adoptive human parents, interaction with newborn infants alter levels of hormones in both brain and plasma, such as cortisol and prolactin (humans: Parsons et al. 2010; rats: Jakubowski and

Terkel 1986; Samuels and Bridges 1983), which can influence subsequent caregiving activities.

10.3.2 Sensory Regulation of Maternal Behavior

After the transient hormonal spurt, the presence of hormones becomes progressively less important. Continued interaction with the offspring further increases their incentive value, assuring the maintenance of maternal responsiveness during the postpartum period. Thus, the sensory experiences acquired by the mother while interacting with the young and processes of learning and reinforcement are the main factors responsible for the maintenance of maternal behaviors during the nonhormonal phase that begins shortly after parturition (for a revision, see Magnusson and Fleming 1995; Stern 1997).

10.3.3 Maternal Circuitry

Among the brain structures critically involved in postpartum maternal responsiveness, the medial preoptic area (mPOA) is believed to act as a primary locus of integration, orchestrating the effective expression of maternal behavior to the developmental stage of the pups across the postpartum period. The mPOA receives converging young-related information from every sensory modality (Simerly and Swanson 1986; Risold et al. 1997) and is a key neural site where the hormones of pregnancy act to synchronize maternal responsiveness to infant-related stimuli at parturition (Bridges 1990). The mPOA has reciprocal important connections with areas of the brain involved in affect, motivation, and cognition, relevant for maternal behavior that provide a means for pup-responsive mPOA neurons to influence the female's responsivity to pups with expression of appropriate maternal behaviors across postpartum. Thus, the mPOA output circuitry plays a critical role in the display not only of maternal behavior per se but also of the accompanying behavioral and physiological adaptations, such as alterations in food intake, stress reactivity, anxiety, and aggressiveness, that also contribute to successful parenting.

Although the maintenance of maternal behavior becomes independent of hormones, it remains dependent on the mPOA and related circuitry. Importantly, across this period, the mother's responsiveness to her developing young changes to match their evolving needs. Recent research has revealed that a substantial functional reorganization of the rat maternal circuitry occurs across postpartum, probably sculpted by the continuous experience of interaction with the pups (Pereira and Morrell 2009, 2011). Specifically, the mPOA has been demonstrated to change its role throughout postpartum, from a necessary facilitatory role for both the onset (Numan et al. 1977; Jacobson et al. 1980; Gray and Brooks 1984; Cohn and Gerall 1989), and the early postpartum expression (Gray and Brooks 1984; Numan et al. 1988; Cohn and Gerall 1989; Lee et al. 2000; Pereira and Morrell 2009) of maternal behavior to an inhibitory role during the late postpartum period (Pereira and Morrell 2009), to allow such changes in the mother's behavior across postpartum attuned to the developing needs of her young.

A unifying system of affect, motivation, and cognition is the mesocorticolimbic dopamine (DA) system. Primary efferents from the mPOA traveling through the lateral hypothalamus interact with components of the mesocorticolimbic DA system to regulate aspects of affective, motivational, and cognitive processes of maternal behavior. This system, which is comprised of DA cell bodies in the midbrain ventral tegmental area (VTA) and its major targets, the nucleus accumbens (NA) and prefrontal cortex (PFC), has been recognized for its central role in several behavioral functions related to motivation and cognition (Berridge 1996 and 2004; Salamone and Correa 2012). The NA receives converging excitatory inputs from most cortical and limbic structures, and hence has long been considered a functional interface between the limbic system and the motor system (Mogenson et al. 1980) that allows the translation of motivation, emotion, and cognition into a coherent action, under the modulatory influence of DAergic inputs from the VTA (LeMoal and Simon 1991; Mogenson et al. 1980).

10.3.4 Neural Mechanisms Underlying Maternal Affect

The close bond between the mother and the infant greatly depend on feelings of love and positive affect. Interestingly, the components of the adult social brain (Brothers 1990), largely correspond to the affective networks identified by affective neuroscience (Panksepp 1998), involving subcortical structures, such as the NA, ventral pallidum, and amygdala, and cortical structures, including the orbitofrontal cortex (OFC), medial prefrontal cortex (mPFC), and insular cortex. It can be proposed that these social and affective regions are likely recruited during the mother–infant interaction because this is the most essential form of interindividual affective interaction.

Besides feelings of love and positive affects, mothers experience preoccupations and obsessive thoughts associated with their infants during the early postpartum period, which are critical for the formation of maternal attachment. These feelings may involve brain circuits that underlie obsessive–compulsive behaviors, such as the dorsolateral PFC and orbito-striatal areas (Leckman et al. 1999; Abramowitz et al. 2010).

This strong emotional commitment with the young is accompanied by affective changes toward the environment, including a generalized reduction in fearfulness, anxiety, and other aspects of emotional reactivity, and aggressiveness toward intruders.

Few studies have examined the brain areas involved in the reduction of fear and anxiety during motherhood. The medial amygdala (MeA), which is implicated in the control of innate fear responses, also modulates anxiety-like behaviors in the mother rat (Fleming 1989). Thus, lesions of this area, as well as, exogenous ovarian hormones, presumably acting on the MeA, reduce anxiety-like behaviors assessed in the open field (Fleming 1989). Conversely, electrical stimulation of the MeA in nulliparous female rats reduces the time spent in the center of the open field, indicative of higher anxiety. Interestingly, a recent study (Smith and Lonstein 2008) shows that some brain areas, traditionally associated with emotional regulation, exhibit lower Fos expression in mothers that have recently interacted with their litter.

Changes along the entire HPA stress axis together with a reduction in the activity of limbic circuits that project to the paraventricular nucleus (PVN) account for the stress-hyporesponsiveness during lactation (Heinrichs et al. 2001; Slattery and Neumann 2008). Such changes have been related primarily to the endocrine profile of the perinatal period and lactation, in particular to the lactogenic hormones oxytocin and prolactin, although the mechanism by which these hormones modify the endocrine stress response is yet unclear.

Areas involved in perception of olfactory cues from the pups, such as the olfactory bulb (OB; Ferreira et al. 1987), or in the contrast between odors from the pups and the intruder, such as the insular PFC and ventromedial hypothalamus (VMH; Ferreira et al. 1987; Lynds 1976), are related to the control of maternal aggression. Also, the ventral stimulation of the nipples and surrounding skin (Stern and Kolunie 1989; Mayer et al. 1987; Factor et al. 1993), as well as, areas that form part of the ascending milk-ejection pathway, such as the peripeduncular nucleus (PPN; Hansen and Ferreira 1986; Factor et al. 1993) and the PVN of the hypothalamus (Olazábal and Ferreira 1997; Giovenardi et al. 1998), modulate the aggressive responses of the mothers toward intruders.

In addition to the abovementioned areas, other cortical and subcortical regions modulate maternal aggression, including the anterior hypothalamus (AHA), the ventromedial hypothalamic nucleus (VMH), the lateral septum (LAS), the amygdala, the periaqueductal gray (PAG), the PFC, and the mPOA (for a review, see Lonstein and Gammie 2002). Several neurotransmitter systems, which include the GABAergic, the serotonergic and the oxitocinergic systems, acting in different receptor subtypes and brain areas, importantly modulate maternal aggression (for a review, see Lonstein and Gammie 2002).

10.3.5 Neural Mechanisms Underlying Motivational Aspects of Maternal Behavior

Current evidence indicates that mPOA connections with the VTA and NA components of the mesocorticolimbic DA system mediate motivational aspects of maternal behavior (Numan and Stolzenberg 2009). The mPOA can influence NA function through a direct projection to it and indirectly through projections to the VTA (Simerly and Swanson 1988; Numan and Numan 1997; Geisler and Zahm 2005). Intact bilateral connections between the mPOA and the VTA and NA are essential for activational aspects of maternal behavior in rats. Severing the dorsolateral connections of the mPOA—but not its other afferents and efferents—disrupt active components of maternal behavior, including retrieval, licking, and nest building, in a manner similar to lesions of the mPOA itself (Terkel et al. 1979; Jacobson et al. 1980; Numan et al. 1985; Arrati et al. 2006; Pereira and Morrell 2009). Moreover, DA is released in the NA of postpartum mother rats during active maternal interaction with pups (Hansen et al. 1993; Champagne et al. 2004; Afonso et al. 2009; Robinson et al. 2011; Pereira et al. 2013). On the other hand, interference with accumbens DA in mother rats by electrolytic or DA-depleting lesions of the VTA, pharmacological manipulation of VTA activity, or intra-accumbens administration of D1 or D2 DA receptor antagonists each selectively and severely disrupt the activational aspects of maternal behaviors in early postpartum rats, while leaving the directional aspects relatively intact (Hansen et al. 1991a, b; Keer and Stern 1999; Silva et al. 2003; Numan et al. 2005, 2009; Seip and Morrell 2009). Of course, DA in the NA does not participate in maternal motivation in isolation, and recent studies, for example, have shown involvement of adenosine and the adenosine A_{2A} receptors in modulating DA-mediated maternal behavior (Pereira et al. 2011).

Interestingly, not only maternal behaviors but also associated affective behaviors have a motivational basis. Our prior studies showed that the D2 DA receptor antagonist haloperidol induced deficits in early postpartum females' affective behaviors, including increased anxiety-like behavior and reduced maternal aggression toward an intruder male. The presence of the pups counteracted haloperidol-induced deficits, restoring maternal anxiolysis (Pereira et al. 2005). In addition, haloperidol-induced low levels of aggression were restored to typical levels exhibited by postpartum females by increasing the incentive salience of the pups (Ferreira et al. 2012).

10.3.6 Neural Mechanisms Underlying Cognitive Aspects of Maternal Behavior

Specific regions of the PFC are activated by offspring cues and parenting behavior in rodents (Fleming and Korsmit 1996; Hernández-González et al. 2005; Febo et al. 2008) and lesions of the mPFC have profound effects on maternal behavior (Afonso et al. 2007; Febo et al. 2010; Pereira and Morrell 2011). The PFC receives polymodal sensory input and has been implicated in stimulus recognition and executive functions contributing to attentional selection, optimal organization and planning, flexibility, and decision making in relation to complex goal-directed behaviors. The mPFC receives DA input from the VTA (mesocortical DA system) and has reciprocal connections with several subcortical structures, including many involved in maternal behavior such as the mPOA, BST, VTA, NA, and the PAG. Importantly, specific subregions of the mPFC appear to be differentially involved in rat maternal behavior across the postpartum period. Permanent lesions and transient inactivation of the anterior cingulate affect the organizational aspects of maternal behavior both during early and late postpartum periods, likely by inducing deficits in attention and behavioral inhibition processes, although mother rats remain interested in pups. Depression of the infralimbic cortex, the most ventral part of the mPFC, severely disrupts maternal behavior only during early postpartum, whereas inactivation of the prelimbic subregion does so only during late postpartum (Pereira and Morrell 2011).

The growing field of cognitive neuroscience, propelled by modern brain imaging techniques and interest in deficits of empathy, has revealed networks of brain activity relating to empathy and emotional mirroring. These imaging studies suggest that two networks are involved in human mothering, a motivational–emotional one, which includes the amygdala, NA, and anterior cingulate cortex, and a social attention network related with social cognition and empathy, which comprises the mPFC, superior temporal gyrus, insula, inferior parietal lobe, and inferior frontal gyrus (Atzil et al. 2011).

Interestingly, the studies of Ruth Feldman show that mothers experiencing synchronous interactions with their infants show a more coherent activation of the amygdala and NA to their infant's stimuli (Atzil et al. 2011). This physiological synchrony can allow the mother to feel and interpret her child signals. Even more interestingly, mothers and fathers show a notable synchrony in the activation of areas related to social cognition, theory of mind, and empathy when they look at their own infant video, which may help parents to make a coincident lecture of nonverbal infant cues (Atzil et al. 2012).

10.4 Conclusion and Perspectives

The maternal behavior of new mothers toward their offspring shows both marked similarities and considerable differences across and within species. In this chapter, our primary interest was to highlight the complexity of motherhood in terms of the emotional concern or affection, motivation, and cognition processes that contribute to individual differences in mothering and their interrelated neural mechanisms. The wide variation in mothering styles observed in nature relies, in part, in individual differences in multiple levels of functioning, including aspects of perception, emotion, attention, and other measures of executive function, that impact mothers' motivation to engage with their young. These systems have developed as a function of the mothers' genetic and early experiences, and are sensitive to the psychobiological context, the developmental maturity of the neonates, and the mothers' affective or mood state.

We would like to see more studies assessing the influence of affect, motivation, and cognition on maternal behavior, and the mediating changes in the brain in an attempt to reveal why mothers differ from one another.

Conflict of Interest The authors declare no conflicts of interest.

References

Abramowitz JS, Meltzer-Brody S, Leserman J, Killenberg S, Rinaldi K, Mahaffey BL, Pedersen C (2010) Obsessional thoughts and compulsive behaviors in a sample of women with postpartum mood symptoms. Arch Womens Ment Health 13(6):523–530

Afonso VM, Sison M, Lovic V, Fleming AS (2007) Medial prefrontal cortex lesions in the female rat affect sexual and maternal behavior and their sequential organization. Behav Neurosci 121(3):515–526

Afonso VM, King S, Chatterjee D, Fleming AS (2009) Hormones that increase maternal responsiveness affect accumbal dopaminergic responses to pup- and food-stimuli in the female rat. Horm Behav 56(1):11–23

Agrati D, Fernández-Guasti A, Ferreira A (2008) The reproductive stage and experience of sexually receptive mothers alter their preference for pups or males. Behav Neurosci 122(5):998–1004

Arrati PG, Carmona C, Dominguez G, Beyer C, Rosenblatt JS (2006) GABA receptor agonists in the medial preoptic area and maternal behavior in lactating rats. Physiol Behav 87(1):51–65

Atzil S, Hendler T, Feldman R (2011) Specifying the brain basis of human attachment: brain, hormones, and behavior in synchronous and intrusive mothers. Neuropsychopharmacology 36:2603–2615

Atzil S, Hendler T, Zagoory-Sharon O, Winetraub Y, Feldman R (2012) Synchrony and specificity in the maternal and the paternal brain: relations to oxytocin and vasopressin. J Am Acad Child Adolesc Psychiatry 51(8):798–811

Barrett J, Fleming AS (2011) Annual research review: all mothers are not created equal: neural and psychobiological perspectives on mothering and the importance of individual differences. J Child Psychol Psychiatry 52(4):368–397

Beach FA (1939) The neural basis of innate behavior. III. Comparison of learning ability and instinctive behavior in the rat. J Comp Psychol 28(2):225–262

Berridge KC (1996) Food reward: brain substrates of wanting and liking. Neurosci Biobehav Rev 20(1):1–20

Berridge KC (2004) Motivation concepts in behavioral neuroscience. Physiol Behav 81(2):179–209

Berridge KC, Kringelbach ML (2008) Affective neuroscience of pleasure: reward in humans and animals. Psychopharmacology (Berl) 199(3):457–480

Berridge KC, Robinson TE (1998) What is the role of dopamine in reward: hedonic impact, reward learning, or incentive salience? Brain Res Rev 28:309–369

Bitran D, Hilvers RJ, Kellogg CK (1991) Ovarian endocrine status modulates the anxiolytic potency of diazepam and the efficacy of aminobutyric acid-benzodiazepine receptor-mediated chloride ion transport. Behav Neurosci 105:653–662

Bridges RS (1990) Endocrine regulation of parental behavior in rodents. In: Krasnegor NA, Bridges RS (ed) Mammalian parenting: biochemical, neurobiological, and behavioral determinants. Oxford, New York, pp 93–117

Brockington IF, Oates J, George S et al (2001) A screening questionnaire for mother-infant bonding disorders. Arch Women's Ment Health 3:133–140

Brothers L (1990) The social brain: a project for integrating primate behaviour and neurophysiology in a new domain. Concepts Neurosci 1:27–251

Carter SC (1998) Neuroendocrine perspectives on social attachment and love. Psychoneuroendocrinology 23:779–818

Carter CS, Altemus M, Chrousos GP (2001) Neuroendocrine and emotional changes in the postpartum period. In: Russell JA, Douglas AJ, Windle RJ, Ingram CD (eds) Progress in brain research. The maternal brain, vol 133. Elsevier, Amsterdam, pp 241–249

Champagne FA, Chretien P, Stevenson CW, Zhang TY, Gratton A, Meaney MJ (2004) Variations in nucleus accumbens dopamine associated with individual differences in maternal behavior in the rat. J Neurosci 24(17):4113–4123

Cohn J, Gerall AA (1989) Pre- and postpuberal medial preoptic area lesions and maternal behavior in the rat. Physiol Behav 46(2):333–336

Factor EM, Mayer AD, Rosenblatt JS (1993) Peripeduncular nucleus lesions in the rat: I. Effects on maternal aggression, lactation, and maternal behavior during pre- and postpartum periods. Behav Neurosci 107:166–185

Fahrbach S, Pfaff DW (1982) Hormonal and neural mechanisms underlying maternal behavior in the rat. In: Pfaff DW (ed) The physiological mechanisms of motivation. Springer, Heidelberg, pp 253–285

Febo M, Stolberg TL, Numan M, Bridges RS, Kulkarni P, Ferris CF (2008) Nursing stimulation is more than tactile sensation: it is a multisensory experience. Horm Behav 54(2):330–339

Febo M, Felix-Ortiz AC, Johnson TR (2010) Inactivation or inhibition of neuronal activity in the medial prefrontal cortex largely reduces pup retrieval and grouping in maternal rats. Brain Res 1325:77–88

Feldman R (2012) Oxytocin and social affiliation in humans. Horm Behav 61(3):380–91

Feldman R, Weller A, Leckman JF, Kuint J, Eidelman AI (1999) The nature of the mother's tie to her infant: maternal bonding under conditions of proximity, separation, and potential loss. J Child Psychol Psychiatry 40(6):929–939

Feldman R, Singer M, Zagoory O (2010) Touch attenuates infants' physiological reactivity to stress. Dev Sci 13:69–71

Ferreira A, Hansen S (1986) Sensory control of maternal aggression in *Rattus norvegicus*. J Comp Psychol 100:173–177

Ferreira A, Dahlof LS, Hansen S (1987) Olfactory mechanisms in the control of maternal aggression, appetite, and fearfulness: effects of lesions to olfactory receptors, mediodorsal thalamic nucleus, and insular prefrontal cortex. Behav Neurosci 101:709–717

Ferreira A, Hansen S, Nielsen M, Archer T, Minor BC (1989) Behavior of mother rats in conflict tests sensitive to anti-anxiety agents. Behav Neurosci 103:193–201

Ferreira A, Pereira M, Agrati D, Uriarte N, FernFernN-Guasti A (2002) Role of maternal behavior on aggression, fear and anxiety. Physiol Behav 77:197–204

Ferreira A, Uriarte N, Pereira M (2012) 8th FENS forum of neuroscience, Barcelona

Fleming AS, Korsmit M (1996) Plasticity in the maternal circuit: effects of maternal experience on Fos-Lir in hypothalamic, limbic, and cortical structures in the postpartum rat. Behav Neurosci 110(3):567–670

Fleming AS, Luebke C (1981) Timidity prevents the virgin female rat from being a good mother: emotionality differences between nulliparous and parturient females. Physiol Behav 27:863–868

Fleming AS, Cheung U, Myhal N, Kessler Z (1989) Effects of maternal hormones on 'timidity' and attraction to pup-related odors in female rats. Physiol Behav 46(3):449–453

Fleming AS, Korsmit M, Deller M (1994) Rat pups are potent reinforcers to the maternal animal: effects of experience, parity, hormones, and dopamine function. Psychobiology 22:491–569

Fleming AS, Ruble D, Krieger H, Wong PY (1997) Hormonal and experiential correlates of maternal responsiveness during pregnancy and the puerperium in human mothers. Horm Behav 31:145–158

Fonagy P, Gergely G, Target M (2007) The parent-infant dyad and the construction of the subjective self. J Child Psychol Psychiatry 48(3-4):288–328

Geisler S, Zahm DS (2005) Afferents of the ventral tegmental area in the rat-anatomical substratum for integrative functions. J Comp Neurol 490(3):270–294

Giovenardi M, Padoin MJ, Cadore LP, Lucion AB (1998) Hypothalamic paraventricular nucleus modulates maternal aggression in rats: effects of ibotenic acid lesion and oxytocin antisense. Physiol Behav 63:351–359

Gonzalez A, Jenkins JM, Steiner M, Fleming AS (2009) The relation between early life adversity, cortisol awakening response and diurnal salivary cortisol levels in postpartum women. Psychoneuroendocrinology 34(1):76–86

Gonzalez A, Jenkins JM, Steiner M, Fleming AS (2012) Maternal early life experiences and parenting: the mediating role of cortisol and executive function. J Am Acad Child Adolesc Psychiatry 51(7):673–682

Gray P, Brooks PJ (1984) Effect of lesion location within the medial preoptic-anterior hypothalamic continuum on maternal and male sexual behaviors in female rats. Behav Neurosci 98(4):703–711

Hahn-Holbrook J, Holt-Lunstad J, Holbrook C, Coyne SM, Lawson ET (2011) Maternal defense: breast feeding increases aggression by reducing stress. Psychol Sci 22(10):1288–1295

Hansen S, Ferreira A (1986) Food intake, aggression, and fear behavior in the mother rat: control by neural systems concerned with milk ejection and maternal behavior. Behav Neurosci 100:64–70

Hansen S, Harthon C, Wallin E, Löfberg L, Svensson K (1991a) Mesotelencephalic dopamine system and reproductive behavior in the female rat: effects of ventral tegmental 6-hydroxydopamine lesions on maternal and sexual responsiveness. Behav Neurosci 105(4):588–885

Hansen S, Harthon C, Wallin E, Löfberg L, Svensson K (1991b) The effects of 6-OHDA-induced dopamine depletions in the ventral or dorsal striatum on maternal and sexual behavior in the female rat. Pharmacol Biochem Behav 39(1):71–77

Hansen S, Bergvall AH, Nyiredi S (1993) Interaction with pups enhances dopamine release in the ventral striatum of maternal rats: a microdialysis study. Pharmacol Biochem Behav 45(3): 673–676

Harlow HF (1958) The nature of love. American Psychologist 13(12):673–685

Hård E, Hansen S (1985) Reduced fear behavior in the lactating rat. Physiol Behav 33:641–664

Harlow HF, Harlow MK, Hansen EW (1963) The maternal affectional system of rhesus monkeys. In: Rheingold HL (ed) Maternal behavior in mammals. Wiley, New York

Heinrichs M, Meinlschmidt G, Neumann I, Wagner S, Kirschbaum C, Ehlert U, Hellhammer DH (2001) Effects of suckling on hypothalamic-pituitary-adrenal axis responses to psychosocial stress in postpartum lactating women. J Clin Endocrinol Metab. 86(10):4798–4804

Hernández-González M, Navarro-Meza M, Prieto-Beracoechea CA, Guevara MA (2005) Electrical activity of prefrontal cortex and ventral tegmental area during rat maternal behavior. Behav Processes 70(2):132–143

Hrdy SB (1977) Infanticide as a primate reproductive strategy. American Scientist 65(1):40–49

Ingram, D (1989) First language acquisition: method, description and explanation. Cambridge: Cambridge University Press

Insel TR, Young LJ (2001) The neurobiology of attachment. Nat Rev Neurosci 2:129–136

Jacobson CD, Terkel J, Gorski RA, Sawyer CH (1980) Effects of small medial preoptic area lesions on maternal behavior: retrieving and nest building in the rat. Brain Res 194(2):471–478

Jakubowski M, Terkel J (1986) Establishment and maintenance of maternal responsiveness in postpartum Wistar rats. Anim Behav 34:256–262

Jay P (1963) Mother-infantrelations in langurs. In: Rheingold HL (ed) Maternal behavior in mammals. Wiley, New York

Keer SE, Stern JM (1999) Dopamine receptor blockade in the nucleus accumbens inhibits maternal retrieval and licking, but enhances nursing behavior in lactating rats. Physiol Behav 67(5):659–669

Kim P, Mayes LC, Leckman JF, Feldman R, Swain JE (2013) Early postpartum parental preoccupation and positive parenting thoughts: Relationship with parent-infant interaction. Infant Mental Health Journal 34:104–116

Kinsley CH, Lambert KG (2008) Reproduction-induced neuroplasticity: natural behavioural and neuronal alterations associated with the production and care of offspring. J Neuroendocrinol 20(4):515–525

Kinsley CH, Trainer R, Stafisso-Sandoz G, Quadros P, Marcus LK, Hearon C, Meyer EA, Hester N, Morgan M, Kozub FJ, Lambert KG (2006) Motherhood and the hormones of pregnancy modify concentrations of hippocampal neuronal dendritic spines. Horm Behav 49(2):131–142

Klaus MH, Trause MA, Kennell JH (1975) Does human maternal behaviour after delivery show a characteristic pattern? Ciba Found Symp (33):69–8

Kolunie JM, Stern JM (1995) Maternal aggression in rats: effects of olfactory bulbectomy, $ZnSO_4$-induced anosmia, and vomeronasal organ removal. Horm Behav 29:492–518

Lambert KG, Berry AE, Griffins G, Amory-Meyers E, Madonia-Lomas L, Love G, Kinsley CH (2005) Pup exposure differentially enhances foraging ability in primiparous and nulliparous rats. Physiol Behav 84(5):799–806

Le Moal M, Simon H (1991) Mesocorticolimbic dopaminergic network: functional and regulatory roles. Physiol Rev 71(1):155–234

Leckman JF, Mayes LC, Feldman R, Evans DW, King RA, Cohen DJ (1999) Early parental preoccupations and behaviors and their possible relationship to the symptoms of obsessive-compulsive disorder. Acta Psychiatr Scand Suppl 396:93–97

Lee A, Clancy S, Fleming AS (2000) Mother rats bar-press for pups: effects of lesions of the mpoa and limbic sites on maternal behavior and operant responding for pup-reinforcement. Behav Brain Res 108(2):215–231

Leibenluft E, Gobbini MI, Harrison T, Haxby JV (2004) Mothers' neural activation in response to pictures of their children and other children. Biol Psychiatry 56(4):225–232 (Erratum Biol Psychiatry 56(8):612)

Leuner B, Gould E (2010) Dendritic growth in medial prefrontal cortex and cognitive flexibility are enhanced during the postpartum period. J Neurosci 30(40):13499–13503

Levine A, Zagoory-Sharon O, Feldman R, Weller A (2007) Oxytocin during pregnancy and early postpartum: individual patterns and maternal-fetal attachment. Peptides 28:1162–1169

Lightman SL, Windle RJ, Wood SA, Kershaw YM, Shanks N, Ingram CD (2001) Peripartum plasticity within the hypothalamo-pituitaryadrenal axis. Prog Brain Res 133:111–129

Lonstein JS (2005) Reduced anxiety in postpartum rats requires recent physical interactions with pups, but is independent of suckling and peripheral sources of hormones. Horm Behav 47(3):241–255

Lonstein JS, Gammie SC (2002) Sensory, hormonal, and neural control of maternal aggression in laboratory rodents. Neurosci Biobehav Rev 26:869–888

Lovic V, Fleming AS (2004) Artificially-reared female rats show reduced prepulse inhibition and deficits in the attentional set shifting task-reversal of effects with maternal-like licking stimulation. Behav Brain Res 148:209–219

Loyd BH, Abidin RR (1985) Revision of the parenting stress index. J Pediatr Psychol 10(2):169–177

Lynds PG (1976) Olfactory control of aggression in lactating female house mice. Physiol Behav 17:157–159

Maestripieri D (1994) Costs and benefits of maternal aggression in lactating female rhesus macaques. Primates 35:443–453

Magnusson JE, Fleming AS (1995) Rat pups are reinforcing to the maternal rat: role of sensory cues. Psychobiology 23:69–75

Mattson BJ, Williams S, Rosenblatt JS, Morrell JI (2001) Comparison of two positive reinforcing stimuli: pups and cocaine throughout the postpartum period. Behav Neurosci 115(3):683–835

Mayer AD, Rosenblatt JS (1993) Contributions of olfaction to maternal aggression in laboratory rats (*Rattus norvegicus*): effects of peripheral deafferentation of the primary olfactory system. J Comp Psychol 107:12–24

Mayer AD, Carter L, Jorge WA, Mota MJ, Tannu S, Rosenblatt JS (1987) Mammary stimulation and maternal aggression in rodents: thelectomy fails to reduce pre- or postpartum aggression in rats. Horm Behav 21:501–510

Mayes LC, Swain JE, Leckman JF (2005) Parental attachment systems: neural circuits, genes, and experiential contributions to parental engagement. Clinical Neuroscience Research 4301–4313

Meins E, Fernyhough C, Wainwright R, Clark-Carter D, Das Gupta M, Fradley E, Tuckey M (2003) Pathways to understanding mind: construct validity and predictive validity of maternal mind-mindedness. Child Dev 74(4):1194–1211

Mogenson GJ, Jones DL, Yim CY (1980) From motivation to action: functional interface between the limbic system and the motor system. Prog Neurobiol 14(2–3):69–97

Newport, EL (1972) Motherese: a study of how mothers speak to young children. Spencer Foundation grant proposal. (cited in Newport 1975)

Newport, EL (1974) Motherese and its relation to the child's acquisition of language. Paper presented at Conference on Language Input and Acquisition, Boston, September 1974. (cited in Newport 1975)

Newport, EL (1975) Motherese: The speech of mothers to young children. Ph.D. dissertation, University of Pennsylvania

Nissen HW (1930) A study of maternal behavior in the white rat by means of the obstruction method. J Genet Psychol 37:377–393

Numan M, Insel TR (2003) The neurobiology of parental behavior. Springer, New York

Numan M, Numan MJ (1997) Projection sites of medial preoptic area and ventral bed nucleus of the stria terminalis neurons that express Fos during maternal behavior in female rats. J Neuroendocrinol 9(5):369–384
Numan M, Stolzenberg DS (2009) Medial preoptic area interactions with dopamine neural systems in the control of the onset and maintenance of maternal behavior in rats. Front Neuroendocrinol 30(1):46–64
Numan M, Rosenblatt JS, Komisaruk BR (1977) Medial preoptic area and onset of maternal behavior in the rat. J Comp Physiol Psychol 91(1):146–164
Numan M, Morrell JI, Pfaff DW (1985) Anatomical identification of neurons in selected brain regions associated with maternal behavior deficits induced by knife cuts of the lateral hypothalamus in rats. J Comp Neurol 237(4):552–564
Numan M, Corodimas KP, Numan MJ, Factor EM, Piers WD (1988) Axon-sparing lesions of the preoptic region and substantia innominata disrupt maternal behavior in rats. Behav Neurosci 102(3):381–396
Numan M, Numan MJ, Pliakou N, Stolzenberg DS, Mullins OJ, Murphy JM, Smith CD (2005) The effects of D1 or D2 dopamine receptor antagonism in the medial preoptic area, ventral pallidum, or nucleus accumbens on the maternal retrieval response and other aspects of maternal behavior in rats. Behav Neurosci 119(6):1588–1604
Numan M, Stolzenberg DS, Dellevigne AA, Correnti CM, Numan MJ (2009) Temporary inactivation of ventral tegmental area neurons with either muscimol or baclofen reversibly disrupts maternal behavior in rats through different underlying mechanisms. Behav Neurosci 123(4):740–751
Olazábal DE, Ferreira A (1997) Maternal behavior in rats with kainic acid-induced lesions of the hypothalamic paraventricular nucleus. Physiol Behav 61(5):779–784
Panksepp J (1998) Affective neuroscience: the foundations of human and animal emotion. Oxford University, New York
Papousek H, Papousek M (1983) Biological basis of social interactions: implications of research for an understanding of behavioural deviance. J Child Psychol Psychiatry 24(1):117–129
Parsons CE, Young KS, Murray L, Stein A, Kringelbach ML (2010) The functional neuroanatomy of the evolving parent infant relationship. Prog Neurobiol 91:220–241
Pawluski JL, Galea LA (2007) Reproductive experience alters hippocampal neurogenesis during the postpartum period in the dam. Neuroscience 149(1):53–67
Pereira M, Morrell JI (2009) The changing role of the medial preoptic area in the regulation of maternal behavior across the postpartum period: facilitation followed by inhibition. Behav Brain Res 205(1):238–248
Pereira M, Morrell JI (2011) Functional mapping of the neural circuitry of rat maternal motivation: effects of site-specific transient neural inactivation. J Neuroendocrinol 23(11):1020–1035
Pereira M, Uriarte N, Agrati D, Zuluaga MJ, Ferreira A (2005) Motivational aspects of maternal anxiolysis in lactating rats. Psychopharmacol (Berl) 180:241–248
Pereira M, Farrar AM, Hockemeyer J, Müller CE, Salamone JD, Morrell JI (2011) Effect of the adenosine A2A receptor antagonist MSX-3 on motivational disruptions of maternal behavior induced by dopamine antagonism in the early postpartum rat. Psychopharmacology (Berl) 213(1):69–79
Pereira M, Farrar AM, Morrell JI, Abercrombie ED (2013) Changes in nucleus accumbens dopamine release during mother-pup interactions at early and late postpartum stages. V Parental brain conference, Germany
Plato (348–328 BC) Republic IV
Risold PY, Thompson RH, Swanson LW (1997) The structural organization of connections between hypothalamus and cerebral cortex. Brain Res Brain Res Rev 24(2–3):197-254
Robinson DL, Zitzman DL, Williams SK (2011) Mesolimbic dopamine transients in motivated behaviors: focus on maternal behavior. Front Psychiatry 9(2):23
Rosenblatt JS (1967) Nonhormonal basis of maternal behavior in the rat. Science 156:152–173
Rosenblatt JS, Lehrman DS (1963) Maternal behavior in the laboratory rat. In: Rheingold HL (ed) Maternal behavior in mammals. Wiley, New York, pp 8–57

Salamone JD, Correa M (2002) Motivational views of reinforcement: implications for understanding the behavioral functions of nucleus accumbens dopamine. Behav Brain Res 137:3–25
Salamone JD, Correa M (2012) The mysterious motivational functions of mesolimbic dopamine. Neuron 76(3):470–485
Samuels MH, Bridges RS (1983) Plasma prolactin concentrations in parental male and female rats: effects of exposure to rat young. Endocrinology 113:1647–1654
Schneirla TC, Rosenblatt JS (1961) Animal research. I. Behavioral organization and genesis of the social bond in insects and mammals. Am J Orthopsychiatry 31:223–253
Seip KM, Morrell JI (2009) Transient inactivation of the ventral tegmental area selectively disrupts the expression of conditioned place preference for pup- but not cocaine-paired contexts. Behav Neurosci 123(6):1325–1338
Silva MR, Bernardi MM, Cruz-Casallas PE, Felicio LF (2003) Pimozide injections into the nucleus accumbens disrupt maternal behaviour in lactating rats. Pharmacol Toxicol 93(1):42–47
Simerly RB, Swanson LW (1986) The organization of neural inputs to the medial preoptic nucleus of the rat. J Comp Neurol 246(3):312–342
Simerly RB1, Swanson LW (1988) Projections of the medial preoptic nucleus: a Phaseolus vulgaris leucoagglutinin anterograde tract-tracing study in the rat. J Comp Neurol 270(2):209–242
Singer LM, Brodzinsky DM, Ramsay D, Steir M, Waters E (1985) Mother-infant attachment in adoptive families. Child Dev 56(6):1543–1551
Slade A (2005) Parental reflective functioning: an introduction. Attach Hum Dev 7(3):269–281
Slattery DA, Neumann ID (2008) No stress please! Mechanisms of stress hyporesponsiveness of the maternal brain. J Physiol 586(2):377–385
Smith CD, Lonstein JS (2008) Contact with infants modulates anxiety-generated c-fos activity in the brains of postpartum rats. Behav Brain Res 190(2):193–200
Steiner JE, Glaser D, Hawilo ME, Berridge KC (2001) Comparative expression of hedonic impact: affective reactions to taste by human infants and other primates. Neurosci Biobehav Rev 25(1):53–74
Stern JM (1997) Offspring-induced nurturance: animal-human parallels. Dev Psychobiol 31(1):19–37
Stern JM, Kolunie JM (1989) Perioral anesthesia disrupts maternal behavior during early lactation in Long-Evans rats. Behav Neur Biol 52(1):20–38
Storey AE, Walsh CJ, Quinton RL, Wynne-Edwards KE (2000) Hormonal correlates of paternal responsiveness in new and expectant fathers. Evol Hum Behav 21:79–91
Strathearn L, Fonagy P, Amico J, Montague PR (2009) Adult attachment predicts maternal brain and oxytocin response to infant Cues. Neuropsychopharmacology 34:2655–2666
Swain JE (2008) Baby stimuli and the parent brain: functional neuroimaging of the neural substrates of parent-infant attachment. Psychiatry 5(8):28–36
Terkel J, Rosenblatt JS (1972) Humoral factors underlying maternal behavior at parturition: cross transfusion between freely moving rats. J Comp Physiol Psychol 80:365–371
Terkel J, Bridges RS, Sawyer CH (1979) Effects of transecting lateral neural connections of the-medial preoptic area onmaternal behavior in the rat: nest building, pup retrieval and prolactin secretion. Brain Res 169(2):369–380
Troisi A, D'Amato FR, Carnera A, Trinca L (1988) Maternal aggression by lactating group-living Japanese macaque females. Horm Behav 22(4):444–452
Uvnäs-Moberg K, Eriksson M (1996) Breastfeeding: physiological, endocrine and behavioural adaptations caused by oxytocin and local neurogenic activity in the nipple and mammary gland. Acta Paediatr 85(5):525–530
van Ijzendoorn MH, Juffer F (2006) The emanuel Miller memorial lecture 2006: adoption as intervention. Meta-analytic evidence for massive catch-up and plasticity in physical, socio-emotional, and cognitive development. J Child Psychol Psychiatry 43:213–223
Weisbard C, Goy RW (1976) Effect of parturition and group composition on competitive drinking order in stumptail macaques (*Macaca arctoides*). Folia Primatol (Basel) 25(2–3):95–121
Wiesner BP, Sheard NM (1933) Maternal behaviour in the rat. Oliver & Boyd, Edinburgh

Wilsoncroft WE (1969) Babies by bar-press: maternal behavior in the rat. Behav Res Meth Instrum 1:229–230
Winnicott DW (1956) Primary maternal preoccupation. In: Winnicott DW (ed) Through paediatrics to psycho-analysis. Hogarth, London, pp 300–305
Wojnar D (2004) Maternal Perceptions of early breastfeeding experiences and breastfeeding outcomes at 6 weeks. Clin Effectiveness Nursing 8:93–100
World Health Organization (2009) Mental health aspects of women's reproductive health: a global review of the literature. http://whqlibdoc.who.int/publications/2009/9789241563567eng.pdf. Accessed Oct 2013

Chapter 11
Role of Sensory, Social, and Hormonal Signals from the Mother on the Development of Offspring

Angel I. Melo

Abstract For mammals, sensory, social, and hormonal experience early in life is essential for the continuity of the infant's development. These experiences come from the mother through maternal care, and have enduring effects on the physiology and behavior of the adult organism. Disturbing the mother–offspring interaction by maternal deprivation (neglect) or exposure to adverse events as chronic stress, maltreatment, or sexual abuse has negative effects on the mental, psychological, physiological, and behavioral health. Indeed, these kinds of negative experiences can be the source of some neuropsychiatric diseases as depression, anxiety, impulsive aggression, and antisocial behavior. The purpose of this chapter is to review the most relevant evidence that supports the participation of cues from the mother and/or littermates during the postnatal preweaning period for the development of nervous system of the offspring. These findings come from the most frequently utilized experimental paradigms used in animal models, such as natural variations in maternal behavior, handling, partial maternal deprivation, and total maternal deprivation and artificial rearing. Through the use of these experimental procedures, it is possible to positively (handling paradigm), or negatively (maternal deprivation paradigms), affect the offspring's development. Finally, this chapter reviews the importance of the hormones that pups ingest through the maternal milk during early lactation on the development of several physiological systems, including the immune, endocrine systems, as well as on the adult behavior of the offspring.

In most mammalian species, social and sensory experiences throughout life affect the individual's physiology and behavior. However, the timing of the experience matters. Experiences acquired during critical or sensitive periods of development often have the most profound and enduring effects in nonhuman primates (Harlow et al. 1965), in rodents (Denenberg et al. 1950; Levine 1957; Weininger 1954), and in sows, ewes, guinea pigs, rabbits, and humans (see reviews; González-Mariscal and Kinsley 2009; González-Mariscal and Melo 2013; Barret and Fleming 2011; Fleming et al. 2002). The lengthy prenatal and postnatal periods provide an opportunity for mothers to influence their offspring through a variety

A. I. Melo (✉)
Centro de Investigación en Reproducción Animal, CINVESTAV-Laboratorio Tlaxcala, Universidad Autónoma de Tlaxcala, Tlaxcala, México
e-mail: meloangel@gmail.com

M. C. Antonelli (ed.), *Perinatal Programming of Neurodevelopment,*
Advances in Neurobiology 10, DOI 10.1007/978-1-4939-1372-5_11,

of mechanisms. During gestation, mothers' influences on fetal development are critical for growth and development and variations in these influences occur as in prenatal stress or maternal malnutrition, and can have long-term consequences for the offspring's physiological and psychological health (Markhan and Koenig 2011; Bosch et al. 2006; Almeida et al. 1996). During the postnatal preweaning period, the main source of maternal influence derives from mothers' milk and the sensory exchanges that occur during mother–infant interactions (Ellis et al. 1996; see Cirulli et al. 2003; Faturi et al. 2010; Caldji et al. 2000; González-Mariscal and Melo 2013). These experiences can influence gene expression and the structure of the emerging brain, as well as the programming of hormonal axes, growth, and the development of organ morphology and/or metabolic pathways.

Mother–offspring interaction involves a synchrony of interchanges of sensory, social, and hormonal stimuli between the mother and her offspring, where the offspring provides the mother with sensory and social stimulation necessary for the establishment and maintenance of maternal behavior and the mother provides to the offspring sensory, social, and hormonal stimulation which promotes continuity in their development (see González-Mariscal and Poindron 2002; Champagne and Meaney 2007; Fleming et al. 2002; Rosenblatt and Lehrman 1963). Mothers adjust their maternal behavior to be consistent with characteristics of the pups and their stage of development. During the first 2 weeks of lactation, when the offspring remain in a "hyporesponsive stress response" (HRSR) stage, they depend completely on the maternal care for their development and survival. The mother approaches the pups frequently, she provides sensory stimulation through body and genital licking (which stimulates urination and defecation), and she groups them into the nest and crouches over them, often aiding their attachment to her teats. In contrast, during the last 2 weeks of lactation the mother spends less time caring for the pups as they become increasingly independent; i.e., they initiate approaches to mother's ventrum from a distance, in order to grasp a nipple and suckle, and they begin to obtain solid foods from mothers snout region and fur (Rosenblatt 1965; Rosenblatt et al. 1985; Alberts and Gubernick 1983). During the period of active mother–litter interactions, the sensory (tactile) and social cues from the mother and/or littermates favor the fine-tuning of specific systems in the nervous system of the young, acting on the genome and the future expression of genes. These observations raise the issue as to (1) how the sensory and social stimuli affect the development of the organism? (2) which specific mechanisms are affected by the early experience? and (3) how are these characteristics transmitted to the next generation? Although there is already a considerable literature that addresses these questions, many questions remain. These are some of the issues that this chapter addresses.

The primary endpoint of this chapter is a review of the experimental evidence that supports the participation of sensory, social, and hormonal cues from the mother and the littermates during the postnatal preweaning period, which is a time window for the effects of early experience on the development of the offspring. This is accomplished through the use of several experimental paradigms that have been devised in order to understand the developmental origins of the physiological, neuroendocrinological, and behavioral phenotype of the offspring. Under laboratory conditions, it is possible to manipulate the environment around the infants in order to change the direction and magnitude of development. Here, we describe the findings from some of the most frequently used experimental procedures that are used to address the above questions. In order to positively affect the offspring's development, the "handling" procedure has been utilized; in contrast, in order to negatively impact on development, maternal separation (MS; early social isolation) has been used. The latter can involve partial separation (3–6 h per day during the first 10–14 postnatal days, PND) or total separation (24 h per day, every day, during PND 3–4 to 22, using the artificial rearing (AR) system). Finally, in addition to the obvious fact that maternal milk is essential for nutrition and growth, the role of bioactive components in milk (for example, hormones) is an emerging issue due to their potential link to metabolic, endocrine, and mental health disorders later in life. Therefore, this chapter ends with a discussion of the hormones (mainly steroids and peptides) that are found in maternal milk, the importance of their levels in milk throughout lactation, and their long-term biological effects on the development of the offspring.

There are vast species differences in the amount of care that mothers provide to their offspring. In some species, mothers spend most of their waking hours caring for their newborn babies, as in many rodents, carnivores, and primates (Rosenblatt 1967; Rosenblatt and Lehrman 1963; Bridges 1996); however, in others, like rabbits, mothers spend at most 3–5 min daily nursing their young. In ungulates, which are precocial, mothers engage in mostly passive behaviors, allowing the young to suckle but without extensive licking or contact stimulation (Poindron et al. 1980; Lévy et al. 1996). In addition to species differences, even within species there are also wide variations in the amount of time mothers spend with their new offspring. In the case of both interspecies and intraspecies variation, mothers' behaviors influence how the offspring will develop. In an elegant series of studies, Meaney and his colleagues have exploited the natural variations that occur in the level of maternal licking behavior within a population of rats, in order to examine the role of gene expression and epigenetics in regulation of variation in the subsequent generation (Meaney 2001; Meaney and Aitken 1985; Meaney et al. 1988, 1989; Champagne et al. 2003). In these studies, a population of dams and their litters are observed using one-zero time-sampling procedures for 8 hours daily across the first 10 PND and the levels of licking and crouching that the mother exhibits is recorded. Mothers and their litters are selected for further analysis based on the level of licking mothers give to their litters. Hence, mothers providing more than one standard deviation above the mean of licking for the entire population are designated as dams that provide high levels of pup body licking and arched back nursing posture (high LG-ABN), whereas mothers that provide licking levels one standard deviation below the mean of the population are designated as dams providing low levels of maternal behavior (low LG-ABN; see later). As reported below, comparisons of young reared by high LG-ABN and low LG-ABN yield fascinating differences in development at the behavioral, molecular, and neuroanatomical levels.

In contrast to studies of natural variations, there are also studies that explore experimentally which features of the early maternal and nest environment contribute to the differences in offspring development. To study the role of early experience on the development of organisms, there are experimental paradigms that manipulate the amount of sensory or social cues that pups receive during the postnatal preweaning period. For instance, under laboratory conditions, there are "positive" effects on the development of the offspring when early stimulation is increased, by (1) reducing the number of pups in the litter (Leigh and Hofer 1973) or (2) separating the pups from the mother for a very brief period of time (15 min; handling which promotes enhanced maternal licking when the pups are returned to the mother, Lee and Williams 1974). On the other hand, it is possible to induce "negative" effects on the offspring by decreasing the amount of stimulation provided to pups by (1) increasing the number of pups within the litter (Fleming et al. 1979), (2) inducing anosmia in the mother (which decreases her licking, Fleming and Rosenblatt 1974; Fleming et al. 1979; Moore 1984), (3) separating the pups from the mother for a portion of each day (3–5 h daily from PND 2–4 to 14; partial maternal separation), or (4) separating the pups from the mother and nest totally, and rearing pups in an AR apparatus (Hall 1975; Total maternal separation, see later). The mechanisms through which these effects are mediated still are not well understood. Furthermore, we know, for instance, that maternal milk from early, but not late, lactation contains bioactive compounds (cytokines, enzymes, and hormones such as prolactin (PRL), leptin, corticosterone, etc.) that pups ingest during in the first days (rodents) or months (primates, humans) of life, and these agents also participate in the development of immune, neuroendocrine, and behavioral profiles of the developing offspring (Melo et al. 2009; Shah et al. 1988; Ellis and Picciano 1995; Ellis et al. 1996; Grosvenor et al. 1992; Hazum 1983; Koldovsky 1980; Brummelte et al. 2006).

Finally, although most of the work in this area has focused on animal models, many of the principles derived from this work apply as well to humans (Teicher et al. 2003, 2006; Weiss et al. 1999; see Barrett and Fleming 2011; González-Mariscal and Melo 2013), where inadequate or abusive care by mothers and fathers puts children at risk for many problems later in life associated with disrupted stress responsiveness, inadequate growth and brain development, as well as problems of affect, cognition, and executive functions; many of these effects of parental neglect have also been described in the rat models (Donatelli et al. 2010; Glaser 2000; Teicher et al. 2003, 2006; Green et al. 2010). In humans, as well as in nonhumans, there is also evidence of transgenerational effects of parenting style or parental separations (see van Ijzendoorn 1992; Champagne 2008).

To summarize, there is now ample evidence that how the young develop in terms of their physiology, brain structure, neural function, and behavior depends on the adequacy of the maternal care received early in life. Included in this care are the sensory, behavioral, and physiological influences whose effects can be sustained through multiple generations. Their effects are varied and are likely mediated through both genomic and nongenomic mechanisms. These are described below.

11.1 Natural Variations in Maternal Behavior

In natural conditions, female virgin rats show an aversion toward pups; however, as pregnancy advances, the neophobia decreases and the pups' attractiveness increases (Fleming and Rosenblatt 1974; Fleming and Luebke 1981). As is well known, in late pregnancy there are changes in the levels of hormones that trigger the beginning of maternal care. These changes consist in a decrease of progesterone, an increase in estrogen, PRL, and oxytocin (OT; Ben-Jonathan et al. 1989; Rosenblatt et al. 1979). The neophobia is decreased by continued exposure to odor derived from pups (pups sensitization paradigm), or by giving a hormonal regimen to an ovariectomized female that mimics the hormonal changes that occur in the latter phase of pregnancy. In rats, inducing anosmia during pregnancy by removing the main olfactory bulbs (Fleming and Rosenblatt 1974; Pollack and Sachs 1975) or by destroying the olfactory epithelium with ZnSO4 spray (Benuck and Rowe 1975) slightly disrupted the expression of maternal care. On the other hand, inducing anosmia in virgin rats by lateral olfactory lesion or by ZnSO4 spray on the olfactory epithelium promoted the expression of maternal care in a sensitization test (Fleming and Rosenblatt 1974; Mayer and Rosenblatt 1977). The most interesting data are that when both the main and the accessory olfactory systems were blocked, a facilitation of maternal care was produced (Fleming et al. 1979). These data suggested that both olfactory systems are involved in the sensitization process through reducing neophobia. Similar effects have been reported in rabbits (Chirino et al. 1999, 2000, 2001). Thus, removing the accessory olfactory bulbs in intact females did not disturb maternal care toward their pups (Chirino et al. 1999). However, the same procedure induced a greater facilitation of maternal behavior in virgins, but not intact female rabbits (Chirino et al. 2000). In contrast with the findings in rats, administration of ZnSO4 to virgin rabbits, a procedure that induces anosmia, did not facilitate maternal behavior (Chirino et al. 2001). Once maternal care is initiated, there are differences among mothers in the intensity of the behavior expressed. As indicated earlier, some of the differences may derive from differences in the extent to which they were cared for by their own mothers, in which pups that were licked a lot by high LG-ABN dams come to lick their offspring more than do pups from low LG-ABN dams (Champagne and Meaney 2001; Champagne et al. 2003, 2006). That differences in hormones of parturition do not mediate these differences is indicated by the fact that adult virgin females that were reared by high LG-ABN and then in adulthood exposed with foster pups (sensitization paradigm) show significantly shorter latencies to display maternal behavior and exhibit higher levels of licking of foster pups, compared to females reared by low LG-ABN mothers (Francis et al. 1999). Moreover, there was a negative correlation between the latency to show pup retrieval in the sensitization paradigm, as virgins, and the frequency of pup licking/grooming toward their own pups after they gave birth. In terms of possible mechanisms mediating these behavioral effects, subsequent studies using the same high low-licking paradigm find that in comparison to their low-licking counterparts, high LG-ABN females showed higher OT receptor levels in the medial preoptic area (MPOA; area involved in the regulation of maternal expression), the lateral septum, the central nucleus of the amygdala, the paraventricular nucleus, and the bed nucleus of the stria terminalis (Champagne and Meaney 2001). Recent work shows that the level of maternal care also affects estrogen and OT receptors in the female offspring. Virgin female offspring of high, as opposed to low LG-ABN mothers show higher levels of estrogen receptor alpha in the MPOA (Champagne et al. 2003). These effects are retained even after cross-fostering, indicating that the effects are experience based: When offspring of high and low LG-ABN mothers are cross-fostered at birth, the biological offspring of low LG-ABN mothers fostered to high LG-ABN show increased estrogen receptor alpha in the MPOA. Epigenetic effects can also be seen at the molecular level, for example, female offspring from high LG-ABN mothers show a higher expression of cyclic AMP-response element binding protein (CREB)-binding protein (CBP), a higher acetylation of histones, and reduced DBA methylation pattern of the gr exon 17 promoter in the hippocampus (Champagne et al. 2003). Moreover, higher levels of cytosine methylation across the ER alpha 1b promoter have been found in the offspring of low LG-ABN, compared to high LG-ABN mothers (Champagne et al. 2006). Based on these results, it is clear that mothers' early experiences of being cared for affect the quality of mothering that they provide to their offspring and that these effects are mediated through experience-based alterations in the neural circuits that underlie mothering in the adult organism.

In addition to having effects on the neural substrates of mothering, early experiences of being mothered also affect other important endocrine and behavioral systems. In contrast to offspring that received low levels of maternal care early in life, offspring that received high levels of LG-ABN also show a reduced stress response, reflected in a moderate hypothalamic–pituitary–adrenal and behavioral response to a stressful situation and lower levels of fear responses. In addition, they show enhanced cognitive ability and more efficient spatial learning (Champagne et al. 2003; Liu et al. 1997, 2000; Caldji et al. 1998). At the cellular level, high LG-ABN animals, show reduced plasma ACTH and CORT responses to restraint stress, increased hippocampal glucocorticoid (GC) receptor messenger ribonucleic acid (mRNA) expression, enhanced GC negative feedback sensitivity, and decreased hypothalamic CRH mRNA levels (see Cirulli et al. 2003; Faturi et al. 2010). Furthermore, these rats also show increased CRH receptor levels in the locus coeruleus and decreased central gamma-aminobutyric acid (GABA)/benzodiazepine receptor levels (Sapolsky and Meaney 1986).

11.2 Handling Paradigm

Neonatal handling is an experimental procedure used to understand how early-life experience can positively affect the fine-tuned neurobehavioral development and place animals on a pathway to prevent pathology. This procedure involves briefly separating pups from the mother and the nest, and placing them in individual cages for 15 min a day during the first 2 weeks of life. Once pups are returned to the nest, mothers show more frequent nursing bouts and spend significantly more time licking them, than do mothers of nonhandled pups. Thus, at the time of weaning, handled pups have experienced more sensory stimulation than nonhandled pups, and their behavior looks similar to what has been reported for the high LG-ABN pups (Meaney et al. 1988, 1989, 1996; Liu et al. 1997; Francis and Meaney 1999; Pryce et al. 2001). Compared to nonhandled animals, handled pups show increased exploratory behavior in novel environments, reduced defecation and urination in an open field, a high degree of exploration in the hole board test, reduced neophobia and startle responsivity, and reduced conditioned taste aversion. They also had a more rapid return of corticosterone to basal levels after restraint stress, increased impulsive behavior in adolescence, reduced helplessness behavior, and reduced emotionalty, and these effects persist throughout life (Levine and Lewis 1959; Hess et al. 1969; Meaney et al. 1996; Costela et al. 1995; Weinberg et al. 1978; Meaney 2001; Beane et al. 2002; Macri et al. 2004; Pryce and Feldon 2003; Kosten et al. 2007). Early handling also has an effect on neural plasticity, producing a greater amplitude of long-term potentiation (LTP) in the hippocampus (Wilson et al. 1986) and on the hypothalamic–pituitary–adrenal (HPA) stress response and its mechanisms (Levine 1957; Meaney et al. 1996). For instance, handled as compared to nonhandled pups, when shocked, show higher levels of GC, but a more rapid return to basal levels, an increased number of GC receptors in the hippocampus and frontal cortex, and lower basal hypothalamic CRH mRNA and protein levels, in comparison with nonhandled rats (Plotsky and Meaney 1993; Meaney and Aitken 1985; Viau et al. 1993). Associated with these changes in the HPA axis, there are elevations as well in GABA-A receptors levels in the locus coeruleus and the nucleus of the tractus solitarius as well as reduced CBZ receptors sites in the central and lateral nucleus of thc amygdala, the frontal cortex, and the LC and NTS (Caldji et al. 2000). Furthermore, pups that had experienced prenatal stress but then later received the handling procedure exhibited decreases in stress behavior and in stress-induced analgesia, compared to nonhandled pups (Sternberg and Ridgway 2003). In summary, handled rats appear to be better adapted to the novel environments, showing reduced exposure to GCs after a stressor and hence reduced risk of incurring its negative effects on the central nervous system. Recently, Pryce and Feldon (2003) proposed a unified nomenclature in order to facilitate discussion on this topic. Thus, they proposed the following: (1) MS should be used to describe the separation of the intact litter from the mother for 1 or more hours per day across the first 14 PND, (2) a single MS as a separation of the intact litter from the mother for a single 24-h period, (3) isolation or maternal deprivation of the pups from the dam and their littermates for one or more hours per day (1, 3, 6, 8, 12 h) across the first 14 PND (SMD). The section that follows describes the findings from SMD and the experimental paradigm of total maternal deprivation and artificial rearing (AR).

11.3 Short or Partial Maternal Deprivation

Basic neurobiological research has established that the mother acts as an external regulator of the infant's states. The mother–infant interaction at least during the first 2 weeks of life is essential for the normal development of the organism, and mainly, of the HPA axis in rats and mice (Gutman and Nemeroff 2002; Macri et al. 2004, 2008; Matthews and Robbins 2003). Sensory and feeding stimulation from the mother and littermates have been shown to sustain the HPA axis of the pups in a hyporesponsive state (Brummelte et al. 2006, 2010; Casolini et al. 1997; Catalani et al. 1993, 2000; Meerlo et al. 2001). When offspring are separated from the mother, or from the mother and littermates, they show behavioral and neurochemical changes immediately after the separation (acute effects) and, if separation is prolonged during the first 14 PND, the effects are negative and can persist for life. Concerning the *acute effects,* it has been reported that rat pups are more active in an unfamiliar area, showing increased locomotor and exploratory behavior, and, when tested as a group, they show more self-grooming, and more defecation and urination (Hofer 1972). They habituate in the box-test at significantly slower rates and the latency to onset of active sleep (REM) is prolonged (Hofer 1973a, b). In addition, the cardiac and respiratory rates of pups separated at 14 PND fell by 40% (Hofer 1973b). All of these separation effects are independent of nutrition and occur despite maintenance of body temperature at nest levels during the separation period. If temperature falls 2–3 °C, activity and other behaviors decrease, and with separation resting cardiac rate and respiratory rates fall 40% within 12–18 h after separation. If nonlactating females or littermates are present during the separation period, most of the behavioral deficits are prevented, and if pups are infused with milk through gastric cannulae, the cardiac effects are prevented (Hofer 1970, 1972, 1973b, c). Other neuroendocrine effects can also be observed within 4–8 h after the initiation of separation, primarily in an enhanced adrenal sensitivity to adrenocorticotropic hormone (ACTH) and a decreased expression of central HPA markers, such as corticotropin-releasing hormone (CRH; Schmidt et al. 2004). By contrast, during the second half of the separation period, negative feedback mechanisms become activated to restrain the continued increase in ACTH and corticosterone release (Schmidt et al. 2004). But, what happens when the MS is repeated during a prolonged period of time during the first 10–14 PND? Concerning the *long-term effects of SMD,* though initially, the main goal of the separation paradigm was to assess the role of maternal care in the development of offspring, and in terms of the handling or very short-term separation, this continues to be a goal. However, the separation paradigm has also been conceived as a chronic stressor designed to affect the development of the offspring's stress susceptibility and onset of stress-related psychiatric disorders in adulthood. There are large variations in the outcomes that this basic paradigm produces which depend on differences in the age of the MD, in the frequency and duration of the separations, subtle differences in procedure, and in animal husbandry. Thus, rats and mice exposed to repeated neonatal SMD show different cognitive, behavioral, and neurodocrine and neurochemical effects, with longer separations being more detrimental than shorter separations (Matthews et al. 1996a, b; Ladd et al. 1996; Hall et al. 1999). Animals exposed to repeated MS of 180–360 min/day for the first 2 weeks of life, in comparison to no separation show significantly increased plasma ACTH and corticosterone (CORT) response to stressful stimuli, and more prolonged responses, an effect opposite to handling (Sapolsky et al. 1984; Anisman et al. 1998). The longer periods of separation also result in decreased GC receptor binding in both the hippocampus and the hypothalamus, and a higher basal CRF mRNA levels in the hypothalamus with higher titers of CRF in the median eminence (Plotsky and Meaney 1993; see Pryce and Feldon 2003; Faturi et al. 2010). However, the effects of prolonged maternal separations are not always consistent. A number of papers also indicate that even 3–4 h daily separations performed from birth until weaning result in behavioral and endocrine changes in the same, rather than in the opposite, direction as produced by handling (see Lehmann and Feldon 2000; Pryce et al. 2001; Pryce and Feldon 2003; Faturi et al. 2010).

11.4 Total Maternal Deprivation and Artificial Rearing (AR)

This paradigm involves a complete maternal deprivation paradigm (24 h daily) and AR of pups isolated in individual cups, without any contact with their mother or littermates (Hall 1998). Separated animals are then compared to their maternally reared (MR) siblings. AR protocol has been used in previous studies (Gonzalez et al. 2001; Lévy et al. 2003; Lovic and Fleming 2004; Lovic et al. 2006; Melo et al. 2006, 2009). In this protocol, pups of 3–4 PND are removed from the nest, and three pups then undergo a stomach or cheek implant to enable formula delivery, while the last pup is marked and returned to the nest to be reared by its mother; this latter treatment comprises the intact MR control group. Two of the pups that had received the implant are reared artificially; one receives minimal anogenital stimulation to induce urination and defecation (AR), and the other receives maximal stimulation, comprising 5–8 tactile stimulation in the body for 45 s (AR-Max) or minimal tactile stimulation along with social stimulation (AR-Social), in which they are reared in the company of two pups of the same age from a donor mother. The third pup has the tube cut off just outside the skin and is returned to the nest to be reared by its mother; this is the sham control treatment (MR-Sham). AR pups are housed individually in open-top plastic cups with corncob bedding, which fit into a second weighted cup floating in a temperature-controlled water bath (36–40 °C). The tube is connected to syringes containing Messer milk formula (Messer et al. 1969; Smart et al. 1986) mounted in a Harvard Apparatus syringe pump programmed to infuse the diet (velocity of infusion is calculated as a function of weight of the all pups) for 10 min every hour, 24 h daily. All pups (AR and MR) are weaned by PND 22, housed in 2–3 animals per acrylic cages and left undisturbed until adulthood.

In comparison to the paradigm of short maternal deprivation, this paradigm has a reduced effect on the HPA axis during the period of separation (Lomanowska et al. 2010; Spear et al. 1989). In addition, the AR or "pup-in-a-cup" procedures do not involve deprivation of nutrients, animals are provided with stable humidity and temperature. Thus, the body weight (Diaz et al. 1981, 1983; Haney et al. 1986; Melo et al. 2009a), liver and kidney tissues (Díaz et al. 1983), and liver protein and DNA content, diacyl glycerolacyl transferase and fatty-acid CoA ligase activities (Haney et al. 1986) are similar to those obtained from pups reared by their mothers. The results obtained to date, enumerated below, are likely attributable to the sensory and/or social deprivation during postnatal preweaning period, rather than to other factors, like undernutrition, loss of body temperature, or chronic stress. Total maternal deprivation and AR produces a wide variety of behavioral and physiologic effects. Behaviorally, animals reared without their mothers (AR) in comparison to mother-reared animals show (1) deficits in maternal behavior toward their own offspring, i.e., they spend less time over their pups and less time licking them (Gonzalez et al. 2001; Lovic and Fleming 2004; Melo et al. 2006); (2) deficits in maternal memory through interaction with pups (Melo et al. 2006; Numan 2006); (3) reduction in attention and increases in activity in adulthood (Lovic and Fleming 2004); (4) hyperactivity and reduced "fear" in a plus maze task (Burton et al. 2006); (5) marked inattention in prepulse inhibition and attention set-shifting tasks (Lovic and Fleming 2004), deficits in social learning (Lévy et al. 2003; Melo et al. 2006), and social behaviors (Gonzalez et al. 2001; Lovic and Fleming 2004); and (6) high levels of maternal and offensive aggressiveness (Melo et al. 2009a). Physiologically, the effects of AR are equally striking. Compared to postpartum MR rats, AR rats show increased basal extracellular dopamine (DA) levels in nucleus accumbens and reduced pup-related DA elevations. These DA changes are related to disrupted maternal behavior (Afonso et al. 2008). In addition, AR rats treated with DA agonists such as amphetamine show dose-dependent increases in locomotor activity above and beyond what is typically observed in MR rats (Lovic et al. 2006). Most of these effects are prevented partially or completely by providing pups with daily additional stroking stimulation with a paintbrush 5–8 times per day designed to simulate mother's licking (Fleming et al. 2002; Gonzalez et al. 2001; Lévy et al. 2003) or by providing peer-derived social stimulation during AR (Melo et al. 2006, 2009a). Interestingly, these behavioral effects of deprivation are similar in many important respects to the institutional inattention/overactivity syndrome seen in infants raised in institutions who are then adopted into enriched homes (Fries and Pollak 2004; Gunnar et al. 2001; Rutter and O'Connor 2004). The effects of AR are also transferred across generations: Daughters of AR mothers show patterns of maternal behavior that are similar to their own mothers (Gonzalez et al. 2001).

The above data strongly suggest that specific cues associated with the interaction between the mother and the offspring, such as sensory (tactile) stimuli, social contact with the mother and littermates, feeding, and rhythmicity of milk delivery, are involved in the development of the offspring. However, when pups are deprived of the above stimuli during AR, they are also deprived of maternal milk. In the section that follows, we describe additional ways that pups can be affected by their mothers, not by maternal behavior, but rather, by aspects of the mother's physiology.

11.5 Hormones in Maternal Milk

Maternal milk provides the young with essential nutrients (fat, proteins, lactose, minerals) to support optimal offspring growth and survival during postnatal life. It is well-known, for well over 50 years, that maternal milk contains a wide array of bioactive compounds such as cytokines, growth factors, vitamins, immunoglobulins, enzymes, and hormones that contribute to the growth, development, and maturation of the newborn (Grosvenor et al. 1992; Hazum 1983; see Koldovsky 1980). The hormones present in milk of several mammalian species are: PRL, gonadotrophins, thyroid-stimulating hormone (TSH), adrenocorticotrophic hormones, oxytocin (OT), thyroid-releasing hormones, luteinizing hormone release hormones (LH-RH), gonadotrophin release hormones (GnRH), thyroid hormones (TH), estradiol, estriol, progesterone, leptin, testosterone, corticosterone, among others (see Grosvenor et al. 1992; Table 11.1). Milk also provides substances that induce immune competence in the offspring (Goldman and Frawley 1996; Goldman 1993; Ellis et al. 1996; Ellis and Picciano 1995).

There are two primary potential sources of the constituents and bioactive components in milk; these are (1) synthetized in the mammary gland from precursors in blood and/or (2) transferred directly from blood to milk through the alveolar–blood barrier. These compounds likely enter into the milk through the alveolar epithelium of the mammary gland by passive diffusion, bound to carrier proteins or by active transport (see Polk 1992; Ellis et al. 1996; Ellis and Picciano 1995).

Once the pups ingest the milk, the milk's components are absorbed through the neonatal gastro intestinal tract (GIT) and are then transferred to the pup circulation in biologically active forms. There is a large literature that reports that the GIT of suckling mammals (mainly in rats and mice) has the ability to absorb several large molecular proteins, as hormones, without changes in its immunological properties (see Koldovsky 1980). Although it is known that lipid-soluble hormones are converted enzymatically to less active forms in the intestinal and hepatic tissues (Baumrucker and Magliaro-Macrina 2011), this is not well established for most of the maternal milk components. Furthermore, this GIT ability appears to be by the low proteolytic activities and a higher permeability for macromolecules in neonates (see Koldovsky 1996). It is possible that components travel throughout the blood to different targets, such as brain, immune tissues, kidney, intestine, and liver, thereby acting on them through their binding to specific receptors.

Although there is little information on the possible biological role of these components, the hypothesis that they influence the development of offspring is often inferred from their relative abundance in early milk compared to mature/late milk and maternal serum. The high concentration of most bioactive compounds in milk declines as lactation progresses while the offspring's milk consumption increases. The level of concentration of the milk's components coincide with the level of absorption in the GIT, i.e., during early lactation there is a high concentration of these components in the milk that is correlated with the level of absorption in the GIT of the offspring, while during late lactation there is a low concentration of these bioactive components, along with a lower capacity for their absorption in the GIT.

Table 11.1 Hormones in milk

Hormones	Species	References
Gonadotropin-releasing hormones (GnRH)	Rats, bovines, mares, humans	McGarrigle and Lachelin (1983); Wolford and Argoudelis (1979); Monk et al. (1975); Heap et al. (1983)
Somatostatin	Rats, bovines, sheep, humans	Holst et al. (1990); Koch et al. (1991); Werner et al. (1988)
Growth hormone-releasing factor (GRF)	Human, bovines, sheep, rats	Amarant et al. (1982); Baram et al. (1977); Sarda and Nair (1981); Werner et al. (1986)
Thyrotropin-releasing hormone (TRH)	Rats, bovine, humans	Amarant et al. (1982); Strbak (1985)
TSH T_4 T_3	Rats	Krulich et al. (1977); Tenore et al. (1981)
Prolactin (PRL)	Rats, bovines, goats, sheep, rabbits, humans	Geschickter and Lewis (1936); Gala et al. (1975); Malven and Mc Murtry (1974); Malven (1977); Grosvenor and Whitworth (1976); Ostrom (1990); Ollivier-Bousquet (1993)
Prostaglandin	Rats, bovines, humans	Chappell et al. (1983); Simmons et al. (1979)
Estrogen 17 β-Estradiol (E2), Estrone (E1), Estriol (E3)	Bovines, humans	Koldovsky and Thornbur (1987); Nilsson et al. (1978); Wolford and Argoudelis (1979)
Progesterone	Bovines, goats	Koldovsky and Thornbur (1987); Hruska and Veznik (1983); Nilsson et al. (1978); Darling et al. (1974); Heap et al. (1973); Zaied et al. (1979)
Oxytocin	Humans	Takeda et al. (1986)
Testosterone	Bovines	Hoffman and Rattenberger (1977); Fritsche and Steinhart (1999)
Glucocorticoids (corticosterone, cortisol)	Bovines, humans	Kulski and Hartmann (1981); Tucker and Schwalm (1977); Ratsimamanga et al. (1956)
Luteinizing hormone-releasing hormone (LH-RH)	Rats	Amarant et al. (1982)
Insulin	Bovines, humans	Ballard et al. (1982); Malven (1977); Cevreska et al. (1975)
Calcitonin	Humans, rats	Werner et al. (1982)
Erythropoietin	Humans, sheep, rats, mice	Grosvenor et al. (1993); Dickson et al. (1985); Bielecki et al. (1972); Grant (1952)

Table 11.1 (continued)

Hormones	Species	References
Melatonin	Humans, bovines, goats	Eriksson et al. (1998); Valtonen et al. (2003)
Epidermal growth factor (EGF)	Rats, humans	Bohuslav et al. (2000); Carpenter (1980); Puccio and Lehy (1988); Pollack et al. (1987)
Insulin-like growth factors (IGFs)	Humans, bovines, rats	Malven et al. (1987); Shing and Klagsbrun (1984)
Leptin	Humans, ruminants, pigs	Chilliard et al. (2001); Wolinski et al. (2014); Savino et al. (2013)
Morphine	Bovines, humans	Henschen et al. (1979); Hazum (1983)
Ghrelin	Humans	Kierson et al. (2006); Aydin et al. (2006)
Cholecystokinin	Humans	Kierson et al. (2006)
Thyroprotein	Bovines	Archibald (1945)

These observations raise the question as to (1) why are these compounds present in maternal milk? (2) why is their concentration higher in early lactation than in late lactation? (3) do they have a biological function? and, (4) if they do, where and how do they act? To consider a milk component to have a biologic function, it has to fulfill several criteria (Peaker and Naville 1991). It has to: (1) be present and active in milk, (2) have a functional effect on the development of the offspring as a neonate and/or later in life, (3) become active or retain activity in the target organs, (4) be present in greater concentration in milk compared to the concentration in serum, and (5) the effect should be eliminated when these components are not present in the milk and restored when it is replaced (Peaker and Neville 1991; Ellis et al. 1996). These criteria of bioactivity can be fulfilled for some milk elements as with PRL, epidermal growth factor, prostaglandins, leptin, and corticosterone (Ellis et al. 1996). However, with others they may be bioactive although they may not fulfill all these criteria. In the following section, we present some evidence of specific hormones present in maternal milk that have biological effects on the development of the neonates.

11.5.1 Prolactin

PRL is a hormone produced in the anterior pituitary gland that functions to maintain milk production during lactation, and it has been identified in the milk of cows, sheep, goats, sows, rats, and humans (Grosvenor et al. 1992). PRL levels are high during early lactation and subsequently decline to the low levels during late lactation (Shyr et al. 1986; Malven and McMurtry 1974; Ellis and Picciano 1995; Ellis et al. 1996; Grosvenor and Whitworth 1976; Gala et al.1980; Kacsoh et al. 1991,

1993). Milk produced during early lactation by rats, bovines, and humans have high PRL bioactivity, which includes its several PRL variants (phosphorylated, glycosylated, and bound; Kacsoh et al. 1993; Ellis and Picciano 1995; Ellis et al. 1996). Milk PRL bioactivity exceeds estimates of PRL immunoreactivity by 2–6 fold in rats (Kacsoh et al. 1993), and by 1.4–4 fold in humans (Ellis and Picciano 1995). The different variants exist in different concentrations in the milk, suggesting either differential transfer from blood to milk or differential synthesis of PRL in the mammary gland itself (Ollivier-Bousquet et al. 1993; Ellis et al. 1996; see also Cox et al. 1996). There are three PRL variants or forms in both rat and human milk: phosphorylated, glycosylated, and bound (Ellis and Picciano 1995). The first one represents 60–80 % of the biologically active PRL only in human milk. The glycosylated PRL variant is also found in milk in humans and rats. There is evidence that the PRL that pups of rats and rabbits ingest through maternal milk is absorbed by the jejunum and ileum, possibly bound by PRL receptors on the epithelial cells (Nagano et al. 1995). The absorption of PRL into the intestine of the pup was confirmed by experiments in which radiolabeled PRL injected to the mother was tracked to her milk and ultimately to the serum of her pups. Once milk-derived PRL is absorbed, it is detected in the plasma 20–25 min later, with an estimated 16 % of the PRL present in milk being absorbed (Gonella et al. 1989; Grosvenor and Whitworth 1976; Whitworth and Grosvenor 1978; Malven and McMurtry 1974). Furthermore, recently it has been reported that ovine PRL injected systemically to lactating mothers on postpartum day 6 was detected 25–30 min later in serum from their pups (Melo et al. 2009b).

What is the effect of this maternal PRL that is absorbed from mothers' milk into the blood of the offspring? There are reports that milk-derived PRL modulates the development of the PRL secretory cells (Hoeffler et al. 1985) and tuberoinfundibular-dopamine (TIDA) system (Shah et al. 1988; Shyr et al. 1986; Shieh and Pan 1999; Ellis et al. 1996) of the pups during early life. Experimentally reducing the levels of PRL in maternal plasma and milk (without abolishing lactation) by administering low doses of bromocriptine (a dopaminergic agonist that inhibits PRL release from anterior pituitary), to lactating rats across lactation days 2–5, resulted in a marked suppression of DA turnover in the median eminence of the offspring, leading to increased plasma PRL concentration (hyperprolactinemia) in the offspring during the juvenile and adult stages. In addition, these females showed normal vaginal opening time but had irregular estrous cycles (Shah et al. 1988), and their ovarian estrogen secretion was disturbed (Crowley et al. 1990). Recently, it has been reported that female adult offspring of bromocriptine-treated mothers induced to display maternal behaviors through a pup sensitization paradigm had lower latencies to respond maternally than did vehicle-treated mothers. Furthermore, in comparison to female offspring of control mothers, female offspring of bromocriptine-treated mothers spent less time hovering over the pups, body licking and in close proximity to pups (Melo et al. 2009b). The simultaneous infusion of ovine PRL by osmotic minipump (Shyr et al. 1986; Shah et al. 1988) or sc injection of ovine PRL (Melo et al. 2009) to the mothers prevented the TIDA

(Shyr et al. 1986; Shah et al. 1988) or maternal (Melo et al. 2009b) effects of bromocriptine.

In addition to the well-established role of immunoglobulins in milk on passive immunity, it has been proposed that PRL in milk early in life participates in the development of acquired immune system in the pup (Ellis et al. 1996). There is evidence that PRL in milk can affect differentiation, proliferation, and function of lymphocytes, macrophages, and natural killer cells involved in the acquired immune system in rats and mice (Berczi et al. 1981; Bour et al. 1991). Using the same schedule of treatment, injection of bromocriptine into mothers during postpartum days 2–5 induced in the offspring at 10, 15, and 21 days of age, a precocious response of splenocyte and thymocyte populations to polyclonal mitogens in vitro was observed. Furthermore, in mice, the administration of PRL-antiserum or bromocriptine to pups on PND 1–3, reduced the percentage of thymocytes and splenocytes that display the CD4+ T-helper phenotype and B cells (Russell et al. 1998). These data suggest that the disruption of lymphocyte suppression early in life is a mechanism for generation of self-reactive lymphocytes involved in autoimmune diseases as Crohn's disease and diabetes. This suggestion derives from epidemiological findings that indicate that the frequency of these diseases is higher in infants fed with formula than those nursed by maternal milk (Goldman 1993).

In summary, the results presented here suggest that milk-borne PRL may be an essential and critical agent in the development of neuroendocrine, reproductive, behavior, immune, and neural function in offspring of mice, rat, and humans.

11.5.2 Glucocorticoids

GC (corticosterone in rodents; cortisol in humans and nonhuman primates) is a steroid hormone secreted by the cortex of the adrenal gland of many mammals (including rodents and primates) that is involved in metabolism, immune responses, and development of stress responses (see Catalani et al. 2000). It has been detected in milk of bovines, humans, rats, mice (Kulski and Hartmann 1981; Tucker and Schwalm 1977; Ratsimamanga 1956), and is reflective of blood GC concentration after physical, pharmacological, or psychological stress. It has been suggested that GC in milk in rodents, in combination with maternal cues, participate in the maintenance of the stress-hyporesponsive period (SHRP) that characterize pups from PND 4–14 when the adrenal response to stress or high levels of ACTH (exogenous) is either minimal or nonexistent, resulting in stable low levels of circulating GC (Sawano et al. 1969; Cirulli et al. 1992; Rosenfeld et al. 1992). The presence of GC in milk further supports the hypothesis that exposure to GC during the postnatal period also induces permanent effects on growth and differentiation in a number of biologic systems, including the central nervous system (CNS). These pituitary–adrenal interrelationships between the mother and offspring during postnatal life via the corticosterone presence in milk and its passage into pups through endothelial cells of intestine have been investigated. Thus, pups of mothers who received daily injections of corticosterone from postpartum day 2 to 22 showed higher corticosterone levels in stomach milk and brain, but not in serum on PND 7, as well as higher corticosterone levels in serum, but not in brain (prefrontal cortex, hypothalamus, or hippocampus) on PND 18 (Brummelte et al. 2010). The experimental paradigms most frequently used to support the above hypothesis involve decreasing the levels of GC below the physiological range in the mother (through adrenalectomy) or increasing levels to the physiological range through injections of corticosterone to lactating mothers (Catalani et al. 1993, 2000; Casolini et al. 1997). Also, GC levels in maternal milk can be manipulated. These studies show that a moderate increase of corticosterone during the postnatal time by mothers drinking corticosterone induced high levels of hippocampal corticosterone receptors and a lower response to stress in 30-day-old offspring (Casolini et al. 1997), as well as a decrease in the adult serotonin 1 A receptor binding in the hippocampus CA1 region (Meerlo et al. 2001). Behaviorally, administration of GC to the neonate results in adults which show good performance in spatial memory tasks (Catalani et al. 1993) reduced anxiety, improved learning, and better capacity to deal with stress (in the case of adult females; Catalani et al. 2000). In contrast, adult male and female offspring of mothers treated with corticosterone (40 mg/kg. sc) during the postnatal period (from birth to 26 postpartum day) showed higher resistance to capture and greater locomotor activity in open field test, and a decreased postnatal cell proliferation in the dentate gyrus (in male, but not female offspring; Brummelte et al. 2006). Finally, males of dams exposed to very high maternal corticosterone during the postpartum period showed more anxiety-like behavior in the elevated plus maze (Brummelte et al. 2012). In rhesus monkeys, naturally occurring variations in endogenous GC concentrations in mothers' milk has been related to variations in infant temperament (Sullivan et al. 2001). In general, these results suggest that early exposure to GC can cause permanent changes in binding capacity of the GC receptors in the hippocampus and in related behavioral and endocrine outcomes (Angelucci et al. 1983).

11.5.3 Thyroid Hormones

It is well known that TH, thyroxin (T4), and 3,5,3′-triiodothyronine (T3) are available to the mammalian embryo from placental transport (Morreale et al. 1987), and to neonate from maternal milk (Oberkotter 1989; Vigouroux et al. 1980; Slebodzinski and Twardon 2004). After birth, in addition to TH produced by its own thyroid gland, offspring ingest a considerable amount of TH in the maternal milk (Oberkotter 1989; Vigouroux et al. 1980; Slebodzinski and Twardon 2004). The levels of T4 in colostrum and milk in rabbits, cows, and women were found to be similar, with about 2 nmol l^{-1}, which, however, represents a very small fraction of levels found in blood serum (Slebodzinski et al. 1986). The physiologic effect of this T4 exposure on the offspring is not known. However, there is some evidence that the other TH, T3, in milk may exert some physiological effects in offspring, particularly during the early postnatal period. The T3 content in milk approximated one-third of the levels found in serum and was about 1.0, 1.8, 0.3, and 0.5 nmol l^{-1} in the rabbit, pig, human, and cow, respectively (Slebodzinski et al. 1986). As with other hormones in milk, concentrations of TH in milk vary across lactation. Although there are species differences in levels of TH, in general there are low levels of T4 and high of T3 milk levels during early lactation, and high milk T4 and low T3 concentrations during late lactation (Slebodzinski and Twardon 2004).

The main source of data supporting the hypothesis that maternal TH participates in the development of altricial offspring during early postnatal period comes from pharmacological studies where serum and milk thyroid content are reduced by administration of goitrogen drugs (e.g., 6-N-propylthiouracil, PTU) to lactating rats (Tamasy et al. 1984, 1986). The experiment consists of administrating PTU to the drinking water (0.1 %) to the mother from parturition and throughout much of lactation. This treatment effectively results in the mother in elevations in serum TSH and reduced serum T4 (Koldovsky 1980; Tamasy et al. 1986). Furthermore, pups' body weights were affected by this regimen, suggesting that the changes in maternal TH levels in serum also occur in the milk (Tamasy et al. 1984). Furthermore, these PTU-treated offspring of 50 PND show severe somatic and motor retardation, decreased exploratory activity, no habituation in the hole-board test, as well as behavioral and hormonal malfunctions (Tamasy et al. 1986). Also, the same treatment significantly decreases Bcl-2 protein and enhances apoptotic cells in the hippocampus neurons at all stages of development and (Huang et al. 2005), as well as inducing a dose-dependent auditory threshold deficits (Golden et al. 1995). Similar effects of PTU on development have also been reported in hamsters, where male hamsters reared by PTU-treated dams showed reduced sexual activity in adulthood (Jansen et al. 2007). There are also studies that suggest a role for TH in the development of offspring. Included among these are (1) thyroidectomy of lactating rats resulted in a transient decrease of T4 in sucklings and (2) the differences in mother milk intake achieved by different litter size also affected the thyroid function of sucklings, where the pups from small size litters (more milk consumption) had a lower thyroid secretion rate, lower plasma TSH, and higher T4 in plasma, a higher growth rate and an accelerated overall maturation (Strbák et al. 1983).

11.5.4 OT Hormone

OT is a small nonapeptide, initially considered to be a reproductive hormone involved in uterine contractions and milk ejection, but recently it has been involved in several central integrative functions, coordinating behavioral and physiological processes and in neural and behavioral development (Carter 2003). OT is abundant in the hypothalamus (supraoptic and paraventricular nucleus), and is transported to the posterior pituitary by neurosecretion, where it is released into the vascular system and acts on the uterus, gonads, heart, and thymus (Gimpl and Fahrenholz 2001). It is released by sensory stimulation during social contact, genital stimulation, sexual performance, and breast stimulation by pups, as demonstrated in several species (Carter 1992; Uvnas-Moberg 1998). OT receptors are found in brainstem and are involved in reproductive, social, affiliative, and adaptive behaviors (Witt 1997). OT has a vagal/parasympathetic function that integrates metabolic, behavioral systems involved in the control of anxiety, obsessiveness, and stress reactivity (Carter 1997; Sawchenko and Swanson 1985). Recently, it has been reported that exposure to OT during the prenatal and/or early postnatal period has lifelong consequences for the development of behavior and physiology of offspring (Boer 1993; Boer et al. 1994; see review Carter 2003). OT receptor binding is increased in female (but not male) offspring that received high levels of maternal licking and grooming (Champagne and Meaney 2001; Francis et al. 2002). Maternal behavioral effects on offspring OT function have, therefore, been demonstrated. However, the presence of OT in maternal milk has only been demonstrated in humans (Takeda et al. 1986). Although the capacity of OT to pass from the mother to the offspring system through the milk and remain stable and with bioactive properties has not been studied in rat models, the behavioral data suggest that neonatal OT can affect offspring development (Hartman et al. 1986). OT administered centrally and peripherally on 6–8 PND results in a reduction in the ultrasonic vocalizations (USV) emitted by isolated pups (Carter 2003), and a single dose of OT administered intracisternally induced an elevation of novelty-induced grooming (Noonan et al. 1989). In addition, the central administration of an OT antagonist blocks the effects of central, but not peripheral, administration of OT (Insel and Winslow 1991). In prairie voles (*Microtus ochrogaster*), OT antagonist (OTA) were administered either on PND 1 or daily from days 1 to 7, and then on day 8, the pups were exposed to social isolation. The single doses of OTA decreased USV during social isolation, but repeated doses increased USV in females (Kramer et al. 2003). Furthermore, early postnatal treatment with OT resulted in increased body weight, changes in nociceptive threshold (Uvnas-Moberg 1998), altered responses of aortic tissue to vasopressin, altered placental and fetal growth during pregnancy (Sohlström et al. 2002), and elevation of novelty-induced grooming (Noonan et al. 1989). When OT is administered early in life (1–14 PND) to offspring from food-restricted mothers, it stimulates postnatal weight gain and decreases the elevated levels of serum corticosterone and blood pressure (Sohlström et al. 2000; Olausson et al. 2003). In prairie voles (*Microtus ochrogaster*), a single injection of OT or OTA on the first day of life affected partner preference formation and aggression (Bales and Carter 2002), alloparental behavior (Pfeifer et al. 2001), and reproductive competency in adults (Bales et al. 2001). Finally, in rats, there is a relation between high levels of licking/grooming received early in life, and OT release and levels of central OT receptors in the pups (Champagne and Meaney 2001; Francis et al. 2002). Moreover, pups that received a high level of licking, as opposed to low levels of licking, showed in adulthood higher levels of maternal care and reduced anxiety-related behaviors (Francis et al. 2002; Pedersen and Boccia 2002).

11.6 Summary

The issues surrounding the roles of "nature versus nurture" in phenotypic development have been debated for many years and continue to the present day. Rather than viewing nature and nurture as dichotomous factors that contribute independently to the traits of the organism, we now know that nature and nurture are not opposites, but that they interact during development of the organism to produce the considerable phenotypic variation that characterizes mammalian behavior we see (Moore 2013). It is undeniable that transmission of many traits across generations results from the inheritance by offspring of genomic information. However, even the highly conserved traits develop, and involve environmental inputs to do so. In many cases, the environmental input is relatively invariant, while in others it shows considerable variation. Moreover, there is a large literature that describes the role of epigenetic mechanisms that are activated when environmental factors or stimuli exert effects on the genome during critical or sensitive periods in the development. Among the critical periods, the postnatal preweaning has acquired importance due to that the sensory, social, and hormonal cues provided through maternal care that can have long-term effects on the development of the offspring. When these influences occur prenatally or during the early postnatal period, they have been referred to as having long-term "programming effects" that suggest the existence of periods of "developmental plasticity" leading to individual differences in physiology and behavior in the offspring.

The use of animal models has been helpful in understanding the mechanisms underlying the association between early life experience and adult phenotypes. In humans, we can talk about correlations and associations; in animal models, we can talk about causes. For example, the short or partial maternal deprivation paradigm provides an excellent model to assess the role of negative (stress) factors on development, whereas the handling paradigm assesses the role of sensory (tactile) cues, and the total separation paradigm permits assessment of both tactile and also social cues on development. Among the cues provided by the mother, the maternal milk is essential not only as a food that furnishes nutrients for offspring growth but also as a vehicle to transfer bioactive elements that participate in the postnatal development of the immature organisms. For example, it has been shown that milk PRL is important during the first days of life for the development of the immune, cognitive, and behavioral systems of rodents. Although it has been known for more than 50 years that there are hormones in the breast milk of human mothers, the role of these hormones on the development of infant is unclear. What is known is that breastfeeding, compared with infant formula feeding, confers protections against several metabolic and physiological problems later in life, such as obesity and related medical complications. Furthermore, in contrast to breast milk, infant formula lacks hormonally active peptides; however, not enough is known to state categorically that specific hormones are essential to the development of the infant. It is hoped that through the use of animal models, where clear effects of hormones in milk have been demonstrated, human studies will be undertaken with specific hypotheses, based on the animal work. Finally, we have yet to fully understand how cues from the mother and littermates operate early in life to modify the genes and their expression. Recent evidence from studies of laboratory rodents illustrate the associations between the experience of prenatal stress, total maternal separation, maternal care, abusive caregiving in infancy, juvenile social housing, and adult social stress and variation in DNA methylation and histone modification. These studies emphasize the possibility that dynamic epigenetic changes (molecular modifications that alter gene expression without altering the underlying DNA sequence) may account for this plasticity and underlie many of the behavioral effects reported in this chapter. Precisely which genes and which environments, at what developmental stages, are important for the early postnatal influences described herein have yet to be examined both in the animal models and in humans.

Acknowledgments The author thanks Prof. Alison S. Fleming for commenting and editing this work and Francisco Corona for his generous help in the preparation of the table and the references.

This study was supported by a CONACYT-México grant to Angel I. Melo (CB-2010-01-156413).

Conflict of Interest The author declare no conflicts of interest.

References

Afonso VM, Grella SL, Chatterjee D, Fleming AS (2008) Previous maternal experience affects accumbal dopaminergic responses to pup-stimuli. Brain Res 1198:115–123

Alberts JR, Gubernick DJ (1983) Reciprocity and resource exchange: a symbiotic model of parent-offspring relation. In: Moltz H, Rosenblum L (eds) Symbiosis in parent-young interactions. Plenum, New York

Almeida SS, Tonkiss J, Galler JR (1996) Prenatal protein malnutrition affects exploratory behavior of female rats in the elevated plus-maze test. Physiol Behav 60:675–680

Amarant T, Fridkin M, Koch Y (1982) Luteinizing hormone-releasing hormone and thyrotropin-releasing hormone in human and bovine milk. Euro J Biochem 127:647–650

Angelucci L, Patacchioli FR, Chierechetti C, Laureti S (1983) Perinatal mother-offspring pituitary-adrenal interrelationship in rats: corticosterone in milk may affect adult life. Endocrinol Exp 17:191–205

Anisman H, Zaharia MD, Meaney MJ, Merali Z (1998) Do early life events permanently alter behavior and hormonal responses to stressors. Int J Dev Neurosci 16:149–164

Archibald JG (1945) Some effects of thyroprotein on the composition of milk. J Dairy Sci 51:1933–1935

Aydin S, Ozkan Y, Kumru S (2006) Ghrelin is present in human colostrum, transitional and mature milk. Peptides 27 :878–882

Bales K, Carter CS (2002) Oxytocin facilitates parental care in female prairie voles (but not in males). Horm Behav 41:456

Bales K, Abdelnabi M, Carter CS (2001) Neonatal injections affects reproductive parameters in male prairie voles. Horm Behav 39:324

Ballard FJ, Nield MK, Francis GL, Dahlenburg GW, Wallace JC (1982) The relationship between the insulin content and inhibitory effects of bovine colostrums on protein breakdown in cultured cells. J Cell Physiol 110:249–254

Baram T, Koch Y, Hazum E, Fridkin M (1977) Gonadotropin-releasing hormone in milk. Science 198:300–302

Barret J, Fleming AS (2011) Annual research review: all mothers are not created equal: neural and psychobiological perspectives on mothering and the importance of individual differences. J Child Psychol Psychiatry 52:368–397

Baumrucker CR, Magliaro-Macrina AL (2011) Hormones in milk. Encyclopedia of dairy sciences, 2nd edn. Academic, New York, pp 565–771
Beane ML, Cole MA, Spencer RL, Rudy JW (2002) Neonatal handling enhances contextual fear conditioning and alters corticosterone stress responses in young rats. Horm Behav 41:33–40
Ben-Jonathan N, Arbogast LA, Hyde JF (1989) Neuroendocrine regulation of prolactin release. Prog Neurobiol 33:399–447
Benuck I, Rowe FA (1975) Centrally and peripherally induced anosmia: influences on maternal behavior in lactating female rats. Physiol Behav 14:439–447
Berczi I, Nagy E, Kovacs K, Horvath E (1981) Regulation of humoral immunitary in rats by pituitary hormones. Acta Endocrinol 98:506–513
Bielecki M, Przala F, Lazewska M (1972) Level of erythropoietin in the woman milk. Acta Physiol 23:435–439
Boer GJ (1993) Chronic oxytocion treatment during late gestation and lactation impairs development of rat offspring. Neurotoxicol Teratol 15:383–389
Boer GJ, Quak, J, de Vries MC, Heinsbroek RPW (1994) Mild sustained effects of neonatal vasopressin and oxytocin treatment on brain growth and behavior of the rat. Peptides 15:229–236
Bohuslav D, Catherine SW, Debra L, McWilliam HS, Dominguez JA, McCuskey RS, Philipps ASF, Koldovsky O (2000) Milk-borne epidermal growth factor modulates intestinal transforming growth factor-α levels in neonatal rats. Pediatr Res 47:194–200
Bosch OJ, Kromer SA, Neumann ID (2006) Prenatal stress: opposite effects on anxiety and hypothalamic expression of vasopressin and corticotropin-releasing hormone in rats selectively bred for high and low anxiety. Eur J Neurosci 23:541–551
Bour DS, Kacsoh B, Mastrol AM (1991) Effect of neonatal milk-prolactin deprivation on the ontogeny of the immune system. Endocr Regul 25:111–119
Bridges R (1996) Biochemical basis of parental behavior in the rat. Adv Stud Behav 25:215–242
Brummelte S, Pawluski JL, Galea LA (2006) High post-partum levels of corticosterone given to dams influence postnatal hippocampal cell proliferation and behavior of offspring: a model of post-partum stress and possible depression. Horm Behav 50:370–382
Brummelte S, Schmidt KL, Taves MD, Soma KK, Galea LA (2010) Elevated corticosterone levels in stomach milk, serum, and brain of male and female offspring after maternal corticosterone treatment in the rat. Dev Neurobiol 70:714–725
Brummelte S, Lieblich SE, Galea LA (2012) Gestational and postpartum corticosterone exposure to the dam affects behavioral and endocrine outcome of the offspring in a sexual-dimorphic manner. Neuropharmacology 62:406–418
Burton C, Lovic V, Fleming AS (2006) Early adversity alters attention and locomotion in adults Sprague-Dawley rats. Behav Neurosci 120:665–675
Caldji C, Tannenbaum B, Sharma S, Francis D, Plotsky PM, Meaney MJ (1998) Maternal care during infancy regulates the development of neural systems mediating the expression of fearfulness in the rat. Proc Natl Acad Sci U S A 95:5335–5340
Caldji C, Diorio J, Meaney MJ (2000) Variations in maternal care in infancy regulate the development of stress reactivity. Biol Psychiatry 48:1164–1174
Carpenter G (1980) Epidermal growth factor is a major growth-promoting agent in human milk. Science 210:198–199
Carter CS (1992) Oxytocin and sexual behavior. Neurosci Biobehav Rev 16:131–144
Carter CS (2003) Developmental consequences of oxytocin. Physiol Behav 79:383–397
Carter CS, Altemus M (1997) Integrative functions of lactional hormones in social behavior and stress management. Ann N Y Acad Sci 807:164–174
Casolini P, Cigliana G, Alemán GS, Ruggiere V, Angelucci L, Catalani A (1997) Effect of increased maternal corticosterone during lactation on hippocampal corticosterone receptors, stress response and learning in offspring in the early stages of life. Neuroscience 79:1005–1012
Catalani A, Marinelli M, Scaccianoce S, Nicolai R, Muscolo LA, Porcu A, Korányi L, Piazza PV, Angelucci L (1993) Progeny of mothers drinking corticosterone during lactation has lower stress-induced corticosterone secretion and better congnitive performance. Brain Res 624:209–215

Catalani A, Casolini P, Scaccianoce S, Patacchioli FR, Spinozzi P, Angelucci L (2000) Maternal corticosterone during lactation permanently affects brain corticosteroid receptors, stress response and behaviour in rat progeny. Neuroscience 1000:319–325

Cevreska S, Kovacev VP, Stankovski M, Kalamaras E (1975) The presence of immunologically reactive insulin in milk of women during the first week of lactation and its relation to changes in plasma insulin concentration. God Medicine Fak Skopje 21:35–41

Champagne FA (2008) Epigenetic mechanisms and the transgenerational effects of maternal care. Front Neuroendocrinol 29:386–397

Champagne FA, Meaney MJ (2001) Like mother, like daughter: evidence for non-genomic transmission of parental behavior and stress responsivity. Prog Brain Res 133:287–302

Champagne FA, Meaney MJ (2007) Transgenerational effects of social environment on variations in maternal care and behavior response to novelty. Behav Neurosci 121:1353–1363

Champagne FA, Francis DD, Mar A, Meaney MJ (2003) Variations in maternal care in the rat as a mediating influence for the effects of environment on development. Physiol Behav 79:359–371

Champagne FA, Weaver ICG, Diorio J, Dymov S, Szyf M, Meaney MJ (2006) Maternal care associated with methylation of the estrogen receptor-alpha1b promoter and estrogen receptor-alpha expression in the medial preoptic area of female offspring. Endocrinology 147:2909–2915

Chappell JE, Clandinin MT, Barbe GJ, Armstrong DT (1983) Prostanoid content of human milk: relationships to milk fatty acid content. Endocrinol Exp 17:351–358

Chilliard Y, Bonnet M, Delavaud C, Faulconnier Y, Leroux C, Djiane J, Bocquier F (2001) Leptin in ruminants. Gene expression in adipose tissue and mammary gland, and regulation of plasma concentration. Domest Anim Endocrinol 21:271–295

Chirino R, González-Mariscal G, Beyer C (1999) Efecto de la lesión del bulbo olfatorio accesorio (BOA) sobre diferentes conductas reproductivas de la coneja. Congreso Nacional de Ciencias. Fisiológicas., 42nd, Zacatecas, México, p 0101

Chirino R, Beyer C, Rosenblatt JS, González-Mariscal G (2000) Virgin female rabbits show pup-induced maternal behavior (MB) only after lesioning the accessory olfactory bulbs. Dev Psychobiol 36:248

Chirino R, Beyer C, Rosenblatt JS, González-Mariscal G (2001) Intranasal zinc sulfate spray induces anosmia but not facilitate maternal behavior in estrous rabbits. Dev Psychobiol 38:197

Cirulli F, Gottlieb S, Levine S (1992) Maternal factors regulate stress responsiveness in the neonatal rat. Psychobiology 20:143–152

Cirulli F, Berry A, Alleva E (2003) Early disruption of the mother-infant relationship: effects on brain plasticity and implications for psychopathology. Neurosci Biobehav Rev 27:73–82

Costela C, Tejedor-Real P, Mico JA, Gilbert-Rahola J (1995) Effect of neonatal handling on learned helplessness model of depression. Physiol Behav 57:407–410

Cox DB, Owens RA, Hartmann PE (1996) Blood and milk prolactin and the rate of milk synthesis in women. Exp Physiol 81:1007–1020

Crowley WR, Shah GV, Watanobe H, Grosvenor CE (1990) Effects of neonatal exposure to estradiol on prolactin secretion and activity of the tuberoinfundibular dopamine system in young adulthood: comparison with neonatal prolactin deficiency. J Neuroendocrinol 2:19–24

Darling J, Laing A, Harkness R (1974) A survey of the steroids in cow's milk. J Endocrinol 62:291–297

Denenberg V, Smith WI, Ross S (1950) A preliminary study of individual and group hoarding in the white rat. J Genet Psychol 77:123–127

Diaz J, Moore E, Petracca F, Schacher J, Stamper C (1981) Artificial rearing of preweanling rat: the effectiveness of direct intragastric feeding. Physiol Behav 27:1103–1105

Diaz J, Moore E, Petracca F, Stamper C (1983) Somatic and central nervous system growth in artificially reared rat pups. Brain Res Bull 11:643–647

Dickson RB, Huff KK, Spencer EM, Lippman ME (1985) Induction of epidermal grow factor-related polypeptides by 17β-estradiol in human breast cancer cells. Endocrinology 118:138–142

Donatelli AL, Seidman LJ, Goldstein JM, Tsuang MT, Buka SL (2010) Children of parents with affective and non-affective psychose: a longitudinal study of behavior problems. Am J Psychiatry 167:1331–1338

Ellis LA, Picciano MF (1995) Bioactive and immunoreactive prolactin variants in human milk. Endocrinology 136:2711–2720
Ellis LA, Mastro AM, Picciano MF (1996) Milk-borne prolactin and neonatal development. J Mammary Gland Biol Neoplasia 1:259–269
Eriksson L, Valtonen M, Laitinen JT, Paananen M, Kaikkonen M (1998) Diurnal rhythm of melatonin in lactating cows and goats. Acta Vet Scand 39:310
Faturi CB, Tiba PA, Kawakami SE, Catallani B, Kerstens M, Suchecki D (2010) Disruptions of the mother-infant relationship and stress-related behaviours: altered corticosterone secretion does not explain everything. Neurosci Biobehav Rev 34:821–834
Fleming AS, Luebke C (1981) Timidity prevents the nulliparous female from being a good mother. Physiol Behav 27:863–868
Fleming AS, Rosenblatt JS (1974) Olfactory regulation of maternal behavior in rats: I. Effects of olfactory bulb removal in experienced and inexperienced lactating and cycling females. J Comp Physiol 87:221–232
Fleming AS, Vaccarino F, Tambosso L, Chee P (1979) Vomeronasal and olfactory system modulation of maternal behavior in the rat. Science 203:327–337
Fleming AS, Kraemer GW, Gonzalez A, Lovic V, Ress A, Melo AI (2002) Mothering begets mothering: the transmission of behavior and its neurobiology across generation. Pharmacol Biochem Behav 73:61–75
Francis DD, Meaney MJ (1999) Maternal care and the development of stress responses. Curr Opin Neurobiol 9:128–134
Francis DD, Dioro J, Liu D, Meaney MJ (1999) Nongenomic transmission across generations of maternal behavior and stress responses in the rat. Science 285:1155–1158
Francis DD, Young LJ, Meaney MJ, Insel TR (2002) Naturally occurring differences in maternal care are associated with the expression of oxytocin and vasopressin (V1a) receptors: gender differences. J Neuroendocrinol 14:349–353
Fries AB, Pollak SD (2004) Emotion understanding in postinstitutionalized Eastern European children. Dev Psychopathol 16:355–369
Fritsche S, Steinhart H (1999) Occurrence of hormonally active compounds in food: a review. Eur Food Res Technol 209:153–179
Gala RR, Singhakowinta A, Brennan MJ (1975) Studies on prolactin in human serum, urine and milk. Horm Res 6:310–320
Gala R, Forsyth I, Turvey A (1980) Milk prolactin is biologically active. Life Sci 26:987–993
Geschickter CF, Lewis D (1936) Lactogenic substance in the human breast: its use in experimental stimulation of mammary secretion and its assay in cases of cystic. Arch Surg 32:598–617
Gimpl G, Fahrenholz F (2001) The oxytocin receptor system: structure, function and regulation. Physiol Behav 81:629–683
Glaser D (2000) Child abuse and neglect and the brain-a review. J Child Psychol Psychiatry 41:97–116
Golden ES, Kehn LS, Rehnberg GL, Crofton KM (1995) Effects of developmental hypothyroidism on auditory and motor function in the rat. Toxicol Appl Pharmacol 135:67–76
Goldman AS (1993) The immune system of human milk: antimicrobial, anti-inflammatory and immunomodulating properties. Pediatr Infect Dis J 12:664–671
Goldman AS, Frawley S (1996) Bioactive components of milk. J Mammary Gland Biol Neoplasia 3:241–242
Gonella P, Harnatz P, Walker QA (1989) Prolactin is transported across the epithelium of the jejunum and ileum of suckling rat. J Cell Physiol 140:138–149
Gonzalez A, Lovic V, Ward GR, Wainwright PE, Fleming AS (2001) Intergenerational effects of complete maternal deprivation and replacement stimulation on maternal behavior and emotionality in female rats. Dev Psychobiol 38:11–32
González-Mariscal G, Kinsley CH (2009) From indifference to ardor: the onset, maintenance, and meaning of the maternal brain. Horm Behav 1:109–136

González-Mariscal G, Melo AI (2013) Parental Behavior. In: Pfaff DW (ed) Neuroscience in the 21st Century. From Basic to Clinical. DOI 10.1007/978-1-4614-1997-6_76, Springer Science+Business Media, Chapter 68, pp 2069–2100
González-Mariscal G, Poindron P (2002) Parental care in mammals: immediate internal and sensory factors of control. Horm Brain Behav 1:215–298
Grant WC (1952) Influence of anoxia of a lactating rat on the blood of normal baby rats. Am J Physiol 171:728–729
Green JG, McLaughlin KA, Berglund PA, Gruber MJ, Sampson NA, Zaslavsky AM (2010) Childhood adversities and adult psychiatric disorders in the national comorbidity survey replication I: associations with first onset of DSM-IV disorders. Arch Gen Psychiatry 67:113–123
Grosvenor CE, Whitworth NS (1976) Incorporation of rat prolactin into rat milk in vivo and in vitro. Endocrinology 70:71–79
Grosvenor CE, Picciano MF, Baumrucker CR (1992) Hormones and growth factors in milk. Endocr Rev 14:710–728
Gunnar MR, Nelson CA, Luciana M (2001) Effects of early deprivation: findings from orphanage-reared infants and children. In: Nelson CA, Luciana M (eds) Handbook of developmental cognitive neuroscience. The MIT Press, Massachusetts, pp 617–629
Gutman DA, Nemeroff CB (2002) Neurobiology of early life stress: rodent studies. Semin Clin Neuropsychiatry 7:89–95
Hall WG (1975) Weaning and growth of artificially reared rats. Science 397:726–735
Hall FS (1998) Social deprivation of neonatal, adolescent, and adults has distinct neurochemical and behavioral consequences. Crit Rev Neurobiol 2:129–162
Hall FS, Wilkinson LS, Humby T, Robbins TW (1999) Maternal deprivation of neonatal rats produces enduring changes in dopamine function. Synapse 32:37–43
Haney PM, Estrin CR, Caliendo A, Patel MS (1986) Precocious induction of hepatic glucokinase and melic enzyme in artificially reared rat pups fed a high-carbohydrate diet. Arch Biochem Biophys 244:787–794
Harlow HF, Dodsworth RO, Harlow MK (1965) Total social isolation in monkeys. Proc Natl Acad Sci U S A 54:90–97
Hartman RD, Rosella-Dampman LM, Emmert SE, Summy-Long JY (1986) Ontogeny of opioid inhibition of vasopressin and oxytocin release in response to osmotic stimulation. Endocrinology 119:1–11
Hazum E (1983) Hormones and neurotransmitters in milk. TIPS 454–456
Heap RB, Gwyn M, Laing J, Walters D (1973) Pregnancy diagnosis in cows: changes in milk progesterone, concentration during the oestrous cycle and pregnancy measured by rapid radioimmunoassay. J Agric Sci 81:151–159
Heap RB, Hamon M, Fleet IR (1983) Factors affecting oestrone sulphate concentrations in milk. Br J Vet 39:79–88
Henschen A, Lottspeich F, Branlt V, Teschemacher H (1979) Novel opioid peptides derived from casein (beta-casomorphins). II. Structure of active components from bovine casein peptone. Hoppe-Seyler´s Zeitschrift für Physiologische Chemie 360:1217–1224
Hess JL, Denenberg VH, Zarrow M, Peiffer WD (1969) Modification of the corticosterone response curve as a function of handling in infancy. Physiol Behav 4:102–109
Hoeffler JP, Boockfor ER, Frawley LS (1985) Ontogeny of prolactin cells in neonatal rats: initial prolactin secretors also release growth hormone. Endocrinology 117:187–195
Hofer MA (1970) Physiological responses of infant rats to separation from their mothers. Science 168:871–873
Hofer MA, Weiner H (1971) Development and mechanisms of cardiorespiratory responses to maternal deprivation in rat pups. Psychosom Med 33:353–362
Hofer MA (1972) Physiological and behavioral processes in early maternal deprivation, in physiology, emotions and psychosomatic illness, a CIBA foundation symposium, London, Amsterdam
Hofer MA (1973a) The role of nutrition in the physiological and behavioral effects of early maternal separation on infant rats. Psychosom Med 35:350–359

Hofer MA (1973b) The effects of brief maternal separation on behavior and heart rates of two week old rat pups. Physiol Behav 10:423–427

Hoffmann B, Rattenberger E (1977) Testosterone concentrations in tissue from veal calves, bull and heifers and in milk samples. J Anim Sci 45:635–641

Holst N, Jenssen TG, Burhol PG (1990) A characterization of immunoreactivesomatostatin in human milk. J Ped Gastroenterol Nutr 10:47–52

Hruska K, Veznik Z (1983) Significance of progesterone determination of milk. Endocrinol Exp 17:207–211

Huang XW, Zhao ZY, Ji C (2005) Effects of hypothyroidism on apoptosis and the expression of Bcl-2 and Bax gene in the neonatal rat hippocampus neurons. Zhonghua Er Ke Za Zhi 43:48–52

Insel TR, Winslow JT (1991) Central administration of oxytocin modulates the infant rat's response to social isolation. Eur J Pharmacol 203:149–152

Jansen HT, Kirby JD, Cooke PS, Arambepola N, Iwamoto GA (2007) Impact of neonatal hypothyroidism on reproduction in the male hamster, *Mesocricetus auratus*. Physiol Behav 23:771–781

Kacsoh B, Toth B, Avery L, Deaver D, Baumrucker C, Grosvenor C (1991) Biological and immunological activities of glycosylated and molecular weigth variants of bovine prolactin in colostrums, and milk. J Anim Sci 69:456

Kacsoh B, Veress Z, Toth BE, Avery LM, Grosvenor CE (1993) Bioactive and immunoreactive variants of prolactin in milk and serum of lactating rats and their pups. J Endocrinol 138:243–257

Kierson JA, Dimatteo DM, Locke RG, Mackley AB, Spear ML (2006) Ghrelin and cholecystokinin in term and preterm human breast milk. Acta Paediatr 95:991–995 (Oslo, Norway 1992)

Koch Y, Werner H, Fridkin M (1991) Hypothalamic hormones in milk. Endocrinol Regulat 25:1–13

Koldovsky O (1980) Hormones in milk. Life Sci 26:1833–1836

Koldovsky O (1996) The potential physiological significance of milk-borne hormonally active substances for the neonate. J Mammary Gland Biol Neoplasia 3:317–323

Koldovsky O, Thornbug W (1987) Hormones in milk. J Pediatr Gastroenterol Nutriol 6:172–196

Kosten T, Lee HJ, Kim JJ (2007) Neonatal handling alters learning in adult male and female rats in a task-specific manner. Res Rep 1154:144–153

Kramer KM, Cushing BS, Carter CS (2003) Developmental effects of oxytocin on stress response: single versus repeated exposure. Physiol Behav 79:775–782

Krulich L, Koldovsky O, Jumawan J, Lau H, Horowitz C (1977) TSH in serum and milk of normal, thryoidectomized, and hyperthyroid lactating rats. Proc Soc Exp Biol Med 155:599–601

Kulski JK, Hartmann PE (1981) Changes in the concentrations of cortisol in milk during different stages of human lactation. Aust J Exp Biol Med Sci 59:769–778

Ladd CO, Owens MJ, Nemeroff CB (1996) Persistent changes in corticotrophin-releasing factor neuronal systems induced by maternal deprivation. Endocrinology 137:1212–1218

Lee MHS, Williams DI (1974) Changes in licking behavior following handling of young. Anim Behav 2:2679–2681

Lehmann J, Feldon J (2000) Long-term biobehavioral effects of maternal separation in the rat: consistent or confusing? Rev Neurosci 11:383–408

Leigh H, Hofer M (1973) Behavioral and physiologic effects of littermate removal on the remaining single pup and mother during the pre-weaning period in rats. Psychosomatics 35:497–508

Levine S (1957) Infantile experience and consummatory behavior in adulthood. J Comp Physiol Psychol 50:609–612

Levine S, Lewis GW (1959) Critical period for the effects of infantile experience on the maturation of a stress response. Science 129:42–43

Lévy F, Kendrick K, Keverne EB, Porter RH, Romeyer A (1996) Physiological, sensory and experiential factors of parental care in sheep. Adv Study Behav 25:385–473

Lévy F, Melo AI, Gale Jr BG, Madden M, Fleming AS (2003) Complete maternal deprivation affects social, but not spatial, learning in adult rats. Dev Psychobiol 43:177–191

Liu D, Diorio J, Tannenbaum B, Caldji C, Francis D, Freedman A et al (1997) Maternal care, hippocampal glucocorticoid receptors and hypothalamic-pituitary-adrenal responses to stress. Science 227:1659–1662
Liu D, Caldji C, Sharma S, Plotsky PM, Meaney MJ (2000) Influence of neonatal rearing conditions on stress-induced adrenocorticotropin responses and norepinepherine release in the hypothalamic paraventricular nucleus. J Neuroendocrinol 12:5–12
Lomanowska AM, Ammari N, Kraemer GW (2010) Interactions between the effects of early isolation rearing and complex housing on adult locomotor activity and sensitivity to amphetamine in rats involve noradrenergic neurotransmission. Pharmacol Biochem Behav 95:100–105
Lovic V, Fleming AS (2004) Artificially-reared female rats show reduced prepulse inhibition and deficits in the attentional set-shifting task-reversal of effects with maternal-like licking stimulation. Behav Brain Res 148:209–219
Lovic V, Fleming AS, Fletcher PJ (2006) Early life tactile stimulation changes adult rat responsiveness to amphetamine. Pharmacol Biochem Behav 84:497–503
Macri S, Mason GJ, Wurbel H (2004) Dissociation in the of neonatal maternal separations on maternal care and the offspring's HPA and fear responses in rats. Eur J Neurosci 20:1017–1024
Macri S, Chiarotti F, Wurbel H (2008) Maternal separation and maternal care act independently on the development of HPA responses in male rats. Behav Brain Res 191:227–234
Malven PV (1977) Prolactin and other protein hormones in milk. J Anim Sci 45:609–616
Malven PV, McMurtry JP (1974) Measurement of prolactin in milk by radio immunoassay. J Dairy Sci 57:411–415
Malven PV, Head HH, Collier RJ, Buonomo FC (1987) Periparturient changes in secretion and mammary uptake of insulin and in concentrations of insulin and insulin-like growth factors in milk of dairy cows. J Dairy Sci 70:2254–2265
Markham JA, Koenig JI (2011) Prenatal stress: role in psychotic and depressive diseases. Psychopharmacology (Berl) 214:89–106
Matthews K, Robbins TW (2003) Early experience as a determinant of adult behavioural responses to reward: the effects of repeated maternal separation in the rat. Neurosci Biobehav Rev 27:45–55
Matthews K, Wilkinson LS, Robbins TW (1996a) Repeated maternal separation of pre-weanling rats attenuates the behavioural responses to primary and conditioned incentives in adulthood. Physiol Behav 59:99–107
Matthews K, Hall FS, Wilkinson LS, Robbins TW (1996b) Retarded acquisition and reduced expression of conditioned locomotor activity in adult rats following repeated early maternal separation: effects of prefeeding, d-amphetamine, dopamine antagonists and clonidine. Psychopharmacology (Berl) 126:75–84
Mayer AD, Rosenblatt JS (1977) Effects of intranasal zinc sulfate on open field and maternal behavior in female rats: experimental effects. J Comp Physiol 89:701–710
McGarrigle HHC, Lachelin GCL (1983) Oestrone, oestradiol and oestriol glucosidurinates and sulphates in human puerperal plasma milk. J Steroid Biochem 18:607–611
McMurtry JP, Malven PV (1974a) Experimental alterations of prolactin levels in goat milk and plasma. Endocrinology 95:559–564
McMurtry JP, Malven PV (1974b) Radioimmunoassay of endogenous and exogenous prolactin in milk of rats. J Endocrinol 61:211–217
Meaney MJ (2001) Maternal care, gene expression, and the transmission of individual differences in stress reactivity across generations. Annu Rev Neurosci 24:1161–1192
Meaney MJ, Aitken DH (1985) The effects of early postnatal handling on hippocampal gluco-corticoid receptor concentration: temporal parameters. Dev Brain Res 22:301–304
Meaney MJ, Aitken DH, Van Berkel C, Bhatnagar S, Sapolsky RM (1988) Effect of neonatal handling on age-related impairments associated with the hippocampus. Science 239:766–768
Meaney MJ, Aitken DH, Viau V, Sharma S, Sarrieau A (1989) Neonatal handling alters adrenocortical negative feedback sensitivity and hippocampal type II glucocorticoid receptor binding in the rat. Neuroendocrinology 50:597–604

Meaney MJ, Diorio J, Francis D, Widdowson J, LaPlante P, Caldji C, Sharma S, Seckl JR, Plotsky PM (1996) Early environmental regulation of forebrain glucocorticoid receptor gene expression: implications for adrenocortical responses to stress. Dev Neurosci 18:49–72

Meerlo P, Horvath KM, Luiten PG, Angelucci L, Catalani A, Koolhaas JM (2001) Increased maternal corticosterone levels in rats: effects on brain 5-HT1A receptors and behavioral coping with stress in adult offspring. Behav Neurosci 115:1111–1117

Melo AI, Lovic V, Gonzalez A, Madden M, Sinopoli K, Fleming AS (2006) Maternal and littermate deprivation disrupt maternal behavior and social-learning of food preference in adulthood: tactile stimulation, nest odor, and social rearing prevent these effects. Dev Psychobiol 48:209–219

Melo AI, Hernández-Curiel M, Hoffman KL (2009a) Maternal and peer contact during the postnatal period participate in the normal development of maternal aggression, maternal behavior, and the behavioral response to novelty. Behav Brain Res 201:14–21

Melo AI, Pérez-Ledezma M, Clapp C, Arnold E, Rivera JC, Fleming AS (2009b) Effects of prolactin deficiency during the early postnatal period on the development of maternal behaviors in female rats: mother's milk makes the difference. Horm Behav 56:281–291

Messer M, Thomas EB, Galofre A, Dallman T, Dallmen PR (1969) Artificial feeding of infants rats by continuous infusion. J Nutr 98:404–410

Monk EL, Erb RE, Mollet TA (1975) Relationships between immunoreactive estrone and estradiol in milk, blood, and urine of diary cows. J Dairy Sci 58:34–40

Moore CL (1984) Maternal contribution to the development of masculine sexual behavior in laboratory rats. Dev Psychobiol 17:347–356

Moore SD (2013) Current thinking about nature and nurture. Hist Philosophy Theory Life Sci 1:629–652

Morreale de Escobar G, Obregón MJ, Calvo R, Escobar del Rey F (1987) Effects of iodine deficiency on thyroid hormone metabolism and the brain in fetal rats: the role of the maternal transfer of thyroxin. Horm Res 26:12–27

Nagano M, Chastre E, Choquet A, Bara J, Gespach C, Kelly PA (1995) Expression of prolactin and growth hormone receptor genes and their isoforms in the gastrointestinal tract. Am Physiol 268:431–442

Nilsson S, Nygren K-G, Johansson EDB (1978) Ethinyl estradiol in human milk and plasma after oral administration. Contraception 17:131–139

Noonan LR, Continell G, Pedersen CA (1989) Neonatal administration of oxytocin Increases novelty-induced grooming in the adult rat. Pharmacol Biochem Behav 33:555–558

Numan M (2006) Hypothalamic neural circuits regulating maternal responsiveness toward infants. Behav Cogn Neurosci Rev 4:163–190

Oberkotter LV (1989) Analysis of term human milk concentration of 3,5,3′-triiodo-L-thyronine by high performance liquid chromatography and radioimmunoassay: correlation with circulating serum levels in lactating women. J Chromatogr 487:445–449

Olausson H, Uvnas-Moberg K, Sohlstrom A (2003) Postnatal oxytocin alleviates adverse effects in adult rat offspring caused by maternal malnutrition. Am J Physiol Endocrinol Metab 284:475–480

Ollivier-Bousquet M, Kann G, Durand G (1993) Prolactin transit through mammary epithelial cells and appearance in milk. Endocr Regul 27:115–124

Ostrom KM (1990) A review of the hormone prolactin during lactation. Prog Food Nutr Sci 14:1–44

Peaker M, Neville MC (1991) Hormones in milk: chemical signals to the offspring? J Endocrinol 131:1–3

Pedersen CA, Boccia ML (2002) Oxytocin links mothering received, mothering bestowed and adult stress responses. Stress 5:259–267

Pfeifer L, Bale K, Carter CS (2001) Neonatal administration of oxytocin affects alloparental behavior in male prairie voles. Horm Behav 40:344

Plotsky PM, Meaney MJ (1993) Early, postnatal experience alters hypothalamic corticotrophin-releasing factor (CRF) mRNA, median eminence CRF content and stress-induced release in adult rats. Mol Brain Res 18:195–200

Poindron P, Le Neindre P, Raksanyi I, Trillat G, Orgeur P (1980) Importance of the characteristics of the young in the manifestation and establishment of maternal behavior in sheep. Reprod Nutr Dev 20:817–826

Polk DH (1992) Do breast milk derived hormones play a role in neonatal development? Early Hum Dev 29:329–331

Pollack EI, Sachs BD (1975) Male copulatory behavior and female maternal behavior in neonatally bulbectomized rats. Physiol Behav 14:337–343

Pollak SD, Nelson CA, Schlaak MF, Roeber BJ, Wewerka SS, Wiik KL, Frenn KA, Loman MM, Gunnar MR (2010) Neurodevelopmental effects of early deprivation in postinstitutional children. Child Dev 81:224–36

Pollack PF, Goda T, Colony PC, Edmond J, Thornburg W, Korc M, Koldovsky O (1987) Effects of enterally fed epidermal growth factor on the small and large intestine of the suckling rat. Regul Pept 17:121–132

Pryce CR, Feldon J (2003) Long-term neurobehavioral impact of the postnatal environments in rats: manipulations, effects and mediating mechanisms. Neurosci Biobehav Rev 27:57–71

Pryce CR, Bettschen D, Bahr NI, Feldon J (2001) Comparison of the effects of infant handling, isolation, and nonhandling on acoustic startle, prepulse inhibition, locomotion, and HPA activity in the adult rat. Behav Neurosci 115:71–83

Puccio F, Lehy T (1988) Oral administration of epidermal growth factor in suckling rats stimulates cell DNA synthesis in fundic and antral gastric mucosae as well as in intestinal mucosa and pancreas. Regul Pept 20:53–64

Ratsimamanga AR, Nigeon-Dureuil M, Rabinowicz M (1956) Presence d´hormone-type cortinique-das le lait de la vache gravide. Comptes Rendus de Societe Biologie (Paris) 150:2179–2182

Rosenblatt JS (1965) The basis of synchrony in the behavioral interaction between the mother and her offspring in the laboratory rat. In: Fos BM (ed) Determinants of infant behavior, vol 3. Methuen, London, pp 6–41

Rosenblatt JS (1967) Nonhormonal basis of maternal behavior in the rat. Science 156:1512–1514

Rosenblatt JS, Lehrman DS (1963) Maternal behavior in the laboratory rat. In: Rheingold HL (ed) Maternal behavior in mammals. Wiley, New York, pp 8–57

Rosenblatt JS, Siegel HI, Mayer AD (1979) Progress in the study of maternal behavior in the rat: hormonal, non-hormonal, sensory, and developmental aspects. Adv Study Behav 10:225–311

Rosenblatt JS, Mayer AD, Siegel HI (1985) Maternal behavior among the nonprimate mammals. In: Adler N, Pfaff D, Goy RW (eds) Handbook of behavioral neurobiology, vol 7. Plenum Press, New York, pp 229–298

Rosenfeld P, Suchecki D, Levine S (1992) Multifactorial regulation of the hypothalamic-pituitary-adrenal axis during development. Neurosci Biobehav Rev 16:553–568

Russell DH, Mills KT, Talamantes FJ, Bern HA (1998) Neonatal administration of prolactin antiserum alters the developmental pattern of T- and B-lymphocytes in the thymus and spleen of BALB/c female mice. Proc Natl Acad Sci U S A 85:7404–7407

Rutter M, O´Connor TG (2004) Are there biological programming effects for psychological development? Findings from a study of Romanian adoptess. Dev Psychol 40:81–94

Sapolsky RM, Krey L, McEwen BS (1984) Glococorticoid sensitive hippocampal neurons are involved in terminating the adrenocortical stress response. Proc Natl Acad Sci U S A 81:6174–6177

Sapolsky RM, Meaney MJ (1986) Maturation of the adrenocortical stress response: neuroendocrine control mechanisms and the stress hyporesponsive period. Brain Res 396:64–76

Sarda A, Nair RMG (1981) Elevated levels of LRH in human milk. J Clin Endocinol Metab 52:826–828

Savino F, Liguori SA, Benetti S, Sorrenti M, Fissore MF, di Montezemolo L (2013) High serum leptin levels in infancy can potentially predict obesity in childhood, especially in formula-fed infants. Acta Paediatr 102:e455–459

Sawano S, Arimura A, Schally AV, Redding TW, Schapiro S (1969) Neonatal corticoid administration: effects upon adult pituitary growth hormone and hypothalamic growth hormone-releasing hormone activity. Acta Endocrinol 61:57–67

Sawchenko PE, Swanson LW (1985) Relationship of oxytocin pathways to the control of neuroendocrine and autonomic function. In: Amico JA, Robinson AG (eds) Oxytocin: clinical and laboratory studies. Elsevier Science Publishers B.V., Amsterdam, pp 87–103

Schmidt MV, Enthoven L, Van Woezik JH, Levine S, De Kloet ER, Oitzl MS (2004) The dynamics of the hypothalamic-pituitary-adrenal axis during maternal deprivation. J Neuroendocrinol 16:52–57

Shah V, Shyr SW, Grosvenor CE, Crowley WR (1988) Hyperprolactinemia after neonatal prolactin (PRL) deficiency in rats: evidence for altered anterior pituitary regulation of PRL secretion. Endocrinology 182:1883–1889

Shieh KR, Pan JT (1999) Stimulatory role of prolactin on the development of tuberoinfundibular dopaminergic neurons in prepubertal female rats: studies with cysteamine and somatostatin. J Neuroendocrinol 11:907–917

Shing YW, Klagsbrun M (1984) Human and bovine milk contain different sets of growth factors. Endocrinology 115:273–282

Shyr SW, Crowley WR, Grosvenor CE (1986) Effect of neonatal prolactin deficiency on prepubertal tuberoinfundibular and tuberohypophyseal dopaminergic neuronal activity. Endocrinology 119:1217–1221

Simmons KR, Moses SC, Perkins BL (1979) Prostaglandin in milk, days open, and estrus detection in dairy cows treated with prostaglandin F2 alpha. J Dairy Sci 62:1443–1448

Slebodzinski AB, Twardon J (2004) Thyroid hormones (TH) and 5′-monodeiodinase (5′-MD) activity in goat's milk from the early, mid- and late lactation period. Acta Vet Hung 52:349–359

Slebodzinski AB, Nowak J, Gawecka H, Sechman A (1986) Thyroid hormones and insulin in milk: a comparative study. Endocrinol Exp 20:247–255

Smart JL, Tonkiss J, Massey RF (1986) A phenomenon: left-biassed asymmetrical eye-opening in artificially reared rat pups. Dev Brain Res 28:134–136

Sohlström A, Carlsson C, Uvnas-Moberg K (2000) Effects of oxytocin treatment in early life on body weight and corticosterone in adult offspring from ad libitum-fed and food-restricted rats. Biol Neonate 78:33–40

Sohlström A, Olausson H, Brismar K, Uvnäs-Moberg K (2002) Oxytocin treatment during early life influences reproductive performance in ad libitum fed and food-restricted female rats. Biol Neonate 81:132–138

Spear LP, Specht SM, Kirstein CL, Kuhn CM (1989) Anterior and Posterior, but not cheek, intraoral cannulation procedures elevate serum corticosterone levels in neonatal rat pups. Dev Psychobiol 22:401–411

Stemberg W, Ridgway CG (2003) Effects of gestational stress and neonatal handling on pain, analgesia, and stress behavior of adult mice. Physiol Behav 78:375–383

Strbak V (1985) The Role of maternal milk in endocrine regulation of sucklings. VEDA, Bratislava, pp 1–108

Strbák V, Macho L, Skultétyová M, Michalicková J, Pohlová G (1983) Thyroid hormones in milk: physiological approach a review. Endocrinol Exp 4:219–235

Sullivan EC, Hinde K, Mendoza SP, Capitanio JP (2001) Cortisol concentrations in the milk of rhesus monkey mothers are associated with confident temperament in sons, but not daughters. Dev Psychobiol 53:96–104

Takeda S, Kuwabara Y, Mizuno M (1986) Concentrations and origin of oxytocin in breastmilk. Endocrinol Jan 33:821–826

Tamasy V, Du J, Vallerga A, Meisami E, Timiras PS (1984) Suckling ability and maternal prolactin levels in hypothyroid rats. Horm Behav 18:457–464

Tamasy V, Meisami E, Du J-Z, Timiras PS (1986) Exploratory behavior, learning ability, and thyroid hormonal responses to stress in female rats rehabilitating from postnatal hypothyroidism. Dev Psychobiol 19:537–553

Teicher MH, Andersen SL, Polcari A, Anderson CM, Navalta CP, Kim DM (2003) The neurobiological consequences of early stress and childhood maltreatment. Neurosci Biobehav Rev 27:33–44

Teicher MH, Tomoda A, Susan L, Andersen SL (2006) Neurobiological consequences of early stress and childhood maltreatment: are results from human and animal studies comparable? Ann N Y Acad Sci 1071:313–323

Tenore A, Oberkotter LV, Koldovsky O, Parks JS, Vanderberg CM (1981) Thyrotropin in human breast milk. Horm Res 14:193–200

Tucker HA, Schwalm JW (1977) Glucocorticoids mammary tissue and milk. J Anim Sci 46:627–634

Uvnas-Moberg K, Alster P, Petersson M, Sohlstro¨m A, Bjorkstrand E (1998) Postnatal oxytocin injections cause sustained weight gain and increased nociceptive thresholds in male and female rats. Pediatr Rev 43:1–5

Valtonen M, Kangas AP, Voutilainen M, Eriksson L (2003) Diurnal rhythm of melatonin in young calves and intake of melatonin in milk. Anim Sci 77:149–154

van Ijzendoornd MH (1992) Intergenerational transmission of parenting: a review of studies in nonclinical populations. Dev Rev 12:76–99

Viau V, Sharmas S, Plotsky PM, Meaney MJ (1993) The hypothalamic-pituitary adrenal response to stress in handled and non-handled rats: differences in stress-induced plasma ACTH secretion are not dependent upon increased corticosterone levels. J Neurosci 13:1097–1105

Vigouroux E, Rastoqui N, Fenerole JM (1980) Estimation of hormonal and nonhormonal iodine uptake from maternal milk in suckling rats. Acta Endocrinol 93:332–338

Weinberg J, Smotherman WP, Levine S (1978) Early handling effects on neophobia and conditioned taste aversion. Physiol Behav 5:589–596

Weininger O (1954) Physiological damage under emotional stress as a function of early experience. Science 119:285–286

Weiss EL, Longhurst JG, Mazure CM (1999) Childhood sexual abuse as a risk factor for depression in women: psychosocial and neurobiological correlates. Am J Psychiatry 156:816–828

Werner S, Widstrom AM, Wahlberg V, Eneroth P, Winberg J (1982) Immunoreactive calcitonin in maternal milk and serum in relation to prolactin and neurotensin. Early Hum Dev 6:77–82

Werner H, Amarant T, Fridkin M, Koch Y (1986) Growth hormone releasing factor-like immunoreactivity in human milk. Biochem Biophys Res Commun 135:1084–1089

Werner H, Katz P, Fridkin M, Koch Y, Levine S (1988) Growth hormone releasing factor and somatostatin concentrations in the milk of lactating women. Eur J Pediatr 147:252–256

Whitworth NS, Grosvenor CE (1978) Transfer of milk prolactin to the plasma of neonatal rats by intestinal absorption. J Endocr 79:191–199

Wilson DA, Willner J, Kurz EM, Nadel L (1986) Early handling increases hippocampal long-term potentiation in young rats. Behav Brain Res 21:223–227

Witt DM (1997) Mechanisms of oxytocin-mediated sociosexual behavior. Ann N Y Acad Sci 807:287–301

Wolford ST, Argoudelis CJ (1979) Measurement of estrogen in cow's milk, human milk, and dietary products. J Dairy Sci 62:1458–1463

Wolinski J, Slupecka M, Romanowicz (2014) Leptin and ghrelin levels in colostrum, milk and blood plasma of sows and pigs neonates during the first week of lactation. Anim Sci J 85:143–149

Zaied AA, Bierschwal CJ, Elmore RG, Youngquist RS, Sharp AJ, Garverick HA (1979) Concentrations of progesterone in milk as a monitor of early pregnancy diagnosis in dairy cows. Theriogenology 12:3–11

Part II
Perinatal Programming: Studies in Humans

Chapter 12
Retrospective Studies

Patrícia Pelufo Silveira and Gisele Gus Manfro

Abstract Large retrospective, epidemiological studies accumulated in the late 1980s, providing increasing evidence to the deeply rooted thought that perinatal events could persistently affect the individual's functioning and health/disease patterns throughout the lifetime. Evidences of such associations can be found in the literature since the beginning of the twentieth century, but studies from Barker, Hales, and colleagues serve as an important hallmark. They proposed the "thrifty phenotype" hypothesis, stating that poor nutrition in fetal and early infant life is detrimental to the development and function of the individuals' organism, predisposing them to the later development of adult chronic diseases. At first used to explain the increased risk for type 2 diabetes in low birth weight individuals, the hypothesis was soon adapted to other systems, becoming one of the core assumptions of the Developmental Origins of Adult Health and Disease (DOHaD) model. The central nervous system is also vulnerable to the effects of environmental variation during fetal or neonatal life. Many researchers have explored the effects of perinatal programming on the human neurodevelopment, and some aspects of the brain structure and/or functioning (such as cognitive function, physiological reactivity to stress, and the risk for behavioral disorders or psychopathology) were shown to be modifiable by the exposure to certain adverse events early in life such as neonatal infections, exposure to gestational psychosocial stress, nutrition during gestation, exposure to drugs, or tobacco smoking during pregnancy. Until recently, most studies focused on birth weight as a strong surrogate of the intrauterine environment, investigating the effects of low birth weight (as a marker of suboptimal fetal environment) on a variety of neurodevelopmental outcomes. Despite the fact that literature reviews on this topic are as old as 1940, the more recent retrospective studies are summarized in this chapter.

P. P. Silveira (✉)
Departamento de Pediatria, Faculdade de Medicina—Hospital de Clínicas de Porto Alegre, Universidade Federal do Rio Grande do Sul, Ramiro Barcelos, 2350, Largo Eduardo Zaccaro Faraco, Porto Alegre, RS 90035-903, Brazil
e-mail: 00032386@ufrgs.br

G. G. Manfro
Departamento de Psiquiatria, Faculdade de Medicina—Hospital de Clínicas de Porto Alegre, Universidade Federal do Rio Grande do Sul, Ramiro Barcelos, 2350, Largo Eduardo Zaccaro Faraco, Porto Alegre, RS 90035-903, Brazil

M. C. Antonelli (ed.), *Perinatal Programming of Neurodevelopment*,
Advances in Neurobiology 10, DOI 10.1007/978-1-4939-1372-5_12,

Retrospective studies are based on existing data that have been recorded for reasons other than research. Therefore, the investigator depends on the availability and accuracy of the medical record (e.g., chart reviews) and there is no follow up of the individuals. There are three general types of retrospective study: case report, case series, and case-control study. A case report describes unusual/instructive case (or multiple cases in the case series). Case-control studies retrospectively compare persons who have an outcome of interest (cases) with others who do not have it (controls), estimating the odds to have the outcome based on the previous exposure to the risk factor of interest (Hess 2004).

Large retrospective, epidemiological studies accumulated in the late 1980s, providing increasing evidence to the deeply rooted thought (Ferreira 1965) that perinatal events could persistently affect the individual's functioning and health/disease patterns throughout the lifetime. After having observed that early twentieth century local infant mortality rates were related to the geographical differences in death rates from ischemic heart disease in England and Wales (Barker and Osmond 1986), Barker and colleagues hypothesized that impaired perinatal growth could be an important risk factor for ischemic heart disease in adult life. To verify this, they studied the records from the east Hertfordshire districts, in England, which were obtained by the attending midwife between 1911 and 1930. She was required to notify every birth to the county medical officer of health within 36 h, as well as to visit the newborn's home periodically throughout the first year of life. Then, using the National Health Service Central Register, they were able to investigate the cause of death of these people in relation to their birth weight. Among 5654 men, those who had had the lowest weights at birth (below 2.5 kg) had the highest death rates from ischemic heart disease as adults (Barker et al. 1989). Soon after that, Hales, Barker, and collaborators also described that reduced growth in early life is strongly linked with impaired glucose tolerance and type II diabetes (Hales et al. 1991), as well as with the prevalence of metabolic syndrome (Barker et al. 1993), with the risk falling progressively in both men and women from those who had the lowest to those who had the highest birth weights. They proposed the "thrifty phenotype" hypothesis (Hales and Barker 1992), which states that poor nutrition in fetal and early infant life is detrimental to the development and function of the individuals' organism, predisposing them to the later development of these adult chronic diseases. At first used to explain the increased risk for type 2 diabetes in low birth weight individuals, the hypothesis was soon adapted to other systems, becoming one of the core assumptions of the Developmental Origins of Adult Health and Disease (DOHaD) model (Barker 2003; Gluckman and Hanson 2004; Silveira et al. 2007; Wadhwa et al. 2009).

The central nervous system is also vulnerable to the effects of environmental variation during fetal or neonatal life. Many researchers have explored the effects of perinatal programming on the human neurodevelopment, and some aspects of the brain structure and/or functioning (such as cognitive function, physiological reactivity to stress, and the risk for behavioral disorders or psychopathology) were shown to be modifiable by the exposure to certain adverse events early in life such as neonatal infection (e.g., Salk et al. 1985), exposure to gestational psychosocial stress (e.g., Selten et al. 1999), nutrition during gestation (e.g., Susser and Lin 1992; Brown et al. 1995), exposure to drugs (e.g., Jacobson et al. 1987), or tobacco smoking during pregnancy (e.g., Mortensen et al. 2005). Until recently, most studies focused on birth weight as a strong surrogate of the intrauterine environment, investigating the effects of low birth weight (as a marker of suboptimal fetal environment) on a variety of neurodevelopmental outcomes. Despite the fact that literature reviews on this topic are as old as 1940 (Benton 1940), the more recent retrospective studies are summarized below.

12.1 Cognitive Functioning

Already in 1949, Asher and Roberts reviewed the birth histories of 4800 children from Britain's ordinary primary and secondary schools, 877 children from the special schools for the educationally subnormal, and 343 mentally defective children, describing that the incidence of very low birth weight was four times greater for the special school and defective children than for the controls (Asher and Roberts 1949). In 1953, Alm studied data obtained through a search of the communal and state records from the year 1947. The study compared 999 boys with a birth weight of 2500 g or less, born at the three largest maternity hospitals in Stockholm between 1902 and 1921, with 1002 control children born with birth weights between 2760 and 3750 g, finding increased rates of institutional care and governmental pensions for spastic paralysis, epilepsy, and educable and uneducable mental deficiency in the low birth weight group (Alm 1953). In 1955, Pasamanick and Lilienfeld used official records identifying information of patients with diagnosed mental deficiency or intelligence quotients under 80 who were born in Baltimore during the period from 1935 to 1952 inclusive, using the subsequent birth as a control, and found associations between mental retardation and complications of pregnancy and parturition (bleeding, toxemia), low birth weight, and abnormal neonatal conditions (seizures, cyanosis, or anoxia; Pasamanick and Lilienfeld 1955). Moore (1965) described that premature children of relatively low IQ had lower birth weight when compared to those of relatively high IQ, especially in males (Moore 1965). Studying 573 institutionalized, mentally retarded children, Katz and Taylor (1967) reported a higher incidence of low birth weight and, especially, of very low birth weight in his samples as compared with the general population of Pennsylvania (Katz and Taylor 1967).

The quarter-century from 1956 to 1981 brought more progress and more promise for the low birth weight/premature infant than any similar previous quantum of time (David and Siegel 1983). Better perinatal medical care accounted for the majority of the improvement, with consequent reduction in mortality rates and a complete change in terms of the previously established epidemiological associations and known risk factors for morbidity in this population (Kessel et al. 1984). As the neonatal intensive care established and improved, interest in the long-term impact of this medical intervention became more and more evident, as it can be detected by the innumerous prospective studies started at the time. Homogeneous, consistent retrospective data from this period are virtually inexistent. An exception is a large series of 12,058 live births from Finland, covering 96% of all children born in the region in 1966, which gathered information from questionnaires and medical records up to the age of 14 years and demonstrated that the incidence of severe mental retardation, lower IQ, and educational problems was significantly higher in the lower birth weight groups (Rantakallio 1985). More recent studies have used birth weight as a continuous variable, in correlation to IQ scores. In a study of 4300 Danish military conscripts born between 1973 and 1975, Sorensen et al. found that the Boerge Prien IQ test increased progressively with the birth weight, leveling out above a birth weight of 4.2 kg (Sørensen et al. 1997). Another large-scale study of 357,768 Swedish military conscripts born as singletons without congenital malformations between 1973 and 1981 showed that poor fetal growth was associated with decreased cognitive functioning in adult life (Lundgren et al. 2004) and this was not explained by differences in socioeconomic status (Bergvall et al. 2006). Recently, a U-shaped association of fetal growth with the risk for intellectual disability in a large Australian retrospective study has been reported (Leonard et al. 2008).

More recent studies are better at discerning the effects of prematurity (less than 37 weeks gestation at birth) from the poor intrauterine growth (low birth weight for a given gestational age). For instance, a study retrospectively divided a total of 1,623,038 live births registered in the Taiwan Birth Registry between 1985 and 1989 into those born at term with low birth weight, those preterm infants with normal birth weight, and the preterm infants with low birth weight; using data from the Basic Competence Test (assessing language, mathematics, and sciences) applied to all junior high school students in Taiwan at age 15 years, the researchers showed that lower birth weight has a negative effect on adolescent learning achievement (Wang et al. 2008). Other retrospective studies have found associations between specific types of prenatal stress and individuals performing less well in cognitive tests. For instance, it was shown a dose–response relationship between maternal smoking during pregnancy and offspring adult intelligence at the mean age of 18.7 years in 3044 singleton males from the Copenhagen Perinatal Cohort at start of the military service (Mortensen et al. 2005).

A very recent study investigated the cognitive abilities of men from the Helsinki Birth Cohort Study who had been separated temporarily from their parent(s) during World War II. These men underwent the Finnish Defense Forces Basic Intellectual Ability Test twice, at 20 years and retest at 70 years. Compared with the men without childhood separation, those who were separated from their parents scored lower on verbal, visuospatial, arithmetic, and general cognitive ability at the ages of 20 and 70 years. All these studies suggest that early-life stress is significantly associated with weaker cognitive performance at different time points during the life course in humans (Pesonen et al. 2013).

12.2 Behavioral Disorders

A relatively consistent literature documents that low birth weight may be associated with attention deficit/hyperactivity disorder (ADHD). An early extensive retrospective study including 1151 children referred to a branch of the Baltimore Board of Education for deviant behavior (lack of concentration, or hyperactive/confused/disorganized behavior) shows that prematurity, with or without other perinatal complications, was found significantly more often among these children than among matched controls (Pasamanick et al. 1956). Data from the 1981 National Health Interview Survey including 11,699 children aged 4–17 years provided information on parent-reported behavior problems and chronic childhood conditions; it shows that low birth weight children were more likely to experience school difficulty (in terms of both repeating grades and requiring special education) as well as to have received higher scores on the hyperactivity subscale of the Behavior Problems Index (McCormick et al. 1990). A case-control study selected a sample of 823 children from lists of newborn discharges occurred between 1983 and 1985 in two major hospitals in Michigan; using data abstracted from medical records at birth and from teachers' ratings on the Teacher Report Form, researchers were able to show that low birth weight children had increased scores on the Attention Problems scale (Breslau et al. 1996). In 1993, a population random sample survey of 1775 children aged 4–16 years in Western Australia, in which linkage with original birth information was performed, used the Child Behavior Checklist (CBCL) and the Teacher Report Form as the behavioral outcomes. It was seen that children who had achieved only 57–72 % of their expected birth weight given their gestational age at delivery were at significant risk to have altered behavior on the CBCL, as well as more likely to be rated as academically impaired (Zubrick et al. 2000).

The 1997 Health Survey for England, besides addressing health issues, demographic and social factors, also collected data on reported birth weight and the Strengths and Difficulties Questionnaire for children aged 4–15 years, a measure that enquires about five dimensions of behavior: conduct problems, hyperactivity, emotional symptoms, peer relationships, and prosocial behavior. In this study, it was found that children with high hyperactivity, peer problem, conduct problem, and total difficulties scores were lighter at birth than those with low/borderline scores. A gender effect was also evident, in which boys with high hyperactivity, peer and conduct problem scores were significantly lighter than boys with low/borderline scores, as well as girls with peer problems were significantly lighter than girls without such problems (Kelly et al. 2001). In a Finnish study, a consecutive series of mothers and their singleton infants who were born healthy in one of the main maternity hospitals in the Helsinki area between March and November 1998 responded to a survey sent to the sample in 2003, in which behavioral symptoms of ADHD were rated by the mother and the father of the child using the ADHD rating scale. The researchers report that a smaller ponderal index, a smaller head circumference, and a smaller head-circumference-to-length ratio at birth predicted significantly higher total and inattention-hyperactivity subscale scores on the ADHD rating scale, even after adjusting for variables such as gender, income, maternal age, length of gestation, maternal alcohol and tobacco use during pregnancy, maternal prepregnancy body mass index (BMI), parity, and the child's BMI at the age of 5–6 (Lahti et al. 2006).

More recently, two retrospective, hospital-based, case-control studies conducted comparing ADHD cases with non-ADHD students show that the number of ADHD cases who had a birth weight below 2500 g was approximately 3–3.6 times the number of control cases who had low birth weight (Sasaluxnanon and Kaewpornsawan 2005; Mick et al. 2002).

In line with the associations with fetal growth, other specific prenatal adverse exposures are prominent risk factors for behavioral problems in the offspring later in life. For example, a study using data from 4704 children 4–15 years of age from the American National Health and Nutrition Examination Survey 1999–2002 demonstrates that prenatal tobacco exposure is significantly associated with ADHD (Braun et al. 2006). In accordance to this finding, a Danish study including all children who were born in Denmark from 1991 through 1994 and recorded in the Danish Psychiatric Central Register until the end of 1999 with hyperkinetic disorders as their main diagnosis (=170 cases), and a random sample of 3765 controls born on the same date shows that maternal smoking during pregnancy is associated a nearly threefold increased risk for hyperkinetic disorder in the offspring; this is still significant after adjustment for years of schooling after basic school, employment status, annual income, cohabitant status, and history of psychiatric disorders in the parents and siblings (Linnet et al. 2005).

The association between maternal smoking during gestation and increased risk for ADHD was also reported in a case-control study performed in 12 public schools in Porto Alegre, Brazil, enrolling 100 subjects with ADHD and 100 non-ADHD controls; this study demonstrated a significantly higher number of cigarettes smoked per day during pregnancy for mothers of subjects with ADHD (Schmitz et al. 2006). Finally, a study performed in twins recruited 2054 pairs at 5–16 years of age, identified from the population-based Greater Manchester Twin Register whose mothers completed a validated set of questionnaires and information on the number of cigarettes smoked during pregnancy, as well as the birth weight of each twin. Teachers were also asked to complete an ADHD rating scale on each twin. The study shows a strong association between maternal smoking during pregnancy and offspring ADHD symptom scores (Thapar et al. 2003). Other studies also report this association between maternal smoking during gestation and the offspring ADHD (Milberger et al. 1998; Milberger et al. 1996). Less consistent evidence also exists linking exposure to stress during gestation and the risk for childhood ADHD (McIntosh et al. 1995).

Psychological stress during pregnancy has long been recognized as a possible risk factor for behavioral and developmental outcomes in humans. An early case-control study involving autistic and nonautistic children demonstrates a higher presence of family discord during the prenatal period in the cases (Ward 1990). Another study enrolling autistic children and controls gathered information on perinatal variables and major life events addressed by the Social Readjustment Rating Scale in

4-week blocks spanning the duration of pregnancy. The study reports an increase in the number of reported prenatal stressors among mothers of autistic children, especially at 21–32 weeks gestation, which is consistent with the embryological age in which brain structures believed to be altered in autism are developing (Beversdorf et al. 2005). Another interesting study linked data from the Louisiana's Department of Health and Hospitals on the birth dates and gender of all individuals seen in the state health system since 1990 who had a diagnosis of autism disease, with data from the National Weather Service on all hurricanes, tropical storms, or floods that struck Louisiana from 1980 to 1995, including the storms' dates, tracks, and degree of destructiveness. They show that the prevalence of autism disease increased in a dose-response fashion with the severity of the prenatal storm exposure, and the effect depends on the period of exposure (higher if the exposure was on mid-gestation), being also more evident in males (Kinney et al. 2008). A case-control study based on children born in Denmark from 1973 to 1999 and including all children ($N=698$) discharged from a Danish psychiatric hospital with a diagnosis of infantile or atypical autism and a set of controls; information about perinatal characteristics was obtained from the Medical Birth Register. Breech presentation, low Apgar score at 5 min (less than 7), low birth weight (less than 2500 g), gestational age at birth of less than 35 weeks, and being small for gestational age and high paternal age were associated with a statistically significantly increased risk of autism (Larsson et al. 2005).

Therefore, many different types of early-life events seem to affect neurodevelopment and consequently behavior. The timing and severity of the insult seem critical to determine the outcome in terms of the severity and morbidity.

12.3 Mental Health

In 1987, a very interesting study linked data from the State Institute of Forensic Medicine collected between 1978 and 1984, and birth data from the Stockholm hospitals in unambiguous cases of suicide victims ($n=281$), alcoholics ($n=66$), and drug addicts ($n=106$). The comparison with 2901 controls showed that suicides involving asphyxiation were closely associated with asphyxia at birth, suicides by violent mechanical means were associated with mechanical birth trauma, and drug addiction was associated with opiate and/or barbiturate administration to mothers during labor (Jacobson et al. 1987). Although at the time it was not possible to foresee any mechanism to explain the findings, the authors were already proposing that birth trauma could be transferred, somehow, to adulthood.

Nonetheless, some evidence of this transmission (or perinatal programming) was already described by then. A 1966 case-control study located birth certificates of 52 schizophrenic adults and their 115 siblings in the Bureau of Vital Statistics at Cleveland City Hall, and compared the birth weight between cases and controls. Despite the fact that most of the schizophrenics were males (who generally have a higher birth weight than females), while the siblings were equally divided between males and females, the schizophrenics weighed significantly less at birth than did their siblings (Lane and Albee 1966).

In a set of 7086 individuals born at the public Helsinki University Central Hospital between 1924 and 1933 that had birth information, mothers' height, weight in late pregnancy, age, parity, and the date of the last menstrual period extracted from birth records together with data on the newborns' length, weight, head circumference, and placental weight, early-life information was linked to the Finnish Hospital Discharge Register, so that diagnosis of broad schizophrenia could be identified. The study reports that mothers' late-pregnancy BMI as well as low birth weight were significantly related to the occurrence of schizophrenia in their offspring (Wahlbeck et al. 2001). Similarly, other subsequent studies also describe an association between perinatal obstetric complications and an increased risk for schizophrenia later in life (Lewis and Murray 1987; Foerster et al. 1991; Rifkin et al. 1994; Verdoux et al. 1997; Kotlicka-Antczak et al. 2001).

In another large retrospective study, 334,188 males born in Sweden between 1973 and 1980 (Swedish Medical Birth Registry) and conscripted into military service in 1990–1997 (Military Service Conscription Registry) were linked to their medical records from then on (Population and Housing Censuses and the Swedish Inpatient Discharge Register), in an attempt to gather incident schizophrenia cases arising in the period following the conscription medical examination. 106 individuals developed schizophrenia and 173 developed other nonaffective psychoses subsequent to their conscription medical examination. Both low birth weight and high birth weight were associated with increased risk for the disease, but the greatest risk was seen in the low birth weight category (Gunnell et al. 2003). However, in a subsequent study in the same population but including women, the authors argue that there was little evidence of an association between birth weight and schizophrenia, although short babies were at an increased risk. In males, low BMI and short height at age 18 were associated with increased risk (Gunnell et al. 2005). The same research group studied the predictors of suicide in 713,370 young adults, born in Sweden between 1973 and 1980, describing a raised risk of attempted suicide for individuals of short birth length, adjusted for gestational age, born fourth or more in birth order, from mothers with lower maternal educational level or aged 19 years or younger. Predictors of suicide were low birth weight adjusted for gestational age and teenage motherhood (Mittendorfer-Rutz et al. 2004). An earlier report using the same strategy and databases describes that adulthood schizophrenia as well as affective and reactive psychosis were associated with specific perinatal variables such as multiparity, maternal bleeding during pregnancy, birth season, and uterine atony; an increased risk for schizophrenia was described in boys who were small for their gestational age at birth as well as whose mothers had had bleeding during late pregnancy (Hultman et al. 1999).

A sharp and time-limited decline in the food intake of the Dutch population following a Nazi blockade in 1944–1945 generated birth cohorts exposed to severe food deprivation; data on these individuals were later linked to the Dutch Psychiatric Registry, which contains information on all inpatients' psychiatric admissions to psychiatric and university hospitals in Holland since 1978. In this study, researchers verified that the offspring exposed to famine in utero had increased relative risks for adulthood schizophrenia in women (Susser and Lin 1992), or in both genders in the most severely undernourished cohort (Susser et al. 1996), suggesting that early prenatal nutrition can have an effect on the risk for this disease. The same association was also found for schizoid and antisocial personality disorders, affective psychosis, and addiction in this cohort (Brown et al. 1995; Hoek et al. 1996; Brown et al. 2000; Franzek et al. 2008; Neugebauer et al. 1999). In a different cohort from the same populational basis, it was shown that maternal exposure to famine prior to conception was associated with poorer mental health and quality of life in her adult offspring (Stein et al. 2009). Interestingly, individuals exposed to a massive famine in China in 1959–1961 experienced similar results, showing that prenatal exposure to famine increases risk of schizophrenia in later life (St Clair et al. 2005; Xu et al. 2009), although this finding was not supported by a large nationally representative sample survey study in China (Song et al. 2009).

Exposure to maternal stress during gestation was also proposed to be involved in the risk for affective and nonaffective psychoses, but the results are less consistent. For instance, in a study performed in the Netherlands, researchers gathered information from the Dutch Psychiatric Registry on admissions to inpatient psychiatric care, selecting data on patients who had been born in one of the 19 municipalities where the impact of a natural disaster (in 1953, a violent gale caused a flood in the southwest of the Netherlands) was high and mortality exceeded 0.25%. The risk of nonaffective psychosis for the exposed birth cohort was found to be increased, but the result was statistically not significant (RR=1.8; 95% CI: 0.9–3.5; Selten et al. 1999). Another Dutch study also investigated the effect of gestational stress (attack and occupation by the German army in May 1940) on the risk for adulthood schizophrenia. Individuals whose mothers were pregnant at the time of the invasion were defined as exposed. Nonexposed cohorts were individuals born in the corresponding periods in the previous and subsequent 2 years. Data on psychiatric admissions were obtained from the National Psychiatric Case Register. Exposed individuals had a small, but statistically significant increased risk of lifetime diagnosis of schizophrenia (van Os and Selten 1998). Finally, a population-based cohort derived from all 92,408 births in 1964–1976 to mothers residing in West Jerusalem gathered information from the Israel's population registry and the Israel's national psychiatric registry. Gestational exposure to the Arab–Israel War of June 1967 was considered as an acute maternal stress exposure. Offspring whose mothers were in the first trimester of pregnancy during the war had a significantly increased relative risk for mood disorders as compared to other trimesters (Kleinhaus et al. 2013).

A study enrolling 52 adolescents whose death before their 20th birthday was deemed a suicide by the Office of the State Medical Examiner, Department of Health, Rhode Island, between 1975 and 1983 and a group of 104 controls (births preceding and following that of the subject, matching sex, race, and hospital of birth in the records of the Division of Vital Statistics of the Department of Health in the State of Rhode Island) shows that several perinatal variables were significantly related to the increased risk for adolescent suicide: previous premature births, bleeding during pregnancy, maternal chronic disease, long duration of labor, decreased fetal heart rate during labor, placenta previa, no antenatal care before 20 weeks, respiratory distress for more than 1 h after birth, and neonatal infection (Salk et al. 1985).

Soon after the proposal of the "thrifty phenotype hypothesis" and the description that low birth weight was associated with ischemic heart disease in the population from the east Hertfordshire districts, Barker and collaborators studied the relationship between early-life variables (birth weight, weight at 1 year of age) registered by the attending midwife between 1911 and 1930 and the incidence of suicide in this population, by comparison to the national rates for men and women of a corresponding age and year of birth. The researchers demonstrate that each kilogram decrease in weight gain between birth and 1 year was associated with an increased risk of suicide of 45 % in men and 31 % in women (Barker et al. 1995). In a subsequent study in the same population, it was demonstrated that, among men but not women, the odds ratios for depression fell with increasing birth weight (Thompson et al. 2001).

Nevertheless, these findings are not a consensus. In a Danish study, data from 10,753 birth certificates were linked to the Danish Psychiatric Central Register, which compiles data on admissions to psychiatric hospitals and to psychiatric departments in general hospitals in Denmark since April 1969, with coverage close to 100 %. A total of 190 men had been discharged from a psychiatric ward with a diagnosis of depression between 1969 and 2002, of which 39 were diagnosed as having a bipolar affective disorder. There was no association between birth weight or ponderal index and risk of depression from age 16 to 49 years in this study (Osler et al. 2005). A subsequent study in the same population also reports no association between indicators of fetal growth and the risk for bipolar disorder (Øgendahl et al. 2006). A larger analysis linking data from the Danish Civil Registration System, Danish Psychiatric Central Register, Cause of Death Register, and the Danish Medical Birth Register covering a period from 1955 to 2005 gathered information in more than 2 million persons and enabled the researchers to calculate relative risks for schizophrenia, bipolar disorder, unipolar depressive disorder, and schizoaffective disorder. The study showed that the loss of a parent (especially by suicide) was a risk factor for all disorders; schizophrenia was related to paternal age and living conditions at birth, and being born small for gestational age was a risk factor for all disorders except for schizophrenia (Laursen et al. 2007).

The Helsinki Birth Cohort Study has provided a large body of evidence concerning the early-life effects on the risk for mental and personality disorders. For instance, a study focused on individuals that had been separated temporarily from their parent(s) during World War II (register in the Finnish National Archives). Of the 12,734 participants, 1487 subjects had a serious mental disorder (registered in the hospital discharge register or the death certificate); after the analysis, it was shown that the separated children had higher risks of substance use mental and personality disorder (Lahti et al. 2012; Räikkönen et al. 2011), as well as depressive symptoms (Pesonen et al. 2007). A study in the same subjects demonstrates that higher trait anxiety is predicted by smaller body size at birth, in infancy and in adulthood. Moreover, faster growth particularly from 7 to 11 years of age and slower growth between 11 and 63 years predicted higher trait anxiety (Lahti et al. 2010).

12.4 Hypothalamic–Pituitary–Adrenal Axis: A Possible Mechanism Linking Early-Life Experiences to the Development of Mood Disorders

The hypothalamic–pituitary–adrenal (HPA) axis is one of the physiological systems that help humans sustain emotional, cognitive, behavioral, and metabolic activity in response to threat. Upon exposure to stress, neurons of the hypothalamus secrete CRH that stimulates the secretion of ACTH from the anterior pituitary. ACTH induces the secretion of glucocorticoids by the adrenal. Thus, the secretion of glucocorticoids is a necessary physiological response to emotional and physical stress.

The loss of a parent during childhood is considered an emotional stress and is associated to increased prevalence of psychiatric disorders such as depression, anxiety, and substance abuse later in life. Rodents and nonhuman primate studies have shown that maternal loss or separation are associated to long-term effects on copying style, social adjustment, cognitive function, and behavioral responses to stress as well as affect latter functioning of HPA axis. Considering this, HPA dysregulation may play a significant role in mediating the pathway between early attachment relationships to adult disorders.

However, few studies have been conducted in humans in order to evaluate the psychobiology of early-life events. An epidemiological community study (Flinn and England 1997) demonstrated that glucocorticoids are associated with health status in infants and also showed that long-term separation from a parent could influence HPA function in children. In accordance to this study, Gunnar et al. (2001) have also described higher cortisol levels in Romanian orphans after 6 years or more of their adoption (Gunnar et al. 2001). But, it is still not known whether these effects may persist into adult life. Some studies have evaluated cortisol levels from adults that have suffered parental loss. Tyrka et al. (2008) studying 44 adults suggested that parental loss is associated to increased cortisol levels at baseline or after stress particularly in males (Tyrka et al. 2008). A study that evaluated 57 adult men demonstrates that parental death during childhood is associated to higher cortisol levels, providing evidence that early-life stressors may have lasting effects on HPA functions, even in the absence of psychopathology (Nicolson 2004). The effect of parental loss on HPA axis later in life may be dependent on gender (DeSantis et al. 2011) and some recent studies have demonstrated that maternal separation in childhood is associated to alterations in the diurnal cortisol pattern in middle age (Kumari et al. 2013).

Many animal studies demonstrate that induced prenatal stress has sizable and long-lasting effects on offspring fear, behavior, neurogenesis, immunity, and stress physiology (Coe et al. 2003). In humans, prenatal anxiety is associated to a wide range of outcomes in children, including decrease in cognitive functions (Davis and Sandman 2010), altered neuropsychological tests (Glover et al. 2004), altered sleep (O'Connor et al. 2007), and behavioral problems (Van den Bergh and Marcoen 2004). It is hypothesized that elevated maternal cortisol (in response to stress exposure, for instance) may cross the placenta and affect obstetric outcomes, as well as child development via HPA functioning (O'Donnell et al. 2012). Although this hypothesis is evident in animal models, less evidence is found in humans. A previous study with 74 10-year-old children demonstrated that prenatal anxiety predicted raised diurnal cortisol (O'Connor et al. 2005). In a study with 889 15-year-old adolescents evaluated regarding their cortisol levels at awakening, 30 min after awakening, at afternoon, and at evening during 3 consecutive days found that prenatal maternal mood was associated with a modest alteration of adolescents' diurnal cortisol levels indexed by reduced cortisol awakening response and flatter diurnal slope (O'Donnell et al. 2013). This study provides a strong evidence that prenatal mood has a small but persistent effect on the HPA axis and human development. In agreement to this study, healthy young adults whose mothers experienced a severe stress during pregnancy (death of a loved one) had lower cortisol levels before a stress test whereas higher cortisol after a stress test as compared to individuals from a control group (Entringer et al. 2009). Depression during pregnancy, as well as during the postpartum period is also associated to HPA dysfunction in the offspring. Prolonged intrauterine exposure to stress may lead to an alteration of the HPA axis functioning, resulting in increased cortisol levels that in turn may negatively affect many systems. Maternal depression can lead to offspring prematurity and low birth weight, mediated throughout maternal higher cortisol levels (Field et al. 2006). Early-life stressors influence trait-like cortisol levels and slope, with both hyper- and hypo-arousal depending on the type of the stressor (Essex et al. 2011).

Children maltreatment has also been associated to the development of psychopathology, but it is not known if it is mediated throughout HPA axis. A recent study demonstrated that women with a history of childhood physical abuse evaluated throughout the Children Trauma Questionnaire displayed a significantly blunted cortisol response to a stress test as compared to women without physical abuse in childhood (Carpenter et al. 2011).

Moreover, recent studies suggested that one of the processes linking intrauterine growth retardation or prenatal stress to long-term adverse outcomes later in life may be the fetal programming of the HPA axis. In adult life, subjects with low birth weight have higher basal morning cortisol levels (Levitt et al. 2000). In a study with 106 healthy young men, the cortisol levels after a stress test were inversely related to birth weight, providing evidence for the concept of fetal programming HPA function (Wüst et al. 2005).

In conclusion, a substantial body of research proposes that early-life experiences can have significant later effects on the behavior and neurodevelopment of offspring, suggesting that the abnormal development of the brain during gestation contributes to many neurological disorders that manifest later in life. Studies aiming at identifying windows of opportunity for intervention (e.g., environmental stimulation) as well as public health measures to improve antenatal care are warranted.

Conflict of Interest The authors declare no conflicts of interest.

References

Alm I (1953) The long-term prognosis for prematurely born children. Acta Paediatr Stockh 42(Suppl. 94):391–394
Asher C, Roberts JA (1949) A study on birth-weight and intelligence. Br J Soc Med 3(2):56–68
Barker D (2003) The midwife, the coincidence, and the hypothesis. BMJ 327(7429):1428–1430
Barker DJP, Osmond C (1986) Infant mortality, childhood nutrition, and ischaemic heart disease in England and Wales. Lancet 1:1077–1081
Barker DJ, Winter PD, Osmond C, Margetts B, Simmonds SJ (1989) Weight in infancy and death from ischaemic heart disease. Lancet 2(8663):577–580
Barker DJ, Hales CN, Fall CH, Osmond C, Phipps K, Clark PM (1993) Type 2 (non-insulin-dependent) diabetes mellitus, hypertension and hyperlipidaemia (syndrome X): relation to reduced fetal growth. Diabetologia 36(1):62–67
Barker DJ, Osmond C, Rodin I, Fall CH, Winter PD (1995) Low weight gain in infancy and suicide in adult life. BMJ 311(7014):1203
Benton AL (1940) Mental development of prematurely born children: a critical review of the literature. Am J Orthopsychiat 10:719–746
Bergvall N, Iliadou A, Tuvemo T, Cnattingius S (2006) Birth characteristics and risk of low intellectual performance in early adulthood: are the associations confounded by socioeconomic factors in adolescence or familial effects? Pediatrics 117(3):714–721
Beversdorf DQ, Manning SE, Hillier A, Anderson SL, Nordgren RE, Walters SE, Nagaraja HN, Cooley WC, Gaelic SE, Bauman ML (2005) Timing of prenatal stressors and autism. J Autism Dev Disord 35(4):471–478
Braun JM, Kahn RS, Froehlich T, Auinger P, Lanphear BP (2006) Exposures to environmental toxicants and attention deficit hyperactivity disorder in U.S. children. Environ Health Perspect 114(12):1904–1909
Breslau N, Brown GG, DelDotto JE, Kumar S, Ezhuthachan S, Andreski P, Hufnagle KG (1996) Psychiatric sequelae of low birth weight at 6 years of age. J Abnorm Child Psychol 24(3):385–400
Brown AS, Susser ES, Lin SP, Neugebauer R, Gorman JM (1995) Increased risk of affective disorders in males after second trimester prenatal exposure to the Dutch hunger winter of 1944–45. Br J Psychiatry 166(5):601–606
Brown AS, van Os J, Driessens C, Hoek HW, Susser ES (2000) Further evidence of relation between prenatal famine and major affective disorder. Am J Psychiatry 157(2):190–195
Carpenter LL, Shattuck TT, Tyrka AR, Geracioti TD, Price LH (2011) Effect of childhood physical abuse on cortisol stress response. Psychopharmacology (Berl) 214(1):367–375
Coe CL, Kramer M, Czéh B, Gould E, Reeves AJ, Kirschbaum C, Fuchs E (2003) Prenatal stress diminishes neurogenesis in the dentate gyrus of juvenile rhesus monkeys. Biol Psychiatry 54(10):1025–1034.
David RJ, Siegel E (1983) Decline in neonatal mortality, 1968–1977: better babies or better care? Pediatrics 71:531–540
Davis EP, Sandman CA (2010) The timing of prenatal exposure to maternal cortisol and psychosocial stress is associated with human infant cognitive development. Child Dev 81(1):131–148
DeSantis SM, Baker NL, Back SE, Spratt E, Ciolino JD, Moran-Santa MM, Dipankar B, Brady KT (2011) Gender differences in the effect of early life trauma on hypothalamic-pituitary-adrenal axis functioning. Depress Anxiety 28(5):383–392
Entringer S, Kumsta R, Hellhammer DH, Wadhwa PD, Wüst S (2009) Prenatal exposure to maternal psychosocial stress and HPA axis regulation in young adults. Horm Behav 55(2):292–298
Essex MJ, Shirtcliff EA, Burk LR, Ruttle PL, Klein MH, Slattery MJ, Kalin NH, Armstrong JM (2011) Influence of early life stress on later hypothalamic-pituitary-adrenal axis functioning and its covariation with mental health symptoms: a study of the allostatic process from childhood into adolescence. Dev Psychopathol 23(4):1039–1058

Ferreira AJ (1965) Emotional factors in prenatal environment: a review. J Nerv Ment Dis 141(1):108–118

Field T, Hernandez-Reif M, Diego M, Figueiredo B, Schanberg S, Kuhn C (2006) Prenatal cortisol, prematurity and low birthweight. Infant Behav Dev 29(2):268–275

Flinn MV, England BG (1997) Social economics of childhood glucocorticoid stress response and health. Am J Phys Anthropol 102(1):33–53

Foerster A, Lewis SW, Owen MJ, Murray RM (1991) Low birth weight and a family history of schizophrenia predict poor premorbid functioning in psychosis. Schizophr Res 5(1):13–20

Franzek EJ, Sprangers N, Janssens AC, Van Duijn CM, Van De Wetering BJ (2008) Prenatal exposure to the 1944-45 Dutch 'hunger winter' and addiction later in life. Addiction 103(3):433–438

Glover V, O'Connor TG, Heron J, Golding J, ALSPAC Study team (2004) Antenatal maternal anxiety is linked with atypical handedness in the child. Early Hum Dev 79(2):107–118

Gluckman PD, Hanson MA (2004) Living with the past: evolution, development, and patterns of disease. Science 305(5691):1733–1736

Gunnar MR, Morison SJ, Chisholm K, Schuder M (2001) Salivary cortisol levels in children adopted from Romanian orphanages. Dev Psychopathol 13(3):611–628

Gunnell D, Rasmussen F, Fouskakis D, Tynelius P, Harrison G (2003) Patterns of fetal and childhood growth and the development of psychosis in young males: a cohort study. Am J Epidemiol 158(4):291–300

Gunnell D, Harrison G, Whitley E, Lewis G, Tynelius P, Rasmussen F (2005) The association of fetal and childhood growth with risk of schizophrenia. Cohort study of 720,000 Swedish men and women. Schizophr Res 79(2–3):315–322

Hales CN, Barker DJ (1992) Type 2 (non-insulin-dependent) diabetes mellitus: the thrifty phenotype hypothesis. Diabetologia 35(7):595–601

Hales CN, Barker DJ, Clark PM, Cox LJ, Fall C, Osmond C, Winter PD (1991) Fetal and infant growth and impaired glucose tolerance at age 64. BMJ 303(6809):1019–1022

Hess DR (2004) Retrospective studies and chart reviews. Respir Care 49(10):1171–1174

Hoek HW, Susser E, Buck KA, Lumey LH, Lin SP, Gorman JM (1996) Schizoid personality disorder after prenatal exposure to famine. Am J Psychiatry 153(12):1637–1639

Hultman CM, Sparén P, Takei N, Murray RM, Cnattingius S (1999) Prenatal and perinatal risk factors for schizophrenia, affective psychosis, and reactive psychosis of early onset: case-control study. BMJ 318(7181):421–426

Jacobson B, Eklund G, Hamberger L, Linnarsson D, Sedvall G, Valverius M (1987) Perinatal origin of adult self-destructive behavior. Acta PsychiatrScand 76(4):364–371

Katz CM, Taylor PM (1967) The incidence of low birthweight in children with severe mental retardation. Am J Dis Child 114:79–87

Kelly YJ, Nazroo JY, McMunn A, Boreham R, Marmot M (2001) Birthweight and behavioural problems in children: a modifiable effect? Int J Epidemiol 30(1):88–94

Kessel SS, Villar J, Berendes HW, Nugent RP (1984) The changing pattern of low birth weight in the United States-1970 to 1980. JAMA 251(15):1978–1982

Kinney DK, Miller AM, Crowley DJ, Huang E, Gerber E (2008) Autism prevalence following prenatal exposure to hurricanes and tropical storms in Louisiana. J Autism Dev Disord 38(3):481–488

Kleinhaus K, Harlap S, Perrin M, Manor O, Margalit-Calderon R, Opler M, Friedlander Y, Malaspina D (2013) Prenatal stress and affective disorders in a population birth cohort. Bipolar Disord 15(1):92–99

Kotlicka-Antczak M, Gmitrowicz A, Sobów TM, Rabe-Jabłonska J (2001) Obstetric complications and Apgar score in early-onset schizophrenic patients with prominent positive and prominent negative symptoms. J Psychiatr Res 35(4):249–257

Kumari M, Head J, Bartley M, Stansfeld S, Kivimaki M (2013) Maternal separation in childhood and diurnal cortisol patterns in mid-life: findings from the Whitehall II study. Psychol Med 43(3):633–643

Lahti J, Räikkönen K, Kajantie E, Heinonen K, Pesonen AK, Järvenpää AL, Strandberg T (2006) Small body size at birth and behavioural symptoms of ADHD in children aged five to six years. J Child Psychol Psychiatry 47(11):1167–1174

Lahti J, Räikkönen K, Pesonen AK, Heinonen K, Kajantie E, Forsén T, Osmond C, Barker DJ, Eriksson JG (2010) Prenatal growth, postnatal growth and trait anxiety in late adulthood-the helsinki birth cohort study. Acta Psychiatr Scand 121(3):227–235

Lahti M, Pesonen AK, Räikkönen K, Heinonen K, Wahlbeck K, Kajantie E, Osmond C, Barker DJ, Eriksson JG (2012) Temporary separation from parents in early childhood and serious personality disorders in adult life. J Pers Disord 26(5):751–762

Lane EA, Albee GW (1966) Comparative birthweights of schizophrenics and their siblings. J Psych 64:227–231

Larsson HJ, Eaton WW, Madsen KM, Vestergaard M, Olesen AV, Agerbo E, Schendel D, Thorsen P, Mortensen PB (2005) Risk factors for autism: perinatal factors, parental psychiatric history, and socioeconomic status. Am J Epidemiol 161(10):916–925

Laursen TM, Munk-Olsen T, Nordentoft M, Bo Mortensen P (2007) A comparison of selected risk factors for unipolar depressive disorder, bipolar affective disorder, schizoaffective disorder, and schizophrenia from a Danish population-based cohort. J Clin Psychiatry 68(11):1673–1681

Leonard H, Nassar N, Bourke J, Blair E, Mulroy S, de Klerk N, Bower C (2008) Relation between intrauterine growth and subsequent intellectual disability in a ten-year population cohort of children in Western Australia. Am J Epidemiol 167(1):103–111

Levitt NS, Lambert EV, Woods D, Hales CN, Andrew R, Seckl JR (2000) Impaired glucose tolerance and elevated blood pressure in low birth weight, nonobese, young south african adults: early programming of cortisol axis. J Clin Endocrinol Metab 85(12):4611–4618

Lewis SW, Murray RM (1987) Obstetric complications, neurodevelopmental deviance, and risk of schizophrenia. J Psychiatr Res 21(4):413–421

Linnet KM, Wisborg K, Obel C, Secher NJ, Thomsen PH, Agerbo E, Henriksen TB (2005) Smoking during pregnancy and the risk for hyperkinetic disorder in offspring. Pediatrics 116(2):462–467

Lundgren M, Cnattingius S, Jonsson B, Tuvemo T (2004) Intellectual performance in young adult males born small for gestational age. Growth Horm IGF Res 14(Suppl. A):S7–S8

McCormick MC, Gortmaker SL, Sobol AM (1990) Very low birth weight children: behavior problems and school difficulty in a national sample. J Pediatr 117:687–693

McIntosh DE, Mulkins RS, Dean RS (1995) Utilization of maternal perinatal risk indicators in the differential diagnosis of ADHD and UADD children. Int J Neurosci 81(1-2):35–46

Mick E, Biederman J, Prince J, Fischer MJ, Faraone SV (2002) Impact of low birth weight on attention-deficit hyperactivity disorder. J Dev Behav Pediatr 23(1):16–22

Milberger S, Biederman J, Faraone SV, Chen L, Jones J (1996) Is maternal smoking during pregnancy a risk factor for attention deficit hyperactivity disorder in children? Am J Psychiatry 153(9):1138–1142

Milberger S, Biederman J, Faraone SV, Jones J (1998) Further evidence of an association between maternal smoking during pregnancy and attention deficit hyperactivity disorder: findings from a high-risk sample of siblings. J Clin Child Psychol 27(3):352–358

Mittendorfer-Rutz E, Rasmussen F, Wasserman D (2004) Restricted fetal growth and adverse maternal psychosocial and socioeconomic conditions as risk factors for suicidal behaviour of offspring: a cohort study. Lancet 364(9440):1135–1140

Moore BC (1965) Relationship between prematurity and intelligence in mental retardates. Am J Ment Defic 70(3):448–453

Mortensen EL, Michaelsen KF, Sanders SA, Reinisch JM (2005) A dose-response relationship between maternal smoking during late pregnancy and adult intelligence in male offspring. Paediatr Perinat Epidemiol 19(1):4–11

Neugebauer R, Hoek HW, Susser E (1999) Prenatal exposure to wartime famine and development of antisocial personality disorder in early adulthood. JAMA 282(5):455–462

Nicolson NA (2004) Childhood parental loss and cortisol levels in adult men. Psychoneuroendocrinology 29(8):1012–1018

O'Connor TG, Ben-Shlomo Y, Heron J, Golding J, Adams D, Glover V (2005) Prenatal anxiety predicts individual differences in cortisol in pre-adolescent children. Biol Psychiatry 58(3):211–217

O'Connor TG, Caprariello P, Blackmore ER, Gregory AM, Glover V, Fleming P, ALSPAC Study Team (2007) Prenatal mood disturbance predicts sleep problems in infancy and toddlerhood. Early Hum Dev 83(7):451–458

O'Donnell KJ, Bugge Jensen A, Freeman L, Khalife N, O'Connor TG, Glover V (2012) Maternal prenatal anxiety and downregulation of placental 11β-HSD2. Psychoneuroendocrinology 37(6):818–826

O'Donnell KJ, Glover V, Jenkins J, Browne D, Ben-Shlomo Y, Golding J, O'Connor TG (2013) Prenatal maternal mood is associated with altered diurnal cortisol in adolescence. Psychoneuroendocrinology. doi:10.1016/j.psyneuen.2013.01.008, pii: S0306-4530(13)00013-9

Øgendahl BK, Agerbo E, Byrne M, Licht RW, Eaton WW, Mortensen PB (2006) Indicators of fetal growth and bipolar disorder: a Danish national register-based study. Psychol Med 36(9):1219–1224

Osler M, Nordentoft M, Andersen AM (2005) Birth dimensions and risk of depression in adulthood: cohort study of Danish men born in 1953. Br J Psychiatry 186:400–403

Pasamanick B, Lilienfeld AM (1955) Association of maternal and fetal factors with development of mental deficiency. 1. Abnormalities in the prenatal and paranatal periods. JAMA 159(3):155–160

Pasamanick B, Rogers ME, Lilienfeld AM (1956) Pregnancy experience and the development of behavior disorders in children. Am J Psychiatry 112(8):613–618

Pesonen AK, Räikkönen K, Heinonen K, Kajantie E, Forsén T, Eriksson JG (2007) Depressive symptoms in adults separated from their parents as children: a natural experiment during World War II. Am J Epidemiol 166(10):1126–1133.

Pesonen AK, Eriksson JG, Heinonen K, Kajantie E, Tuovinen S, Alastalo H, Henriksson M, Leskinen J, Osmond C, Barker DJ, Räikkönen K (2013) Cognitive ability and decline after early life stress exposure. Neurobiol Aging 34:1674–1679

Räikkönen K, Lahti M, Heinonen K, Pesonen AK, Wahlbeck K, Kajantie E, Osmond C, Barker DJ, Eriksson JG (2011) Risk of severe mental disorders in adults separated temporarily from their parents in childhood: the Helsinki birth cohort study. J Psychiatr Res 45(3):332–338

Rantakallio P (1985) A 14-year follow-up of children with normal and abnormal birth weight for their gestational age. A population study. Acta Paediatr Scand 74(1):62–69

Rifkin L, Lewis S, Jones P, Toone B, Murray R (1994) Low birth weight and schizophrenia. Br J Psychiatry 165(3):357–362

Salk L, Lipsitt LP, Sturner WQ, Reilly BM, Levat RH (1985) Relationship of maternal and perinatal conditions to eventual adolescent suicide. Lancet 1(8429):624–627

Sasaluxnanon C, Kaewpornsawan T (2005) Risk factor of birth weight below 2,500 grams and attention deficit hyperactivity disorder in Thai children. J Med Assoc Thai 88(11):1514–1518

Schmitz M, Denardin D, Laufer Silva T, Pianca T, Hutz MH, Faraone S, Rohde LA (2006) Smoking during pregnancy and attention-deficit/hyperactivity disorder, predominantly inattentive type: a case-control study. J Am Acad Child Adolesc Psychiatry 45(11):1338–1345

Selten JP, van der Graaf Y, van Duursen R, Gispen-de Wied CC, Kahn RS (1999) Psychotic illness after prenatal exposure to the 1953 Dutch flood disaster. Schizophr Res 35(3):243–245

Silveira PP, Portella AK, Goldani MZ, Barbieri MA (2007) Developmental origins of health and disease (DOHaD). J Pediatr (Rio J) 83(6):494–504

Song S, Wang W, Hu P (2009) Famine, death, and madness: schizophrenia in early adulthood after prenatal exposure to the Chinese great leap forward famine. Soc Sci Med 68(7):1315–1321

Sørensen HT, Sabroe S, Olsen J, Rothman KJ, Gillman MW, Fischer P (1997) Birth weight and cognitive function in young adult life: historical cohort study. BMJ 315(7105):401–403

St Clair D, Xu M, Wang P, Yu Y, Fang Y, Zhang F, Zheng X, Gu N, Feng G, Sham P, He L (2005) Rates of adult schizophrenia following prenatal exposure to the Chinese famine of 1959–1961. JAMA 294(5):557–562

Stein AD, Pierik FH, Verrips GH, Susser ES, Lumey LH (2009) Maternal exposure to the Dutch famine before conception and during pregnancy: quality of life and depressive symptoms in adult offspring. Epidemiology 20(6):909–915

Susser ES, Lin SP (1992) Schizophrenia after prenatal exposure to the Dutch Hunger Winter of 1944–1945. Arch Gen Psychiatry 49(12):983–988

Susser E, Neugebauer R, Hoek HW, Brown AS, Lin S, Labovitz D, Gorman JM (1996) Schizophrenia after prenatal famine. Further evidence. Arch Gen Psychiatry 53(1):25–31

Thapar A, Fowler T, Rice F, Scourfield J, van den Bree M, Thomas H, Harold G, Hay D (2003) Maternal smoking during pregnancy and attention deficit hyperactivity disorder symptoms in offspring. Am J Psychiatry 160(11):1985–1989

Thompson C, Syddall H, Rodin I, Osmond C, Barker DJ (2001) Birth weight and the risk of depressive disorder in late life. Br J Psychiatry 179:450–455

Tyrka AR, Wier L, Price LH, Ross N, Anderson GM, Wilkinson CW, Carpenter LL (2008) Childhood parental loss and adult hypothalamic-pituitary-adrenal function. Biol Psychiatry 63(12):1147–1154

Van denBBR, Marcoen A (2004) High antenatal maternal anxiety is related to ADHD symptoms, externalizing problems, and anxiety in 8- and 9-year-olds. Child Dev 75(4):1085–1097

van Os J, Selten JP (1998) Prenatal exposure to maternal stress and subsequent schizophrenia. The May 1940 invasion of The Netherlands. Br J Psychiatry 172:324–326

Verdoux H, Geddes JR, Takei N, Lawrie SM, Bovet P, Eagles JM, Heun R, McCreadie RG, McNeil TF, O'Callaghan E, Stöber G, Willinger MU, Wright P, Murray RM (1997) Obstetric complications and age at onset in schizophrenia: an international collaborative meta-analysis of individual patient data. Am J Psychiatry 154(9):1220–1227

Wadhwa PD, Buss C, Entringer S, Swanson JM (2009) Developmental origins of health and disease: brief history of the approach and current focus on epigenetic mechanisms. Semin Reprod Med 27(5):358–368

Wahlbeck K, Forsén T, Osmond C, Barker DJ, Eriksson JG (2001) Association of schizophrenia with low maternal body mass index, small size at birth, and thinness during childhood. Arch Gen Psychiatry 58(1):48–52

Wang WL, Sung YT, Sung FC, Lu TH, Kuo SC, Li CY (2008) Low birth weight, prematurity, and paternal social status: impact on the basic competence test in Taiwanese adolescents. J Pediatr 153(3):333–338

Ward AJ (1990) A comparison and analysis of the presence of family problems during pregnancy of mothers of "autistic" children and mothers of normal children. Child Psychiatry Hum Dev 20(4):279–288

Wüst S, Entringer S, Federenko IS, Schlotz W, Hellhammer DH (2005) Birth weight is associated with salivary cortisol responses to psychosocial stress in adult life. Psychoneuroendocrinology 30(6):591–598

Xu MQ, Sun WS, Liu BX, Feng GY, Yu L, Yang L, He G, Sham P, Susser E, St Clair D, He L (2009) Prenatal malnutrition and adult schizophrenia: further evidence from the 1959-1961 Chinese famine. Schizophr Bull 35(3):568–576

Zubrick SR, Kurinczuk JJ, McDermott BM, McKelvey RS, Silburn SR, Davies LC (2000) Fetal growth and subsequent mental health problems in children aged 4 to 13 years. Dev Med Child Neurol 42(1):14–20

Chapter 13
Prenatal Stress and Its Effects on the Fetus and the Child: Possible Underlying Biological Mechanisms

Vivette Glover

Abstract Many prospective studies have shown that if a mother is depressed, anxious or stressed while pregnant, this increases the risk for her child having a wide range of adverse outcomes including emotional problems, symptoms of attention deficit hyperactivity disorder (ADHD) or impaired cognitive development. Although genetics and postnatal care clearly affect these outcomes, evidence for a prenatal causal component also is substantial. Prenatal anxiety/depression may contribute 10–15 % of the attributable load for emotional/behavioural outcomes.

The mechanisms underlying these changes are just starting to be explored. One possible mediating factor is increased exposure of the fetus to cortisol, as has been shown in animal studies. However, the human hypothalamic–pituitary–adrenal (HPA) axis which makes cortisol functions differently in human pregnancy from in most animals. The maternal HPA axis becomes gradually less responsive to stress as pregnancy progresses. And there is only a weak, if any, association between a mother's prenatal mood and her cortisol level, especially later in pregnancy. Cytokines are alternative possible mediators. An additional explanation is that stress or anxiety causes increased transfer of maternal cortisol across the placenta to the fetus. The placenta plays a crucial role in moderating fetal exposure to maternal factors and presumably in preparing the fetus for the environment in which it is going to find itself. There is some evidence in both rat models and in humans that prenatal stress can reduce placental 11β-HSD2, the enzyme which metabolises cortisol to inactive cortisone. The level of cortisol in the amniotic fluid, surrounding the baby in the womb, has been shown to be inversely correlated with infant cognitive development. However, several other biological systems are likely to be involved. Serotonin is another possible mediator of prenatal stress induced programming effects on offspring neurocognitive and behavioural development. The role of epigenetic changes in mediating alterations in offspring outcome following prenatal stress is likely to be important and starting to be explored.

V. Glover (✉)
Institute of Reproductive and Developmental Biology, Imperial College London,
Hammersmith Campus, Du Cane Road, London W12 0NN, UK
e-mail: v.glover@imperial.ac.uk

M. C. Antonelli (ed.), *Perinatal Programming of Neurodevelopment*,
Advances in Neurobiology 10, DOI 10.1007/978-1-4939-1372-5_13,

13.1 Prenatal Stress and Child Outcomes

Stress is a generic term, which includes a wide range of different exposures, and many different types of prenatal stress have been shown to be associated with altered outcome for the child. These include symptoms of maternal anxiety and depression (O'Connor et al. 2003; Van den Bergh et al. 2008), pregnancy-specific anxiety (Huizink et al. 2002), bereavement (Khashan et al. 2008), life events, including a bad relationship with the partner (Bergman et al. 2007) and exposure to acute disasters such as a Canadian ice storm (Laplante et al. 2008), 9/11 (Yehuda et al. 2005), Chernobyl (Huizink et al. 2007) or a hurricane in Louisiana (Kinney et al. 2008a). It is clear that it is not just a diagnosable mental illness or very extreme or "toxic stress" that can alter the outcome. Exposures which can have an effect vary from the very severe, such as the death of an older child, to quite mild stresses, such as daily hassles.

It has been suggested that mild-to-moderate stress may actually improve some outcomes. Mild prenatal stress has been shown in some studies to accelerate motor development and cognitive ability (DiPietro et al. 2006). This is an interesting idea and deserves further investigation. Other studies have found a linear dose response between prenatal maternal anxiety and emotional/behavioural outcomes for the child (O'Connor et al. 2002). It is possible that prenatal stress has different patterns and direction of effect for different outcomes. For example, mild-to-moderate stress may accelerate physical maturation and cognitive function while also increasing symptoms of anxiety.

Many independent prospective studies have now shown that if the mother is anxious, depressed or stressed while she is pregnant her child is at increased risk of a wide range of problems (Van den Bergh et al. 2005; Talge et al. 2007; Glover 2011; see Table 13.1). These include both neurodevelopmental, such as emotional and behavioural disorders (O'Connor et al. 2002), and physical problems, such as asthma (Khashan et al. 2012). It is important to note that in all these studies, the findings show only an increased risk. Most children are not affected. But an increased risk, for example, a doubling of a probable mental disorder, from about 6–12 %, if the mother in the top 15 % of anxiety or depression as shown in a normal population (O'Donnell et al. 2014), is of real clinical significance.

Different studies have examined children at times from birth until adulthood. Many have shown that prenatal stress is associated with somewhat lower birthweight and reduced gestational age (Wadhwa et al. 2011; Rice et al. 2010). Studies have found an increased proportion of children who are mixed handed, rather than right handed, after prenatal stress (Glover et al. 2004; Rodriguez and Waldenstrom 2008), and also altered fingerprint patterns (King et al. 2009). These physical alterations are of interest because they are features that are known to develop in utero. Being mixed handed is not a problem in itself, but it is known that people with a range of neurodevelopmental problems such as autism and schizophrenia are also more likely to be mixed handed. Recent studies have shown prenatal stress is associated with reduced telomere length (Entringer et al. 2011, 2013). This is an intriguing finding, as well as of concern, as reduced telomere length is associated with a reduced life span.

Table 13.1 Studies showing prenatal anxiety, depression or stress is associated with an increased risk of the following conditions

Psychological/behavioural/cognitive	References
Worse function on the Brazelton test in newborns	(Rieger et al. 2004)
More sleep problems in infants	(O'Connor et al. 2007)
More anxiety in infants	(Bergman et al. 2007)
More difficult temperament in infants	(Austin et al. 2005; Werner et al. 2007; Davis et al. 2007; Blair et al. 2011)
Worse cognitive ability in infancy Increased cognitive ability in infancy	(Huizink et al. 2003; Bergman et al. 2007; Laplante et al. 2004; DiPietro et al. 2006)
ADHD in childhood	(O'Connor et al. 2002, 2003; Van Den Bergh and Marcoen 2004; Rodriguez and Bohlin 2005; Li et al. 2010)
Emotional problems in childhood and adolescence	(O'Connor et al. 2002, 2003; Van Den Bergh and Marcoen 2004; Pawlby et al. 2009; Barker et al. 2011)
Conduct disorder in childhood	(O'Connor et al. 2002, 2003; Rice et al. 2010)
Decreased cognitive ability in childhood	(Laplante et al. 2008; Barker et al. 2011)
Autism or autism spectrum disorder No increased risk of autism	(Beversdorf et al. 2005; Kinney et al. 2008b; Class et al. 2013; Li et al. 2009)
Vulnerability to bullying at school	(Lereya and Wolke 2012)
Increased risk of schizophrenia in adulthood	(van Os and Selten 1998; Khashan et al. 2008)
Physical	
Lower birthweight and/or gestational age	(Wadhwa et al. 2011; Rice et al. 2010)
Reduced telomere length	(Entringer et al. 2011, 2013)
Oral cleft	(Ingstrup et al. 2013)
Altered fingerprint pattern	(King et al. 2009)
Mixed handedness	(Glover et al. 2004; Rodriguez and Waldenstrom 2008)
Altered immune function	(O'Connor et al. 2013)
Asthma	(Khashan et al. 2012)
Obesity	(Entringer 2013)

Some investigators have looked at the newborns of mothers who report stress during pregnancy and found a poorer performance on the Neonatal Behavioral Assessment Scale relative to newborns of mothers who do not report stress during pregnancy (Rieger et al. 2004), showing that adverse behavioural outcomes are also observable from the very beginning. Studies of infants and toddlers have shown more difficult temperament (Austin et al. 2005; Werner et al. 2007; Davis et al. 2007; Blair et al. 2011), sleep problems (O'Connor et al. 2007), lower cognitive performance and increased fearfulness associated with higher maternal stress during pregnancy (Bergman et al. 2007). Other studies have examined the association between prenatal stress and neurodevelopmental outcomes in children aged 3–16 years. Many independent groups have shown that prenatal stress increases child emotional problems, especially symptoms of anxiety and depression, and symptoms of attention deficit hyperactivity disorder (ADHD) and conduct disorder (O'Connor et al. 2002, 2003; Van Den Bergh and Marcoen 2004; Rodriguez and Bohlin 2005;

Li et al. 2010; Pawlby et al. 2009; Barker et al. 2011; Rice et al. 2010). Studies have also shown a reduction in cognitive performance associated with prenatal stress (Laplante et al. 2008; Barker et al. 2011).

Some research (Beversdorf et al. 2005; Kinney et al. 2008b; Class et al. 2013), although not all (Li et al. 2009), has found an association between prenatal stress and increased risk of autism or autistic spectrum disorder. Two studies have found an increased risk of schizophrenia in adults born to mothers who experienced stress during pregnancy (van Os and Selten 1998; Khashan et al. 2008). Both showed effects with severe stress, the death of a relative or exposure to the invasion of the Netherlands in 1940. One group, using magnetic resonance imaging (MRI), has shown associations between prenatal stress and specific regional reductions in grey matter density in the brain (Buss et al. 2010). Such altered grey matter may be associated with neurodevelopmental and psychiatric disorders as well as cognitive and intellectual impairment.

There is little consistency in the literature as to the most sensitive time in gestation for the influence of prenatal stress. It is likely that there are different times of sensitivity dependent on the outcome studied and the stage of development of the relevant brain structures. The two studies of schizophrenia found the most sensitive period was in the first trimester (van Os and Selten 1998; Khashan et al. 2008). This is when neuronal cells migrate to their eventual site in brain, a process previously suggested to be disrupted in schizophrenia. In contrast, two studies of conduct disorder, or antisocial behaviour, found associations with stress in mid or late pregnancy (O'Connor et al. 2003; Rice et al. 2010).

Many human studies, as discussed above, have shown that there is an association between maternal stress during pregnancy and an altered outcome for the child. The evidence for this is very strong and has been shown in many independent prospective studies from around the world. What is harder to establish is that the association is causal. If a mother is stressed while she is pregnant, she may well be stressed postnatally and this could affect her parenting. There can be other associated confounding factors such as smoking or alcohol consumption, which may affect her child, for example, in behaviour and birthweight. There also could be genetic continuity. The mother may have certain genes which make her more likely to become anxious or depressed and she may pass these genes on to her child, which in turn makes them more prone to emotional or behavioural problems.

Several studies have tried to address these points but the first evidence to consider is that from animals. With animal studies, it is much easier to establish that prenatal stress has a direct effect on the outcome for the offspring. Newborn rat pups of prenatally stressed mothers can be cross-fostered to non-stressed mothers on the 1st day after birth, with control pups of unstressed mothers cross-fostered also (Weinstock 2001; Maccari et al. 2003). This can establish that any differences in outcome are caused by stress in the prenatal period. Many such studies have shown that there are definite programming effects of prenatal stress on behaviour, cognitive development and brain structure of the offspring. The nature of the effects can be affected by the timing of the exposure in gestation, the type of the stress, the strain of the animal and the age at which the offspring was tested (Weinstock 2007).

Some altered outcomes are not observed in the youngest offspring, but become apparent as they mature. The effects can also depend on the sex of the offspring. In general, although not always, prenatal stress increases anxiety and depressive behaviour to a greater extent in female offspring and impairs learning and cognition more in the males (Weinstock 2007).

In humans there are several types of evidence which also suggest that prenatal stress is causing fetal programming, but it is harder to be definitive. There is good evidence for an association between the mother's emotional state and the behaviour or heart rate of her fetus. Experiments in which a pregnant mother is asked to carry out a stressful computer task, while the fetal heart rate is monitored, showed that the fetal heart rate went up during the task, but only in mothers who rated themselves as anxious (Monk et al. 2003). Thus, even before birth, the fetus can be affected by the maternal emotional state, although we do not know what the mechanism is for this (it is too quick to be caused by the stress hormone cortisol). There is evidence for continuity between fetal behaviour and neurological maturation at 2 years of age (DiPietro et al. 2007). The fact that maternal stress during pregnancy is associated with altered outcomes at birth including reduced birthweight (Wadhwa et al. 1993), reduced scores on a neonatal assessment (Rieger et al. 2004) and epigenetic changes in the glucocorticoid receptor in cord blood (Hompes et al. 2013) is evidence for some prenatal, independent of postnatal, effects. Findings of altered fingerprint patterns (King et al. 2009) and handedness (Glover et al. 2004) are also strong evidence for prenatal effects as the pattern for both of these is set in utero.

Another approach to establishing that the associations between the maternal emotional state and long-term outcome for the child are, at least in part, causal is by controlling for confounding factors such as prenatal smoking and alcohol consumption and for postnatal maternal mood. Several studies have done this and found a strong signal remaining for prenatal anxiety or depression, thus controlling for impaired parenting due to postnatal depression or anxiety, e.g. O'Connor et al. (2002). If the observed associations are primarily due to an anxious mother passing on predisposing genes to her child, it would not be expected to be specifically associated with prenatal as opposed to postnatal mood. In a recent study (O'Donnell et al. 2014) we have shown that the associations with child emotional and behavioural problems last until 13 years of age. In this study, we also show that allowing for paternal prenatal mood makes little difference to the associations with prenatal maternal anxiety or depression, thus adding further evidence for maternal prenatal effects independent of genetics.

An interesting study compared the association between prenatal stress and child outcome in children born after in vitro fertilisation, in those who were genetically related to the mother with those who were not (Rice et al. 2010). They showed that there was an association between maternal stress in pregnancy and child symptoms of ADHD and conduct disorder, and that the association with conduct disorder was apparent in the unrelated mothers. This gives strong support to the idea that the association between prenatal stress and child conduct disorder can be independent of genetic factors. However, the fact that the increase in symptoms of ADHD was apparent only in those with related mothers does not conclusively rule out a prenatal environmental component. There may be a gene–environment interaction. Prenatal stress may only have the effect of increasing symptoms of ADHD in the genetically vulnerable mother and child pairs. More research is needed to disentangle the role of genetic factors for all outcomes.

One further indication that the effects of prenatal stress are not just due to genetic continuity is the group of studies that have shown children of mothers exposed to acute disasters, such as the Canadian ice storm (Laplante et al. 2008). With these "natural experiments", the level of stress was objectively assessed, and the exposure was of a specific duration. This reduces the confounding effects of pre-existing emotional problems and genetic continuity, and also postnatal emotional and parenting effects.

There clearly are additional effects of both postnatal maternal mood and parenting. For example, the association between prenatal anxiety and child fearfulness was found to be greater in those children with an insecure attachment to their mother (Bergman et al. 2008). In our recent study (O'Donnell et al. in press), we found that the magnitude of the effect of prenatal maternal mood was similar to that of postnatal and that the two were additive.

Although there is some room for scepticism, the evidence is mounting that in humans, as in animals, prenatal stress has a direct causal effect on fetal development including neurodevelopment. However, the early postnatal environment is equally important, and can either exacerbate or ameliorate the prenatal effects.

It is important to consider the clinical magnitude of these effects. When we compared the outcome for children of the 15 % most prenatally anxious or depressed women with the rest, in a large normal population (O'Donnell et al. in press), we found that the rate of probable mental disorder doubled from about 6–12 % at age 13 years, after allowing for a very wide range of possible confounders. This is of clear clinical and public health significance but also shows that most children are not affected. Those children that are affected are often affected in different ways (Bergman et al. 2007). One reason for this may be differential genetic vulnerabilities.

13.2 Mechanisms

In animal studies, there has been much research on underlying mechanisms with an especial focus on the hypothalamic–pituitary–adrenal (HPA) axis (Weinstock 2005; Harris and Seckl 2011; Khulan and Drake 2012). The effects of prenatal stress on the offspring can be partially mimicked by giving the pregnant animal a synthetic glucocorticoid such as dexamethasone (Matthews and Phillips 2011; Crudo et al. 2013) or adrenocorticotropic hormone (ACTH) to stimulate the production of cortisol (or corticosterone in rodents), and at least partially blocked by adrenalectomy (Weinstock 2008).

A model of some potential underlying mechanisms, based on the animal data, is shown in Fig. 13.1, with references in Table 13.2. The hypothesis is that prenatal stress causes an increase in maternal cortisol; this then crosses the placenta in a

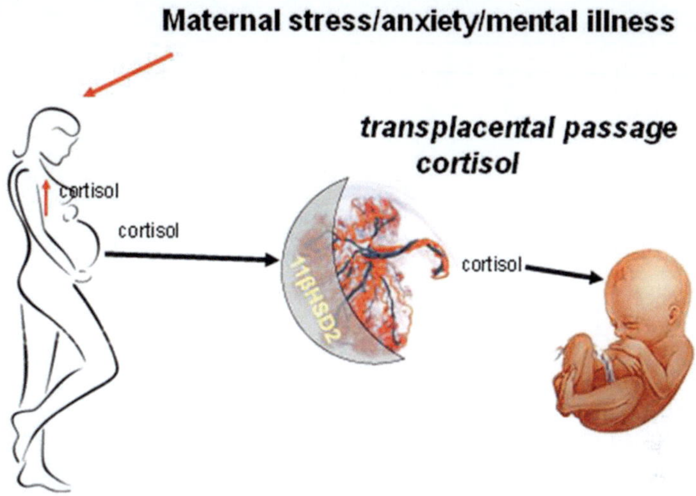

Fig. 13.1 A model of some potential underlying mechanisms, based on the animal data

quantity sufficient to affect the development of the fetal brain. However, each stage of this needs to be tested in humans, and this is only just starting.

There is some evidence that in humans, as in animal models, prenatal administration of dexamethasone is associated with more behavioural and emotional problems in the child (Khalife et al. 2014), lasting at least until adolescence, where a thinning of the cortex has been shown (Davis et al. 2013). There is also evidence for the potentially widespread role for exposure to increased cortisol in human fetal brain development by a study showing, by microarray analysis, that increased cortisol exposure affects the expression of over a thousand genes in fetal brain cells (Salaria et al. 2006).

However, there is either a weak or no correlation found in many studies between a mother's symptoms of anxiety and depression while pregnant and her level of cortisol (Sarkar et al. 2006; O'Donnell et al. 2009; Davis and Sandman 2010; Baibazarova et al. 2013). In human pregnancy, the placenta produces increasing concentrations of corticotrophin-releasing hormone (CRH) which stimulates maternal production of cortisol; towards the end of pregnancy, plasma levels reach those found in melancholic depression. This in turn is associated with a dampened cortisol response to stress (Kammerer et al. 2002). Maternal plasma cortisol levels correlate strongly with cord blood (Gitau et al. 1998) and amniotic fluid (Sarkar et al. 2007; Baibazarova et al. 2013), and prenatal maternal cortisol levels can be a predictor of child outcome independent of maternal mood (Davis and Sandman 2010). However, the maternal mediator between prenatal stress, anxiety and depression and altered child outcome is currently not known.

One possible biological group of maternal mediators could be those associated with the immune system and inflammation, such as the pro-inflammatory cytokines. There is a growing literature associating them with depression (Hepgul et al.

Table 13.2 Possible mechanisms

Mother	References
Cortisol	(Sarkar et al. 2006)
Cytokines	(Coussons-Read et al. 2007)
Placenta	
11β- HSD2	(O'Donnell et al. 2012)
Amniotic fluid	
Cortisol	(Bergman et al. 2010)
Child	
HPA axis function	(O'Donnell et al. 2013; Entringer et al. 2009)
Brain structure	(Buss et al. 2010)
Epigenetic changes	(Hompes et al. 2013)

2013). Increased cytokines have been associated with psychosocial stress during pregnancy (Coussons-Read et al. 2007). Elevated stress was related to higher serum interleukin-6 (IL-6) both in early and late pregnancy. No relationships between stress and cytokines were apparent during the second trimester. However, elevated stress levels across pregnancy were predictive of elevated production of the pro-inflammatory cytokines IL-1B and IL-6 by stimulated lymphocytes in the third trimester, suggesting that stress during pregnancy affects the function of immune system cells. A recent study has confirmed that depressed pregnant women have higher levels of IL-6 in the first trimester (Haeri et al. 2013). However, another study has failed to find any association between maternal symptoms of anxiety and depression during pregnancy and levels of IL-6 (Blackmore et al. 2011), at 18 or 32 weeks. This is clearly an area that needs further exploration.

Increase in activity of the sympathetic system may also be important although it has been much less studied than the HPA axis. However, noradrenaline, unlike cortisol which is only partially metabolised (Gitau et al. 1998), is totally metabolised by the placenta (Giannakoulopoulos et al. 1999).

The placenta plays a crucial part in fetal programming. Dependent on the chemical signals it receives from the mother, it can alter its filtering capacity and thus alter the exposure of the fetus to specific chemicals (Jansson and Powell 2007). Animal studies have shown that prenatal stress can have an effect on placental function, including on the regulation of 11β-hydroxysteroid dehydrogenase 2 (11β-HSD2), the enzyme that breaks down cortisol (corticosterone in rodents) to inactive product. Two studies have shown that prenatal stress in the last week of gestation caused a downregulation of expression of this enzyme (Mairesse et al. 2007; Jensen Pena et al. 2012). However, acute stress on day 20 of gestation caused an upregulation (Welberg et al. 2005).

We have shown that with increasing maternal anxiety the correlation between maternal plasma and amniotic fluid cortisol increased significantly (Glover et al. 2009). We have more recently shown directly, in women having an elective caesarean section, that maternal symptoms of anxiety on the previous day were associated with a downregulation of 11β-HSD2 (O'Donnell et al. 2012). This would be compatible with some chemical signal from the mother causing an alteration in the filtering capacity of the placenta, allowing more cortisol to pass through to the fetus. Another study has failed to find this downregulation of placental 11β-HSD2 in association with prenatal maternal symptoms of anxiety or depression, (Ponder et al. 2011), although they did find alterations in the noradrenaline transporter. This may be due to differences in the design of the study, and especially in the inclusion of women who have experienced labour. There is some evidence that labour alters the expression of 11β-HSD2 (personal communication).

In addition to cortisol, another factor which may be important in altering fetal brain development is 5-hydroxytryptamine (5-HT), which acts as a trophic factor to regulate fetal neuronal cell division, differentiation and synaptogenesis (Gaspar et al. 2003). 5-HT has a different role during early development from that in adulthood (Oberlander 2012). Whilst selective serotonin reuptake inhibitors can alleviate anxiety and depression later in life, treatment of newborn mice with these drugs causes an increase in these symptoms (Ansorge et al. 2004). Recent work has identified an endogenous serotonin biosynthetic pathway within the human placenta, which plays a role in offspring neurodevelopment (Bonnin et al. 2011).

A major mechanism for removing 5-HT is its metabolism to inactive 5-hydroxyindoleacetic acid by the enzyme monoamine oxidase A (MAO A). This is the enzyme that metabolises a range of monoamine neurotransmitters including noradrenaline and dopamine. The placenta is a very rich source of MAO A suggesting its importance in the regulation of fetal monoamine exposure. We have recently shown that prenatal maternal depression is associated with a downregulation of expression of placental MAO A (Blakeley et al. 2013), suggesting that another mechanism underlying the effects of prenatal mood on fetal brain development may be via increased exposure to 5-HT. This is a promising area for future research.

There is little direct human evidence yet that fetal overexposure to specific chemicals is mediating the effects of prenatal stress. However, amniotic fluid cortisol levels have been shown to be inversely correlated with cognitive development in the infant (Bergman et al. 2010), but only in those children who were insecurely attached. It is clear that at least some prenatal effects can be buffered or modified by the postnatal environment.

Animal studies have shown that many of the long-term effects of the early environment, including the psychosocial, are due to epigenetic changes, which can be maintained through the life span and even the grandchild generation (Meaney and Szyf 2005; Bale et al. 2011; Monk et al. 2012; Gudsnuk and Champagne 2012). These are changes "on top of the DNA" which alter whether a specific gene is turned on or off, and if turned on, how much of it is expressed. Prenatal stress has been shown to cause, through microRNA regulation, certain epigenetic signatures of psychiatric and neurological diseases in the offspring (Zucchi et al. 2013).

In human studies too, epigenetic changes in the child are starting to be found after prenatal stress, with an initial focus on the glucocorticoid receptor, the receptor that responds to cortisol (Harris and Seckl 2011). Methylation in the promoter of the glucocorticoid receptor NR3C1 in the newborn has been found to be altered after prenatal stress, in a cohort from the Congo, and the changes were associated with reduced birthweight (Mulligan et al. 2012). High pregnancy-specific anxiety has also been shown to be associated with epigenetic changes in the promoter for this receptor in the newborn (Hompes et al. 2013). And maternal exposure to intimate partner violence has been shown to be associated with epigenetic changes to the promoter for the same receptor in the blood of their adolescent children (Radtke et al. 2011).

In animal models, it has been found that prenatal stress can have a long-term effect on the function of the HPA axis in the offspring, although the patterns are quite complex (Weinstock 2005). There has been little work so far in humans but our group has shown that prenatal anxiety was associated with raised morning cortisol in 10-year-old children (O'Connor et al. 2005), but that the pattern had changed by adolescence (O'Donnell et al. 2013). In 15-year-old children, there were modest but significant effects, with the morning rise being reduced and a flatter diurnal slope. It is unlikely that these changes in the diurnal cortisol pattern underlie any of the emotional, behavioural or cognitive changes seen in older children as they are much too small.

A notable finding of all the prenatal stress and child outcome studies is that most of the children are not affected. This is probably due at least in part to different genetic vulnerabilities and gene–environment interactions (Caspi et al. 2003). Although no interactions have been found between prenatal anxiety, genetic variation in the 5-HT transporter and child outcome (Braithwaite et al. 2013) we are finding small interactions between prenatal anxiety and variants of the COMT and BDNF genes (unpublished observations). This is certainly an area where further research is warranted.

13.3 Conclusion

There is good evidence that various forms of prenatal stress contribute to long-term neurodevelopmental changes in the child. The underlying mechanisms are just starting to be understood, and probably include the HPA axis, changes in the filtering capacity of the placenta and epigenetic changes in the child. However, much work is needed before we understand these underlying mechanisms and are able to evaluate and target different interventions properly.

Conflict of Interest The author declares no conflict of interest.

References

Ansorge MS, Zhou M, Lira A, Hen R, Gingrich JA (2004) Early-life blockade of the 5-HT transporter alters emotional behavior in adult mice. Science 306:879–881

Austin MP, Hadzi-Pavlovic D, Leader L, Saint K, Parker G (2005) Maternal trait anxiety, depression and life event stress in pregnancy: relationships with infant temperament. Early Hum Dev 81:183–190

Baibazarova E, van de Beek C, Cohen-Kettenis PT, Buitelaar J, Shelton KH, van Goozen SH (2013) Influence of prenatal maternal stress, maternal plasma cortisol and cortisol in the amniotic fluid on birth outcomes and child temperament at 3 months. Psychoneuroendocrinology 38:907–915

Bale TL, Baram TZ, Brown AS, Goldstein JM, Insel TR, Mccarthy MM, Nemeroff CB, Reyes TM, Simerly RB, Susser ES, Nestler EJ (2011) Early life programming and neurodevelopmental disorders. Biol Psychiatry 68:314–319

Barker ED, Jaffee SR, Uher R, Maughan B (2011) The contribution of prenatal and postnatal maternal anxiety and depression to child maladjustment. Depress Anxiety 28:696–702

Bergman K, Sarkar P, O'Connor TG, Modi N, Glover V (2007) Maternal stress during pregnancy predicts cognitive ability and fearfulness in infancy. J Am Acad Child Adolesc Psychiatry 46:1454–1463

Bergman K, Sarkar P, Glover V, O'Connor TG (2008) Quality of child-parent attachment moderates the impact of antenatal stress on child fearfulness. J Child Psychol Psychiatry 49:1089–1098

Bergman K, Sarkar P, Glover V, O'Connor TG (2010) Maternal prenatal cortisol and infant cognitive development: moderation by infant-mother attachment. Biol Psychiatry 67:1026–1032

Beversdorf DQ, Manning SE, Hillier A, Anderson SL, Nordgren RE, Walters SE, Nagaraja HN, Cooley WC, Gaelic SE, Bauman ML (2005) Timing of prenatal stressors and autism. J Autism Dev Disord 35:471–478

Blackmore ER, Moynihan JA, Rubinow DR, Pressman EK, Gilchrist M, O'Connor TG (2011) Psychiatric symptoms and proinflammatory cytokines in pregnancy. Psychosom Med 73:656–663

Blair MM, Glynn LM, Sandman CA, Davis EP (2011) Prenatal maternal anxiety and early childhood temperament. Stress 14:644–651

Blakeley PM, Capron LE, Jensen AB, O'Donnell KJ, Glover V. (2013) Maternal prenatal symptoms of depression and down regulation of placental monoamine oxidase A expression. J Psychosom Res.75(4):341–5

Bonnin A, Goeden N, Chen K, Wilson ML, King J, Shih JC, Blakely RD, Deneris ES, Levitt P (2011) A transient placental source of serotonin for the fetal forebrain. Nature 472:347–350

Braithwaite EC, Ramchandani PG, O'Connor TG, van Ijzendoorn MH, Bakermans-Kranenburg MJ, Glover V, Netsi E, Evans J, Meaney MJ, Murphy SE (2013) No moderating effect of 5-HTTLPR on associations between antenatal anxiety and infant behavior. J Am Acad Child Adolesc Psychiatry 52:519–526

Buss C, Davis EP, Muftuler LT, Head K, Sandman CA (2010) High pregnancy anxiety during mid-gestation is associated with decreased gray matter density in 6–9-year-old children. Psychoneuroendocrinology 35:141–153

Caspi A, Sugden K, Moffitt TE, Taylor A, Craig IW, Harrington H, McClay J, Mill J, Martin J, Braithwaite A, Poulton R (2003) Influence of life stress on depression: moderation by a polymorphism in the 5-HTT gene. Science 301:386–389

Class QA, Abel KM, Khashan AS, Rickert ME, Dalman C, Larsson H, Hultman CM, Langstrom N, Lichtenstein P, D'Onofrio BM (2013) Offspring psychopathology following preconception, prenatal and postnatal maternal bereavement stress. Psychol Med 44:1–14

Coussons-Read ME, Okun ML, Nettles CD (2007) Psychosocial stress increases inflammatory markers and alters cytokine production across pregnancy. Brain Behav Immun 21:343–350

Crudo A, Suderman M, Moisiadis VG, Petropoulos S, Kostaki A, Hallett M, Szyf M, Matthews SG (2013) Glucocorticoid programming of the fetal male hippocampal epigenome. Endocrinology 154:1168–1180

Davis EP, Sandman CA (2010) The timing of prenatal exposure to maternal cortisol and psychosocial stress is associated with human infant cognitive development. Child Dev 81:131–148

Davis EP, Glynn LM, Schetter CD, Hobel C, Chicz-Demet A, Sandman CA (2007) Prenatal exposure to maternal depression and cortisol influences infant temperament. J Am Acad Child Adolesc Psychiatry 46:737–746

Davis EP, Sandman CA, Buss C, Wing DA, Head K (2013) Fetal glucocorticoid exposure is associated with preadolescent brain development. Biol Psychiatry 74:647–655

DiPietro JA, Novak MF, Costigan KA, Atella LD, Reusing SP (2006) Maternal psychological distress during pregnancy in relation to child development at age two. Child Dev 77:573–587
DiPietro JA, Bornstein MH, Hahn CS, Costigan K, Achy-Brou A (2007) Fetal heart rate and variability: stability and prediction to developmental outcomes in early childhood. Child Dev 78:1788–1798
Entringer S (2013) Impact of stress and stress physiology during pregnancy on child metabolic function and obesity risk. Curr Opin Clin Nutr Metab Care 16:320–327
Entringer S, Kumsta R, Hellhammer DH, Wadhwa PD, Wust S (2009) Prenatal exposure to maternal psychosocial stress and HPA axis regulation in young adults. Horm Behav 55:292–298
Entringer S, Epel ES, Kumsta R, Lin J, Hellhammer DH, Blackburn EH, Wust S, Wadhwa PD (2011) Stress exposure in intrauterine life is associated with shorter telomere length in young adulthood. Proc Natl Acad Sci U S A 108:E513–E518
Entringer S, Epel ES, Lin J, Buss C, Shahbaba B, Blackburn EH, Simhan HN, Wadhwa PD (2013) Maternal psychosocial stress during pregnancy is associated with newborn leukocyte telomere length. Am J Obstet Gynecol 208:134.e1–e7
Gaspar P, Cases O, Maroteaux L (2003) The developmental role of serotonin: news from mouse molecular genetics. Nat Rev Neurosci 4:1002–1012
Giannakoulopoulos X, Teixeira J, Fisk N, Glover V (1999) Human fetal and maternal noradrenaline responses to invasive procedures. Pediatr Res 45:494–499
Gitau R, Cameron A, Fisk NM, Glover V (1998) Fetal exposure to maternal cortisol. Lancet 352:707–708
Glover V (2011) Annual research review: prenatal stress and the origins of psychopathology: an evolutionary perspective. J Child Psychol Psychiatry 52:356–367
Glover V, O'Connor TG, Heron J, Golding J (2004) Antenatal maternal anxiety is linked with atypical handedness in the child. Early Hum Dev 79:107–118
Glover V, Bergman K, Sarkar P, O'Connor TG (2009) Association between maternal and amniotic fluid cortisol is moderated by maternal anxiety. Psychoneuroendocrinology 34:430–435
Gudsnuk K, Champagne FA (2012) Epigenetic influence of stress and the social environment. ILAR J 53:279–288
Haeri S, Baker AM, Ruano R (2013) Do pregnant women with depression have a pro-inflammatory profile? J Obstet Gynaecol Res 39:948–952
Harris A, Seckl J (2011) Glucocorticoids, prenatal stress and the programming of disease. Horm Behav 59:279–289
Hepgul N, Cattaneo A, Zunszain PA, Pariante CM (2013) Depression pathogenesis and treatment: what can we learn from blood mRNA expression? BMC Med 11:28
Hompes T, Izzi B, Gellens E, Morreels M, Fieuws S, Pexsters A, Schops G, Dom M, Van Bree R, Freson K, Verhaeghe J, Spitz B, Demyttenaere K, Glover V, Van den Bergh B, Allegaert K, Claes S (2013) Investigating the influence of maternal cortisol and emotional state during pregnancy on the DNA methylation status of the glucocorticoid receptor gene (NR3C1) promoter region in cord blood. J Psychiatr Res 47:880–891
Huizink AC, de Medina PG, Mulder EJ, Visser GH, Buitelaar JK (2002) Psychological measures of prenatal stress as predictors of infant temperament. J Am Acad Child Adolesc Psychiatry 41:1078–1085
Huizink AC, Robles deMPG, Mulder EJ, Visser GH, Buitelaar JK (2003) Stress during pregnancy is associated with developmental outcome in infancy. J Child Psychol Psychiatry 44:810–818
Huizink AC, Dick DM, Sihvola E, Pulkkinen L, Rose RJ, Kaprio J (2007) Chernobyl exposure as stressor during pregnancy and behaviour in adolescent offspring. Acta Psychiatr Scand 116:438–446
Ingstrup KG, Liang H, Olsen J, Nohr EA, Bech BH, Wu CS, Christensen K, Li J (2013) Maternal bereavement in the antenatal period and oral cleft in the offspring. Hum Reprod 28:1092–1099
Jansson T, Powell TL (2007) Role of the placenta in fetal programming: underlying mechanisms and potential interventional approaches. Clin Sci (Lond) 113:1113
Jensen Pena C, Monk C, Champagne FA (2012) Epigenetic effects of prenatal stress on 11beta-hydroxysteroid dehydrogenase-2 in the placenta and fetal brain. PLoS ONE 7:e39791

Kammerer M, Adams D, Castelberg BV, Glover V (2002) Pregnant women become insensitive to cold stress. BMC Pregnancy Childbirth 2:8
Khalife N, Glover V, Taanila A, Ebeling H, Järvelin MR, Rodriguez A (2014) Prenatal glucocorticoid treatment and later mental health in children and adolescents. PLoS One. 22;8(11):e81394
Khashan AS, Abel KM, McNamee R, Pedersen MG, Webb RT, Baker PN, Kenny LC, Mortensen PB (2008) Higher risk of offspring schizophrenia following antenatal maternal exposure to severe adverse life events. Arch Gen Psychiatry 65:146–152
Khashan AS, Wicks S, Dalman C, Henriksen TB, Li J, Mortensen PB, Kenny LC (2012) Prenatal stress and risk of asthma hospitalization in the offspring: a Swedish population-based study. Psychosom Med 74:635–641
Khulan B, Drake AJ (2012) Glucocorticoids as mediators of developmental programming effects. Best Pract Res Clin Endocrinol Metab 26:689–700
King S, Mancini-Marie A, Brunet A, Walker E, Meaney MJ, Laplante DP (2009) Prenatal maternal stress from a natural disaster predicts dermatoglyphic asymmetry in humans. Dev Psychopathol 21:343–353
Kinney DK, Miller AM, Crowley DJ, Huang E, Gerber E (2008a) Autism prevalence following prenatal exposure to hurricanes and tropical storms in Louisiana. J Autism Dev Disord 38:481–488
Kinney DK, Munir KM, Crowley DJ, Miller AM (2008b) Prenatal stress and risk for autism. Neurosci Biobehav Rev 32:1519–1532
Laplante DP, Barr RG, Brunet A, Galbaud duFG, Meaney ML, Saucier JF, Zelazo PR, King S (2004) Stress during pregnancy affects general intellectual and language functioning in human toddlers. Pediatr Res 56:400–410
Laplante DP, Brunet A, Schmitz N, Ciampi A, King S (2008) Project ice storm: prenatal maternal stress affects cognitive and linguistic functioning in 5 1/2-year-old children. J Am Acad Child Adolesc Psychiatry 47:1063–1072
Lereya ST, Wolke D (2012) Prenatal family adversity and maternal mental health and vulnerability to peer victimisation at school. J Child Psychol Psychiatry 54:644–652
Li J, Vestergaard M, Obel C, Christensen J, Precht DH, Lu M, Olsen J (2009) A nationwide study on the risk of autism after prenatal stress exposure to maternal bereavement. Pediatrics 123:1102–1107
Li J, Olsen J, Vestergaard M, Obel C (2010) Attention-deficit/hyperactivity disorder in the offspring following prenatal maternal bereavement: a nationwide follow-up study in Denmark. Eur Child Adolesc Psychiatry 19:747–753
Maccari S, Darnaudery M, Morley-Fletcher S, Zuena AR, Cinque C, Van Reeth O (2003) Prenatal stress and long-term consequences: implications of glucocorticoid hormones. Neurosci Biobehav Rev 27:119–127
Mairesse J, Lesage J, Breton C, Breant B, Hahn T, Darnaudery M, Dickson SL, Seckl J, Blondeau B, Vieau D, Maccari S, Viltart O (2007) Maternal stress alters endocrine function of the fetoplacental unit in rats. Am J Physiol Endocrinol Metab 292:E1526–E1533
Matthews SG, Phillips DI (2011) Minireview: transgenerational inheritance of the stress response: a new frontier in stress research. Endocrinology 151:7–13
Meaney MJ, Szyf M (2005) Environmental programming of stress responses through DNA methylation: life at the interface between a dynamic environment and a fixed genome. Dialogues Clin Neurosci 7:103–123
Monk C, Myers MM, Sloan RP, Ellman LM, Fifer WP (2003) Effects of women's stress-elicited physiological activity and chronic anxiety on fetal heart rate. J Dev Behav Pediatr 24:32–38
Monk C, Spicer J, Champagne FA (2012) Linking prenatal maternal adversity to developmental outcomes in infants: the role of epigenetic pathways. Dev Psychopathol 24:1361–1376
Mulligan CJ, D'Errico NC, Stees J, Hughes DA (2012) Methylation changes at NR3C1 in newborns associate with maternal prenatal stress exposure and newborn birth weight. Epigenetics 7:853–857
O'Connor TG, Heron J, Golding J, Beveridge M, Glover V (2002) Maternal antenatal anxiety and children's behavioural/emotional problems at 4 years. Report from the Avon longitudinal study of parents and children. Br J Psychiatry 180:502–508

O'Connor TG, Heron J, Golding J, Glover V (2003) Maternal antenatal anxiety and behavioural/emotional problems in children: a test of a programming hypothesis. J Child Psychol Psychiatry 44:1025–1036

O'Connor TG, Ben-Shlomo Y, Heron J, Golding J, Adams D, Glover V (2005) Prenatal anxiety predicts individual differences in cortisol in pre-adolescent children. Biol Psychiatry 58:211–217

O'Connor TG, Caprariello P, Blackmore ER, Gregory AM, Glover V, Fleming P, ALSPAC Team (2007) Prenatal mood disturbance predicts sleep problems in infancy and toddlerhood. Early Hum Dev 83:451–458

O'Connor TG, Winter MA, Hunn J, Carnahan J, Pressman EK, Glover V, Robertson-Blackmore E, Moynihan JA, Lee FE, Caserta MT (2013) Prenatal maternal anxiety predicts reduced adaptive immunity in infants. Brain Behav Immun 32:21–28

O'Donnell K, O'Connor TG, Glover V (2009) Prenatal stress and neurodevelopment of the child: focus on the HPA axis and role of the placenta. Dev Neurosci 31:285–292

O'Donnell KJ, Bugge Jensen A, Freeman L, Khalife N, O'Connor TG, Glover V (2012) Maternal prenatal anxiety and downregulation of placental 11beta-HSD2. Psychoneuroendocrinology 37:818–826

O'Donnell KJ, Glover V, Barker ED, O'Connor TG (2014) The persisting effect of maternal mood in pregnancy on childhood psychopathology. Dev Psychopathol 26(2):393–403

O'Donnell KJ, Glover V, Jenkins J, Browne D, Ben-Shlomo Y, Golding J, O'Connor TG (2013) Prenatal maternal mood is associated with altered diurnal cortisol in adolescence. Psychoneuroendocrinology 38:1630–1638

Oberlander TF (2012) Fetal serotonin signaling: setting pathways for early childhood development and behavior. J Adolesc Health 51:S9–S16

Pawlby S, Hay DF, Sharp D, Waters CS, O'Keane V (2009) Antenatal depression predicts depression in adolescent offspring: prospective longitudinal community-based study. J Affect Disord 113:236–243

Ponder KL, Salisbury A, McGonnigal B, Laliberte A, Lester B, Padbury JF (2011) Maternal depression and anxiety are associated with altered gene expression in the human placenta without modification by antidepressant use: implications for fetal programming. Dev Psychobiol 53:711–723

Radtke KM, Ruf M, Gunter HM, Dohrmann K, Schauer M, Meyer A, Elbert T (2011) Transgenerational impact of intimate partner violence on methylation in the promoter of the glucocorticoid receptor. Transl Psychiatry 1:e21

Rice F, Harold GT, Boivin J, van den Bree M, Hay DF, Thapar A (2010) The links between prenatal stress and offspring development and psychopathology: disentangling environmental and inherited influences. Psychol Med 40:335–345

Rieger M, Pirke KM, Buske-Kirschbaum A, Wurmser H, Papousek M, Hellhammer DH (2004) Influence of stress during pregnancy on HPA activity and neonatal behavior. Ann N Y Acad Sci 1032:228–230

Rodriguez A, Bohlin G (2005) Are maternal smoking and stress during pregnancy related to ADHD symptoms in children? J Child Psychol Psychiatry 46:246–254

Rodriguez A, Waldenstrom U (2008) Fetal origins of child non-right-handedness and mental health. J Child Psychol Psychiatry 49:967–976

Salaria S, Chana G, Caldara F, Feltrin E, Altieri M, Faggioni F, Domenici E, Merlo-Pich E, Everall IP (2006) Microarray analysis of cultured human brain aggregates following cortisol exposure: implications for cellular functions relevant to mood disorders. Neurobiol Dis 23:630–636

Sarkar P, Bergman K, Fisk NM, Glover V (2006) Maternal anxiety at amniocentesis and plasma cortisol. Prenat Diagn 26:505–509

Sarkar P, Bergman K, Fisk NM, O'Connor TG, Glover V (2007) Ontogeny of foetal exposure to maternal cortisol using midtrimester amniotic fluid as a biomarker. Clin Endocrinol (Oxf) 66:636–640

Talge NM, Neal C, Glover V (2007) Antenatal maternal stress and long term effects on child neurodevelopment: how and why? J Child Psychol Psychiatry 48:245–261

Van den Bergh BR, Marcoen A (2004) High antenatal maternal anxiety is related to ADHD symptoms, externalizing problems, and anxiety in 8- and 9-year-olds. Child Dev 75:1085–1097

Van den Bergh BR, Mulder EJ, Mennes M, Glover V (2005) Antenatal maternal anxiety and stress and the neurobehavioural development of the fetus and child: links and possible mechanisms. A review. Neurosci Biobehav Rev 29:237–258

Van den Bergh BR, Van Calster B, Smits T, Van Huffel S, Lagae L (2008) Antenatal maternal anxiety is related to HPA-axis dysregulation and self-reported depressive symptoms in adolescence: a prospective study on the fetal origins of depressed mood. Neuropsychopharmacology 33:536–545

van Os J, Selten JP (1998) Prenatal exposure to maternal stress and subsequent schizophrenia. The May 1940 invasion of The Netherlands. Br J Psychiatry 172:324–326

Wadhwa PD, Sandman CA, Porto M, Dunkel-Schetter C, Garite TJ (1993) The association between prenatal stress and infant birth weight and gestational age at birth: a prospective investigation. Am J Obstet Gynecol 169:858–865

Wadhwa PD, Entringer S, Buss C, Lu MC (2011) The contribution of maternal stress to preterm birth: issues and considerations. Clin Perinatol 38:351–384

Weinstock M (2001) Alterations induced by gestational stress in brain morphology and behaviour of the offspring. Prog Neurobiol 65:427–451

Weinstock M (2005) The potential influence of maternal stress hormones on development and mental health of the offspring. Brain Behav Immun 19:296–308

Weinstock M (2007) Gender differences in the effects of prenatal stress on brain development and behaviour. Neurochem Res 32:1730–1740

Weinstock M (2008) The long-term behavioural consequences of prenatal stress. Neurosci Biobehav Rev 32:1073–1086

Welberg LA, Thrivikraman KV, Plotsky PM (2005) Chronic maternal stress inhibits the capacity to up-regulate placental 11beta-hydroxysteroid dehydrogenase type 2 activity. J Endocrinol 186:R7–R12

Werner EA, Myers MM, Fifer WP, Cheng B, Fang Y, Allen R, Monk C (2007) Prenatal predictors of infant temperament. Dev Psychobiol 49:474–484

Yehuda R, Engel SM, Brand SR, Seckl J, Marcus SM, Berkowitz GS (2005) Transgenerational effects of posttraumatic stress disorder in babies of mothers exposed to the World Trade Center attacks during pregnancy. J Clin Endocrinol Metab 90:4115–4118

Zucchi FC, Yao Y, Ward ID, Ilnytskyy Y, Olson DM, Benzies K, Kovalchuk I, Kovalchuk O, Metz GA (2013) Maternal stress induces epigenetic signatures of psychiatric and neurological diseases in the offspring. PLoS ONE 8:e56967

Chapter 14
Using Natural Disasters to Study Prenatal Maternal Stress in Humans

Suzanne King and David P. Laplante

Abstract Animal studies of prenatal maternal stress permit random assignment of pregnant animals to stress and no-stress groups, and allow total control of the type, severity, and timing of the stressor in utero. Human studies have obvious constraints that make the use of experimental methods nearly impossible. Studying pregnant women who experience natural disasters during pregnancy, however, approximates the random assignment to groups enjoyed by animal studies, and can characterize the timing of the stressor in utero with great precision. In this chapter, we briefly describe our three ongoing prospective longitudinal studies of children exposed to prenatal maternal stress from natural disasters. We present results from Project Ice Storm in detail, showing effects of prenatal maternal stress on cognitive and neurodevelopment. We contrast these results with preliminary findings from the Iowa Flood Study and introduce the QF2011 Queensland Flood Project. In the "Discussion" section, we present conclusions to date and discuss the relative effects of the severity of maternal objective disaster exposure and maternal subjective distress levels, the moderating effects of fetal sex and the timing of the stressor in utero, and the longevity of the effects. Finally, we discuss some possible mechanisms that may mediate the effects of prenatal maternal stress on the neurodevelopment of children.

14.1 Introduction

Retrospective epidemiological studies suggest that maternal exposure to a severe stressor during pregnancy (e.g., divorce, death of a relative, foreign invasion) increases the fetus' risk for suboptimal growth and for developing a variety of neurodevelopmental disorders later in life, such as autism and schizophrenia. The subsequent challenge is to understand the process that is responsible for these effects: how much of the effect of maternal exposure to an event is due to the objective severity of the event itself, how much is due to her subjective level of distress, how

S. King (✉)
Department of Psychiatry, McGill University, Montreal, QC, Canada
e-mail: suzanne.king@mcgill.ca

S. King · D. P. Laplante
Douglas Hospital Research Centre, Verdun, QC, Canada

M. C. Antonelli (ed.), *Perinatal Programming of Neurodevelopment*, Advances in Neurobiology 10, DOI 10.1007/978-1-4939-1372-5_14,

much to her hormonal response, and how might these aspects of the stress experience interact to affect the unborn child? And how might the mother's genetic profile moderate her subjective and hormonal response to a potentially stressful event? Similarly, how might the timing of the stressor in utero, the sex of the child, and the child's own genetic profile moderate the effects of the mother's distress on his or her own neurodevelopment?

When testing the "fetal programming hypothesis" prospectively, animal studies randomly assign pregnant animals to stress or nonstress conditions and find that maternal glucocorticoids (GCs) pass the placenta and alter fetal brain development, particularly the hypothalamic–pituitary–adrenal (HPA) axis and immune system, causing changes in a wide variety of outcomes for the offspring. Using experimental animals, the type, severity, and timing of the prenatal stress can be varied and controlled. In humans, however, the criteria for the gold standard of the rigorous experimental method are difficult to meet because we cannot randomly assign pregnant women to different stress conditions to determine the effects on the fetus. Without such rigor in the research design, the internal validity of a study is called into question, and the biopsychosocial mechanisms responsible for the wide variety of consequences of prenatal maternal stress (PNMS) seen in human offspring remain obscure.

Natural experiments offer an opportunity to circumvent the obstacles to random assignment in human research on PNMS. Many forms of disaster distribute their harm in quasi-random fashion, affect sizable communities that include large numbers of pregnant women, and have sudden onsets that allow the researcher to determine the moderating effects of the timing of the event in gestation. In human research, it is also important to distinguish between the pregnant woman's objective exposure to an event and her degree of subjective distress in response to it. Although one may assume that two rats of the same strain will not differ appreciably in their level of distress at having their tails pinched, for example, the human stress response is highly individual. Appraisal theory, which describes the intrapsychic processes that explain why two people experiencing the same event may have completely different emotional reactions to it (Folkman and Lazarus 1988; Lazarus 1991; Smith and Lazarus 1990), suggests that the objective degree of exposure to a stressor and the subsequent subjective distress are relatively independent of each other.

The goal of our research program is to capitalize on natural disasters in order to complement current understanding of how stress to the pregnant woman impacts her unborn child. We aim to determine the relative effects of several objective, subjective, and hormonal components of the stress experience by tracking these aspects of stress as they cascade onto biological processes in mother, placenta, and fetus, and onto development through early childhood and beyond.

In this chapter, we will summarize our three ongoing studies of women exposed to natural disasters during pregnancy and the findings on the child that are relevant to neurodevelopment, and discuss the general trends in the results.

14.2 Background

Disasters and other independent life events have been used in retrospective PNMS research and have shown effects on the development of psychopathology in later life. PNMS during specific weeks of gestation significantly increases the risk of developing schizophrenia and major depression: This has been shown in retrospective studies of the death of the mother's husband (Huttenen and Niskanen 1978) or other relatives (Khashan et al. 2008) during the pregnancy, of hurricanes and tornadoes (Kinney et al. 1999a), and foreign invasion (van Os and Selten 1998). Similar conclusions have been made for autism. In a historical analysis of ten severe storms in Louisiana, Kinney et al. (1999b; 2008) report that rates of autism in the state were significantly increased in a dose–response manner according to the severity of the storms. For children exposed during weeks 20–24 or 36–40 of pregnancy, for storms classified by the National Weather Service as moderate or high intensity, the prevalence rates for autism rose above baseline rates of 5.2/10,000 to 9.7, and 26.6/10,000 for moderate- and high-intensity storms, respectively. Although these retrospective studies demonstrate the power of a stressor in pregnancy to disrupt development in the fetus, they are incapable of elucidating the active ingredient in the prenatal stress experience.

Prospective studies of independent stressors, on the other hand, can be designed so as to disentangle the effects of any genetic transmission of maternal traits, the intrauterine environment, and the postnatal rearing environment. Our approach to PNMS research for the past 15 years has been to study pregnant women experiencing natural disasters. A disaster is any event that "causes disruption exceeding the adjustment capacity of the affected community" (Lechat 1979). By their nature, disasters tend to have sudden onsets and to be independent of the control of individuals (i.e., "independent stressors"). Using disasters for natural experiments in PNMS presents an approximation of the randomization afforded by true experiments and capitalizes on the relatively large potential subject pool following a disaster to a large community.

We currently have three such studies under way: Project Ice Storm, begun after the January 1998 Quebec Ice Storm in Canada; The Iowa Flood Study, including pregnant women exposed to the floods of June 2008 in the USA; and QF2011, the Queensland Flood Study of pregnant women in and around Brisbane, Australia. Although the samples are relatively small, with initial sample sizes between 200 and 300 women, statistical power is enhanced by having a range of exposure levels from mild to severe; by having a deep level of information on the women, their exposures, and their distress gathered soon after the events; by the fact that the degrees of exposure to the disasters were independent of family psychosocial attributes; and by the acute onsets of the events which allow exact dating of the stressors in the pregnancy. The independent nature of these stressful events allows us to worry less about controlling for potential confounds and allows us to disentangle the effects of objective degrees of exposure to the event from the women's subjective distress and hormonal responses.

As such, the overarching aim of our research program is to increase understanding of the ways in which different aspects of PNMS influence the development of the unborn child. We began this endeavor with the preconceived, underlying hypothesis that the more severe the woman's objective exposure to a disaster, the greater would be her subjective distress, which would result in higher levels of basal cortisol, which should then be the active ingredient in the effects on the fetus.

14.2.1 Project Ice Storm

Project Ice Storm was conceived following one of Canada's worst natural disasters in history: the January 1998 Quebec ice storm. During the first week of the year, a series of freezing rain storms passed through southern Quebec. The weight of the ice toppled high-tension power lines and utility poles, collapsing the power grid to the region. Resulting power outages ranged from a few hours to as long as 6 weeks for 3 million people in the province of Quebec. On Friday, January 9, the downtown core of Montreal was blacked out, leaving the city in darkness and commuters stranded in metro cars. The military were called in to assist local forces in removing broken trees and other debris from roads. Cold fronts followed the mild weather, plunging the region into seasonal temperatures of −10 °C to −20 °C. Without electricity to power central heating, pumps for well water, farm equipment, or factories, the personal and financial costs of the ice storm crisis left a significant impact on the population. Security forces went door to door to rescue isolated individuals in danger from cold and hypothermia, asphyxiation from unconventional heating devices, and fire due to blocked chimneys. There were more than 27 deaths attributed to the ice storm.

We identified physicians who deliver babies in the four main hospitals in the region southeast of Montreal (the Montérégie) which is typified by bedroom communities and rural areas. Physicians who agreed to collaborate indicated the number of patients in their practice who met our inclusion criteria: over 18 years of age, French-speaking, and pregnant on January 9,1998, or became pregnant between January 9 and April 9 (the preconception-exposed group). We delivered the correct number of postal questionnaires to each physician whose office staff addressed and mailed them on June 1,1998. Of the 1,440 questionnaires that were sent, 224 women responded—a 15.5 % response rate. Of the 224 initial responders, 178 women gave us their names and addresses, and permission to recontact them. Some of these women's pregnancies ended in miscarriage or stillbirth, and other families were lost to follow-up over the years. Our largest sample for a single testing was 110 children at age 5½ years. Our most recent assessment, at the age of 13½, included approximately 80 families.

The first questionnaire in June 1998 included a series of objective items about the women's exposure to the ice storm crisis. We constructed items to reflect four main categories of disaster exposure: threat, loss, change, and scope. To create a total score, we attributed a maximum of 8 points to each category and then summed them to create the Storm32 objective stress score. Table 14.1 presents the Storm32 items and their weightings. Women also completed the Impact of Event Scale-Revised (IES-R; Brunet et al. 2003; Weiss and Marmar 1997) which reflects the severity of their posttraumatic stress symptoms relative to the ice storm: hyperarousal, avoidance, and intrusive thoughts or images. We included saliva sampling kits for the assessment of cortisol levels at seven times points over 24 h. The questionnaire also included scales of current anxiety and depression. At 6 months after each woman's due date, we sent a second questionnaire about the birth and the newborn, and queried about other life events in the preceding year. Mental health and life event scales were repeated at every assessment.

The goal of Project Ice Storm has been to determine the effects of PNMS on the cognitive, behavioral, physical, and motor behavior of the children as they develop. Over the years, and as funding permitted, Project Ice Storm mothers completed postal questionnaires 6 months after their due dates, and when the children were 2, 4, 5, 6, 8, 9, 11, and 13 years of age. The children's school teachers completed rating scales of the children throughout elementary school, and the children have completed self-report questionnaires since the age of 11. We assessed the children's development through direct testing at ages 2, 5, 8, 11, and 13 years, including structural brain magnetic resonance imaginings (MRIs) at age 11, and a social stress test at age 13. DNA, diurnal and reactive cortisol, and blood for assessing glucose tolerance and immune functioning have also been collected from the children at various points. Beginning in 2013, further developmental assessments and brain imaging will continue between the ages of 15 and 18 years, and perhaps beyond as funding permits.

The Sample

Approximately half of the sample had given birth by June 1, the date the first questionnaire was mailed. The sample had, and maintains, a fairly equal balance of women who were exposed to the ice storm during the first, second, and third trimesters and in the 3-month preconception period. In addition, their infants were in equal numbers of boys and girls. Rates of preterm birth (8 %) and postpartum depression (17 %) were within the normal range. The 224 initial responders to the Ice Storm questionnaire were significantly better educated and had higher family incomes, than the averages for their region; 70.8 % of the participants were from households in the upper-middle class or above, and only 4.5 % were from lower and lower-middle classes. This socioeconomic bias in the sample explains the high cognitive functioning of their children (see sections below).

Stress Levels

On average, the women in the sample were without electricity for 14 days, with some going without power for as long as 45 days. The families were without phone

Table 14.1 Storm32: questions used to assess the four dimensions (threat, loss, scope, and change) of our objective stress questionnaire that the mothers completed after the ice storm

Threat		Loss		Scope		Change	
1	Were you injured? No=0 Yes=1	1	Did your residence suffer damage as a result of the ice storm? No=0 Yes=2	1	How many days were you without electricity? 0=0–5 days 1=6–13 days 2=14–19 days 3=20–21 days 4=> 22 days	1	Did your fam-ily stay together for the duration of the ice storm? Yes=0 No=1
2	Was anyone close to you injured? No=0 Yes=1	2	Did you experience a loss of personal income? No=0 Yes=2	2	How many days were you without the use of your telephone? 0=0 days 1=0.01–1 day 2=2–4.5 days 3=5–7 days 4=8+ days	2	Did you spend any time in a temporary shelter? No=0 Yes=1
3	Were you ever in danger due to:	3	How much was the total financial loss including income, food, damage to home? 0= < CAD 100 1=CAD 100–CAD 1000 2=CAD 1000–CAD 10,000 3=CAD 10,000–CAD 100,000 4=> CAD 100,000			3	How often were you required to change residence during the ice storm? 0=0 1=1 time 2=2+ times
3.1	…the cold No=0 Yes=1					4	Did you take in guests during the ice storm? No=0 Yes=1

Table 14.1 (continued)

Threat		Loss	Scope	Change	
3.2	…exposure to downed electrical power lines No=0 Yes=1			5	Did you experience an increase in physical work during the ice storm? 0=less or same 1=little or lot more
3.3	…exposure to carbon monoxide No=0 Yes=1			6	Number of nights away from home: 0=none 1=1–7.5 nights 2=8+nights
3.4	…lack of potable water No=0 Yes=1				
3.5	…lack of food No=0 Yes=1				
3.6	…falling branches and ice No=0 Yes=1				
8 points		8 points	8 points	8 points	

service for an average of 4 days, and some for as long as 34 days. Only one-third of the families in the study never left their own home, while most families were forced to move to friends', neighbors' and relatives' homes; half the sample moved once or twice, and 15 % moved up to five times during the crisis.

The initial sample scored between 0 and 24 on the Storm32 objective stress questionnaire, with an average score of 10.6 (standard deviation (SD) = 4.7) and a fairly normal distribution of scores (Fig. 14.1). The distribution of subjective stress levels, as assessed by the IES-R, was positively skewed (Fig. 14.2), and mean levels were moderate (M = 11.9; SD = 12.5), with 16.4 % of women scoring above the cutoff of 22 often used for screening for posttraumatic stress disorder (PTSD). In general, cortisol levels for the women showed the usual diurnal patterns and, as expected,

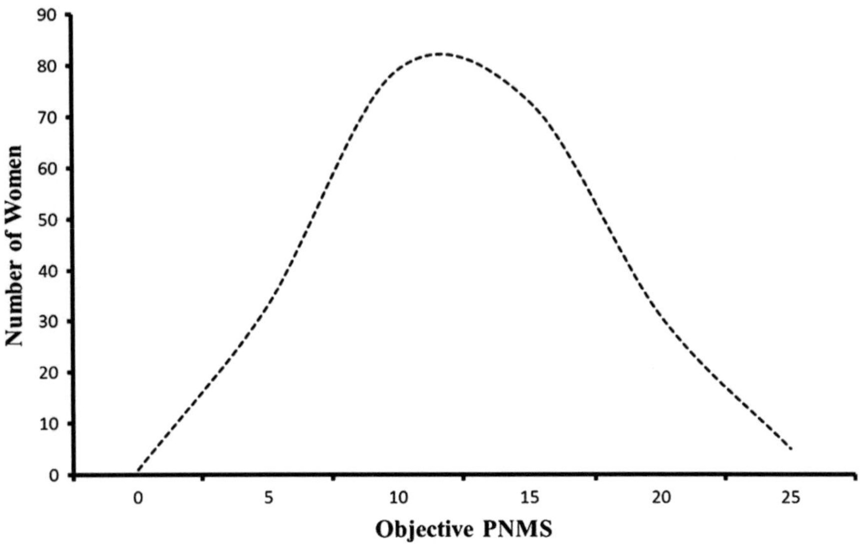

Fig. 14.1 Distribution of Storm32 objective prenatal maternal stress (*PNMS*) scores for Project Ice Storm cohort

were significantly higher in women who were still pregnant at the time they took the samples. This difference in basal levels according to pregnancy status has made the inclusion of maternal cortisol in statistical analyses untenable for much of our study.

The three forms of stress (objective, subjective, and cortisol) are relatively uncorrelated with each other. Storm32 and IES-R scores tend to be correlated at less than 0.30, and Storm32 and diurnal cortisol levels are uncorrelated ($r<0.20$). In the subsample that was still pregnant at saliva sampling, IES-R and integrated diurnal cortisol correlated −0.26 ($p<0.05$); although statistically significant, these two measures of stress cannot be considered highly associated, and the association that is present is negative: the more severe the women's PTSD symptoms the lower their basal cortisol levels, echoing findings from the PTSD literature (Zoladz and Diamond 2013). Socioeconomic status (SES), as reflected in the Hollingshead Index which takes both maternal and paternal education and occupation into account, has been uncorrelated with both objective and subjective stress ($r<0.20$).

Cognitive Development

General Intelligence Perhaps one of the most obvious clues to any effects of PNMS on neurodevelopment is the effect on general intelligence. As noted earlier, the families in Project Ice Storm are better educated and have higher incomes than the regional averages. Not surprisingly, then, their children, even those from high-stress families, have consistently scored above average on all tests of cognitive development. Statistically, this presents an inherent obstacle to explaining the

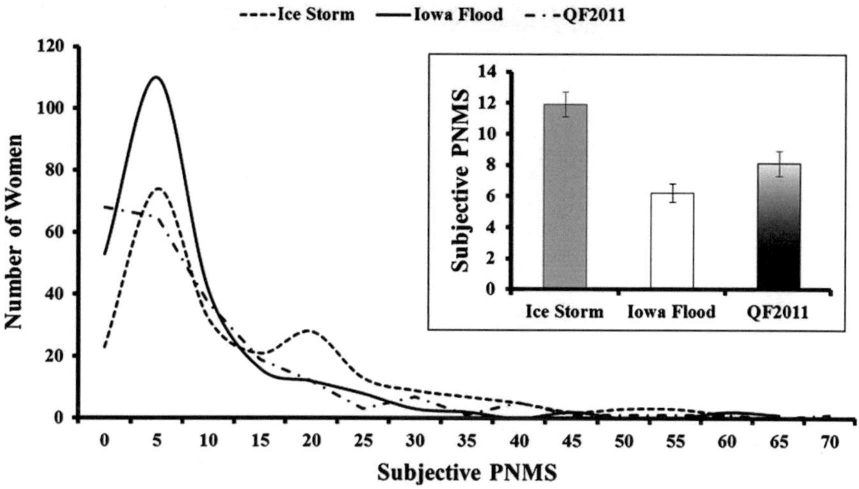

Fig. 14.2 Distribution and means (with standard errors) of subjective prenatal maternal stress (*PNMS*) levels (IES-R) for women in all three cohorts

limited variance in outcome variables. Psychosocially, this limits the study's generalizability to more disadvantaged families who would have had fewer financial resources for dealing with the ice storm crisis.

We obtained our first small grant for Project Ice Storm in time to test a subset of Project Ice Storm children at age 2 years (Laplante et al. 2004, 2007). To select the sample, we included only children whose mothers were exposed to the ice storm in the first (n=21), second (n=14), or third (n=23) trimesters (that is, we excluded the preconception-exposed group). We also excluded mothers who reported cigarette or alcohol use in pregnancy, and those who reported any other major life events while pregnant. In order to limit the sample size while maximizing the variance between groups, we split the sample into three equal groups according to the Storm32 objective stress scale and invited the low- and high-stress groups for face-to-face testing, but not the moderate-stress group. Children were tested with the Bayley Scales for Infant Development for general intelligence (the Mental Development Index: MDI), the MacArthur Communicative Developmental Inventory (MCDI) for receptive and productive language, and they were videotaped during a 15-min free play session to assess the maturity of their play style.

At this 2-year assessment, as illustrated in Fig. 14.3, we found that children from the high-objective stress group had, on average, approximately one full SD lower Bayley scores than those from the low-stress group, even controlling for birth weight; SES was unrelated to the Bayley MDI, perhaps because of the limited range in SES levels. The large effect of PNMS was obtained with the Storm32 objective stress score; the women's subjective distress level, reflected in the IES-R posttraumatic stress symptom score, had no effect on Bayley scores. At this age, the results demonstrated a significant moderating effect of the timing in utero of the ice storm:

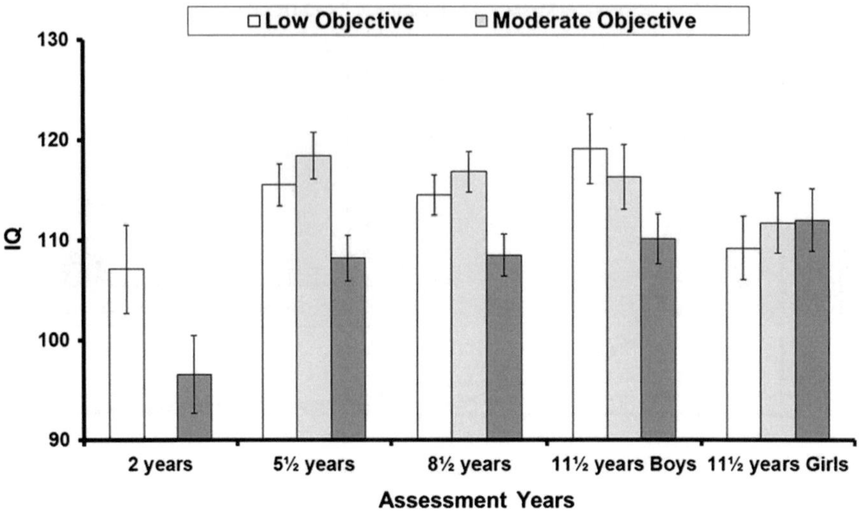

Fig. 14.3 Means and standard errors for Bayley MDI scores at 2 years and WPPSI/WISC-III IQ scores at 5½, 8½, and 11½ years

The difference between low and high Storm32 groups was significant for children exposed to the ice storm in the first or second trimester, but not in the third.

Given that both animal and human research suggest that prenatal stress or maternal pregnancy anxiety predicts a more difficult temperament in children (Weinstock 2008), we were concerned that scores on the Bayley may reflect effects of PNMS on the children's behavior with the experimenter rather than their cognitive development per se. By observing the children during a free play session while their mothers sat nearby, however, we could circumvent potential confounding by the child's temperament and attempt a replication of the results found by the more structured Bayley test using a more implicit test of cognitive development (Laplante et al. 2007). The experimenter presented all of the children with the same array of toys arranged identically. Raters coded videos of each child's play behaviors, moment by moment, into three categories: stereotypical, such as waving or mouthing a toy; relational, when touching two or more toys together without any specific function; and functional, in which a toy is used according to its intended function, such as rolling the truck along the floor or making pouring motions with the teapot. Results from these ratings replicated the effects of objective stress from the ice storm, as well as the moderation of the effect by the timing in utero: for children exposed to the ice storm in the first or second trimester, the low Storm32 objective stress group engaged in significantly less of the immature stereotypical play, and in significantly more of the more mature functional play, than the high-stress group. On the other hand, although the effect of objective stress was again highly significant, subjective stress (IES-R) levels were also significantly correlated with functional play outcomes.

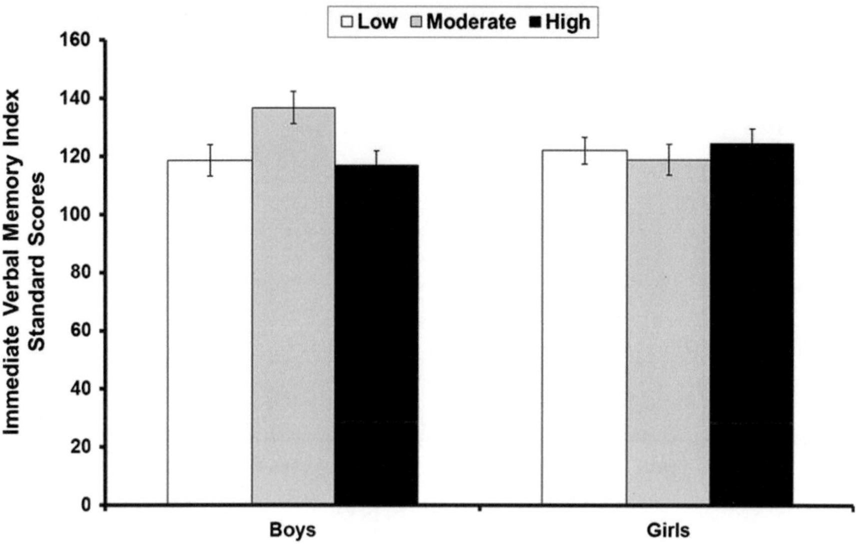

Fig. 14.4 Immediate Verbal Memory Index standard scores for boys and girls at 5½ years of age (means and standard errors)

The results obtained at age 2 years encouraged the Canadian Institutes of Health Research (CIHR) to support the project, and more extensive cognitive assessments were done at the ages of 5½, 8½, and 11½ years, including Wechsler intelligence scales, memory, and attention measures. Given the new funding, we were able to include all children from the study, including the moderate-stress group. Once again, at ages 5½ (Laplante et al. 2008) and 8½ we found that the high objective stress group had significantly lower full scale IQ than the low-stress group (5½: 108.2 vs. 115.5; 8½: 108.5 vs. 114.5); what we had not anticipated, however, was that the moderate-stress group obtained higher IQ scores than the low-stress group (5½, 118.4; 8½, 116.8). These curvilinear results were confirmed with a significant quadratic term when using the Storm32 score as a continuous measure. We could not have predicted these curvilinear results from the findings at age 2 since we had decided not to include the moderate-stress group in the testing at that age (Fig. 14.3).

Verbal Memory In similar fashion, the results using the Immediate Verbal Memory Index of the Children's Memory Scales with the children at 5½ years of age once again show this curvilinearity effect of Storm32 objective stress, although this time in boys only (Fig. 14.4).

Language Development The susceptibility of language development to the effects of PNMS has been a recurring theme in Project Ice Storm results. During the assessment at 2 years, we administered the MCDI which presents the mother with a list of words. She indicated, for each word, whether the child understood it (receptive vocabulary) and/or used the word (expressive/productive vocabulary). At this age, the high objective stress group understood 10% fewer words than the low-

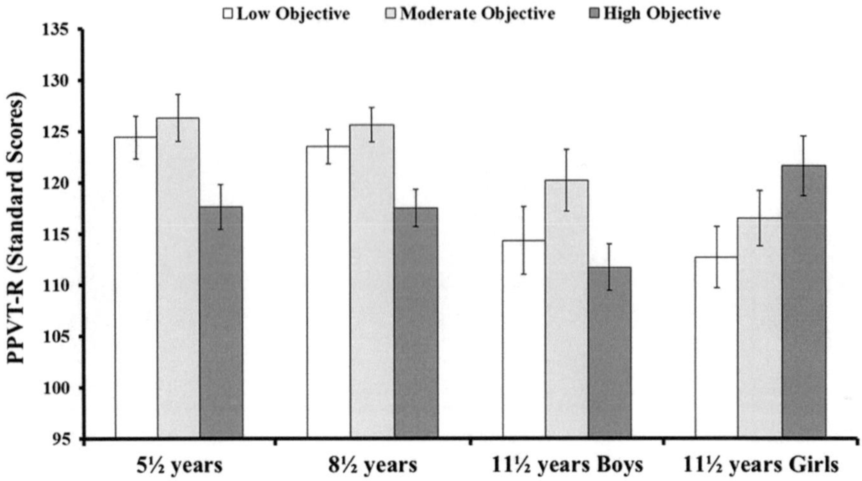

Fig. 14.5 Receptive vocabulary standard scores as a function of year of assessment (means and standard errors). *PPVT-R* Peabody Picture Vocabulary Test

stress group (82.4 vs. 90.6 words), and used 27% fewer words than their low stress peers (54.7 vs. 74.9; Laplante et al. 2004). As with the results for the Bayley scales, there was no effect of the mothers' subjective stress scores. However, unlike the results for the Bayley and for the play style, there was no timing effect on language development.

Beginning with the assessment at age 5½ years, we have used the Peabody Picture Vocabulary Test (PPVT-R) to assess receptive language, and have used the verbal tests from the Wechsler scales to assess other components of language development. At age 5½, we found significant effects of Storm32 objective stress, but not subjective stress (IES-R), on PPVT-R scores and replicated the curvilinear effects of PNMS on language that we saw in the full-scale Wechsler IQ scores, with a slight increase in vocabulary with moderate stress but a sharp drop in vocabulary with higher stress levels (Fig. 14.5; Laplante et al. 2008). Within the Wechsler itself, we found similar significant and curvilinear effects of objective stress on the two verbal tests we administered (Information and Similarities), but no effect on the nonverbal Block Design test.

Similar IQ and receptive vocabulary outcomes were observed in boys and girls at 8½ years of age: Children exposed to moderate levels of objective PNMS exhibited the highest scores while those exposed to high levels had the lowest scores. At age 11½, however, the results differed by sex: For boys, the curvilinear pattern was replaced by a significant, negative, linear effect of objective stress on IQ, while for girls there was no longer any effect of prenatal stress (Fig. 14.3). For PPVT-R scores at 11½ years, boys still showed the curvilinear effect, while for girls there was now a significant, linear, positive effect of objective stress on vocabulary. Thus, the effects of PNMS on cognitive outcomes appear to become sexually dimorphic at the onset of puberty (Fig. 14.5).

Physical Development

Conventional wisdom would suggest that any effects of PNMS on cognitive outcomes would be the result of a direct neurodevelopmental insult during gestation, probably due to the effects of maternal stress hormones passing through placenta. Indeed, a review of the literature (Charil et al. 2010) supports the notion that PNMS has important, negative effects on brain development, at least in laboratory animals. The hippocampus appears to be especially sensitive to PNMS. Unfortunately, these studies are difficult to replicate in humans.

Dermatoglyphic Asymmetry Neurodevelopmental insults may be evident in parts of the human body outside of the brain itself. For example, individuals diagnosed with schizophrenia, which is presumed to be neurodevelopmental in origin, exhibit higher rates of some physical traits than do normal controls. Schizophrenia patients have, on average, more minor physical anomalies (Lane et al. 1997) and have greater left–right asymmetry in the number of ridges in their fingerprints (Markow and Wandler 1986). Fingerprints develop out of the same fetal ectoderm as the brain between weeks 14 and 22 which overlaps with the timing of crucial hippocampal development. There was considerable debate within the schizophrenia literature about the causes of this dermatoglyphic asymmetry, whether genetic or neurodevelopmental in origin.

To address this debate, we tested the hypothesis that maternal stress from the ice storm would predict dermatoglyphic asymmetry in the children, but only in those for whom the onset of the ice storm occurred at some point between weeks 14 and 22 of pregnancy (King et al. 2009). Results showed that mere exposure to the ice storm during the key time period resulted in half a standard deviation greater asymmetry than in children exposed at other times of gestation. Within the group exposed during weeks 14–22, greater asymmetry was associated with both more severe objective stress ($r=0.348, p<0.10$) and subjective stress ($r=0.500, p<0.05$). Finally, for the 17 children in the 14–22-week exposure target group whose mothers were still pregnant when they took diurnal salivary cortisol samples, higher levels of cortisol were associated with less asymmetry ($r=-0.557, p<0.05$), that is, greater evidence of neurodevelopmental insult was associated with lower levels of maternal cortisol. Although we would have predicted that higher maternal cortisol would have been associated with greater evidence of neurodevelopmental insult, our cortisol assessments were taken 5–6 months after the ice storm crisis and may be poor reflections of actual acute levels of cortisol at the time of the storm. The stress literature is debating the meaning of the low cortisol levels in people diagnosed with PTSD; there is support for the notion that these low cortisol levels pre-existed the trauma exposure (a difficult hypothesis to test) and represent a vulnerability for developing symptoms of PTSD in the face of trauma rather than a concomitant effect of the exposure (Yehuda et al. 2000).

Brain Structure The effect of ice storm stress on dermatoglyphic asymmetry suggested the possibility that an effect would also be seen on the development of the

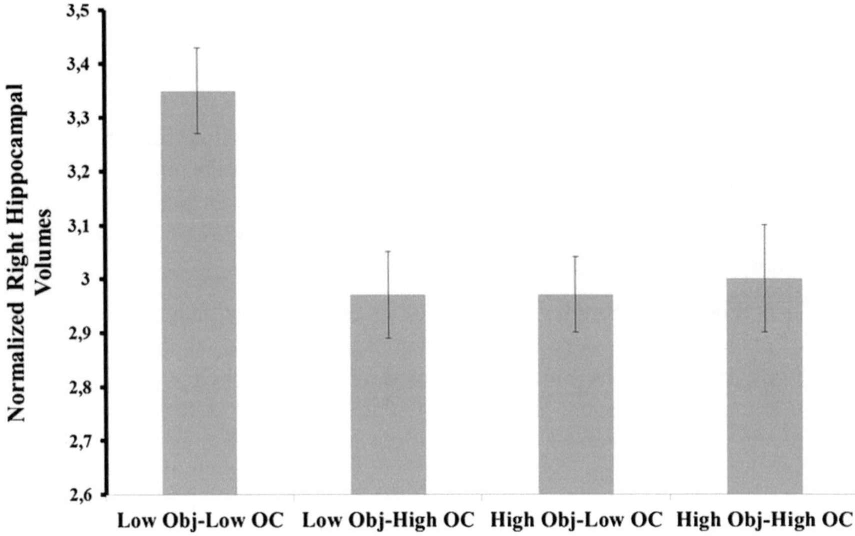

Fig. 14.6 Normalized right hippocampal volumes as a function of objective PNMS and obstetric complications (OC) (means and standard errors)

hippocampus. Structural brain scans (MRI) were conducted when the children were aged 11½ years; for comparison, we also scanned a matched sample of children who were born in the year before the ice storm. Unpublished results show that in both boys and girls, smaller normalized right and left hippocampal volumes were associated with being exposed to a greater number of maternal obstetric complications. Moreover, boys exposed to higher levels of objective PNMS had smaller normalized right hippocampal volumes. However, an observed objective PNMS × obstetric complication interaction suggests that smaller normalized right hippocampal volumes were observed in boys who were exposed to either high levels of objective stress, irrespective of the level of maternal obstetric complications, or low levels of objective PNMS and high levels of maternal obstetric complications (Fig. 14.6).

Thus, exposure to a major stressor in utero appears to have implications for brain development. But what of our comparison cohort of postnatally exposed children? Near the end of our scanning sessions, we began to wonder whether the neurodevelopment of children in our comparison group, who had been infants during the ice storm, might be influenced by the stress of their mothers and, if so, by what mechanism? Apart from having a stressed mother, or being cold, it occurred to us that maternal stress might be transmitted to infants directly via breast milk. Indeed, studies show that maternal cortisol does reach the infant via breast milk (Patacchioli et al. 1992). For the 15 children who were breastfed during the ice storm crisis, we created a composite score, multiplying the mother's objective stress score times the number of days that her child was breastfed during the disaster. The higher this composite stress exposure score, the smaller the child's left hippocampal volume ($r=-0.581$, $p<0.10$); examination of the scatter plot ruled out the effects of outliers (Fig. 14.7; King et al. 2012).

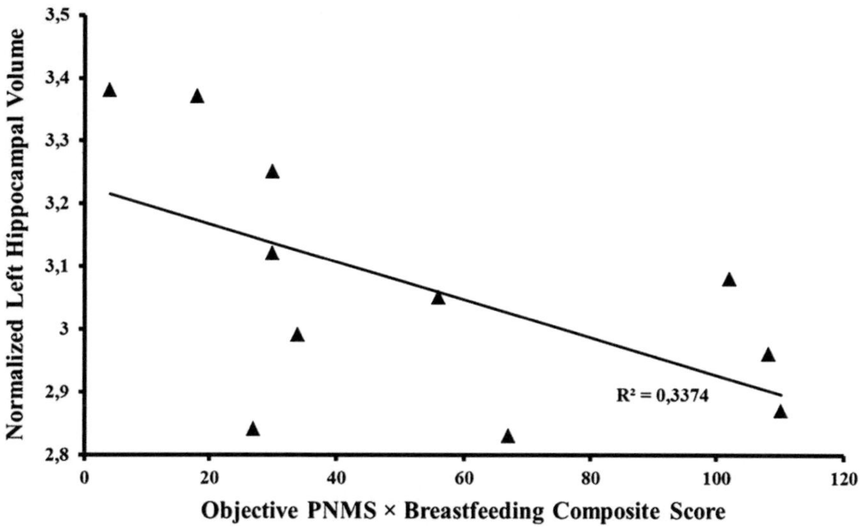

Fig. 14.7 Scatterplot of the control children's normalized left hippocampal volumes as a function of their mother's objective prenatal maternal stress (*PNMS*) × duration of breastfeeding composite scores

HPA Axis Functioning Because the animal literature suggests that PNMS alters HPA functioning in the offspring, we have obtained salivary cortisol samples from our subjects before and after potentially stressful situations on several occasions. At about the age of 5 years, children in Quebec are scheduled to receive vaccinations. We traveled to the medical clinics in order to be present for the vaccinations of 34 children who provided saliva samples while in the waiting room before the injection, and 15 min post injection. Results suggest that neither maternal objective nor subjective stress was related to the children's acute cortisol reaction to the vaccination. In all children, however, lower acute stress response predicted more severe internalizing (e.g., depression and anxiety) and externalizing (e.g., aggression, destructiveness) at age 5½ on the maternal-rated Child Behavior Checklist (CBCL).

We used a similar protocol to determine the children's stress reactivity from the MRI scans at age 11½. Preliminary results from the Project Ice Storm children suggest there is no association between either objective or subjective PNMS and pre- or post-MRI cortisol levels.

For the subjects from the infancy-exposed MRI comparison group who had been breastfed during the ice storm, the results suggest that higher composite stress scores were associated with smaller left hippocampal volumes, which were then associated with smaller acute cortisol responses to the stress of the MRI scan ($r=0.632$, $p<0.05$), which were then associated with more severe CBCL internalizing problems at the same age (King et al. 2012). This latter association is consistent with the results from the prenatally exposed Ice Storm cohort for whom more blunted cortisol responses to vaccination were associated with more severe internalizing problems. The clinical implications of these results are that the failure to mount a stress response is a greater risk factor for psychopathology than an increase in cortisol, something seen in conditions such as schizophrenia (Brenner et al. 2009).

Finally, at age 13½ years, 63 Project Ice Storm teens completed the Trier Social Stress Test (Kirschbaum et al. 1993). Data from that protocol have not been analyzed to date, but will provide useful information on the longevity of the effects of PNMS on the HPA axis into early adolescence. The data collection protocol at age 15½ (2013–2014) will include salivary cortisol awakening responses in association with daily stressors during the school year.

Project Ice Storm Limitations

Project Ice Storm was the first prospective longitudinal study of a natural disaster that teased apart the objective and subjective aspects of the PNMS experience. As such, it provides unique data about the nature of PNMS and its effects. However, there are several limitations to the project that can neither be overlooked nor remedied.

These limitations are related to the numerous challenges involved in initiating a large study of pregnant women in the immediate aftermath of a natural disaster. First, despite the significant effects that both objective exposure and subjective distress have on a variety of outcomes, we have no data that enlighten us about the biological mechanisms involved in these effects. Part of the problem is that pregnant women in a target population will not all stay pregnant for long; fetuses will not wait for researchers to prepare a protocol that would permit them to collect precious biological samples that may contain clues about the cascade of effects from maternal exposure, through the placenta and umbilical cord to the fetus. In Project Ice Storm, half the sample had already given birth by the time of the first postal survey. Any delay between the disaster and initial data collection may also compromise the prospective nature of the data since the difficulty of the women's birthing experiences may color their recall of the disaster. In addition, the longer the delay between the disaster and the collection of biological data, such as cortisol, the more difficult the interpretation of the hormonal data; we have found it difficult to conduct analyses combining cortisol samples from pregnant women and from women who have already given birth and who may or may not be breastfeeding. Even the subjective stress data must be interpreted within the context of a relatively prolonged stress response: Do high levels of distress 6 months after the ice storm necessarily reflect similarly high levels during the storm itself, and how many of the women with low distress levels at 6 months had had high levels at the height of the disaster? Thus, speed is of the essence in setting up a disaster study.

The second challenge is to initiate a study (including protocol development, ethics approval, subject recruitment, and assessment) in record time in the absence of dedicated funds. When funds are limited, it is difficult to conduct the most effective recruitment methods. Project Ice Storm relied upon inexpensive postal questionnaires for recruitment and obtained a 15 % response rate from a potential sample of 1,400 women, rather than using more direct, personal, and costly face-to-face methods. Carrying a sample of fewer than 180 cases at the beginning of the study, and fewer than 100 cases 15 years later, means the sample size limits the complexity of the statistical analyses that can be conducted; once the sample is separated by child sex and trimester of exposure, additional interactions and nonlinear associations push the sample size beyond its limits.

Another side effect of the cost-saving recruitment method is that the sample is biased towards the highly educated family. This bias limits the generalizability of the results to similarly advantaged families, and also (in theory) reduces statistical power due to the restricted variance in the children's outcomes in the upper ranges of functioning.

Finally, the occasional critic of Project Ice Storm has noted that it is limited by not having evaluated the pregnant women in the study before the ice storm struck. Not being blessed with the gift of prophecy, this limitation seemed quite impossible to overcome.

14.2.2 The Iowa Flood Study

In 2008, we discovered that we could conduct a pre-post disaster study without necessarily being able to predict future disasters. In early June of that year, the Midwest of the USA was hit with back-to-back storms. The record-breaking levels of rainfall lead to the flooding of rivers and breeching of levees. Iowa lay at the heart of the destruction, suffering its worst flooding in more than 50 years. The Cedar and Iowa rivers exceeded record high water levels set back in the early twentieth century. In Iowa alone, 83 of 99 counties were declared federal disaster areas. In Cedar Rapids and Linn County, 1300 city blocks (9.2 square miles) were covered in water, engulfing the city hall, Linn County jail, fire department, and public library. The severe flooding in Johnson County, Iowa City, and the University of Iowa forced the evacuation of 35,000–40,000 people from their homes. The disastrous flooding lasted from early June to early July damaging 5238 homes and 940 businesses. The agricultural industry suffered a US$ 2 billion loss as many corn and soybean crops were destroyed. Approximately two dozen people were killed and 150 injured as a result of the storms and flooding. Overall, the 2008 Midwestern flood ranks among the top ten disasters in US history.

As the flood waters were still rising, we contacted Dr. Michael O'Hara, the well-known postpartum depression researcher at the University of Iowa, to discuss a collaboration. We learned that Dr. O'Hara's doctoral student, Kimberly Nylen (now Hart), was in the midst of conducting a study linking stress in pregnancy to obstetric and birth complications (Nylen et al. 2013a, b). We invited them to collaborate on a replication of Project Ice Storm by piggybacking our disaster protocol onto their birth outcomes project.

The Sample

The 134 women who had already been recruited into Nylen's doctoral research were invited to join the Iowa Flood Study. Additional participants continued to be recruited for both studies via media advertisements distributed throughout Eastern Iowa, in-person recruiting in the Department of Obstetrics and Gynecology at the University of Iowa Hospitals and Clinics, letters distributed to patients of obstetrics and gynecology practices in Eastern Iowa, and the clinic associated with the Johnson County Department of Public Health.

When our final recruitment phase was completed, we had flood stress data from 268 women who had been pregnant before or during the flooding, of whom 74 had been assessed by Nylen before the flooding for psychopathology, life events and daily hassles, social support, coping style, and other psychosocial variables. Despite efforts to recruit a wider range of participants, the majority of subjects were from either upper-middle class (55 %) or upper class (20 %), with only one woman in the lower-class category, and 25 % in the lower-middle- and middle-class groups—not appreciably more representative than the Project Ice Storm sample. Although the Iowa Flood Study failed to include a preconception-exposed group of any size ($n=18$), 34 women from Nylen's project, who had given birth before the floods, constituted a postpartum-exposed group that may be followed into the future to determine the effects of maternal stress during infancy and, perhaps, the role of breastfeeding.

Stress Levels

Much of the initial assessment protocol was similar to that of Project Ice Storm. However, the objective stress questionnaire from the ice storm needed major revisions to be relevant to a flood disaster. The final version included old and new items in the four categories of threat, loss, scope, and change. Weighting each category equally, with a maximum of 25 points each, our Iowa Flood Study scale is named IF100. As expected, a large percentage of women experienced very little hardship from the flood, and a small percentage experienced catastrophic loss and danger (Fig. 14.8).

Although Storm32 and IF100 scores cannot be compared directly between the two events, both studies used the IES-R to measure subjective distress. As in Project Ice Storm, the IES-R distribution is highly skewed, with 6.7 % of women scoring above the cutoff of 22 for potential PTSD. Despite the fact that several women from Iowa lost their homes in the floods, something unheard of in the ice storm sample, the average IES-R scores were significantly lower in the Iowa Flood Study than in Project Ice Storm (Fig. 14.2).

In the hope of circumventing the difficulties noted in Project Ice Storm of assessing persistent PTSD symptoms 6 months post-disaster, in Iowa we introduced two "peritraumatic" measures: the Peritraumatic Distress Inventory and the Peritraumatic Dissociative Experiences Questionnaire. Both measures ask the respondent

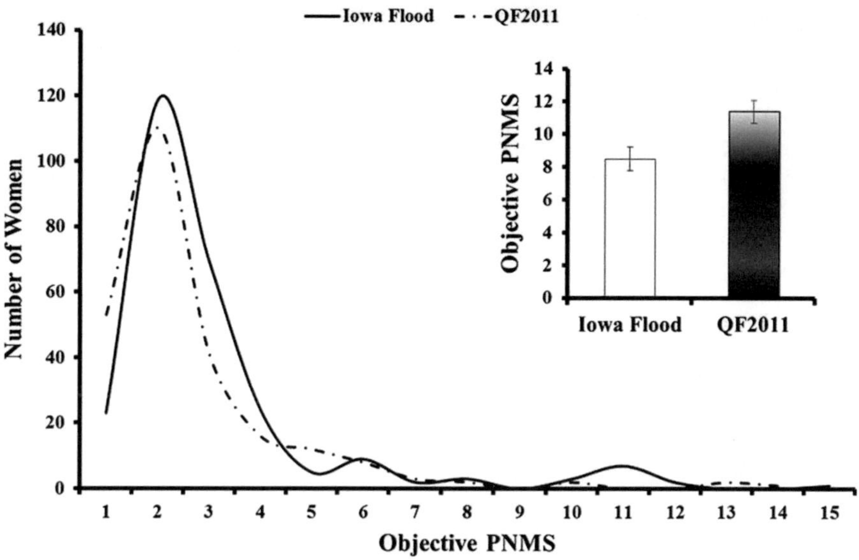

Fig. 14.8 Distribution and means of objective prenatal maternal stress (*PNMS*) levels for women in the Iowa Flood (IF100) and QF2011 (QIF100) cohorts

to recall and report what they had experienced at the time of the trauma. At recruitment, these peritraumatic measures correlate between 0.60 and 0.70 with the IES-R total score. Future analyses will elucidate the relative contributions these three distress measures make in explaining variance in child outcomes.

Assessments of Child Development

Funded by the CIHR, face-to-face evaluations of the children's cognitive, behavioral, motor, and physical development have been conducted at ages 2½ ($n=161$) and 4 years ($n=130$) with a 5½-year assessment to begin in 2013–2014. To date, only the 2½-year assessment has preliminary results that are sufficiently advanced to present here.

Cognitive Development

General Intelligence and Language As in Project Ice Storm, maternal stress from the disaster explained significant amounts of variance in the children's composite scores from the Bayley. Unlike in the Ice Storm cohort, however, there was no effect of maternal objective stress exposure (IF100). Instead, a significant interaction suggested that high levels of maternal subjective distress (IES-R scores) were associated with lower Bayley scores for those exposed during the first trimester.

Thus, the Iowa results support the idea of a certain vulnerability in early gestation for cognitive development assessed at this age, but suggest that the smoking gun may be maternal subjective distress and not objective hardship in this cohort.

Mothers completed the MCDI during this assessment, checking words on a list indicating their children's receptive and productive vocabulary. Although it appears that the language development of boys and girls were affected in different ways by their mother's flood stress, it appears that for all children exposed to the flood during the first or second trimesters the greater their mother's subjective distress, the lower their productive language at this age.

HPA Axis Functioning

The Iowa Flood Study included a relatively controlled test of stress reactivity in the toddlers at age 2½. Mothers left the testing room for up to 3 min leaving their toddlers alone (but under surveillance by research staff via camcorder and one-way mirror). Saliva samples were taken before and after the separation. The children's cortisol reactions to the separation were mixed, with some showing an increase in cortisol by 25 % or more (~40 %), others a decrease by at least 25 % (~20 %), and still others showed little change (~40 %). Preliminary analyses of those data, explaining approximately 27 % of the variance in pre-post separation cortisol change, suggest that the association between the mothers' objective stress exposure and the child's cortisol response differed for boys and girls, with girls appearing to be more sensitive to the effects of the stress. Timing also appears to play a role, with greater cortisol responses in children exposed to objective maternal stress in early gestation and lower responses (even decreases in cortisol) in those exposed later in utero.

There was no stress induction included in the 4-year-old assessment, but we plan to follow the effects of maternal stress from the Iowa floods on the children's HPA axis functioning at later ages.

Iowa Flood Study Limitations

The Iowa Flood Study is the world's first pre-post disaster study of PNMS. Although the data are still being analyzed, it promises to yield unique and crucial results that will further increase our understanding of PNMS. In particular, the ability to control for pre-trauma psychosocial functioning, coping styles, and social support will permit us to better understand pregnant women within their own contexts, and to learn which psychosocial factors may buffer them and their unborn children from adverse outcomes, and which may increase risk.

We had hoped that the Iowa study would provide a larger, more representative sample than the Ice Storm study, and that the range of exposure severity would be greater than that in Project Ice Storm which lacks a "no stress" comparison group. Although the range of hardship is probably greater in Iowa than in Ice Storm, the average IES-R scores were lower. We also failed to recruit a sample with a significantly greater range of socioeconomic levels, despite more direct recruitment in sites such as community health clinics. Finally, we had developed a protocol for collecting placentas, umbilical cords, and cord blood in Iowa, but we did not obtain the required funding in time, limiting our ability to study a complete picture of the cascading stress process from mother to placenta to fetus.

14.2.3 QF2011 The Queensland Flood Study

In 2011, after years of severe drought, the Australian state of Queensland experienced some of its worst summer flooding in living memory. Starting late December 2010, the state experienced torrential rainfall, leading to severe flooding throughout much of January 2011. The rains lead to the state's second largest flood in the past 100 years. Severe flooding was recorded throughout most of the Brisbane River catchment, affecting at least 70 towns, including the state capital of Brisbane. Brisbane is the third largest city in Australia and home to more than 1 million people. The flooding in Brisbane alone affected 200,000 people, caused residences on 2100 streets to be evacuated, inundated 18,000 properties, and left 100,000 homes and businesses without electricity. In total, three-quarters of Queensland was declared a disaster zone. Altogether, 35 deaths were directly attributed to the flooding. The flooding severely damaged many roads and bridges leaving many individuals stranded in their homes until they could be rescued by the army. Economic losses were enormous, amounting to around Australian $ 1.5 billion.

Working through our network of contacts, we were referred to Dr. Sue Kildea, director of midwifery research at the Mater Mother's Hospital and at the Mater Medical Research Institute in Brisbane. In Australia, nearly all babies are brought into the world with the aid of midwives, working in hospitals, and 10,000 babies per year are born within the walls of the Mater. Dr. Kildea was coinvestigator on a multisite randomized control trial of two forms of midwifery care, the M@NGO Trial (Tracy et al. 2011), which was in mid-recruitment at the time of the floods. With our invitation to collaborate, she saw an opportunity to determine the extent to which the experimental condition, Group Midwifery Practice, might buffer the effects of stress from the flooding on the pregnant women and their unborn children. She agreed to append a new flood study onto the existing Brisbane arm of the M@NGO Trial. Although flooded out of her home herself, she rallied a team on-site to prepare ethics documents and to put in place procedures for collection of the birth biological samples. Our Montreal team sent a young placentology student to Brisbane to be on-call 24/7 to receive the needed tissues at the time of birth and a research assistant to help the Mater team with the coordination. The number of individuals in Montreal and in Brisbane who were involved in making the QF2011 Queensland Flood Study happen is too great (in quality and in quantity) to describe here. A grant application was submitted to the CIHR on March 1, the first placenta was collected in April, and notification of funding arrived in July of that year.

One advantage of adding QF2011 to the stable of PNMS studies is that it will allow a more direct replication of the Iowa Flood Study—two studies of similar disasters during the same season of the year.

The Sample

Women who were taking part in the M@NGO trial, and who were still pregnant during the flooding, were recontacted to invite them into QF2011. Additional women meeting the inclusion criteria were recruited in person and by flyer through the Mater Hospital. At this writing, 2½ years post-flood, we are beginning to finalize the description of the QF2011 sample. Objective and subjective stress levels are available for 228 women. Following the birth of the children, we were able to obtain dates of birth and sex of the children from 201 women. Thirty-five women were in their third trimester of pregnancy, 81 in their second trimester, and 75 in the first trimester of pregnancy at the peak of the flooding. An additional 10 women became pregnant within 3 months of the peak of the flooding. The women gave birth to 96 daughters and 105 sons.

Stress Levels

With the Iowa Flood Study protocol already developed and tested, relatively little adaptation was required to prepare the objective stress questionnaire for the Australian context. Nonetheless, we created more detailed questions about insurance coverage, business losses of the spouse, and other specifics. Once the laborious task of creating the scoring scheme was completed, the Queensland Flood Objective Stress Scale (QFOSS) had a possible maximum of 200 points: 50 points each for the threat, loss, change, and scope scales; we also created an additional scale on nutrition but which does not figure into the QFOSS score.

As in Iowa, the QFOSS scale has a positively skewed distribution, unlike Storm32 which was normally distributed. In both flood studies, a large percentage of women experienced very little hardship as a result of the flood, while a non-negligible percentage lost nearly all of their worldly goods.

Because the QFOSS was essentially an extension of the IF100, we could pull an IF100 score out of the Queensland questionnaire (which we called QIF100). In so doing, we saw that the average objective PNMS score was significantly higher in the Brisbane cohort ($M=11.4$; SD=10.8) relative to the Iowa cohort ($M=8.6$; SD=11.4; Fig. 14.8).

The IES-R was administered at recruitment, again with the expected skewed distribution. When the IES-R scores were compared across all three studies, women who experienced the ice storm had higher IES-R scores compared to the women who experienced either flood. When comparing the flood victims, women in the Brisbane cohort had higher IES-R scores relative to women in the Iowa cohort (Fig. 14.2).

When comparing the peritraumatic experiences of the flood victims, women in the Brisbane cohort had higher peritraumatic distress levels (recruitment: $M=11.9$, SD=8.9; follow-up survey: $M=9.5$, SD=6.3) compared to women in the Iowa cohort (recruitment: $M=10.4$, SD=7.6; follow-up survey: $M=7.5$, SD=7.2). The women's peritraumatic dissociative experiences did not differ at either assessment point.

As in Project Ice Storm and Iowa, QF2011 subjects provided diurnal cortisol samples, allowing us to study both the cortisol awakening response and the complete diurnal pattern of cortisol secretion. These data have yet to be analyzed.

Infant Assessments

The first face-to-face assessment of the QF2011 children was conducted at 16 months of age (2012–2013). The Bayley scales, the VMI Visual Motor Integration, the cognitive free play session, and the MCDI (completed by the mother) were included in the protocol. In addition, the brief separation from the mother, identical to the task used in Iowa, was conducted to assess the child's HPA axis response to stress. This protocol was identical to the one we used in the Iowa Flood Study at 2½ years and will be repeated in the QF2011 2½-year assessments (2013–2014).

To date, there are no results to report. The demands of working out the bureaucratic regulations of multiple institutions, ethical requirements protecting the data, while at the same time keeping up with data collection, keep the production of results from prospective studies of this nature moving at a snail's pace.

14.3 Conclusions and Discussion

After 15 years of studying PNMS, we would like to have definitive conclusions to present here. A certain number of patterns have emerged in Project Ice Storm over time, only to be contradicted by our subjects as they approach adolescence and to be challenged by preliminary results from Iowa. It seems that conclusions are moving targets, changing over time, across geographic boundaries and across disasters of different varieties.

14.3.1 Objective Exposure, Subjective Distress, and Cortisol

We began our research program with a preconceived model of how PNMS would work: A pregnant woman would experience an objective stressor, the severity of which would predict her subjective response to it, which would then drive her hormonal response, and the resulting increase in cortisol levels would overwhelm the 11-β-HSD2 in the placenta, sending noxious cortisol to the fetus and interrupting otherwise optimal neurodevelopment of whichever system was in ascendance at that moment in gestation.

In Project Ice Storm, however, the correlations among objective and subjective stress measures and cortisol have been low (<0.30) suggesting that the three elements of the PNMS experience are relatively independent, at least in the Quebec winter of 1998. For Iowa, these same correlations are in the range of 0.40–0.60. There may well be something about the nature of the different disasters that explain these discrepancies: The 1998 Quebec ice storm created disruptions in daily life for the entire population of the region, even for those who never lost power, which is reflected in the normal distribution of objective stress (Fig. 14.1). Thus, although there was relatively little loss of house and home from the ice storm compared to the Iowa floods, there could also be no "no stress" group in Project Ice Storm as there could be in the Iowa Flood Study and in QF2011 in which some families were never inconvenienced and others lost everything they owned (Fig. 14.8). We also wonder about the possibility of additional physical stress caused by exposure to cold—an oversight in the Storm32 questionnaire—that may have its own effects that are independent of those from psychosocial stress.

Our elaboration of a measure of objective stress exposure in Project Ice Storm was initially intended as a simple control measure, the backdrop against which maternal subjective distress would emerge as the smoking gun in causing suboptimal neurodevelopmental outcomes in the unborn child. Yet, our results from this first study suggest that objective exposure is a powerful agent in influencing the child's cognitive development (IQ, verbal memory, language) and physical development (hippocampal volumes, obesity, insulin secretion) in the absence of a concomitant effect of maternal subjective distress. Similarly, preliminary results suggest that, in both Ice Storm and Iowa, children's HPA axis responses to their own exposure to stress is driven more by maternal objective exposure than by subjective distress. Thus, the strength of the effects of objective stress exposure have surprised us and, to date, we have no data upon which to base a hypothesis about how mere objective exposure to a stressor by the pregnant woman can influence fetal outcomes while bypassing maternal subjective distress.

Much of the effect of maternal objective exposure on child cognitive outcomes in Project Ice Storm appears to be, surprisingly, curvilinear. Thus, 5- and 8-year-old children whose mothers had objective stress scores in the mid-range had higher Wechsler IQs and language and memory scores than those with lower maternal objective stress, although high stress still predicted the worst performance. One might apply the Yerkes–Dodson Law of Optimal Arousal (Yerkes and Dodson 1908) to explain this effect, but the application breaks down when we consider that Yerkes–Dodson would apply to the stress and arousal of the pregnant mother, which would seem to have limited relevance to the performance of her child on an IQ test at age 5 or 8 years. One might rather invoke the notion of hormesis (Calabrese 2008), a concept from toxicology that describes the observation that small doses of some toxins have beneficial effects while effects become catastrophic at larger doses.

Maternal subjective distress from a disaster also has a role to play in the children's neurodevelopment. Although in Project Ice Storm, performance on the Bayley scales of infant cognitive development at age 2 years was related to objective stress, and not subjective distress, the opposite was true in Iowa with subjective stress explaining the bulk of the variance in general intelligence and language development. Even within Project Ice Storm, there are some inconsistencies in the effects of PNMS on cognitive development. Although at age 2 years it was objective stress predicting Bayley scores, objective and subjective stress explained equal amounts of variance in the maturity of the children's free play. And although objective stress explained the lion's share of variance in cognitive development at ages 5, 8, and 11 years, we are starting to see a shift at age 13 years towards a greater effect of maternal subjective stress, rather than objective. Thus, differences between disasters (winter/summer, ice storm/flood) and between the ages at which assessments occur (childhood, adolescence) seem to influence the relative roles of prenatal exposures to objective and subjective maternal stress.

Thus, one general conclusion from our work is that the degrees of objective exposure and subjective distress of a pregnant woman going through a natural disaster have differential effects on a variety of developmental trajectories of their children. An additional complexity emerges in much of our data, however. Although not mentioned in our review above, for some outcomes (e.g., fetal growth, Dancause et al. 2011; motor skills, Cao et al. 2014), the worst child outcomes are associated with mothers who exhibited a "mismatch" between their objective exposure and their subjective distress; in Project Ice Storm, the shortest birth lengths and the lowest bilateral coordination and visual motor integration performances belonged to children whose mothers either had high subjective stress in the face of mild objective exposure or had low distress in the face of high exposure levels. Preliminary analyses from Iowa suggest that it is these over- and under-reacting mothers who are also at greatest risk of postpartum depression (Brock et al., unpublished data). Further analyses are required to determine whether maternal psychopathology triggered by a mismatch in disaster reaction mediates the effect of the disaster on child outcomes.

Logistic difficulties with maternal cortisol assessment, being too long after the disaster itself to tell us about immediate maternal response to the event, and difficulties equating samples from pregnant and postpartum women, prevent us from drawing conclusions about the role of maternal GCs in these PNMS effects. To date, we have seen that lower (rather than higher) levels of maternal diurnal cortisol, sampled 5–6 months after the crisis, are associated with greater dermatoglyphic asymmetry which is presumed to reflect neurodevelopmental insult in utero. Other research paradigms are required to elucidate the mechanisms by which this unexpected association occurs.

14.3.2 Timing in Gestation and Sex Effects

Another important consideration in PNMS research is the timing in gestation of the stressor and how this might moderate the stress effects. Thanks to the use of sudden onset natural disasters, we have been able to date the stressors with accuracy.

The patterns of results suggest that every moment in pregnancy presents a window of vulnerability for some form of development. Objective exposure (in Project Ice Storm) or subjective distress (in Iowa) predicted lower cognitive development scores at age 2 years, but only when exposure occurred in early gestation. In Project Ice Storm, maternal subjective distress during the second trimester predicted greater dermatoglyphic asymmetry. On the other hand, for motor development, we showed that the later in gestation girls were exposed to the ice storm the lower their bilateral coordination and visual motor integration scores at age 5½. The third trimester also appears to be a sensitive period for the development of attention problems, according to preliminary analyses from the ice storm. These timing effects appear to be fleeting, however. In Project Ice Storm, the timing effects on cognitive development are only seen at age 2 years, and disappear in assessments between ages 5½ and 13½.

The timing of the stressor in gestation is not the only significant moderator of the effects of PNMS on neurodevelopment. The literature presents an inconsistent picture of whether males or females are more sensitive to the effects of PNMS, and our results are similarly irregular. In Project Ice Storm, sex does not moderate effects of maternal stress on cognitive outcomes until age 11½ when the effect becomes negative and linear for boys, and nonexistent in girls; these results are echoed in the hippocampal volume data at the same age. Girls, on the other hand, appear more vulnerable than boys to the effects of late gestation exposure on motor functioning. We have also found girls to be more vulnerable to the effects of maternal subjective stress in predicting childhood asthma (Turcotte-Tremblay et al. 2014). As such, neither sex appears to be more or less at risk of neurodevelopmental consequences of PNMS, but results suggest that we must continue to be vigilant in testing hypotheses about sex as a potential moderator of the effects.

14.3.3 The Longevity of Effects

Children who were born preterm begin life with certain developmental disadvantages compared to their term-born peers. But by elementary school, they tend to catch up to their classmates with few long-term consequences. We suspected to find a similar effect of prenatal maternal disaster exposure. Results suggest, however, that significant effects can still be seen at later ages as preliminary analyses of Project Ice Storm results at age 13½ continue to demonstrate. The magnitude of the effects seems to diminish with age, but remain in the small-to-moderate range at age 13.

Apart from a general weakening of effect over time, the age of the child at the time of the assessment also seems to be an important consideration in the study of PNMS as noted in our review of animal research on PNMS and brain development (Charil et al. 2010). We have seen timing effects at age 2 years that disappear at later ages, and patterns of objective exposure effects on cognition that held stable through childhood suddenly change at age 11½. The onset of puberty may well change many of our conclusions from younger assessment ages.

14.3.4 Possible Mediators of PNMS

The conventional wisdom implicates the maternal HPA axis, and the noxious effects of GCs that invade the placenta, as the mechanism by which PNMS exerts its influence in programming the fetus. Several investigators are testing complementary hypotheses about mediation of PNMS by other agents such as the immune system, epigenetics, and androgens. All three of our disaster studies provide genotyping data with which to study gene-by-environment interactions in the years to come. We anticipate furthering knowledge about these mechanisms through analysis of the biological specimens from QF2011.

In the meantime, our preliminary analyses of blood from a subset of children from Project Ice Storm at age 13 suggest that the degree of maternal objective exposure explains: significant amounts of variance in immune system measures such as cell counts and cytokines; epigenetic signals (Cao-Lei et al., Under review); and (in girls) testosterone levels (Veru et al, Under review). Thus, we anticipate that the field will be expanding in its elaboration of the mechanisms of action of PNMS over the coming years.

14.3.5 Limitations of Prenatal Maternal Stress Work in Disasters

The use of natural disasters as a PNMS paradigm has a number of advantages, as noted in the introduction. There are several disadvantages, however, that render this approach difficult to manage and the results difficult to interpret. Most of these limitations are a function of the disaster itself, as noted above in the section on Project Ice Storm Limitations. Because disasters are unpredictable, there is an inevitable delay between the event and the collection of the first data point: potential on-site collaborators, who may themselves be victims of the disaster, must be found and convinced of the importance of the project; the protocol must be elaborated and adapted to the particular disaster; ethics approval must be obtained; and all of these must be done often at some geographic distance and (as with QF2011) at great time zone differences.

The advantages of having a single, sudden-onset disaster as the PNMS paradigm are tempered by the disadvantage of a potential confound of the timing in gestation of the event by the season of the event. In "normal" years, there may be seasonal patterns in outcomes such as preterm birth, metabolism, or temperament. Without a control group, matched by birth month, one cannot be entirely certain that significant timing effects of a natural disaster are not the result of annual patterns by date of birth rather than a function of fetal windows of vulnerability. Having a suite of three natural disaster studies, one in winter and two in summer, helps us to circumvent this particular challenge.

14.3.6 Conclusion

Despite these limitations and challenges, the advantages associated with studying an independent, sudden-onset stressor that is applied to large populations of pregnant women give added value to these studies and provide data that are not available elsewhere. Project Ice Storm, the Iowa Flood Study, and QF2011 are well positioned to complement the PNMS data generated from other study paradigms. Together, all of these approaches will fill in the gaps in knowledge about the effects of PNMS and its mechanisms and, we hope, guide the development of interventions to circumvent maladaptive fetal programming.

The reader interested in staying abreast of the results from Project Ice Storm, The Iowa Flood Study, and QF2011 may wish to consult our website called "SPIRAL" (Stress in Pregnancy International Research Alliance) at www.mcgill.ca/spiral/.

Conflicts of Interest The authors declare no conflicts of interest.

References

Brenner K, Liu A, Laplante DP, Lupien S, Pruessner JC, Ciampi A et al (2009) Cortisol response to a psychosocial stressor in schizophrenia: blunted, delayed, or normal? Psychoneuroendocrinology 34(6):859–868

Brunet A, St-Hilaire A, Jehel L, King S (2003) Validation of a French version of the impact of event scale—revised. Can J Psychiatry 48:55–60

Calabrese EJ (2008) Hormesis: why it is important to toxicology and toxicologists. Environ Toxicol Chem 27(7):1451–1474

Cao-Lei L, Massart R, Suderman M, Machnes Z, Laplante D, Szyf M, King S (Under review) DNA methylation signatures of prenatal maternal objective stress exposure to a natural disaster: project Ice Storm

Cao X, Laplante DP, Brunet A, Ciampi A, King S. (2014) Prenatal maternal stress affects motor function in 5½-year-old children: Project Ice Storm. Dev Psychobiol 56:117–25

Charil A, Laplante DP, Vaillancourt C, King S (2010) Prenatal stress and brain development. Brain Res Rev 65(1):56–79

Dancause KN, Laplante D, Oremus C, Fraser S, Brunet A, King S (2011) Disaster-related prenatal maternal stress influences birth outcomes: project Ice Storm. Early Hum Dev 87:813–820. doi:10.1016/j.earlhumdev.2011.06.007

Folkman S, Lazarus RS (1988) Coping as a mediator of emotion. J Pers Soc Psychol 54(3):466–475

Huttenen MO, Niskanen P (1978) Prenatal loss of father and psychiatric disorders. Arch Gen Psychiatry 35(4):429–431

Khashan AS, Abel KM, McNamee R, Pedersen MG, Webb RT, Baker PN et al (2008) Higher risk of offspring schizophrenia following antenatal maternal exposure to severe adverse life events. Arch Gen Psychiatry 65(2):146–152

King S, Mancini-Marïe A, Brunet A, Walker E, Meaney MJ, Laplante DP (2009) Prenatal maternal stress from a natural disaster predicts dermatoglyphic asymmetry in humans. Dev Psychopathol 21:343–353

King S, Charil A, Laplante DP, Paus T, Pruessner J, Schmitz N, Brunet A (2012) Breastfeeding under stress due to a natural disaster affects children's hippocampus, stress reactivity, and internalizing problems. Paper presented at the international society for the study of behavioral development, Edmonton, Alberta

Kinney DK, Hyman W, Greetham C, Tramer S (1999a) Increased relative risk for schizophrenia and prenatal exposure to a severe tornado. Schizophr Res 36:45

Kinney DK, Hyman W, LeBlanc D, Vernooy A, Tramer S (1999b). Prenatal exposure to severe storms and risk for schizophrenia and autism. Paper presented at the society for research in psychopathology, Montreal, 18–21 Nov 1999

Kinney DK, Miller AM, Crowley DJ, Huang E, Gerber E (2008) Autism prevalence following prenatal exposure to hurricanes and tropical storms in Louisiana. J Autism Dev Disord 38:481–488

Kirschbaum C, Pirke KM, Hellhammer DH (1993) The 'Trier Social Stress Test'-a tool for investigating psychobiological stress responses in a laboratory setting. Neuropsychobiology 28(1-2):76–81

Lane A, Kinsella A, Murphy P, Bryne M, Keenan J, Colgan K et al (1997) The anthropometric assessment of dysmorphic features as an index of its developmental origins. Psychol Med 27:1155–1164

Laplante DP, Barr RG, Brunet A, Galbaud duFG, Meaney M, Saucier J-F et al (2004) Stress during pregnancy affects intellectual and linguistic functioning in human toddlers. Pediatr Res 56:400–410

Laplante DP, Zelazo PR, Brunet A, King S (2007) Functional play at 2 years of age: effects of prenatal maternal stress. Infancy 12(1):69–93

Laplante DP, Brunet A, Schmitz N, Ciampi A, King S (2008) Project Ice Storm: prenatal maternal stress affects cognitive and linguistic functioning in 5½-year-old children. J Am Acad Child Adolesc Psychiatry 47(9):1063–1072

Lazarus RS (1991) Emotion and adaptation. Oxford University Press, New York

Lechat MF (1979) Disasters and public health. Bull World Health Organ 57:11–17

Markow TA, Wandler K (1986) Fluctuating dermatoglyphic asymmetry and the genetics of liability to schizophrenia. Psychiatry Res 19(4):323–328

Nylen KJ, O'Hara MW, Engeldinger J (2013a) Perceived social support interacts with prenatal depression to predict birth outcomes. J Behav Med 36(4):427–440. doi:10.1007/s10865-012-9436-y

Nylen KJ, Williamson JA, O'Hara MW, Watson D, Engeldinger J (2013b) Validity of somatic symptoms as indicators of depression in pregnancy. Arch Womens Ment Health 16(3):203–210. doi:10.1007/s00737-013-0334-2

Patacchioli FR, Cigliana G, Cilumbriello A, Perrone G, Capri O, Alema GS et al (1992) Maternal plasma and milk free cortisol during the first 3 days of breast-feeding following spontaneous delivery or elective cesarean section. Gynecol Obstet Invest 34(3):159–163

Smith CA, Lazarus RS (1990) Emotion and adaption. In: Pervin LA (ed) Handbook of personality: theory and reserach. Guilford, New York pp 609-637

Tracy S, Hartz D, Hall B, Allen J, Forti A, Lainchbury A et al (2011) A randomised controlled trial of caseload midwifery care: M@NGO (Midwives @ New Group practice Options). BMC Pregnancy Childbirth 11(1):82

Turcotte-Tremblay A-M, Lim R, Laplante DP, Kobzik L, Brunet A, King S. (2014) Prenatal Maternal Stress Predicts Childhood Asthma in Girls: Project Ice Storm. BioMed Research International 2014:10

van Os J, Selten JP (1998) Prenatal exposure to maternal stress and subsequent schizophrenia. The May 1940 invasion of The Netherlands. Br J Psychiatry 172:324–326

Veru F, Dancause KN, Laplante DP, King S, Luheshi GN. (Submitted) Prenatal maternal stress predicts reductions in CD4+ lymphocytes, increases in innate-derived cytokines and Th2 shift in adolescents: A Project Ice Storm Pilot Study

Weinstock M (2008) The long-term behavioural consequences of prenatal stress. Neurosci Biobehav Rev 32(6):1073–1086

Weiss DS, Marmar CR (1997) The impact of event scale—revised. In: Wilson JP, Keane TM (eds) Assessing psychological trauma and PTSD: a practitioner's handbook. Guilford, New York, pp 399–411

Yehuda R, Bierer LM, Schmeidler J, Aferiat DH, Breslau I, Dolan S (2000) Low cortisol and risk for PTSD in adult offspring of holocaust survivors. Am J Psychiatry 157(8):1252–1259

Yerkes RM, Dodson JD (1908) The relation of strength of stimulus to rapidity of habit formation. J Comp Neurol Psychol 18:459–482

Zoladz PR, Diamond DM (2013) Current status on behavioral and biological markers of PTSD: a search for clarity in a conflicting literature. Neurosci Biobehav Rev 37(5):860–895

Chapter 15
Early Life Influences on Cognition, Behavior, and Emotion in Humans: From Birth to Age 20

Bea R. H. Van den Bergh, Eva M. Loomans and Maarten Mennes

Abstract The long-lasting effects of fetal exposure to early life influences (ELI) such as maternal anxiety, stress, and micronutrient deficiencies as well as mediating and moderating factors are quite well established in animal studies, but remain unclear in humans. Here, we report about effects on cognition, behavior, and emotion in offspring aged 5–20 years old in two prospective longitudinal birth cohorts.

The long-lasting effects of fetal exposure to early life influences (ELI) such as maternal anxiety, stress, and micronutrient deficiencies as well as mediating and moderating factors are quite well established in animal studies, but remain unclear in humans. Here, we report about effects on cognition, behavior, and emotion in offspring aged 5–20 years old in two prospective longitudinal birth cohorts.

Maternal anxiety in the first and second trimester of pregnancy was associated with more variable performance during a simple reaction time task at age 5–6. In addition, children of women who reported high levels of anxiety (state anxiety > 90th percentile) were slower and performed more variable on a choice reaction time task. Moreover, prenatal maternal anxiety was associated with problem behavior (e.g., hyperactivity/inattention problems, emotional symptoms) in 5–6-year-olds. Our findings suggest a heightened vulnerability to developmental modulation of programming effects of maternal anxiety during pregnancy in boys. Maternal caffeine intake during the first and second trimester of pregnancy was not associated with a higher risk for behavior problems in 5–6-year-olds. No evidence was found

B. R. H. Van den Bergh (✉) · E. M. Loomans
Department of Psychology, Tilburg University, Warandelaan 2, PO Box 90153, 5000 LE Tilburg, The Netherlands
e-mail: Bea.vdnBergh@uvt.nl

B. R. H. Van den Bergh
Department of Psychology, KU Leuven, Leuven, Belgium

E. M. Loomans
Department of Epidemiology and Health Promotion, Public Health Service, Amsterdam, The Netherlands

M. Mennes
Donders Institute for Brain, Cognition and Behavior, Department of Cognitive Neuroscience, Radboud University, Nijmegen, The Netherlands

M. C. Antonelli (ed.), *Perinatal Programming of Neurodevelopment*,
Advances in Neurobiology 10, DOI 10.1007/978-1-4939-1372-5_15,

for mediation by fetal growth restriction or gestational age or for effect modification by the child's gender. Higher maternal concentrations of omega-3 fatty acids (docosahexaenoic acid, DHA, and eicosapentaenoic acid, EPA) during early pregnancy decreased the risk for overall problem behavior and emotional symptoms (mothers rating their child). In contrast, higher maternal concentrations of omega-6 fatty acid (archidonic acid, AA) and a higher omega-6 to omega-3 ratio during gestation increased the risk for overall problem behavior, hyperactivity/inattention problems, and peer relationship problems. Evidence was found neither for mediation by preterm birth, and being small for gestational age, nor for effect modification by the child's gender.

In 15-year-old offspring, hypothalamo–pituitary–adrenocortical (HPA) axis function was measured through establishing a saliva diurnal cortisol profile and depressive symptoms were measured with the Children's Depression Inventory. Maternal anxiety at 12–22 weeks of pregnancy was associated with an attenuated diurnal cortisol profile in both boys and girl due to elevated cortisol secretion in the evening. Moreover, in female adolescents this flattened cortisol profile mediated the link between prenatal maternal anxiety and depressive symptoms. Prenatal maternal anxiety during week 12–22 of pregnancy was also associated with attention-deficit/hyperactivity disorder (ADHD)-related symptoms, lower intelligence scores, and impairments in endogenous cognitive control (i.e., the ability from within one self to control actions, strategies, and thoughts) as measured with neuropsychological tasks in 15–17-year-old offspring. Importantly, results of event-related potentials (ERPs) at age 17 and functional magnetic resonance imaging (fMRI) at age 20 confirmed a less optimal endogenous cognitive control function and indicated aberrant brain functioning in adolescents of mothers reporting high levels of prenatal maternal anxiety.

Results from both prospective cohorts indicate that ELI (e.g., prenatal maternal anxiety and fatty acid status) enhance neurobiological vulnerability and influence offspring's cognitive, emotional, and behavioral functioning well into adolescence. A possible mechanism might be modulation of programming of the offspring's (neuro)physiology and brain structure–function relationships.

15.1 Atypical Early Influences During Fetal Life are Associated with Less Favorable Neurodevelopmental Outcomes in Later Life

The fetal programming hypothesis (Barker and Osmond 1986), Developmental Origins of Health and Disease (DOHaD; Gluckman and Hanson 2004; Gluckman et al. 2008), and Developmental Origins of Behavior, Health and Disease (DOBHaD; Van den Bergh 2011a) propose that human health and development have their origin in early life, in the womb. The fetus responds to its uterine environment and to changes and disturbances in this environment, e.g., those elicited by maternal stress, micronutrient deficiencies, or placental dysfunction. The induced alterations in fetal physiology and metabolic responses may modulate the trajectory of developmental processes, i.e., "modulate the programming" of the developmental pattern within key tissues and organ systems (Gluckman et al. 2008). We deliberately choose to term these processes developmental modulation of developmental programming, because we presume that developmental modulation of developmental programming by means of early environmental cues in order to shape an organism's development is a fundamental part of the trajectory in typical development across species (Van den Bergh 2007, Van den Bergh 2011a).

Animal research starting around 1950 demonstrated that experimentally manipulated exposure to adverse environmental factors in early life (e.g., (induced) stress, stress hormones administered to the pregnant animal or fetus, cytokines, alcohol, hypoxia, cocaine) are causally related to several short- and long-term effects on offspring behavior, affects developing brain areas (i.e., hippocampus, amygdala, and frontal lobes), and is associated with changes in neuronal circuits that are involved in cognitive and emotional processing and in modulating stress responses (Bock et al. 2005; de Kloet et al. 2005; Seckl and Meaney 2004; Weinstock 2008; Stanwood et al. 2001; Stanwood and Levitt 2004; Son et al. 2006; Baier et al. 2012). In humans, since 1990, an increasing number of prospective studies shows that exposure to atypical ELI during pregnancy is associated with adverse birth outcomes and a range of less favorable child neurobehavioral outcomes. Most studies looked at the association between prenatal exposure to maternal stress, anxiety, or depression and emotional and behavioral problems (i.e., temperamental reactivity, externalizing behavior) measured with self-report questionnaires or behavioral observation scales. A small number of studies examined specific aspects of cognitive function; these studies revealed evidence for an association between prenatal exposure to atypical prenatal environmental factors and reduced attention, lowered IQ scores, and lowered linguistic competence. These findings from human studies are in accordance with results from animal studies, including inconsistent findings that probably result from genetic differences, differences in intensity/severity, duration/chronicity, controllability/coping, the developmental timing of ELI, and differences in postnatal environment (e.g., in adversity or in maternal care-giving style; for reviews, see Glover et al. 2010; O'Donnell et al. 2009; Van den Bergh et al. 2005b; Weinstock 2008; Räikkönen et al. 2011; Beijers et al. 2014; Graignic-Philippe et al. 2014).

Our studies were aimed at a better understanding of developmental modulation of programming of cognition, behavior, and emotion in humans: from birth to age 20. In the following paragraphs, we will present results from two prospective, longitudinal birth cohort studies. The first cohort: The Amsterdam-Born Children and their Development Study (ABCD study) started in 2003 and aims to examine a broad range of factors during pregnancy and in early life that are potentially related to the child's health and development at birth and in later life (van Eijsden et al. 2011). To date, mothers and children have been assessed until children were 5–6 years old. The second cohort of pregnant women which was started in 1986 in Leuven, Belgium, was followed up from 12–22 weeks of gestation until their offspring was 20 years old (Mennes 2008).

15.2 Results from the Amsterdam-Born Children and their Developmental Study

In the following paragraphs, we will briefly describe the study design and procedures of the ABCD study. Thereafter, we will present results with regard to the association between three ELI that are highly prevalent in pregnant women (i.e., experience of negative emotions, caffeine intake, and a suboptimal fatty acid status) and children's cognitive functioning and behavior.

15.2.1 ABCD-Study Design and Procedures

Pregnant women (n=8266) filled out a questionnaire covering sociodemographic, obstetric, lifestyle, and psychosocial conditions on average at 16 weeks gestation and an extra blood sample was taken during routine blood collection for prenatal screening purposes (n=4389). Pregnancy outcomes were obtained through Youth Health Care of the Public Health Service Amsterdam, and from the Dutch Perinatal Registry. Three months after delivery, mothers received a questionnaire concerning the course of pregnancy and delivery, the health and development of their baby, and questions about their own lifestyle (n=5131 returned). When the children were 5 years old, mothers received three questionnaires (n=6161 sent). The first questionnaire covered the child's health, medical conditions, family sociodemographics, and children's problem behavior (n=4488 returned). The second was a food frequency questionnaire (n=2851 returned). The third, which was addressed to the child's teacher, concerned school performance and problem behavior (n=3588 returned). Furthermore, children were invited for a health check at school (n=3321) where they took part in cognitive testing.

15.2.2 Maternal Anxiety during Pregnancy and Children's Neurocognitive Functioning and Behavior at the Age of 5–6

Analyses in a group of 922 mothers with a large range of anxiety scores (mean state-anxiety score=36) and children revealed that there indeed is an association between maternal anxiety during the first and second trimester (mean 16th week) and alterations in children's neurocognitive functioning at the age of 5–6. Children of anxious pregnant mothers were more variable in their performance during a simple reaction time test than children of less anxious women, but no associations were found between antenatal anxiety and the children's mean reaction time during both a simple and a choice reaction time task. Examination of nonlinear associations revealed a significant nonlinear association between antenatal anxiety and the children's variability in reaction time in the incompatible part (i.e., incompatible stimulus-response mode) of the choice reaction time task. Visual inspection of the data showed that higher levels of maternal anxiety were related to a stronger than linear increase in children's variability in reaction time. This finding suggested that the modulation of programming effects of antenatal anxiety become stronger when reported anxiety levels rise. Subsequent analyses in a highly anxious subsample (state-anxiety score > 90th percentile, mean state anxiety = 54.7, n = 100) showed that higher levels of antenatal anxiety were more strongly associated with longer reaction times and more intraindividual variability in reaction time in the incompatible part of the choice reaction time task. The child's sex moderated the relation between antenatal anxiety and intraindividual variability in the simple reaction time task in the highly anxious subsample. Boys performed more variables on the simple reaction time task, but no significant associations were found in girls (Loomans et al. 2012b).

In addition, children of mothers who reported higher levels of state anxiety (mean state–anxiety score = 36.7) during the first and second trimester (mean 16th week) showed more overall problem behavior, hyperactivity/inattention problems, emotional symptoms, peer relationship problems, and conduct problems and showed less pro-social behavior when mothers had rated their child's behavior (n = 3446). When child behavior was rated by their primary school teacher (n = 3520), maternal anxiety during pregnancy was related to more overall problem behavior and less pro-social behavior. The child's sex moderated the relation between antenatal anxiety with overall problem behavior and hyperactivity/inattention problems in children when reported by their mother. Higher levels of antenatal anxiety were more strongly related to overall problem behavior in boys than in girls. Furthermore, antenatal anxiety was significantly associated with hyperactivity/inattention problems in boys, while this was not the case in girls (Loomans et al. 2011).

15.2.3 Maternal Caffeine Intake During Pregnancy is not Associated with Problem Behavior in 5–6-Year-Old Children

Analyses in 3439 mothers and children indicated that maternal dietary caffeine intake (self-reported coffee, tea, and soft drink consumption) during the first and second trimester (mean 16th week) pregnancy was not associated with a higher risk for hyperactivity/inattention problems, emotional symptoms, conduct problems, peer relationship problems, suboptimal pro-social behavior, and overall problem behavior in their 5–6-year-old children. In addition, no evidence was found for moderation by the child's sex or for mediation by fetal growth restriction and gestational age. Given the fact that a relatively large group of women consumed considerable quantities of caffeine (n = 862 > 4 cups per day), we were able to fully explore the effect of high doses of caffeine intake; nevertheless, our findings did not provide evidence for a dose-response effect of intrauterine caffeine exposure. Because nausea, a common symptom in the first trimester of healthy pregnancies (n = 1.586

women reported nausea), did reduce caffeine intake significantly in our sample, we repeated analyses in a subsample of only nonnauseous women that revealed no associations between caffeine consumption and children's problem behavior (Loomans et al. 2012a).

15.2.4 Maternal Omega-3 and Omega-6 Fatty Acid Status During Pregnancy is Related to Children's Risk of Problem Behavior at Age 5–6

We found evidence for long-term developmental modulation of programming influences of maternal long-chain polyunsaturated fatty acid (LCPUFA) status during the first and second trimester of pregnancy (gestational week at blood sampling, mean = 13) on offspring's problem behavior at age 5–6 years when rated by mothers (n=2502). Higher concentrations of omega-3 fatty acid DHA in maternal plasma phospholipids decreased the risk for offspring's overall problem behavior. Higher concentrations of omega-3 fatty acid EPA decreased the risk for overall problem behavior and emotional symptoms. In contrast, higher concentrations of omega-6 fatty acid AA increased the risk for peer relationship problems and a higher omega-6 to omega-3 ratio increased the risk for overall problem behavior, hyperactivity/inattention problems, and peer relationship problems. No associations were found when teachers rated children's behavior. No evidence was found for mediation by preterm birth and being born small for gestational age. The child's sex did not modify the association between maternal LCPUFA status and children's behavioral outcome (Loomans et al. 2014).

15.3 Results from the 1986 Leuven Cohort

Next, we describe the results of a second longitudinal study on the neurobehavioral effects of prenatal exposure to maternal anxiety and stress (PEMAS) on human offspring. Initiated in 1986, the Leuven Cohort investigated effects of PEMAS on fetal, infant, and childhood development, continuing into adolescence with follow-up investigations in 14–17- and 20-year-old adolescents. To uncover biological mechanisms underlying effects of PEMAS on depressed mood, we examined the offspring's pituitary HPA axis activity. In addition to HPA axis functioning, we examined cognitive processing and associated neural activity using ERPs and fMRI (Van den Bergh 2011a).

Maternal state anxiety, measured with the State Trait Anxiety Inventory (STAI), was assessed during weeks *12–22, 23–31,* and *32–40* of pregnancy in 86 pregnant women who completed the STAI as part of a larger battery of standardized psychological questionnaires probing anxiety and stress. It is important to note that we had access to the full range of anxiety scores in our sample. The mother also completed the same battery of questionnaires at all postnatal research phases, allowing to control for a possible influence of postnatal maternal anxiety. Importantly, although postnatal maternal anxiety was associated with many outcome measures, it did not explain effects related to PEMAS. In addition, we were able to show that most of our findings were (statistically) independent of confounding factors including smoking and alcohol use during pregnancy or birth weight.

15.3.1 PEMAS is Associated with Altered Fetal, Infant, and Child Neurobehavioral Development

The full cohort participated in fetal ultrasound measurements (120 min) that focused on body movements, eye movements, and heart rhythm, resulting in the identification of behavioral states (i.e., sleep–wake cycles). State-dependent motor activity was significantly higher and the percentage of time spent in quiet (or deep) sleep significantly lower in fetuses from pregnant women reporting high levels of anxiety compared to fetuses from pregnant women reporting low levels of anxiety. A mother's anxiety during pregnancy explained between 10 and 25 % of the differences in irritability, excessive crying, irregularities in biological functions, and (difficult) temperament in the offspring during the first 7 months after birth. In contrast, we observed no significant relationship with PEMAS for (clinical) observations about the neurological condition, general cognitive and motor development, and feeding behavior (Van den Bergh et al. 1989, Van den Bergh 1990).

In a second research phase of the Leuven Cohort, we examined neurobehavioral development in the 8–9-year-old children ($n=72$; 38 boys). Children from mothers reporting high PEMAS showed significantly more behavioral self-regulation problems compared to children from mothers reporting lower levels of PEMAS (explained variance between 17 and 22 %). Importantly, problem behavior was reported at home (reported by the mother), in class (reported by the child's teacher), and during experimental test settings (reported by an observer). Reported problem behavior reflected impaired regulation of emotion and cognition, ADHD symptoms, increased externalizing behavior, and augmented self-reported feelings of anxiety (Van den Bergh and Marcoen 2004).

15.3.2 HPA Mediates the Link Between PEMAS and Adolescent Depressed Mood

In a third follow-up of the cohort, when children were 14–15 years old, we were able to examine whether offspring HPA axis mediated the link between PEMAS and reported emotional problems in the offspring. Such link was hypothesized based on the implication of HPA axis functioning in depressive symptomatology (Claes 2004; Swaab et al. 2005; Heim et al. 2000) and evidence from animal studies that early life stress can alter HPA axis functioning (Maccari et al. 2003; Macrì et al.

2007). Using cortisol measurements in the 15-year-old offspring ($n=58$; 29 boys), we showed that in both, boys and girls, PEMAS during week 12–22 of pregnancy was associated with a high, flattened diurnal cortisol profile showing elevated cortisol secretion in the evening. However, only in girls we were able to partly explain the effect of PEMAS on depressed mood by the effect of the flattened cortisol profile on depressed mood (Van den Bergh et al. 2008). These results suggested that PEMAS can prenatally modulate the programming HPA axis functioning and induce a vulnerable phenotype in the offspring. A plausible, albeit untested, underlying mechanism might operate through a "resetting" of HPA axis set points by antenatal exposure to maternal anxiety during critical periods resulting in a hyperactive HPA axis (de Kloet et al. 2005; Seckl and Meaney 2004). However, PEMAS-related modulation of programming is likely not limited to the HPA axis. For instance, evidence from animal studies suggests PEMAS-related sequelae in neural circuits involved in emotional and cognitive processing.

The fundamental brain architecture of the fetus is established during the first two trimesters of pregnancy, and its sensory organs gain functionality during the third trimester. From the first month of gestation, brain development proceeds as a continuous dialogue between the fetus's genome and its environment. Accordingly, environmental factors (such as PEMAS) that disturb the expression of genes involved in cellular proliferation, migration, and differentiation may impact early brain development, thereby constraining sensory and cognitive/emotional development even in the absence of structural brain alterations (Meaney 2010; Van den Bergh et al. 2005b).

15.3.3 Neuropsychological, ERP, and fMRI Measures Reveal Evidence for Impaired Endogenous Cognitive Control Associated with PEMAS

In addition to assessing associations between PEMAS and emotional functioning, we included measures that enabled us to assess associations between PEMAS and neural functioning related to cognitive processes (e.g., inhibition, planning, memory). The field of cognitive neuroscience has produced several robust (now mostly computerized) tasks that target specific aspects of cognitive functioning. Classic examples are Go/NoGo paradigms assessing inhibition or N-back memory tasks that index memory capacity. By administering these tasks during ERP measurements and fMRI scans, neuroimagers have been able to assess the timing and speed of cognitive functions as well as make inferences about the involvement of specific brain regions or larger functional networks (e.g., Whelan et al. 2012). Just as the occipital cortex is essential to visual processing, the prefrontal cortex is given a central role in cognitive processing (Miller and Cohen 2001), including attentional control, error monitoring, and executive functions. In the context of our earlier findings suggesting altered attention-related, externalizing behaviors following high levels of

PEMAS, we expanded our research to investigate the association between PEMAS and prefrontal cortical functioning.

In follow-up phases of the Leuven cohort at ages 15 (n=64; 33 boys) and 17 (n=49; 29 boys), we investigated cognitive functioning with IQ tests and computerized neuropsychological tasks assessing working memory, sustained attention, and response control. These neuropsychological domains were specifically targeted based on the type of problem behavior shown to be associated with PEMAS in infancy and childhood. Specifically, at age 15 we included three tasks that assess functions closely tied to aspects of self-regulation and that have a well-established relationship with prefrontal cortex: (1) visual attention control and working memory (Encoding), (2) response control (Stop Task paradigm) and (3) sustained attention (Continuous Performance Task). For the follow-up at age 17 we extended our behavioral task battery to five tasks, taxing five prefrontal functions: (1) the ability to orient attention (Cued Attention), (2) working memory (N-back), (3) external response inhibition (Go/NoGo), (4) the ability to perform two tasks simultaneously (Dual Tasks), and (5) the ability to switch between response sets (Response Shifting).

By evaluating the cognitive functions needed to successfully perform in each of these tasks, we observed a specific pattern of cognitive deficits in adolescents of mothers reporting high levels of anxiety during weeks 12–22 of their pregnancy. Children of mothers reporting high levels of PEMAS were impaired in the Continuous Performance Task and Response Shifting tasks. Both tasks require the ability to endogenously (i.e., autonomously, from within oneself and without external sources) inhibit reactions to interfering and distracting stimuli or inhibit a learned response. In contrast, adolescents of mothers reporting high levels of PEMAS performed adequately in tasks that triggered response inhibition in an exogenous, external manner (e.g., through presentation of a sound in the Stop Signal Task). Next, to a deficit in endogenous cognitive control, adolescents of mothers reporting high levels of PEMAS during weeks 12–22 of their pregnancy exhibited lowered scores on Vocabulary and Block Design, two subtests of the Wechsler Intelligence Scale for Children-Revised (WISC-R) at age 15, and decreased performance when the cognitive load of the task was increased (e.g., in Dual Tasks) at age 17. Finally, we observed no association between PEMAS and performance on working memory tasks (Van den Bergh et al. 2005a; Van den Bergh et al. 2006; Mennes et al. 2006).

To further investigate the relationship between PEMAS and neuronal functioning associated with these cognitive functions, we measured ERPs during some tasks administered at age 17. ERPs are small changes in the electrical activity of the brain caused by an internal or external event, and they are interpreted as reflections of brain activity related to the processing of that event. Using small electrodes placed on the scalp of a participant, we can measure this electrical activity. Accordingly, when ERPs are measured during cognitive tasks, we can make inferences about the cognitive processes underlying the changes seen in the electrical currents and functional significance can be ascribed to the observed waveforms. We recorded ERPs with 19 scalp electrodes during 4 tasks. These included a Go/NoGo, N-back, and switching paradigm similar to those used in the behavioral assessments, as well as a gambling paradigm specifically requiring high levels of endogenous cognitive control. In accordance with the behavioral results, we observed no relationship between PEMAS and ERPs measured during the Go/NoGo paradigm. This result confirmed that offspring of mothers reporting high levels of PEMAS have likely no behavioral or neuronal impairments in exogenous cognitive control. In contrast, we observed an association between PEMAS and decision making during the gambling task. This association was evident in both a less optimal behavioral performance and aberrant brain activity in adolescents of mothers reporting high levels of PEMAS during weeks 12–22 of their pregnancy. Particularly the early frontal P2a ERP component measured during endogenous cognitive trials was related to the level of PEMAS. As the P2a is thought to be related to the task relevance of a stimulus (Potts 2004), we speculated that adolescents of mothers reporting high levels of PEMAS based their decisions during the gambling tasks on different features of the gambling stimulus compared to adolescents of mothers reporting lower levels of PEMAS, thereby focusing on the most obvious features of the stimulus (e.g., color), while missing the intricacies of the stimulus present in less evident stimulus features (e.g., the numbers indicating a possible amount of gain/loss). As such, these results showed a link between brain activity of the offspring and the level of anxiety experienced by their mother during pregnancy, specifically targeting endogenous cognitive control processes (Mennes et al. 2009).

ERPs are measured with millisecond accuracy and have an excellent temporal resolution regarding the ongoing cognitive processes. However, ERPs lack spatial specificity, preventing tying specific cognitive processes to specific brain regions. Therefore, to complement the ERP results with spatial information about which areas in prefrontal cortex show differences in functionality related to PEMAS, we assessed endogenous cognitive control using fMRI in the follow-up phase of the Leuven Cohort at age 20. During fMRI scans, the blood oxygenation level dependent (BOLD) response is measured while performing a (cognitive) task. Based on the ratio of deoxy- and oxygenated hemoglobin in the blood, the BOLD response is thought to be coupled with oxygen consumption in the context of neuronal processes. As such, fMRI is said to provide an indirect measurement of neuronal activity (Menon 2012). By relating variations in the BOLD response over time to the timing of events in an experimental paradigm, neuroimagers can relate cognitive processes to specific areas of the brain. For the follow-up phase at age 20, we invited all ten boys of the high anxious group present in the Leuven Cohort as well as ten boys, matched on IQ, of the low-average anxious group (final $n=18$). All participants performed four cognitive tasks during fMRI scanning, again similar to the tasks included in the follow-up phases at ages 15 and 17: response switching, N-back, Go/NoGo, and gambling. Confirming the previous results, we observed that offspring of mothers reporting high levels of PEMAS during weeks 12–22 of pregnancy exhibited altered, endogenously controlled decision making. This behavioral difference was complemented by altered patterns of brain activation in regions involved in cognitive control including inferior frontal junction (e.g., Zysset et al. 2006) and areas in the middle frontal gyrus. Again, we observed no association between PEMAS and brain activity during the Go/NoGo task requiring exogenous cognitive control (Mennes 2008).

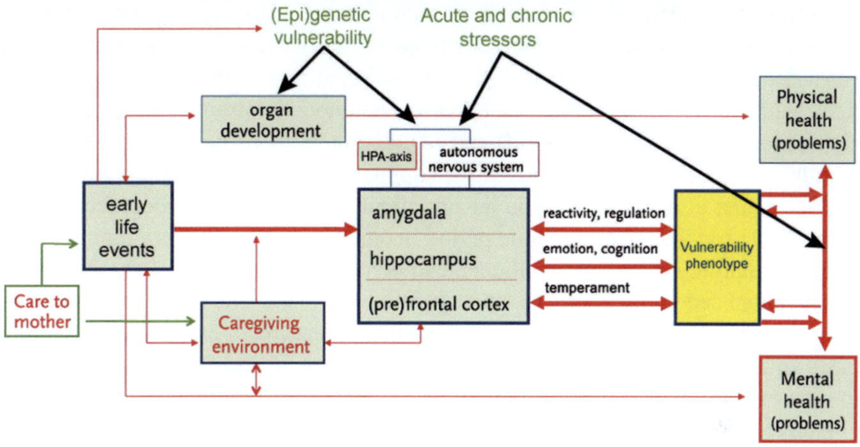

Fig. 15.1 Developmental programming of early brain and behavior development and mental health and physical health (problems). (Adapted from Van den Bergh 2011b)

15.4 Discussion

Results from both our prospective longitudinal birth cohorts corroborate and extend results from preclinical studies as they provide evidence for adverse neurodevelopmental consequences of atypical ELI from early childhood until late adolescence, and are in accordance with the DOBHaD paradigm proposed by Van den Bergh (2011a). Especially high levels of maternal anxiety during the first and second trimester of pregnancy were associated with self-regulation problems, (externalizing) behavior problems, and enhanced self-reported feelings of anxiety in the offspring, taking important confounders (e.g., postnatal maternal anxiety) into account. Neuropsychological, ERP, and fMRI measures provided evidence for impaired cognitive functioning and altered brain activity patterns related to prenatal maternal anxiety. In addition, data from the ABCD study showed that maternal omega-3 and omega-6 fatty acid status during early pregnancy are related to children's problem behavior at ages 5–6, whereas maternal caffeine intake was not associated with offspring's behavior.

15.4.1 Modulation of Developmental Programming of Early Brain and Behavior Development and Mental Health

In Fig. 15.1, Van den Bergh (2011a, b) attempts to integrate results from preclinical, clinical, neurobehavioral, developmental, and epidemiological research that revealed direct or indirect evidence for the DOBHaD hypothesis. Importantly, early brain and behavior development are integrated and put in a central place in the figure. Research in animals has convincingly shown that developmental exposure to excess glucocorticoids or stress may modulate the programming the peripheral and central nervous system (CNS) involved in the two coacting stress-regulating subsystems: (1) the hypothalamic–pituitary–adrenal axis (with the hormones corticotrophin-releasing hormone, vasopressin, adrenocorticotrophic hormone, mineralocorticoid, and glucocorticoid) and (2) the autonomous nervous system (with noradrenaline and adrenaline). Effective coping with stress is important throughout life (de Kloet et al. 2005) and is an important behavioral regulation factor. The perinatal period is seen as a unique period in ontogeny where the fine-tuning of the stress-regulating system and resilience can be permanently modulated which may lead to enhanced vulnerability to develop diseases in later life. Evidence for alterations in neuronal circuits was shown in preclinical research. For instance, in limbic brain structures (hippocampus, amygdala) and prefrontal cortex, both are involved in stress reactivity and regulation patterns, in emotional (e.g., anxiety, anger) and cognitive (e.g., learning, memory) processing, and in temperamental variation in behavior (e.g., novelty seeking, harm avoidance, reactive temperament) (Gluckman and Hanson 2004; Seckl and Holmes 2007; Barker 1998; Räikkönen et al. 2011; Oitzl et al. 2010; Lupien et al. 2009). These changes may influence how an individual "behaves" (i.e., perceives, interprets, and reacts) to its environment and to situations of acute and chronic stress. In concert with physiological activity (the hypothalamic–pituitary–adrenal axis and the autonomous nervous system), these processes may underlie behavioral problems and psychopathology, or more in general, mental health problems. In accordance, recent theories hold that individuals vary in their biological sensitivity (Boyce and Ellis 2005) or in their susceptibility (Belsky and Pluess 2009) to environmental influences. These theories predict that some individuals are more susceptible than others to both the adverse and beneficial effects of, respectively, unsupportive and supportive environments. The nature of the environment and this difference in sensitivity or susceptibility will influence how mental health or mental health problems are shaped; these processes covary with physical health and health problems (Van den Bergh 2011, p. 21–22).

15.4.2 Potential Underlying Mechanisms

While the mediating and moderating factors and the long-lasting effect of ELI on offspring are quite well established in animal studies (for reviews, see other chapters in this book), in humans they are only beginning to be understood (Schlotz and Phillips 2009; Van den Bergh 2011). In studies focusing on offspring biological systems possibly altered by ELI, some studies examined HPA axis function. While animal research showed that exposure to excess glucocorticoids during specific sensitive periods in utero might alter the homeostasis of the fetal HPA axis (see Weinstock 2008), this hypothesis is not tested in a direct way in humans. However, altered basal or stress-related cortisol secretion were seen in infants and adolescent offspring subjects of anxious (e.g., O'Connor et al. 2005; O'Donnell et al. 2009) or highly stressed women developing posttraumatic stress disorder (Yehuda et al.

2005). Accordingly, results from the Leuven cohort showed that the offspring HPA axis mediates the link between prenatal maternal anxiety and offspring's emotional problems.

While in humans the role of the maternal and fetal HPA axis in mediating the transmission of stress from mother to fetus are still not clear, current data indicate that key targets for programming may include not only cortisol secretion itself but also glucocorticoid receptor (GR) and 11β-hydroxysteroid dehydrogenase type 2 (11βHSD2) gene expression in a range of tissues (Bertram and Hanson 2002; McGowan et al. 2009; Oberlander et al. 2008; Hompes et al. 2013). For instance, preclinical research has shown that early life experiences may induce epigenetic modifications in the expression of the GR gene (*NR3C1*) and in other key players in the biological stress response (Weaver et al. 2004; Zhang et al. 2010, Meaney and Ferguson-Smith 2010; Meaney et al. 2007). In humans, McGowan et al. (2009) showed that childhood abuse leads to increased promoter methylation and decreased expression at the *NR3C1* gene in hippocampal brain tissue. Oberlander et al. (2008) observed a link between prenatal emotional stress and methylation of the *NR3C1* gene 1_F promoter in DNA from human cord blood. Furthermore, an altered stress response reactivity at 3 months of age correlated with a higher methylation status of the *NR3C1* gene. Hompes et al. (2013) investigated the association between prenatal stress, in combination with HPA axis functioning, and the methylation pattern of the *NR3C1* 1_B, 1_D, and 1_F promoter regions in cord blood mononuclear cells. Their results indicate that prenatal maternal emotional state, particularly pregnancy related anxiety, is associated with the methylation state of the *NR3C1* gene in the infant.

The few studies that focused on CNS structures or on the structure-functioning relationship found associations between ELI and altered brainstem auditory evoked potential (DiPietro et al. 2010), auditory ERPs (Harvison et al. 2009), and gray matter volume reductions in several parts of the brain, such as the prefrontal and premotor cortex, the medial temporal lobe, and cerebellum (Buss et al. 2010). As indicated above, the results of the Leuven cohort indicated that altered patterns of brain activation in regions involved in cognitive control such as inferior frontal junction and areas in the middle frontal gyrus were involved. However, it is clear that more studies confirming these results are needed before firm conclusions can be reached.

It is known that a wide spectrum of micronutrients during pregnancy may have long-lasting effects on offspring's physical health (Godfrey and Barker 2001), as well as cognitive functioning and behavior (Monk et al. 2013). Findings from the ABCD cohort suggest that maternal LCPUFA concentrations during pregnancy might modulate the programming offspring's neurodevelopmental outcomes. Despite the fact that long-chain polyunsaturated fatty acids are essential components in brain development, it is possible that the alterations in cognitive functioning and increased risk for problem behavior could be attributed to deficiencies in other nutrients relevant to brain development, such as iron, zinc, selenium, iodine, folate, and vitamin A (Georgieff 2007; Monk et al. 2013). In addition, fetal exposure to maternal bacterial or viral infections during pregnancy, as a result of stress-induced immune dysregulation (Coussons-Read et al. 2007; Meyer and Feldon 2009; Meyer et al. 2009; Bilbo 2013), could be an alternative explanation for the increased prevalence in problem behavior and altered cognitive functioning in our studies.

15.5 Conclusion

The DOBHaD hypothesis has led to convergence of knowledge from different fields and has stimulated interdisciplinary research. There is clear evidence that prenatal and early postnatal adversity may affect lifelong behavior and both mental and physical health. Our chapter may further stimulate research examining the influence of ELI that are highly prevalent in women in the childbearing age (e.g., negative emotions, micronutrient deficiencies, environmental toxins) on offspring's neurodevelopmental outcomes, with the use of objective, sensitive assessment techniques.

Both the prenatal and early postnatal periods are targets for innovative, preventative, and intervention strategies. Because maternal lifestyle and stress are modifiable, expected potential, societal, and economic returns on investment, e.g., by improving behavior and health of the next generation, are substantial (Shonkoff 2010). The pregnancy and the early postnatal period are times of both great opportunity and considerable risk, and their influence can extend over a lifetime.

Conflict of Interest The authors declare no conflicts of interest.

References

Baier C, Katunar M, Adrover E, Pallarés M, Antonelli MC (2012) Gestational restraint stress and the developing dopaminergic system: an overview. Neurotox Res 22:16–32

Barker DJ (1998) In utero programming of chronic disease. Clin Sci 95:115–128

Barker DJ, Osmond C (1986) Infant mortality, childhood nutrition, and ischaemic heart disease in England and Wales. Lancet 1:1077–1081

Beijers R, Buitelaar JK, de Weerth C (2014). Mechanisms underlying the effects of prenatal psychosocial stress on child outcomes: beyond the HPA axis. European child & adolescent psychiatry, 1–14.

Belsky J, Pluess M (2009) Beyond diathesis stress: differential susceptibility to environmental influences. Psychol Bull 135:885–908

Bertram C, Hanson M (2002) Prenatal programming of postnatal endocrine responses by glucocorticoids. Reproduction 124:459–467

Bilbo SD (2013) Frank A. Beach award: programming of neuroendocrine function by early-life experience: a critical role for the immune system. Horm Behav 63:684–691

Bock J, Gruss M, Becker S, Braun K (2005) Experience-induced changes of dendritic spine densities in the prefrontal and sensory cortex: correlation with developmental time windows. Cereb Cortex 15:802–808

Boyce Wt, Ellis BJ (2005) Biological sensitivity to context: I. An evolutionary-developmental theory of the origins and functions of stress reactivity. Dev Psychopathol 17:271–301

Buss C, Davis EP, Muftuler LT, Head K, Sandman CA (2010) High pregnancy anxiety during mid-gestation is associated with decreased gray matter density in 6–9-year-old children. Psychoneuroendocrinology 35:141 − 53

Claes SJ (2004) CRH, stress, and major depression: a psychobiological interplay. Vitam Horm 69:117–150

Coussons-Read ME, Okun ML, Nettles CD (2007) Psychosocial stress increases inflammatory markers and alters cytokine production across pregnancy. Brain Behav Immun 21:343–350

de Kloet ER, Joels M, Holsboer F (2005) Stress and the brain: from adaptation to disease. Nat Rev Neurosci 6:463–475
DiPietro JA, Kivlighan KT, Costigan KA, Rubin SE, Shiffler DE, Henderson JL, Pillion JP (2010) Prenatal antecedents of newborn neurological maturation. Child Dev 81:115–130
Georgieff MK (2007) Nutrition and the developing brain: nutrient priorities and measurement. Am J Clin Nutr 85:614S–620S
Glover V, O'connor TG, O'donnell K (2010) Prenatal stress and the programming of the HPA axis. Neurosci Biobehav Rev 35:17–22
Gluckman PD, Hanson MA (2004) Living with the past: evolution, development, and patterns of disease. Science 305:1733–1736
Gluckman PD, Hanson MA, Cooper C, Thornburg KL (2008) Effect of in utero and early-life conditions on adult health and disease. N Engl J Med 359:61–73
Godfrey KM, Barker DJ (2001) Fetal programming and adult health. Public Health Nutr 4:611–624
Graignic-Philippe R, Dayan J, Chokron S, Jacquet AY, Tordjman S (2014) Effects of prenatal stress on fetal and child development: A critical literature review. Neurosci Biobehav Rev 43C:137–162. doi: 10.1016/j.neubiorev.2014.03.022
Harvison KW, Molfese DL, Woodruff-Borden J, Weigel RA (2009) Neonatal auditory evoked responses are related to perinatal maternal anxiety. Brain Cogn 71:369–374
Heim C, Ehlert U, Hellhammer DH (2000) The potential role of hypocortisolism in the pathophysiology of stress-related bodily disorders. Psychoneuroendocrinology 25:1–35
Hompes T, Izzi B, Gellens E, Morreels M, Fieuws S, Pexsters A, Schops G, Dom M, Van Bree R, Freson K, Verhaeghe J, Spitz B, Demyttenaere K, Glover V, Van den Bergh B, Allegaert K, Claes S (2013) Investigating the influence of maternal cortisol and emotional state during pregnancy on the DNA methylation status of the glucocorticoid receptor gene (NR3C1) promoter region in cord blood. J Psychiatr Res 47:880–891
Loomans EM, van der Stelt O, van Eijsden M, Gemke RJBJ, Vrijkotte T, van den Bergh BRH (2011) Antenatal maternal anxiety is associated with problem behaviour at age five. Early Hum Dev 87:565–570
Loomans EM, Hofland L, van der Stelt O, van der WAL MF, Koot HM, van den Bergh BRH, Vrijkotte TGM (2012a) Caffeine intake during pregnancy and risk of problem behavior in 5- to 6-year-old children. Pediatrics 130:e305–e313
Loomans EM, van der Stelt O, van Eijsden M, Gemke RJBJ, Vrijkotte T, van den Bergh BRH (2012b) High levels of antenatal maternal anxiety are associated with altered cognitive control in five-year-old children. Dev Psychobiol 54:441–450
Loomans EM, Van den Bergh BRH, Schelling M, Vrijkotte TGM, van Eijsden M (2014) Maternal long-chain polyunsaturated fatty acid status during early pregnancy and children's risk of problem behavior at age 5–6 years. J Pediatr 164:762–768
Lupien SJ, Mcewen BS, Gunnar MR, Heim C (2009) Effects of stress throughout the lifespan on the brain, behaviour and cognition. Nat Rev Neurosci 10:434–445
Maccari S, Darnaudery M, Morley-Fletcher S, Zuena AR, Cinque C, Van Reeth O (2003) Prenatal stress and long-term consequences: implications of glucocorticoid hormones. Neurosci Biobehav Rev 27:119–127
Macrì S, Spinelli S, Adriani W, Dee Higley J, Laviola G (2007) Early adversity and alcohol availability persistently modify serotonin and hypothalamic-pituitary-adrenal-axis metabolism and related behavior: what experimental research on rodents and primates can tell us. Neurosci Biobehav Rev 31:172–180
McGowan PO, Sasaki A, D'alessio AC, Dymov S, Labonte B, Szyf M, Turecki G, Meaney MJ (2009) Epigenetic regulation of the glucocorticoid receptor in human brain associates with childhood abuse. Nat Neurosci 12:342–348
Meaney MJ (2010) Epigenetics and the biological definition of gene × environment interactions. Child Dev 81:41–79
Meaney MJ, Ferguson-Smith AC (2010) Epigenetic regulation of the neural transcriptome: the meaning of the marks. Nat Neurosci 13:1313–1318

Meaney MJ, Szyf M, Seckl JR (2007) Epigenetic mechanisms of perinatal programming of hypothalamic-pituitary-adrenal function and health. Trends Mol Med 13:269–277

Mennes M (2008) Longitudinal study on the effects of maternal anxiety during pregnancy: neuropsychological and neurophysiological examination of cognitive control in the adolescent offspring. Unpublished doctoral dissertation. Katholieke Universiteit Leuven, Belgium

Mennes M, Stiers P, Lagae L, Van den Bergh B (2006) Long-term cognitive sequelae of antenatal maternal anxiety: involvement of the orbitofrontal cortex. Neurosci Biobehav Rev 30:1078–1086

Mennes M, Van den Bergh B, Lagae L, Stiers P (2009) Developmental brain alterations in 17 year old boys are related to antenatal maternal anxiety. Clin Neurophysiol 120:1116–1122

Menon RS (2012) The great brain versus vein debate. Neuroimage 62:970–974

Meyer U, Feldon J (2009) Prenatal exposure to infection: a primary mechanism for abnormal dopaminergic development in schizophrenia. Psychopharmacology (Berl) 206:587–602

Meyer U, Feldon J, Fatemi SH (2009) In-vivo rodent models for the experimental investigation of prenatal immune activation effects in neurodevelopmental brain disorders. Neurosci Biobehav Rev 33:1061–1079

Miller EK, Cohen JD (2001) An integrative theory of prefrontal conrtex function. Annu Rev Neurosci 24:167–202

Monk C, Georgieff MK, Osterholm EA (2013) Research review: maternal prenatal distress and poor nutrition—mutually influencing risk factors affecting infant neurocognitive development. J Child Psychol Psychiatry 54:115–130

O'Donnell K, O'connor TG, Glover V (2009) Prenatal stress and neurodevelopment of the child: focus on the HPA axis and role of the placenta. Dev Neurosci 31:285–292

O'Connor TG, Ben-Shlomo Y, Heron J, Golding J, Adams D, Glover V (2005) Prenatal anxiety predicts individual differences in cortisol in pre-adolescent children. Biol Psychiatry 58:211–217

Oberlander TF, Weinberg J, Papsdorf M, Grunau R, Misri S, Devlin AM (2008) Prenatal exposure to maternal depression, neonatal methylation of human glucocorticoid receptor gene (NR3C1) and infant cortisol stress responses. Epigenetics 3:97–106

Oitzl MS, Champagne DL, van der Veen R, de Kloet ER (2010) Brain development under stress: Hypotheses of glucocorticoid actions revisited. Neurosci Biobehav Rev 34:853–866

Potts GF (2004) An ERP index of task relevance evaluation of visual stimuli. Brain Cogn 56:5–13

Räikkönen K, Seckl JR, Pesonen A-K, Simons A, Van den Bergh BRH (2011) Stress, glucocorticoids and liquorice in human pregnancy: programmers of the offspring brain. Stress 14:590–603

Schlotz W, Phillips DIW (2009) Fetal origins of mental health: evidence and mechanisms. Brain Behav Immun 23:905–916

Seckl JR, Holmes MC (2007) Mechanisms of disease: glucocorticoids, their placental metabolism and fetal 'programming' of adult pathophysiology. Nat Clin Pract Endocrinol Metab 3:479–488

Seckl JR, Meaney MJ (2004) Glucocorticoid programming. Ann N Y Acad Sci 1032:63–84

Shonkoff JP (2010) Building a new biodevelopmental framework to guide the future of early childhood policy. Child Dev 81:357–367

Son GH, Geum D, Chung S, Kim EJ, Jo J-H, Kim C-M, Lee KH, Kim H, Choi S, Kim HT, Lee C-J, Kim K (2006) Maternal stress produces learning deficits associated with impairment of nmda receptor-mediated synaptic plasticity. J Neurosci 26:3309–3318

Stanwood GD, Levitt P (2004) Drug exposure early in life: functional repercussions of changing neuropharmacology during sensitive periods of brain development. Curr Opin Pharmacol 4:65–71

Stanwood GD, Washington RA, Shumsky JS, Levitt P (2001) Prenatal cocaine exposure produces consistent developmental alterations in dopamine-rich regions of the cerebral cortex. Neuroscience 106:5–14

Swaab DF, Bao A-M, Lucassen PJ (2005) The stress system in the human brain in depression and neurodegeneration. Ageing Res Rev 4:141–194

Van den Bergh BRH (1990) The influence of maternal emotions during pregnancy on fetal and neonatal behavior. J Prenat Perinat Psychol Health 5:119–130

Van den Bergh B (2007) De prenatale oorsprong van welvaartsziekten en gedragsproblemen. [Prenatal origin of lifestyle diseases and behavioural problems] In: van Houdenhove B (ed) Stress, het lijf, en het brein. Lannoo, Tielt

Van den Bergh BR (2011a) Developmental programming of early brain and behaviour development and mental health: a conceptual framework. Dev Med Child Neurol 53(Suppl 4):19–23

Van den Bergh BRH (2011b) Prenatal programming of cognitive functioning, stress responsiveness and depression in humans: From birth to age 20. In A. Plagemann (Ed.), *Perinatal programming. The state of the art.* (pp. 199–205). Berlin: De Gruyter. (ISBN978-3-11-024944-6; e-ISBN 978-3-11-024945-3)

Van den Bergh BRH, Marcoen A (2004) High antenatal maternal anxiety is related to adhd symptoms, externalizing problems, and anxiety in 8- and 9-year-olds. Child Dev 75:1085–1097

Van den Bergh BRH, Mulder EJH, Visser GHA, Poelmann-Weesjes G, Bekedam DJ, Prechtl HFR (1989) The effect of (induced) maternal emotions on fetal behaviour: a controlled study. Early Hum Dev 19:9–19

Van den Bergh BR, Mennes M, Oosterlaan J, Stevens V, Stiers P, Marcoen A, Lagae L (2005a) High antenatal maternal anxiety is related to impulsivity during performance on cognitive tasks in 14- and 15-year-olds. Neurosci Biobehav Rev 29:259–269

Van den Bergh BR, Mulder EJ, Mennes M, Glover V (2005b) Antenatal maternal anxiety and stress and the neurobehavioural development of the fetus and child: links and possible mechanisms. A review. Neurosci Biobehav Rev 29:237–258

Van den Bergh BRH, Mennes M, Stevens V, Van der Meere J, Borger N, Stiers P, Marcoen A, Lagae L (2006) ADHD deficit as measured in adolescent boys with a continuous performance task is related to antenatal maternal anxiety. Pediatr Res 59:78–82

Van den Bergh BR, Van Calster B, Smits T, Van Huffel S, Lagae L (2008) Antenatal maternal anxiety is related to HPA-axis dysregulation and self-reported depressive symptoms in adolescence: a prospective study on the fetal origins of depressed mood. Neuropsychopharmacology 33:536–545

van Eijsden M, Vrijkotte TG, Gemke RJ, Van Derwal MF (2011) Cohort profile: the Amsterdam born children and their development (ABCD) study. Int J Epidemiol 40:1176–1186

Weaver ICG, Diorio J, Seckl JR, Szyf M, Meaney MJ (2004) Early environmental regulation of hippocampal glucocorticoid receptor gene expression: characterization of intracellular mediators and potential genomic target sites. Ann N Y Acad Sci 1024:182–212

Weinstock M (2008) The long-term behavioural consequences of prenatal stress. Neurosci Biobehav Rev 32:1073–1086

Whelan R, Conrod PJ, Poline J-B, Lourdusamy A, Banaschewski T, Barker Gj, Bellgrove MA, Buchel C, Byrne M, Cummins TDR, Fauth-Buhler M, Flor H, Gallinat J, Heinz A, Ittermann B, Mann K, Martinot J-L, Lalor EC, Lathrop M, Loth E, Nees F, Paus T, Rietschel M, Smolka MN, Spanagel R, Stephens DN, Struve M, Thyreau B, Vollstaedt-Klein S, Robbins TW, Schumann G, Garavan H (2012) Adolescent impulsivity phenotypes characterized by distinct brain networks. Nat Neurosci 15:920–925

Yehuda R, Engel SM, Brand SR, Seckl J, Marcus, SM, Berkowitz, SG (2005) Transgenerational effects of posttraumatic stress disorder in babies of mothers exposed to the world trade center attack during pregnancy. J Clin Endocrinol Metab 90:4115–4118

Zhang T-Y, Hellstrom IC, Bagot RC, Wen X, Diorio J, Meaney MJ (2010) Maternal care and DNA methylation of a glutamic acid decarboxylase 1 promoter in rat hippocampus. J Neurosci 30:13130–13137

Zysset S, Wendt CS, Volz KG, Neumann J, Huber O, Von Cramon, DY (2006) The neural implementation of multi-attribute decision making: a parametric fMRI study with human subjects. Neuroimage 31:1380–1388

Part III
Epigenetic And Translational Studies

Chapter 16
Perinatal Programming of Neurodevelopment: Epigenetic Mechanisms and the Prenatal Shaping of the Brain

Paula A. Desplats

Abstract The recent years have witnessed an exponential growth in the knowledge of epigenetic mechanisms, and piling evidence now links DNA methylation and histone modifications with a wide range of physiological processes from embryonic development to memory formation and behavior. Not surprisingly, deregulation of epigenetic modifications is associated with human diseases as well.

An important feature of epigenetics is the ability of transducing environmental input into biological signaling, mainly by modulation of the transcriptome in response to a particular scenario. This characteristic generates developmental plasticity and allows the manifestation of a variety of phenotypes from the same genome.

The early-life years represent a period of particular susceptibility to epigenetic alteration, as active changes in DNA methylation and histone marks are occurring as part of developmental programs and in response to environmental cues, which notably include psychosocial stimulation and maternal behavior. Memory formation and storage, response to stress in adult life, behavior, and manifestation of neurodegenerative conditions can all be imprinted in the organism by epigenetic modifications that contribute to shape the brain during prenatal or early postnatal life. Moreover, if these epigenetic alterations are preserved in the germ line, changes induced in one generation are likely inherited by future offspring. Programming by transgenerational inheritance thus represents a central mechanism by which environmental conditions may influence disease risk across multiple generations.

As novel techniques emerge and as genome-wide profiling of disease-associated methylomes is achieved, epigenetic marks open a new source for biomarker discovery.

P. A. Desplats (✉)
Department of Neurosciences, University of California San Diego,
La Jolla, CA, USA
e-mail: pdesplat@ucsd.edu

M. C. Antonelli (ed.), *Perinatal Programming of Neurodevelopment*,
Advances in Neurobiology 10, DOI 10.1007/978-1-4939-1372-5_16,

16.1 Introduction

The sequencing of the human genome and the fast development of molecular biology techniques in the recent years had a profound impact on the analysis of human disease. Psychiatry, developmental psychology, and related fields have also incorporated the analysis of the genome and the study of the effects of nucleotide variations in the DNA on cell biology, physiology, and emotional states as part of their efforts on understanding individual differences in brain development, function, and disease (Meaney 2010).

During the past recent years, we have witnessed an exponential growth in the interest and knowledge of epigenetics. The capacity of profiling epigenetic marks at the genome-wide level ignited the research across the lifespan of the organisms, and what was before considered of exclusive domain of development, is now recognized to play roles in health and disease during the postnatal periods and on adulthood.

From the most current definition of epigenetics as "the study of mitotically and/or meiotically heritable changes in gene function that cannot be explained by changes in DNA sequence" (Riggs et al. 1996) to the theory of the developmental origin of adult onset diseases (Barker and Clark 1997), epigenetic mechanisms have been reported to mediate discordant tween phenotypes; the responses to the environment; plasticity and memory formation and to underlie a myriad of neurological diseases from autism, Rett, Angelman, and Prader–Willi syndromes, to adult onset disorders including Parkinson's and Alzheimer's diseases.

The perinatal environment in particular has profound effects on modulation of epigenetic mechanisms that might mediate long-term health effects. Maternal exposure to stress, alcohol, drugs, chemicals, and poor nutrition are being recognized to impact on brain function and learning.

While most of our knowledge is derived from cellular and animal models, the emerging field of epigenetic epidemiology might provide mechanistic insights and will likely guide potential interventions aimed at improving health outcomes of "exposed" individuals.

16.2 Epigenetics 101

The "central dogma of modern biology" summarizes the processes involved in maintaining and translating the genetic code necessary to sustain life. In its basic form, it implies the self-propagation of the DNA, the transcription of the genetic code into a messenger RNA, and the translation of the message to synthesize proteins (Crick 1970). As molecular biology grew deeper, new equations were added to the dogma, which now accounts for DNA synthesis from and RNA template by reverse transcription in retroviruses and retrotransposons (Kim et al. 2004a; Medstrand et al. 2005) and also includes the replicative ability of self-aggregating prions (Cohen and Prusiner 1998; Shorter and Lindquist 2005). More recently, the discovery of noncoding RNAs, including microRNAs and long noncoding RNAs, added a new level of complexity to the flow of genetic information, with the recognition of the role of these molecules in the regulation of gene expression (McManus and Sharp 2002; Magistri et al. 2012).

The term "epigenetics" was coined by Waddington early in 1942 to conceptualize the interaction of the gens with their environment to produce a phenotype (Waddington 1942). As the word implies, the Greek prefix *epi* states that these marks operate "above" the DNA sequence and might be seen as a new addition to the dogma as proteins now feedback into the flow between DNA and RNA. Beyond semantics, the concept of epigenetics was used to explain certain anomalous phenotypes that could not be accounted by Mendelian inheritance considering non-Mendelian traits inherited by the maternal line like mitochondria. Epigenetic mechanisms are believed to be the first line of response of the organisms to the physical environment, including the intrauterine milieu, exposure to chemicals, endocrine disruptors, alcohol, drugs of abuse, nutrients, and also to psychosocial stimulus, particularly to early-life stress.

16.2.1 The Chromatin and the Histone Code

The scenario for epigenetic reactions is the DNA molecule, which resides in the cellular nucleus in a complex with histone proteins, forming the chromatin. The basic unit of the chromatin is the nucleosome, which contains a segment of DNA of 147 base pairs (bp) wrapped around an octamer of histones (composed by two molecules of each H2A, H2B, H3, and H4; Kornberg 1974). The chromatin is a dynamic molecule that can adopt either an open and relaxed conformation known as euchromatin, most favorable for gene expression, or that can exist in a closed, tightly packed structure named heterochromatin, which is nonpermissive for transcription. Transition between these states is highly determined by epigenetic mechanisms including histone modifications and DNA methylation and is crucial for the regulation of the transcriptome (Fig. 16.1).

The N-terminal tails of histone proteins are targets for a wide arrange of modifications, including acetylation, methylation, phosphorylation, ubiquitylation, sumoylation, and adenosine triphosphate (ADP)-ribosylation, and which are crucial determinants of the chromatin structure (Vaquero et al. 2003). Histone covalent modifications occur at precise sites and are catalyzed by specialized enzymes. Histone acetyltransferases (HATs), which acetylate histones H3 and H4 (Roth et al. 2001), and histone deacetylases (HDACs), which deacetylase histone tails (Kuo and Allis 1998; Grozinger and Schreiber 2002), are the groups most extensively investigated. In addition, histone lysine methyl transferases (HKMTs) and the more recently described histone demethylases (Tsukada et al. 2006) are also important chromatin modifiers. Lysine modifications can be present as mono-, di- or tri-methylated states, adding complexity to what is called "the histone code," or the specific pattern of histone modifications that outlines which portions of the genome will be expressed at a certain time and in a certain cell type (Jenuwein and Allis 2001).

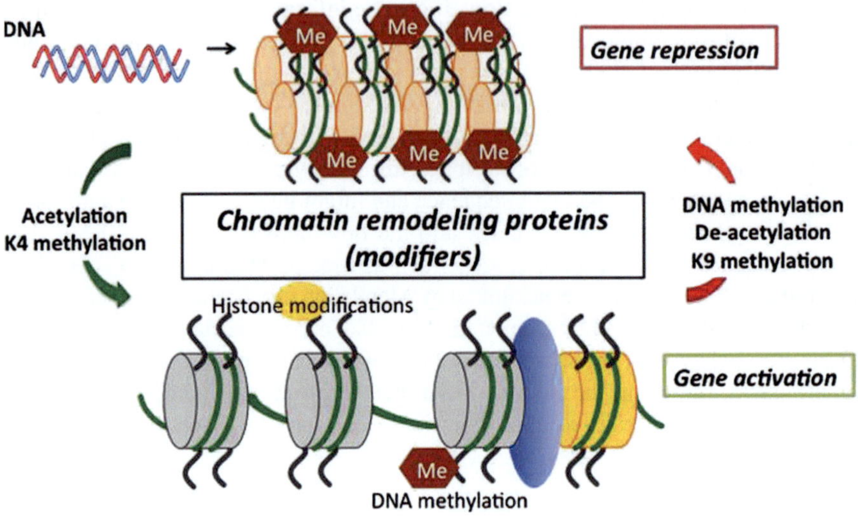

Fig. 16.1 A simplified schematic of the epigenetic modulation of gene expression. The overall balance between DNA methylation and the degree of acetylation/methylation of specific histone residues predict the transcriptional status of a gene. Heavily methylated heterochromatin (*upper*) is associated with repression of gene expression. Increase in histone acetylation and methylation of the lysine residue at position 4 of the histone 3 (H3K4) is associated with transition to chromatin-relaxed states that are permissive for transcription (*lower*)

While the combinatorial possibilities of the histone code and the association of chromatin modifier activities with other transcription factors render a quite complex array of regulatory scenarios, a simplified vision considers that histone acetylation and mono-methylation on specific lysine residues are signals for predominantly active chromatin configuration (Perry and Chalkley 1982; Lee et al. 1993), while marks such as certain phosphorylated residues, lost of acetylation and di- and tri-methylation of lysine residues 9 and 27 are associated with heterechromatin conformations that are nonpermissive for transcription (Allis et al. 2006). It is important to note that all these modifications are not gene specific, in contradiction with the fine-tuning of transcription attributed to epigenetic regulation. Gene-specific targeting is achieved by chromatin remodeling complexes, whereby transcription factors that bind DNA at specific recognition sites recruit histone-modifying enzymes inducing a local change in chromatin conformation and thus determining the expression status of particular genes (Jenuwein and Allis 2001).

16.2.2 DNA Methylation

DNA methylation is one of the oldest known epigenetic modifications whose changes are highly correlated with gene expression (Razin and Riggs 1980). This variation consists in the addition of a methyl group, from S-adenosyl-L-methionine donor, to the 5′ position of a cytosine ring present on a CpG dinucleotide on the DNA template. The reaction is catalyzed by a family of specialized enzymes, the DNA methyltransferases (DNMTs) and since this new 5-methyl-cytosine (5mC) residue is part of the chemical structure of the DNA itself, it is more stable than other epigenetic marks and holds a high potential for diagnosis and biomarker development (Beck et al. 1999). A characteristic of DNA methylation in mammals is that not all CpG sites are modified in any cell type, therefore this epigenetic mark confers cell-specific identity (Razin 1998). Distribution of methylated regions along the genome shows enrichment at noncoding regions and interspersed repetitive elements. Regions with a high density of CpGs are known as "CG islands" and these clusters that are normally found at gene promoters tend to be unmethylated (Weber et al. 2007). Recent evidence is showing that DNA methylation in gene regions other than islands might be important for transcriptional regulation. Methylation at introns might regulate noncoding RNAs that contribute to chromatin silencing and RNA degradation (McGowan et al. 2009). In the brain, the distribution of DNA methylation regarding CpG neighborhood context and genomic location shows that loci with decreased methylation are more likely to locate at CpG islands and associate with promoter regions located around 1500 or 2000 bp upstream of the transcription start sites (TSS1500, TSS200) and also at first exon sites, while CpG sites located farther away from islands (open sea) and at the gene bodies are more likely to present increased methylation (Davies et al. 2012).

DNMTs can be divided according to the methylation status of the DNA template on which they catalyze the methyl transfer. Thus, "de novo" methylation enzymes, including DNMT3A and B, catalyze methyl transfer onto a "naked" or unmethylated DNA template and are mainly responsible for changes in methylation during embryogenesis and development. DNMT1, on the other hand, is the "maintenance" methylation enzyme, which copies DNA marks from a hemi-methylated template and which has major importance preserving methylation patterns in adult tissues, including postmitotic cells in the brain (Okano et al. 1998).

The potential roles of DNA methylation in cellular functions after development were obscured for a long time for the lack of evidence of a "demethylation" enzyme, which would confer plasticity as a quality needed to respond to environmental stimulus. A growing line of evidence now supports the idea that, similar to chromatin modification, DNA methylation is potentially reversible (Ramchandani et al. 1999). New reports link demethylation with further modification of the 5-methylcytosine to produce a variety of chemical modifications. Among these, a potential regulatory role for 5-hydroxymethyl-C (5hmC) in neurons and embryonic stem cells was suggested after the discovery of high levels of 5hmC in the brain (Kriaucionis and Heintz 2009; Tahiliani et al. 2009). In mammals, the members of the Ten-Eleven-Translocation (TET) protein family, which are expressed in tissues with active DNA demethylation, catalyze the hydroxymethylation of 5mC into 5hmC (Tahiliani et al. 2009; Ito et al. 2010). TET1 expression is mainly confined to embryonic stem cells, whereas TET2 and TET3 are more ubiquitous (Tahiliani et al. 2009; Szwagierczak et al. 2010). Conversion to 5hmC shifts molecular recognition by reducing the recruitment of methyl-binding proteins, including MeCP2 and DNMT1, and resulting in passive demethylation (Valinluck and Sowers 2007). In addition, TET proteins could further oxidize 5hmC to 5-formylcytosine (5fC) and 5-carboxylcytosine (5caC; He et al. 2011; Ito et al. 2011), raising the possibility that these enzymes actively induce DNA demethylation by successive oxidation of 5mC. To date, no enzyme has been identified that can lead to complete demethylation of 5mC to C (Franchini et al. 2012).

An important feature of epigenetic modifications is their correlation to chromatin structure, in particular in the case of DNA methylation. Thus, active regions of chromatin permissive for gene expression are generally associated with low levels of DNA methylation, whereas compacted heterochromatin that represses transcription contains hyper methylated DNA regions (Razin 1998; Fig. 16.1). If DNA methylation sites occur at regulatory gene areas, they determine the transcriptional status of the gene. If CpG residues lie in the recognition/binding site of a regulatory factor, the presence of the methyl group might directly block the binding of transcription factors by steric interference (McGowan et al. 2009). Alternatively, a higher density of DNA methylation in a defined region can recruit methylated-DNA binding proteins such as methyl CpG-binding protein 2 (MeCP2), which in turn might complex with Sin3A and histone modifying enzymes to transition to heterochromatin, repressive states (Nan et al. 1997).

In sum, the overall balance between DNA methylation and varied histone modifications dictates chromatin conformation and ultimately determines the expression of a particular set of genes at a due moment.

16.3 The Role of Epigenetic Regulation in the Brain

16.3.1 The Dynamic Epigenetic Machinery and its Role in Neuronal Function

Albeit fundamental knowledge about epigenetic mechanisms came from developmental studies, a growing body of evidence supports the notion that regulation of chromatin structure through histone acetylation and DNA methylation might play an important role in long-lasting behavioral changes in the context of learning and memory. Thus, while sequence of DNA contains the cellular memory, epigenetic modulation that alters chromatin structure might be viewed as a novel way to store physiological and behavioral memories that can be preserved until adulthood or even beyond generations (Levenson and Sweatt 2005).

The relevance of epigenetic mechanisms for postmitotic terminally differentiated neurons in the adult central nervous system (CNS) was a conundrum. Indeed the epigenetic machinery is necessary to preserve the neuronal identity itself. Neuronal cells posses a unique specialized proteome populated by proteins involved in excitability, transmitter release, and maintenance of transmembrane potential, which if expressed in other cellular types might result highly toxic.

Thus, genes that are to be expressed only in neurons have a neuron-restrictive silencer element (NRSE) in their promoters (Li et al. 1993). The transcription factor REST binds to NRSEs and represses gene expression outside the CNS (Chong et al. 1995). REST-mediated gene silencing requires the modulation of chromatin structure and seems to involve reductions in histone acetylation. REST/SIN3A repressor complexes are associated with HDAC1, whereas REST/CoREST complexes are associated with HDAC2 (Huang et al. 1999; Naruse et al. 1999; Ballas et al. 2001).

The long-term storage of memories that shape behavior and cognition is perhaps one of the most fascinating functions of the human brain. Long-standing research established that transcription and translation are necessary steps for memory formation and storage. More recently, novel evidence showed epigenetic mechanisms play a crucial role in memory formation (Levenson et al. 2004).

Contextual fear conditioning, whereby an animal learns to associate a novel context with an aversive stimulus, is a hippocampus-dependent task that requires specific H3 acetylation mediated by mitogen-activated protein kinase/extracellular-signal-regulated kinase (MAPK/ERK) kinase signaling (Levenson et al. 2004). Interestingly, latent inhibition, another form of long-term memory, is associated with altered H4 acetylation (Levenson et al. 2004), suggesting that a histone code might exist for memory formation, where precise patterns of histone modification are linked to specific types of memory.

The formation of long-term memory is dependent upon the ability of the neurons to adjust the strength of synapses, a process known as synaptic plasticity. Long-term potentiation (LTP) is a form of synaptic plasticity that requires the activation of N-methyl-D-aspartate (NMDA) receptors and the MEK–ERK/MAPK signaling cascade, which also results in increased H3 acetylation (Levenson et al. 2004). Similarly, the activation of dopaminergic, cholinergic, and glutamatergic pathways in the hippocampus is associated with ERK-dependent H3 phosphorylation (Crosio et al. 2003).

A crucial event in neurobiology is the actual retrieval of stored memories, a process that induces a window of plasticity during which perturbations of the molecular signaling involved in memory formation can promote enduring behavioral changes (Lattal and Wood 2013). The effects of retrieval-induced plasticity are critical in two events: during reconsolidation of the original memory, when molecular events are needed to stabilize the memory and during extinction, in which learning during the retrieval event creates an additional memory of a changed environment (Torregrossa and Taylor 2013). Epigenetics provide a plausible cellular and molecular mechanism for maintaining the persistent loss of behavior induced by extinction (Lattal and Wood 2013). The use of a selective HDAC3 inhibitor (HDAC3i) facilitates extinction of a previously established cocaine-induced conditional place preference, while simultaneously enhancing the long-term formation of a concurrent object location memory in the same subjects. Importantly, during extinction consolidation, HDAC3i promoted a distinct pattern of histone acetylation and gene expression in the infralimbic cortex, hippocampus, and nucleus accumbens (Malvaez et al. 2013).

Novel evidence also links DNA methylation on the regulation of associative learning. An elegant study, using a Pavlovian reward conditioning model in which animals form an association between reward-paired cues and future reward, showed that learning events produce changes in DNA methylation in genes upregulated in dopamine neurons (Day et al. 2013). Furthermore, blockage of DNA methylation in the ventral tegmental area that mediates reward learning prevented memory formation. These results have a broad implication in the understanding of the molecular mechanisms of reward-related learning, which are critical in human behavior as they relate to decision making and addiction.

An important feature of neuronal physiology is rhythmicity. Circadian rhythms are generated endogenously by a biological timekeeping mechanism known as the circadian clock, which resides in the suprachiasmatic nuclei (SCN) of the brain and synchronize its cycles with subsequent rounds of transcription and translation. The acetylation of histones H3 and H4 associated with the promoters of core molecular clock genes is differentially regulated during the circadian cycle (Naruse et al. 2004). Infusion of the HDAC inhibitor (HDACi) trichostatin A into the SCN increases the expression of clock genes Period 1 (Per1) and Period 2 (Per2) in the mouse, suggesting the involvement of epigenetic mechanisms on circadian regulation (Naruse et al. 2004). In addition, adjustment of phases to environmental stimulus also requires transcription and is highly regulated by light. Discrete pulses of light have been shown to induce increase H3 and H4 acetylation at Per1 and Per2 promoters (Naruse et al. 2004), as well to result in a significant increase in H3 phosphorylation (Crosio et al. 2000).

16.3.2 Deregulation of Epigenetic Mechanisms in Neurodevelopment and Neurodegeneration

Due to their crucial roles in neuronal function, it is not surprising that deviations on the fine-tuned regulation of gene expression by the epigenetic machinery result in disease states. While the role of disrupted epigenetic mechanisms either as causative or consequence of pathology needs to be clarified yet, abundant evidence links DNA methylation and histone modifications with neurodegenerative disorders.

Alzheimer's disease (AD), the prevalent neurodegenerative condition of the older adults, is characterized by amyloid β deposition, resulting from the cleavage of the amyloid precursor protein (APP) by presenilins (Masliah et al. 1994). The cleavage of APP also generates the APP C-terminal peptide (AICD) that might interact with the histone acetyltransferase Tip60, increasing acetylation of histones H3K14 and H4K5 in vitro (Kim et al. 2004c). Moreover, the elegant work of L.H. Tsai and her group recently demonstrated that neuron-specific overexpression of HDAC2 in the hippocampus decreases dendritic spine density, synapse number and plasticity, and impairs memory formation in a mouse model, with some of those deficits also identified in postmortem brain tissue from AD patients (Guan et al. 2009). In a recent study analyzing DNA methylation at the temporal neocortex in subjects ranging from 17 weeks of gestation to 104 years old, AD cases included in the analyzed cohort showed an acceleration of age-related changes in DNA methylation (Siegmund et al. 2007). Yet another study on the entorhinal cortex layer II, a region that exhibits substantial pathology in AD, showed decreased levels of DNA methylation and methylation maintenance factors in AD (Mastroeni et al. 2010), highlighting a previously unsuspected role of epigenetic mechanisms on pathology.

Parkinson's disease (PD) is the second most common neurodegenerative disorder in the elderly, affecting 2% of the population over 60 years old (Dauer and Przedborski 2003). PD is manifested clinically by resting tremor and postural instability, and neuropathologically by the loss of dopaminergic neurons and accumulation of α-synuclein in Lewy bodies, primarily in the striatonigral system (Hirsch et al. 1999). At the molecular level, disruption of several physiological pathways contribute to pathology, including mitochondrial alterations, deregulation of autophagy and clearance mechanisms (Levy et al. 2009) and reduced adult neurogenesis (Desplats et al. 2012). PD is a multifactorial disease and majority of the cases are of idiopathic origin, likely due to the joint contribution of genetic susceptibility and environmental triggers. We recently reported that aberrant interaction between α-synuclein and DNMT1, the main enzyme involved in maintaining the DNA methylation pattern in adult cells, results in the hypomethylation of specific targets in the brains of PD patients (Desplats et al. 2011).

Abnormal epigenetic regulation has also been reported for other neurodegenerative and neurological diseases including Huntington, multiple sclerosis, epilepsy, and amyotrophic lateral sclerosis, among others (Urdinguio et al. 2009).

More relevant for the present chapter is the role of epigenetic alterations in neurodevelopmental disorders (NDD). While detailed description of the broad spectra of NDD is out of scope of the present chapter, an excellent portrayal of disorders and their clinical, molecular, and epigenetic factors can be found at the review by Millan (Millan 2013). In general, NDD are characterized by impaired or delayed CNS development, leading to disruption of discrete cerebral functions or generalized impairment in multiple domains, particularly in cognition, language, emotion, motor control, and intellectual disability (Millan 2013). NDD can be further grouped as 1) disorders characterized by discrete genetic anomalies: including Down; Rett; Fragile X; alpha-thalassemia X-linked intellectual disability (ATRX) syndromes and also imprinting disorders like Angelman and Prader–Willi syndromes; or as 2) heterogeneous, multifactorial, and polygenic disorders: including autism and schizophrenia, where environmental factors like perinatal infection, malnutrition, and trauma have a direct role in etiology (Millan 2013).

Importantly, many NDD are accompanied by aberrant epigenetic regulation. As selected examples, ATRX is characterized by perturbations of higher-order chromatin organization, due to mutations on ATRX, a member of the SW1/SFN family of chromatin-remodeling proteins, and resulting in altered coordination of DNA methylation patterns and H3K9me3 binding, which is manifested clinically as mental retardation, alpha thalassemia, and pronounced intellectual disability (Berube 2011). On the other hand, alterations on imprinting associated with Angelman syndrome, are due to maternal deletion imprinting on $GABA_A$ receptor subunits or UBE3A ubiquitin ligase, which result in cognitive deficits, motor dysfunction, and seizures (Jana 2012). Similarly, in Prader–Willi syndrome, a paternal deletion of imprinting results in the loss of paternal Necdin, Magel2, and snoRNA genes, triggering a series of symptoms including difficulties in feeding during infancy, poor social function, and cognitive deficits (Cassidy et al. 2012). Perhaps, the most obvious link between NDD and epigenetic alterations is represented by Rett syndrome, an X-linked developmental disorder caused by mutations on the methyl-DNA binding protein MeCP2, that results in anomalous patterns of gene transcription and which is clinically manifested by cerebral hypotrophy, intellectual disability at various degrees, autistic features, and motor dysfunction (Grillo et al. 2012).

A notable example of NDD associated with environmentally and genetically induced changes is the group of autism spectrum disorders (ASD), which encompass syndromes characterized by deficits in social behavior and communication and accompanied by intellectual disability, restricted interests, and repetitive stereotyped behaviors (Manning-Courtney et al. 2013). The complex genetics of sporadic ASD are not yet fully elucidated, still some studies on cerebral tissue support the idea that epigenetic processes are disrupted in sporadic ASD in a tissue- and time-dependent fashion, particularly on patterning of H3K4me3 on several loci considered as risk factors for ASD and known to regulate neuronal development and synaptic communication (Shulha et al. 2012). In addition, increased methylation of MeCP2 promoter, which also correlates with reduced levels of MeCP2 in the frontal cortex, was reported in ASD patients (Nagarajan et al. 2006). Hypermethylation of oxytocin receptor, involved in promoting social communication was also observed (Gregory et al. 2009), pointing out to the complex interplay of epigenetic mechanisms and its contribution to disease.

Finally, schizophrenia deserves special attention, as several epigenetic abnormalities have been associated to this NDD. Schizophrenic patients present a reduction in GAD67, the rate-limiting enzyme for GABA synthesis, in the cortex and hippocampus (Grayson and Guidotti 2013). In addition, reduced levels of reelin, which controls dendritic spine formation, axonal development, glutamatergic transmission, and synaptic plasticity was also reported in the frontal cortex of schizophrenic individuals (Grayson and Guidotti 2013). Both deficits seem to be rooted in the same mechanism: activation of mGluR2 receptors reduces DNA methylation reversing the transcriptional downregulation of these genes (Gavin and Akbarian 2012). In addition, H3K4 methylation seems to be involved in GAD67 regulation (Grayson and Guidotti 2013). Thus, intricate epigenetic regulation, involving histone acetylation and methylation, MeCP2 availability, and DNA methylation underlie complex neurochemical changes associated with schizophrenia, bringing together a myriad of environmental and genetic factors that vary among individuals, disease progression, and tissue (Millan 2013).

16.4 Genome, Epigenome, and Environment

Developmental plasticity transduces environmental cues into cellular pathways during gestation, enabling a single genotype to produce a broad diversity of adult phenotypes. Epigenetic mechanisms are the first molecular responders to environmental inputs and sometimes the epygenotype shaped under certain particular conditions might contribute to disease susceptibility later in life or even in the next generations (Dolinoy et al. 2007b).

A wide range of environmental factors have been documented to impact on the neonatal epigenome, from chemical compounds, like endocrine disruptors bisphenol A (Dolinoy et al. 2007a) or vinclozolin (Skinner et al. 2008); to the effects of maternal diet including nutrition and folate availability (Foley et al. 2009), maternal metabolic status (Lehnen et al. 2013), and alcohol consumption (Perkins et al. 2013). But while the deleterious effects of exposure to a range of substances is easier to anticipate, the effects of psychosocial factors encompassed by parental behavior and prenatal stress (PS) are far more intriguing and can take a deep toll on shaping the epigenetic landscape of the perinatal brain. Thus, the fetal basis of adult disease hypothesis postulates that environmental factors acting during prenatal and early postnatal life influence developmental plasticity, which might predispose the organism to adult chronic diseases.

16.5 Epigenetic Programming of the Brain During Early Life

The process of "fetal programming" is mediated by the impact of prenatal experience on the developing hypothalamic–pituitary–adrenal (HPA) axis, a dynamic metabolic system that regulates homeostatic mechanisms, including the ability to respond to stressors (Van den Hove et al. 2006), and which is highly sensitive to adverse early-life experiences (Meaney 2001). Animal studies show that PS leads to vulnerability to anxiety and decreased learning, memory, and locomotor dysfunction (Weinstock 2001; Huizink et al. 2004), effects that are mediated by increased responsiveness of the HPA axis to stress, and to reductions in the expression of the glucocorticoid receptor (GR) in the hippocampus of adult offspring (Zuena et al. 2008). Postnatal stimulation might counteract the biological effects of PS, as changes in the postnatal environment including handling and early adoption improve the performance of adult offspring in cognitive tasks (Meaney et al. 1988) and reduce stress-induced corticosterone secretion (Vallee et al. 1999). In humans, prenatal depressed/stressed maternal mood is associated with higher rates of preterm delivery and lower birth weight (Wadhwa et al. 1993; Van den Bergh et al. 2005), elevated cortisol (Field et al. 2004), subsequent working memory performance in young women (Entringer et al. 2009), and changes in GR regulation (Mulligan et al. 2012).

Several studies support the hypothesis that parental programming is mediated by epigenetic mechanisms that stably alter gene transcription affecting physiology and behavior (Weaver et al. 2004; Caldji et al. 2011; Mulligan et al. 2012).

16.5.1 The Paradigm of the Glucocorticoid Receptor

A key determinant of early-postnatal life is the interaction of the immature individual with his/her parents and the presence and sensory input from the mother is crucial during neonatal and infancy development (McClelland et al. 2011).

The remarkable studies by Meany and coworkers unveiled the epigenetic mechanisms underlying the HPA response to stress, which involves the regulation of GR expression. The organization of the GR gene (*Nr3c1*) is similar in rodents and humans and comprises nine exons, where exons 2 to 9 are coding regions while exon 1 is a regulatory segment encompassing several promoters. This modular structure enables the activation of different promoters by specific transcription factors to confer tissue-specific patterns of GR expression patterns (Meaney 2010). Activation of GRs in the mammalian forebrain is associated with a decrease in neurogenesis (Uno et al. 1989) and synaptic plasticity (McEwen 2007; de Kloet et al. 2008).

Studies in rodents demonstrated that postnatal handling decreased the magnitude of behavioral and HPA responses to stress in adulthood (Denenberg 1964; Levine 1970), effects that were later linked with changes in maternal care (Levine 1970). Postnatal handling of rat pups increase the licking/grooming (LG) behavior of the mother (Liu et al. 1997), which is a major source of tactile stimulation with effects on endocrine and cardiovascular functions (Levine 1994).

The adult offspring of naturally high-LG mothers show more modest behavioral and endocrine responses to stress compared to animals reared by low-LG mothers (Liu et al. 1997; Francis et al. 1999; Weaver et al. 2004). Cross-fostering studies, where pups born to high-LG mothers are fostered at birth to low-LG mothers (and vice versa), confirmed a direct relation between maternal care and postnatal behavioral and HPA responses to stress (Francis et al. 1999; Caldji et al. 2000, 2003; Weaver et al. 2004), with the rearing mother determining the phenotype of the offspring. The effects of maternal care involve alterations in the function of the corticotrophin-releasing factor (CRF) systems in selected brain regions (Plotsky et al. 1989; Bale and Vale 2004). The adult offspring of high-LG mothers show decreased CRF expression in the hypothalamus and reduced glucocorticoid responses to acute stress in comparison with the adult offspring of low-LG mothers (Liu et al. 1997; Francis et al. 1999; Weaver et al. 2004, 2005). The high-LG offspring showed significantly increased hippocampal GR expression and decreased hypothalamic CRF levels. Moreover, altered maternal care also affects the behavioral responses to stress (Francis et al. 1999; Caldji et al. 2003). The adult offspring of high-LG mothers showed decreased startle responses, and substantially less fearfulness in the presence of stressors. Stress during pregnancy decreases maternal responsivity in lactating rats (Fride et al. 1985; Moore and Power 1986) and, as expected, the effects on maternal behavior are apparent in the development of the offspring. As adults, the offspring of high-LG/gestationally stressed mothers were comparable to those of low-LG dams on their own maternal behavior, and also in fear behavior and hippocampal GR gene expression. These effects are also apparent in subsequent litters even in the absence of the stressor (Champagne and Meaney 2006). Thus, the behavior of the mother "program" stable changes in gene expression in the offspring that might set the basis for individual differences in behavioral and neuroendocrine responses to stress in adulthood.

The fundamental question posed by these enthralling studies is how maternal care might stably affect gene expression and which are the mechanisms that might translate social interactions into biological effects. The answer to this question holds one of the most exquisite examples of the epigenetic response to the social environment.

Tactile stimulation in the form of pup LG increases serotonin (5-HT) metabolism in the hippocampus, which in turn elevates the expression of nerve growth factor-inducible factor A (NGFI-A) and CREB-binding protein (CBP; Meany et al. 2000). NGFI-A binds to the exon 1_7 promoter of GR to modulate its expression (Weaver 2007). Remarkably, mother–pup interactions actively regulate NGFI-A binding, as changes in DNA binding affinity are detected immediately after nursing and decay after 25 min without mother-pup contact (Meaney 2010).

Epigenetic mechanisms come into play as NGFI-A and CBP form a complex that binds to exon 1_7 promoter and reshapes the methylation at this region, setting a new epigenomic landscape that will persist into adulthood (Weaver et al. 2004; Weaver 2007). The adult offspring of high-LG mothers show low levels of methylation at specific CpG residues located at 5′- end of the NGFI A consensus binding sequence (Weaver et al. 2004). The 5′- and the 3′-CpG residues of this binding site are unmethylated on the hippocampus of fetal rats, although both sites appear heavily methylated on postnatal day 1 (PD1) regardless of maternal interaction. Between PD1 and PD7 epigenetic programming is imprinted by the maternal interaction, as the 5′ CpG is only demethylated in pups reared by high-LG while it appears methylated in low-LG offspring (Caldji et al. 1998; Champagne et al. 2003).

Furthermore, CBP has histone acetyltransferase activity, and its levels are increased in the hippocampus of high-LG offspring. Binding of CBP to exon 1_7 promoter contributes to modify the local chromatin structure enabling further binding of NGFI-A and other transcription factors that regulate GR expression (Szyf et al. 2005). Notably, blockage of HDACs by direct infusion of an inhibitor into the hippocampus of the adult offspring eliminates the maternal imprinting differences at the exon 1_7 promoter by increasing histone acetylation on the offspring of low-LG mothers and thus erasing maternal effects on NGFI-A binding and GR expression and, most importantly, reversing the differences in the HPA response to stress (Weaver et al. 2004). Another elegant study showed subsequently that the reverse pattern of results could be obtained in response to a dietary manipulation that increases the availability of methyl-donor groups: increased methylation of the 5′ -CpG in the offspring of high-LG mothers decreased NGFI-A binding and GR expression, with concomitant increase in HPA responses to stress (Weaver et al. 2005).

The precise epigenetic alterations that can convert sensory impressions received from the environment into biological outcomes are expected to function in transducing the signals from a myriad of stimulus and on a range of targets as well. The methylation status of the promoter of the brain-derived neurotrophic factor (*Bdnf*) gene can also be reshaped by maternal influence during early life (Roth et al. 2009) but moreover it can be altered by contextual fear conditioning in the adult rat (Lubin et al. 2008). BDNF is a crucial regulator of neuronal physiology implicated in learning and memory (Yamada et al. 2002) and whose alterations have been associated with AD, depression (Nagahara and Tuszynski 2011), and schizophrenia (Martinotti et al. 2012). Changes in methylation in response to external stimulus, including but not limited to chronic stress, drugs of abuse, chemical exposure, and psychosocial interactions, might therefore induce alterations on mental health later in life (Tsankova et al. 2007; Jiang et al. 2008; Akbarian and Huang 2009).

16.5.2 Neonatal Stress and Epigenetic Alterations on Learning and Memory

As discussed in previous sections, prenatal and early postnatal life is a critical window for epigenetic modulation of programming, whose outcomes might endure into adult life. Beyond altered responses to stress, numerous clinical studies report a strong association between early-life experience and cognitive functions. Importantly, chronic childhood stress due to emotional/social factors, as those triggered by extreme poverty, loss of a parent, social deprivation, or abuse, seem to have a bigger impact on learning and memory impairments later in life (Chugani et al. 2001; Kaplan et al. 2001; Wilson et al. 2007). Noteworthy, improving the social environment of institutionalized children by placement in functional families significantly improves learning and memory in the long term. Remarkably, the timing of placement in foster care appears to be crucial, as highest improvements are observed when placement occurs before the age of 2 years (Nelson et al. 2007). These data stress the plasticity of the processes that shape cognitive function in response to early-life experiences, making epigenetic mechanisms likely players in this modulation of programming.

Enriched early-life experience is associated with persistent attenuation of the stress response and with improved learning and memory (Meaney et al. 1991; Fenoglio et al. 2005), effects mediated at the molecular level by reduced expression of corticotropin releasing hormone (CRH) in the hypothalamus and increased levels of GR in the hippocampus (Plotsky and Meaney 1993). Similar to what was described earlier for GR, regulation of CRH involves epigenetic mechanisms. Expression of CRH gene is modified in response to circadian rhythm and stress-provoked stimulus, as well as by MeCP2 (McClelland et al. 2011). The CRH promoter contains a regulatory NRSE element that recruits chromatin-modifying factors (Naruse et al. 1999). It has been previously reported that enriched early-life experience reduces the number of excitatory synapses and input to CRH-expressing hypothalamic neurons, without altering GABAergic neurotransmission, a phenomenon that might signal the initiation of the epigenetic cascade that will endure those changes throughout life (Korosi et al. 2010). On the contrary, chronic early-life stress results in persistent elevation of CRH in the hippocampus, which impairs dendritic spines, synaptic plasticity, and memory function (McClelland et al. 2011).

16.5.3 Psychosocial Stress as Environmental Input

Social interaction is considered the major source of stress for humans in their daily life and from a neurobiological standpoint, the study of chronic social stress models in rodents is accepted as highly relevant to the pathological stress response in humans (Stankiewicz et al. 2013). Psychosocial stress seems to evoke epigenetic modifications. Hinwood et al. reported that rats subjected to chronic social defeat presented increased H3 acetylation in neurons and glia in the prefrontal cortex (Hinwood et al. 2011), and similar changes were observed in the nucleus accumbens of mice exposed to chronic social defeat stress or to social isolation, both conditions serving as depression models (Wilkinson et al. 2009). Moreover, social defeat and social isolation also result in altered histone methylation in genes related to cellular plasticity, inflammation, and transcriptional regulation, thus suggesting that histone modification seems to be the common transducer of social stress phenotypes (Stankiewicz et al. 2013).

On the other hand, acute restraint—suggested as one of the most potent and harmful psychogenic stressors—is also linked to epigenetic alterations. Experimental paradigms using male rats showed that soon after a 30-min period of restraint, the hippocampus displays changes on H3 methylation that correlate with transcriptional repression (Hunter et al. 2009). While acute stress, including forced swim and predator stress results in H3 phosphorylation in rodents (Bilang-Bleuel et al. 2005), voluntary exercise, on the contrary, significantly reduces neuronal H3 phosphorylation on mice brains (Binder et al. 2004), in agreement with its role in stress alleviation. Moreover, increased H3 phospho-acetylation in the hippocampus has also been associated with novelty stress in rats, with changes being evident as early as 30 min after exposure, revealing the rapid formation of epigenetic patterns (Chandramohan et al. 2007).

Changes in the neuronal network in response to adaptive and maladaptive responses to chronic stress are highly relevant to human health, and can be modeled by chronic restraint paradigms (Stankiewicz et al. 2013). In adult rats, a 7-day restraint stress results in a significant decrease of H3 trimethylation of the lysine residue 27 (H3K27me3), an epigenetic mark associated with gene silencing (Hunter et al. 2009). Interestingly, these epigenetic changes were subtler than those triggered by acute stressors, in agreement with the gradual decrease in transcription reported for chronic stress and associated with habituation mechanisms (Hunter et al. 2009).

An intriguing study tested for the first time in humans the idea that extreme maternal psychosocial stressors may modify locus-specific epigenetic marks in the newborn resulting in altered health outcomes. The cohort consisted in mothers and their newborns from the Democratic Republic of Congo, who suffered a variety of psychosocial stressors ranging from socio-economic status to extreme war situations. The authors reported a significant correlation between culturally relevant measures of maternal PS, newborn birth weight, and newborn methylation at the GR promoter (Mulligan et al. 2012). These striking results provided strong evidence of the effect of early-life stress on epigenetic modulation of programming in humans, but importantly, opened the possibility of measuring concrete methylation marks as a potential biomarker of stress-related modulation of programming.

Childhood maltreatment is yet another source of extreme stress that likely influences fundamental biological processes engraving long-lasting epigenetic marks. A study on individuals who had experienced at least two types of trauma other than childhood abuse, half of which showed symptoms of post-traumatic stress disorder (PTSD) while the rest did not manifest PTSD in adulthood, reported a significant larger proportion of DNA methylation changes in the group with childhood abuse versus nonabused individuals (Mehta et al. 2013).

An impressive study suggested recently that parent–infant interactions might also result in significant epigenetic modifications persistent for life. Moreover, these abnormal DNA methylation patterns were associated with suicidal behavior in adulthood. Analysis of hippocampal samples of suicidal victims showed increased DNA methylation on the exon 1_F promoter of the GR (the human homologous for the exon 1_7 promoter in the rat), when compared with control individuals who died suddenly from other causes (McGowan et al. 2009). The most striking observation was that abnormal methylation was only detected when suicide was accompanied with a developmental history of child maltreatment.

Another measure of early-life social conditions—the parental socioeconomic status (SES)—has been also associated with chronic disease risk in the adult life (Galobardes et al. 2006). A recent study analyzed a US birth cohort of women to investigate whether indicators of early-life and adult SES were associated with differential methylation at repetitive genomic elements Sta2, Alu, and LINE-1, commonly used to profile global methylation. Low family income at birth was associated with increased Sat2 methylation in adulthood, whereas a single-parent family resulted in higher Alu methylation. Lower adult education associated with lower Sat2 methylation (Tehranifar et al. 2013). Although the impact of these epigenetic drifts on health remains to be determined, global DNA methylation can contribute to genomic instability, which is a common cause of human cancers and observed in white blood cells in association with disease (Terry et al. 2011).

Lastly, a recent investigation analyzing genome-wide methylation on blood samples from children raised since birth in institutional care in comparison to children raised by their parents, reported greater levels of DNA methylation in the institutionalized group, mostly associated with genes involved in immune response, cellular signaling, neuronal communication, and brain development and functioning (Naumova et al. 2012). Institutional care may represent one of the most intense deprivations in human children, and has been associated with development deficits in every domain examined, including physical growth, motor development, cognitive functions, and social behavior (O'Connor and Rutter 2000). Fortunately, human studies also suggest that there is a rapid catch-up in development after placement on enriching environments (Nelson et al. 2007). Linkage of plastic epigenetic mechanisms to the enduring effects of early stress also supports the successful intervention of adoptive placement and its positive outcome in children cognition and development from a molecular standpoint.

16.6 Transgenerational Epigenetic Inheritance

The risk of developing several human diseases is determined by a heritable component. Novel evidence suggests that epigenetic processes might also contribute with these heritable origins of disease with effects that extend for more than one generation. In evolutionary terms, the passage of environmental information to the immediate offspring contributes with a better fit progeny. If heritable memory persists for several generations each facing a constantly changing background, these marks might result in maladaptive phenotypes contributing to disease.

A fascinating new topic in the field is the study of the transgenerational inheritance of epigenetic traits, whereby an altered ancestral methylome is passed through several generations without being altered or erased (Reik and Walter 2001).

Embryonic exposure to endocrine disrupting chemicals, like vinclozolin, during the window of gonadal sex determination induces a transgenerational effect on male reproduction and sperm production (Anway et al. 2005). When prenatally exposed animals reach adulthood, they manifest a variety of diseases including breast tumors, prostate disease, and immune abnormalities, but, importantly, these abnormalities are still evident in subsequent generations (Anway and Skinner 2006). This transgenerational phenotype appears to be transmitted through the male germ line and paternal allele (Anway et al. 2005; Anway and Skinner 2006), as although females develop the disease (Nilsson et al. 2008), they do not transmit the abnormal traits. Importantly, this phenomenon is associated with alterations in the germ-line epigenome induced by environmental exposure (Anway et al. 2005).

The biological process involved in the transgenerational epigenetic programming occurs during a critical period of development. During gonadal sex determination, the primordial germ cells migrate to the genital ridge and are subjected to waves of demethylation and re-methylation (Morgan et al. 2005). This period of abrupt changes in global methylation represents a highly sensitive window where alterations in the methylome of the germ line might become permanently fixed and transmitted to future generations. All the somatic cells derived from this germ line will have a shift in their epigenomes and transcriptomes, that in some tissues will promote disease states (Skinner 2011). This mechanism of transmission of environmental exposure effects has been related to adult onset diseases including spermatogenic defects, prostate and kidney diseases, immune abnormalities, and female reproductive defects (Anway et al. 2005, 2006; Guerrero-Bosagna et al.

2010; Skinner et al. 2010; Nilsson et al. 2012). Moreover, a recent study reported that transgenerational inheritance events affect the epigenetic programming of the brain transcriptome and alter anxiety behavior (Skinner et al. 2008), as far as three generations after removal of exposure.

The significance of these discoveries on epidemiology and human health are vast, as the environment experienced by great-grandparents might induce alterations in the health of their adult descendants almost 100 years later.

16.7 DNA Methylation as a Potential Biomarker

The increasing number of genes that show epigenetic alterations in association with disease highlights the potential application of these modifications—particularly DNA methylation—in diagnosis, prognosis, and therapy monitoring assays.

DNA methylation is currently used as a successful biomarker in cancer (Kim et al. 2004b; Belinsky et al. 2005; He et al. 2010), and holds promise for neurodegenerative disorders as well. The advantages of epigenetic markers are many: (a) novel platforms allow easier detection and quantification of DNA methylation with high reproducibility rates; (b) DNA methylation occurs mainly at defined and rather small genomic regions, facilitating the design of targeted probes (Anglim et al. 2008), in comparison to genetic mutations that might occur along the whole gene body; (c) as methylation changes represent rapid response mechanisms that precede gene expression, detection of DNA methylation might potentially be an indicator of early pathological changes, and, finally (d) it can be assessed from blood, enabling sampling by noninvasive procedures that minimize patient risk and discomfort.

A crucial question on the feasibility of an epigenetic biomarker approach monitoring alterations from peripheral blood is whether blood might be a reasonable surrogate for brain tissue. Recent reports raised some concern about the use of whole blood DNA on methylation studies because the heterogeneity of the various white cell types, each having their own particular epigenome, has the potential to confound DNA methylation measurements and disease associations (Adalsteinsson et al. 2012; Reinius et al. 2012). As our knowledge on epigenetics grows deeper, improved analysis methods have emerged incorporating data on the proportional number of single white cell types to correct for confounders (Adalsteinsson et al. 2012) and to exclude nonvariable CpG sites to improve association signals (Meng et al. 2010), features that will aid in the analysis of complex biological systems. While the mechanisms that might underlie the covariation of methylation across different tissues are largely unknown, recent reports show that despite cell-specific variability, interindividual variation is reflected across brain and blood (Davies et al. 2012). Taken together, these studies strongly suggest that blood might represent an adequate proxy to interrogate alterations on the brain methylome associated with disease.

Identification of differentially methylated markers might aid to diagnose disorders for which only few genetic mutations have been unequivocally linked, such as AD. A recent comparison of genome-wide methylation on homozygotic twins discordant for AD manifestations unveiled a global loss of methylation associated with AD (Mastroeni et al. 2009). In addition, methylation patterns in whole blood DNA were recently associated with schizophrenia symptoms. In particular, the methylation levels at 11 CpG sites significantly correlated with reality distortion symptoms (Liu et al. 2013), a discovery that can represent a big stride towards a translational application of DNA methylation in psychiatry diagnosis.

We recently investigated genome-wide methylation on postmortem frontal cortex and matched blood samples from PD patients and age-matched healthy control subjects. Comparison between Parkinson's-associated and control methylomes led us to identify differentially methylated regions (DMRs). Overall, methylation profiles were similar among tissues and the distribution of DMRs regarding CpG context (islands, shores, shelves, or open sea) did not significantly differ between brain and blood. Majority of altered sites in Parkinson's samples appeared hypomethylated, and corresponded to 277 autosomal loci in brains and 380 loci in blood. Importantly, we detected covariation of 65 hypermethylated and 61 hypomethylated genes between brain and blood. Functional analysis of covariating genes showed that higher than 60% of the genes were previously reported as deregulated in PD or linked to PD by genome-wide association studies (GWAS; Masliah et al. 2013). These results suggested that blood might indeed serve as an appropriate source for biomarker discovery in neurodegenerative disorders.

The promising study by Mulligan and coworkers also contributes with the idea of epigenetic biomarkers. The authors showed distinctive changes in the methylation at precise CpG residues on the NR3C1 promoter associated with PS, that were reliable quantitated from cord blood samples (Mulligan et al. 2012), opening the tantalizing idea of detecting early-life epigenetic modulation of programming in newborns which might dictate early interventions.

16.8 Final Remarks

Epigenetic modifications provide another layer of molecular complexity to the precise mechanisms that contribute to shape the brain during development and early postnatal life. The most prominent feature of these processes is their ability to modulate and transduce signals received from the environment, whether it is in the form of physic/chemical input or psychosocial interactions. Thus, epigenetic modifications generate a source of variation that contributes to developmental plasticity. While this adaptive mechanism could be favorable in preparing the offspring for an unfriendly landscape, it can also represent a source of increased disease susceptibility.

The variety of input signals that directly alter DNA methylation marks or histone modifications on the genome might prompt the perception that most environmental scenarios are likely to induce detrimental alterations. The beauty of the knowledge of epigenetic plasticity relies in the possibility of modulating molecular changes by one of the most salient human conditions: social interaction and parental bonding.

Conflict of Interest The author declares no conflict of interest.

References

Adalsteinsson BT, Gudnason H, Aspelund T, Harris TB, Launer LJ, Eiriksdottir G, Smith AV, Gudnason V (2012) Heterogeneity in white blood cells has potential to confound DNA methylation measurements. PloS ONE 7:e46705

Akbarian S, Huang HS (2009) Epigenetic regulation in human brain-focus on histone lysine methylation. Biol Psychiatry 65:198–203

Allis CD, Jenuwein T, Reinberg D (2006) Overview and concepts. In: Allis CD, Jenuwein T, Reinberg D (eds) Epigenetics. Cold Spring Harbor Laboratory, New York, pp 23–61

Anglim PP, Alonzo TA, Laird-Offringa IA (2008) DNA methylation-based biomarkers for early detection of non-small cell lung cancer: an update. Mol Cancer 7:81

Anway MD, Skinner MK (2006) Epigenetic transgenerational actions of endocrine disruptors. Endocrinology 147:S43–S49

Anway MD, Cupp AS, Uzumcu M, Skinner MK (2005) Epigenetic transgenerational actions of endocrine disruptors and male fertility. Science 308:1466–1469

Anway MD, Leathers C, Skinner MK (2006) Endocrine disruptor vinclozolin induced epigenetic transgenerational adult-onset disease. Endocrinology 147:5515–5523

Bale TL, Vale WW (2004) CRF and CRF receptors: role in stress responsivity and other behaviors. Annu Rev Pharmacol Toxicol 44:525–557

Ballas N, Battaglioli E, Atouf F, Andres ME, Chenoweth J, Anderson ME, Burger C, Moniwa M, Davie JR, Bowers WJ, Federoff HJ, Rose DW, Rosenfeld MG, Brehm P, Mandel G (2001) Regulation of neuronal traits by a novel transcriptional complex. Neuron 31:353–365

Barker DJ, Clark PM (1997) Fetal undernutrition and disease in later life. Rev Reprod 2:105–112

Beck S, Olek A, Walter J (1999) From genomics to epigenomics: a loftier view of life. Nat Biotechnol 17:1144

Belinsky SA, Klinge DM, Dekker JD, Smith MW, Bocklage TJ, Gilliland FD, Crowell RE, Karp DD, Stidley CA, Picchi MA (2005) Gene promoter methylation in plasma and sputum increases with lung cancer risk. Clin Cancer Res 11:6505–6511

Berube NG (2011) ATRX in chromatin assembly and genome architecture during development and disease. Biochem Cell Biol 89:435–444

Bilang-Bleuel A, Ulbricht S, Chandramohan Y, De Carli S, Droste SK, Reul JM (2005) Psychological stress increases histone H3 phosphorylation in adult dentate gyrus granule neurons: involvement in a glucocorticoid receptor-dependent behavioural response. Euro J Neurosci 22:1691–1700

Binder E, Droste SK, Ohl F, Reul JM (2004) Regular voluntary exercise reduces anxiety-related behaviour and impulsiveness in mice. Behav Brain Res 155:197–206

Caldji C, Tannenbaum B, Sharma S, Francis D, Plotsky PM, Meaney MJ (1998) Maternal care during infancy regulates the development of neural systems mediating the expression of fearfulness in the rat. Proc Natl Acad Sci U S A 95:5335–5340

Caldji C, Francis D, Sharma S, Plotsky PM, Meaney MJ (2000) The effects of early rearing environment on the development of GABAA and central benzodiazepine receptor levels and novelty-induced fearfulness in the rat. Neuropsychopharmacology 22:219–229

Caldji C, Diorio J, Meaney MJ (2003) Variations in maternal care alter GABA(A) receptor subunit expression in brain regions associated with fear. Neuropsychopharmacology 28:1950–1959

Caldji C, Hellstrom IC, Zhang TY, Diorio J, Meaney MJ (2011) Environmental regulation of the neural epigenome. FEBS Lett 585:2049–2058

Cassidy SB, Schwartz S, Miller JL, Driscoll DJ (2012) Prader-Willi syndrome. Genet Med 14:10–26

Champagne FA, Meaney MJ (2006) Stress during gestation alters postpartum maternal care and the development of the offspring in a rodent model. Biol Psychiatry 59:1227–1235

Champagne FA, Weaver IC, Diorio J, Sharma S, Meaney MJ (2003) Natural variations in maternal care are associated with estrogen receptor alpha expression and estrogen sensitivity in the medial preoptic area. Endocrinology 144:4720–4724

Chandramohan Y, Droste SK, Reul JM (2007) Novelty stress induces phospho-acetylation of histone H3 in rat dentate gyrus granule neurons through coincident signalling via the N-methyl-D-aspartate receptor and the glucocorticoid receptor: relevance for c-fos induction. J Neurochem 101:815–828

Chong JA, Tapia-Ramirez J, Kim S, Toledo-Aral JJ, Zheng Y, Boutros MC, Altshuller YM, Frohman MA, Kraner SD, Mandel G (1995) REST: a mammalian silencer protein that restricts sodium channel gene expression to neurons. Cell 80:949–957

Chugani HT, Behen ME, Muzik O, Juhasz C, Nagy F, Chugani DC (2001) Local brain functional activity following early deprivation: a study of postinstitutionalized Romanian orphans. Neuroimage 14:1290–1301

Cohen FE, Prusiner SB (1998) Pathologic conformations of prion proteins. Annu Rev Biochem 67:793–819

Crick F (1970) Central dogma of molecular biology. Nature 227:561–563

Crosio C, Cermakian N, Allis CD, Sassone-Corsi P (2000) Light induces chromatin modification in cells of the mammalian circadian clock. Nat Neurosci 3:1241–1247

Crosio C, Heitz E, Allis CD, Borrelli E, Sassone-Corsi P (2003) Chromatin remodeling and neuronal response: multiple signaling pathways induce specific histone H3 modifications and early gene expression in hippocampal neurons. J Cell Sci 116:4905–4914

Dauer W, Przedborski S (2003) Parkinson's disease: mechanisms and models. Neuron 39:889–909

Davies MN, Volta M, Pidsley R, Lunnon K, Dixit A, Lovestone S, Coarfa C, Harris RA, Milosavljevic A, Troakes C, Al-Sarraj S, Dobson R, Schalkwyk LC, Mill J (2012) Functional annotation of the human brain methylome identifies tissue-specific epigenetic variation across brain and blood. Genome Biol 13:R43

Day JJ, Childs D, Guzman-Karlsson MC, Kibe M, Moulden J, Song E, Tahir A, Sweatt JD (2013) DNA methylation regulates associative reward learning. Nat Neurosci 16:1445–1452

de Kloet ER, Karst H, Joels M (2008) Corticosteroid hormones in the central stress response: quick-and-slow. Front Neuroendocrinol 29:268–272

Denenberg VH (1964) Critical periods, stimulus input, and emotional reactivity: a theory of infantile stimulation. Psychol Rev 71:335–351

Desplats P, Spencer B, Coffee E, Patel P, Michael S, Patrick C, Adame A, Rockenstein E, Masliah E (2011) Alpha-synuclein sequesters Dnmt1 from the nucleus: a novel mechanism for epigenetic alterations in Lewy body diseases. J Biol Chem 286:9031–9037

Desplats P, Spencer B, Crews L, Patel P, Morvinski-Friedmann D, Kosberg K, Roberts S, Patrick C, Winner B, Winkler J, Masliah E (2012) Alpha-synuclein induces alterations in adult neurogenesis in Parkinson's disease models via p53-mediated repression of Notch1. J Biol Chem 287:31691–31702

Dolinoy DC, Huang D, Jirtle RL (2007a) Maternal nutrient supplementation counteracts bisphenol A-induced DNA hypomethylation in early development. Proc Natl Acad Sci U S A 104:13056–13061

Dolinoy DC, Weidman JR, Jirtle RL (2007b) Epigenetic gene regulation: linking early developmental environment to adult disease. Reprod Toxicol 23:297–307

Entringer S, Buss C, Kumsta R, Hellhammer DH, Wadhwa PD, Wust S (2009) Prenatal psychosocial stress exposure is associated with subsequent working memory performance in young women. Behav Neurosci 123:886–893

Fenoglio KA, Brunson KL, Avishai-Eliner S, Stone BA, Kapadia BJ, Baram TZ (2005) Enduring, handling-evoked enhancement of hippocampal memory function and glucocorticoid receptor expression involves activation of the corticotropin-releasing factor type 1 receptor. Endocrinology 146:4090–4096

Field T, Diego M, Hernandez-Reif M, Vera Y, Gil K, Schanberg S, Kuhn C, Gonzalez-Garcia A (2004) Prenatal maternal biochemistry predicts neonatal biochemistry. Int J Neurosci 114:933–945

Foley DL, Craig JM, Morley R, Olsson CA, Dwyer T, Smith K, Saffery R (2009) Prospects for epigenetic epidemiology. Am J Epidemiol 169:389–400

Franchini DM, Schmitz KM, Petersen-Mahrt SK (2012) 5-Methylcytosine DNA demethylation: more than losing a methyl group. Annu Rev Genet 46:419–441

Francis D, Diorio J, Liu D, Meaney MJ (1999) Nongenomic transmission across generations of maternal behavior and stress responses in the rat. Science 286:1155–1158

Fride E, Dan Y, Gavish M, Weinstock M (1985) Prenatal stress impairs maternal behavior in a conflict situation and reduces hippocampal benzodiazepine receptors. Life Sci 36:2103–2109

Galobardes B, Davey Smith G, Jeffreys M, McCarron P (2006) Childhood socioeconomic circumstances predict specific causes of death in adulthood: the Glasgow student cohort study. J Epidemiol Community Health 60:527–529

Gavin DP, Akbarian S (2012) Epigenetic and post-transcriptional dysregulation of gene expression in schizophrenia and related disease. Neurobiol Dis 46:255–262

Grayson DR, Guidotti A (2013) The dynamics of DNA methylation in schizophrenia and related psychiatric disorders. Neuropsychopharmacology 38:138–166

Gregory SG, Connelly JJ, Towers AJ, Johnson J, Biscocho D, Markunas CA, Lintas C, Abramson RK, Wright HH, Ellis P, Langford CF, Worley G, Delong GR, Murphy SK, Cuccaro ML, Persico A, Pericak-Vance MA (2009) Genomic and epigenetic evidence for oxytocin receptor deficiency in autism. BMC Med 7:62

Grillo E, Villard L, Clarke A, Ben Zeev B, Pineda M, Bahi-Buisson N, Hryniewiecka-Jaworska A, Bienvenu T, Armstrong J, Roche-Martinez A, Mari F, Veneselli E, Russo S, Vignoli A, Pini G, Djuric M, Bisgaard AM, Mejaski Bosnjak V, Polgar N, Cogliati F, Ravn K, Pintaudi M, Melegh B, Craiu D, Djukic A, Renieri A (2012) Rett networked database: an integrated clinical and genetic network of Rett syndrome databases. Hum Mutat 33:1031–1036

Grozinger CM, Schreiber SL (2002) Deacetylase enzymes: biological functions and the use of small-molecule inhibitors. Chem Biol 9:3–16

Guan JS, Haggarty SJ, Giacometti E, Dannenberg JH, Joseph N, Gao J, Nieland TJ, Zhou Y, Wang X, Mazitschek R, Bradner JE, DePinho RA, Jaenisch R, Tsai LH (2009) HDAC2 negatively regulates memory formation and synaptic plasticity. Nature 459:55–60

Guerrero-Bosagna C, Settles M, Lucker B, Skinner MK (2010) Epigenetic transgenerational actions of vinclozolin on promoter regions of the sperm epigenome. PLoS ONE 5

He Q, Chen HY, Bai EQ, Luo YX, Fu RJ, He YS, Jiang J, Wang HQ (2010) Development of a multiplex MethyLight assay for the detection of multigene methylation in human colorectal cancer. Cancer Genet Cytogenet 202:1–10

He YF, Li BZ, Li Z, Liu P, Wang Y, Tang Q, Ding J, Jia Y, Chen Z, Li L, Sun Y, Li X, Dai Q, Song CX, Zhang K, He C, Xu GL (2011) Tet-mediated formation of 5-carboxylcytosine and its excision by TDG in mammalian DNA. Science 333:1303–1307

Hinwood M, Tynan RJ, Day TA, Walker FR (2011) Repeated social defeat selectively increases deltaFosB expression and histone H3 acetylation in the infralimbic medial prefrontal cortex. Cereb Cortex 21:262–271

Hirsch EC, Hunot S, Faucheux B, Agid Y, Mizuno Y, Mochizuki H, Tatton WG, Tatton N, Olanow WC (1999) Dopaminergic neurons degenerate by apoptosis in Parkinson's disease. Mov Disord 14:383–385

Huang Y, Myers SJ, Dingledine R (1999) Transcriptional repression by REST: recruitment of Sin3A and histone deacetylase to neuronal genes. Nat Neurosci 2:867–872

Huizink AC, Mulder EJ, Buitelaar JK (2004) Prenatal stress and risk for psychopathology: specific effects or induction of general susceptibility? Psychol Bull 130:115–142

Hunter RG, McCarthy KJ, Milne TA, Pfaff DW, McEwen BS (2009) Regulation of hippocampal H3 histone methylation by acute and chronic stress. Proc Natl Acad Sci U S A 106:20912–20917

Ito S, D'Alessio AC, Taranova OV, Hong K, Sowers LC, Zhang Y (2010) Role of Tet proteins in 5mC to 5hmC conversion, ES-cell self-renewal and inner cell mass specification. Nature 466:1129–1133

Ito S, Shen L, Dai Q, Wu SC, Collins LB, Swenberg JA, He C, Zhang Y (2011) Tet proteins can convert 5-methylcytosine to 5-formylcytosine and 5-carboxylcytosine. Science 333:1300–1303
Jana NR (2012) Understanding the pathogenesis of Angelman syndrome through animal models. Neural Plast 2012:710943
Jenuwein T, Allis CD (2001) Translating the histone code. Science 293:1074–1080
Jiang Y, Langley B, Lubin FD, Renthal W, Wood MA, Yasui DH, Kumar A, Nestler EJ, Akbarian S, Beckel-Mitchener AC (2008) Epigenetics in the nervous system. J Neurosci 28:11753–11759
Kaplan GA, Turrell G, Lynch JW, Everson SA, Helkala EL, Salonen JT (2001) Childhood socioeconomic position and cognitive function in adulthood. Int J Epidemiol 30:256–263
Kim FJ, Battini JL, Manel N, Sitbon M (2004a) Emergence of vertebrate retroviruses and envelope capture. Virology 318:183–191
Kim H, Kwon YM, Kim JS, Lee H, Park JH, Shim YM, Han J, Park J, Kim DH (2004b) Tumor-specific methylation in bronchial lavage for the early detection of non-small-cell lung cancer. J Clin Oncol 22:2363–2370
Kim HS, Kim EM, Kim NJ, Chang KA, Choi Y, Ahn KW, Lee JH, Kim S, Park CH, Suh YH (2004c) Inhibition of histone deacetylation enhances the neurotoxicity induced by the C-terminal fragments of amyloid precursor protein. J Neurosci Res 75:117–124
Kornberg RD (1974) Chromatin structure: a repeating unit of histones and DNA. Science 184:868–871
Korosi A, Shanabrough M, McClelland S, Liu ZW, Borok E, Gao XB, Horvath TL, Baram TZ (2010) Early-life experience reduces excitation to stress-responsive hypothalamic neurons and modulate the programming the expression of corticotropin-releasing hormone. J Neurosci 30:703–713
Kriaucionis S, Heintz N (2009) The nuclear DNA base 5-hydroxymethylcytosine is present in Purkinje neurons and the brain. Science 324:929–930
Kuo MH, Allis CD (1998) Roles of histone acetyltransferases and deacetylases in gene regulation. Bioessays 20:615–626
Lattal KM, Wood MA (2013) Epigenetics and persistent memory: implications for reconsolidation and silent extinction beyond the zero. Nat Neurosci 16:124–129
Lee DY, Hayes JJ, Pruss D, Wolffe AP (1993) A positive role for histone acetylation in transcription factor access to nucleosomal DNA. Cell 72:73–84
Lehnen H, Zechner U, Haaf T (2013) Epigenetics of gestational diabetes mellitus and offspring health: the time for action is in early stages of life. Mol Hum Reprod 19:415–422
Levenson JM, Sweatt JD (2005) Epigenetic mechanisms in memory formation. Nat Rev Neurosci 6:108–118
Levenson JM, O'Riordan KJ, Brown KD, Trinh MA, Molfese DL, Sweatt JD (2004) Regulation of histone acetylation during memory formation in the hippocampus. J Biol Chem 279:40545–40559
Levine S (1970) The pituitary-adrenal system and the developing brain. Prog Brain Res 32:79–85
Levine S (1994) The ontogeny of the hypothalamic-pituitary-adrenal axis. The influence of maternal factors. Ann N Y Acad Sci 746:275–288 (discussion 289–293)
Levy OA, Malagelada C, Greene LA (2009) Cell death pathways in Parkinson's disease: proximal triggers, distal effectors, and final steps. Apoptosis 14:478–500
Li L, Suzuki T, Mori N, Greengard P (1993) Identification of a functional silencer element involved in neuron-specific expression of the synapsin I gene. Proc Natl Acad Sci U S A 90:1460–1464
Liu D, Diorio J, Tannenbaum B, Caldji C, Francis D, Freedman A, Sharma S, Pearson D, Plotsky PM, Meaney MJ (1997) Maternal care, hippocampal glucocorticoid receptors, and hypothalamic-pituitary-adrenal responses to stress. Science 277:1659–1662
Liu J, Chen J, Ehrlich S, Walton E, White T, Perrone-Bizzozero N, Bustillo J, Turner JA, Calhoun VD (2013) Methylation patterns in whole blood correlate with symptoms in schizophrenia patients. Schizophr Bull 40:769–776
Lubin FD, Roth TL, Sweatt JD (2008) Epigenetic regulation of BDNF gene transcription in the consolidation of fear memory. J Neurosci 28:10576–10586
Magistri M, Faghihi MA, St Laurent G, 3rd, Wahlestedt C (2012) Regulation of chromatin structure by long noncoding RNAs: focus on natural antisense transcripts. Trends Genet 28:389–396

Malvaez M, McQuown SC, Rogge GA, Astarabadi M, Jacques V, Carreiro S, Rusche JR, Wood MA (2013) HDAC3-selective inhibitor enhances extinction of cocaine-seeking behavior in a persistent manner. Proc Natl Acad Sci U S A 110:2647–2652

Manning-Courtney P, Murray D, Currans K, Johnson H, Bing N, Kroeger-Geoppinger K, Sorensen R, Bass J, Reinhold J, Johnson A, Messerschmidt T (2013) Autism spectrum disorders. Curr Probl Pediatr Adolesc Health Care 43:2–11

Martinotti G, Di Iorio G, Marini S, Ricci V, De Berardis D, Di Giannantonio M (2012) Nerve growth factor and brain-derived neurotrophic factor concentrations in schizophrenia: a review. J Biol Regul Homeost Agents 26:347–356

Masliah E, Mallory M, Hansen L, DeTeresa R, Alford M, Terry R (1994) Synaptic and neuritic alterations during the progression of Alzheimer's disease. Neurosci Lett 174:67–72

Masliah E, Dumaop W, Galasko D, Desplats P (2013) Distinctive patterns of DNA methylation associated with Parkinson disease: Identification of concordant epigenetic changes in brain and peripheral blood leukocytes. Epigenetics 8:1030–1038

Mastroeni D, McKee A, Grover A, Rogers J, Coleman PD (2009) Epigenetic differences in cortical neurons from a pair of monozygotic twins discordant for Alzheimer's disease. PloS ONE 4:e6617

Mastroeni D, Grover A, Delvaux E, Whiteside C, Coleman PD, Rogers J (2010) Epigenetic changes in Alzheimer's disease: decrements in DNA methylation. Neurobiol Aging 31:2025–2037

McClelland SKA, Cope J, Ivy A, Baram T (2011) Emerging roles of epigenetic mechanisms in the enduring effects of neonatal stress and experience on lerning and memory. Neurobiol Learn Mem 96:79–88

McEwen BS (2007) Physiology and neurobiology of stress and adaptation: central role of the brain. Physiol Rev 87:873–904

McGowan PO, Sasaki A, D'Alessio AC, Dymov S, Labonte B, Szyf M, Turecki G, Meaney MJ (2009) Epigenetic regulation of the glucocorticoid receptor in human brain associates with childhood abuse. Nat Neurosci 12:342–348

McManus MT, Sharp PA (2002) Gene silencing in mammals by small interfering RNAs. Nat Rev Genet 3:737–747

Meaney MJ (2001) Maternal care, gene expression, and the transmission of individual differences in stress reactivity across generations. Annu Rev Neurosci 24:1161–1192

Meaney MJ (2010) Epigenetics and the biological definition of gene x environment interactions. Child Dev 81:41–79

Meaney MJ, Mitchell JB, Aitken DH, Bhatnagar S, Bodnoff SR, Iny LJ, Sarrieau A (1991) The effects of neonatal handling on the development of the adrenocortical response to stress: implications for neuropathology and cognitive deficits in later life. Psychoneuroendocrinology 16:85–103

Meaney MJ, Aitken DH, van Berkel C, Bhatnagar S, Sapolsky RM (1988) Effect of neonatal handling on age-related impairments associated with the hippocampus. Science 239:766–768

Meany MDJ, Donaldson L, Yau J, Chapman K, Seckl J (2000) Handling alters the expression of messenger RNAs for AP-2, NGFI-A and NGFI-B in the hippocampus of neonatal rats. J Neuroscience 20:3936–3945

Medstrand P, van de Lagemaat LN, Dunn CA, Landry JR, Svenback D, Mager DL (2005) Impact of transposable elements on the evolution of mammalian gene regulation. Cytogenet Genome Res 110:342–352

Mehta D, Klengel T, Conneely KN, Smith AK, Altmann A, Pace TW, Rex-Haffner M, Loeschner A, Gonik M, Mercer KB, Bradley B, Muller-Myhsok B, Ressler KJ, Binder EB (2013) Childhood maltreatment is associated with distinct genomic and epigenetic profiles in posttraumatic stress disorder. Proc Natl Acad Sci U S A 110:8302–8307

Meng H, Joyce AR, Adkins DE, Basu P, Jia Y, Li G, Sengupta TK, Zedler BK, Murrelle EL, van den Oord EJ (2010) A statistical method for excluding non-variable CpG sites in high-throughput DNA methylation profiling. BMC Bioinformatics 11:227

Millan MJ (2013) An epigenetic framework for neurodevelopmental disorders: from pathogenesis to potential therapy. Neuropharmacology 68:2–82

Moore CL, Power KL (1986) Prenatal stress affects mother-infant interaction in Norway rats. Dev Psychobiol 19:235–245
Morgan HD, Santos F, Green K, Dean W, Reik W (2005) Epigenetic modulation of programming in mammals. Hum Mol Genet 14 (Spec No 1):R47–R58
Mulligan CJ, D'Errico NC, Stees J, Hughes DA (2012) Methylation changes at NR3C1 in newborns associate with maternal prenatal stress exposure and newborn birth weight. Epigenetics 7:853–857
Nagahara AH, Tuszynski MH (2011) Potential therapeutic uses of BDNF in neurological and psychiatric disorders. Nat Rev Drug Discov 10:209–219
Nagarajan RP, Hogart AR, Gwye Y, Martin MR, LaSalle JM (2006) Reduced MeCP2 expression is frequent in autism frontal cortex and correlates with aberrant MECP2 promoter methylation. Epigenetics 1:e1–e11
Nan X, Campoy FJ, Bird A (1997) MeCP2 is a transcriptional repressor with abundant binding sites in genomic chromatin. Cell 88:471–481
Naruse Y, Aoki T, Kojima T, Mori N (1999) Neural restrictive silencer factor recruits mSin3 and histone deacetylase complex to repress neuron-specific target genes. Proc Natl Acad Sci U S A 96:13691–13696
Naruse Y, Oh-hashi K, Iijima N, Naruse M, Yoshioka H, Tanaka M (2004) Circadian and light-induced transcription of clock gene Per1 depends on histone acetylation and deacetylation. Mol Cell Biol 24:6278–6287
Naumova OY, Lee M, Koposov R, Szyf M, Dozier M, Grigorenko EL (2012) Differential patterns of whole-genome DNA methylation in institutionalized children and children raised by their biological parents. Dev Psychopathol 24:143–155
Nelson CA 3rd, Zeanah CH, Fox NA, Marshall PJ, Smyke AT, Guthrie D (2007) Cognitive recovery in socially deprived young children: the Bucharest Early Intervention Project. Science 318:1937–1940
Nilsson EE, Anway MD, Stanfield J, Skinner MK (2008) Transgenerational epigenetic effects of the endocrine disruptor vinclozolin on pregnancies and female adult onset disease. Reproduction 135:713–721
Nilsson E, Larsen G, Manikkam M, Guerrero-Bosagna C, Savenkova MI, Skinner MK (2012) Environmentally induced epigenetic transgenerational inheritance of ovarian disease. PLoS ONE 7:e36129
O'Connor TG, Rutter M (2000) Attachment disorder behavior following early severe deprivation: extension and longitudinal follow-up. English and Romanian adoptees study team. J Am Acad Child Adolesc Psychiatry 39:703–712
Okano M, Xie S, Li E (1998) Cloning and characterization of a family of novel mammalian DNA (cytosine-5) methyltransferases. Nat Genet 19:219–220
Perkins A, Lehmann C, Lawrence RC, Kelly SJ (2013) Alcohol exposure during development: Impact on the epigenome. Int J Dev Neurosci 31:391–397
Perry M, Chalkley R (1982) Histone acetylation increases the solubility of chromatin and occurs sequentially over most of the chromatin. A novel model for the biological role of histone acetylation. J Biol Chem 257:7336–7347
Plotsky PM, Meaney MJ (1993) Early, postnatal experience alters hypothalamic corticotropin-releasing factor (CRF) mRNA, median eminence CRF content and stress-induced release in adult rats. Brain Res Mol Brain Res 18:195–200
Plotsky PM, Cunningham ET Jr, Widmaier EP (1989) Catecholaminergic modulation of corticotropin-releasing factor and adrenocorticotropin secretion. Endocr Rev 10:437–458
Ramchandani S, Bhattacharya SK, Cervoni N, Szyf M (1999) DNA methylation is a reversible biological signal. Proc Natl Acad Sci U S A 96:6107–6112
Razin A (1998) CpG methylation, chromatin structure and gene silencing-a three-way connection. Embo J 17:4905–4908
Razin A, Riggs AD (1980) DNA methylation and gene function. Science 210:604–610
Reik W, Walter J (2001) Genomic imprinting: parental influence on the genome. Nat Rev Genet 2:21–32

Reinius LE, Acevedo N, Joerink M, Pershagen G, Dahlen SE, Greco D, Soderhall C, Scheynius A, Kere J (2012) Differential DNA methylation in purified human blood cells: implications for cell lineage and studies on disease susceptibility. PloS ONE 7:e41361

Riggs ADM, Martienssen RA, Russo VEA (1996) Introduction in epigenetic mechanisms of gene regulation. Cold Spring Harbor Laboratory, Cold Spring Harbor, pp 1–4 (ed, V.E.A. Russo et al.)

Roth SY, Denu JM, Allis CD (2001) Histone acetyltransferases. Annu Rev Biochem 70:81–120

Roth TL, Lubin FD, Funk AJ, Sweatt JD (2009) Lasting epigenetic influence of early-life adversity on the BDNF gene. Biol Psychiatry 65:760–769

Shorter J, Lindquist S (2005) Prions as adaptive conduits of memory and inheritance. Nat Rev Genet 6:435–450

Shulha HP, Cheung I, Whittle C, Wang J, Virgil D, Lin CL, Guo Y, Lessard A, Akbarian S, Weng Z (2012) Epigenetic signatures of autism: trimethylated H3K4 landscapes in prefrontal neurons. Arch Gen Psychiatry 69:314–324

Siegmund KD, Connor CM, Campan M, Long TI, Weisenberger DJ, Biniszkiewicz D, Jaenisch R, Laird PW, Akbarian S (2007) DNA methylation in the human cerebral cortex is dynamically regulated throughout the life span and involves differentiated neurons. PLoS ONE 2:e895

Skinner MK (2011) Environmental epigenetic transgenerational inheritance and somatic epigenetic mitotic stability. Epigenetics 6:838–842

Skinner MK, Anway MD, Savenkova MI, Gore AC, Crews D (2008) Transgenerational epigenetic programming of the brain transcriptome and anxiety behavior. PloS ONE 3:e3745

Skinner MK, Manikkam M, Guerrero-Bosagna C (2010) Epigenetic transgenerational actions of environmental factors in disease etiology. Trends Endocrinol Metab 21:214–222

Stankiewicz AM, Swiergiel AH, Lisowski P (2013) Epigenetics of stress adaptations in the brain. Brain Res Bull 98C:76–92

Szwagierczak A, Bultmann S, Schmidt CS, Spada F, Leonhardt H (2010) Sensitive enzymatic quantification of 5-hydroxymethylcytosine in genomic DNA. Nucleic Acids Res 38:e181

Szyf M, Weaver IC, Champagne FA, Diorio J, Meaney MJ (2005) Maternal programming of steroid receptor expression and phenotype through DNA methylation in the rat. Front Neuroendocrinol 26:139–162

Tahiliani M, Koh KP, Shen Y, Pastor WA, Bandukwala H, Brudno Y, Agarwal S, Iyer LM, Liu DR, Aravind L, Rao A (2009) Conversion of 5-methylcytosine to 5-hydroxymethylcytosine in mammalian DNA by MLL partner TET1. Science 324:930–935

Tehranifar P, Wu HC, Fan X, Flom JD, Ferris JS, Cho YH, Gonzalez K, Santella RM, Terry MB (2013) Early life socioeconomic factors and genomic DNA methylation in mid-life. Epigenetics 8:23–27

Terry MB, Delgado-Cruzata L, Vin-Raviv N, Wu HC, Santella RM (2011) DNA methylation in white blood cells: association with risk factors in epidemiologic studies. Epigenetics 6:828–837

Torregrossa MM, Taylor JR (2013) Learning to forget: manipulating extinction and reconsolidation processes to treat addiction. Psychopharmacology (Berl) 226:659–672

Tsankova N, Renthal W, Kumar A, Nestler EJ (2007) Epigenetic regulation in psychiatric disorders. Nat Rev Neurosci 8:355–367

Tsukada Y, Fang J, Erdjument-Bromage H, Warren ME, Borchers CH, Tempst P, Zhang Y (2006) Histone demethylation by a family of JmjC domain-containing proteins. Nature 439:811–816

Uno H, Tarara R, Else JG, Suleman MA, Sapolsky RM (1989) Hippocampal damage associated with prolonged and fatal stress in primates. J Neurosci 9:1705–1711

Urdinguio RG, Sanchez-Mut JV, Esteller M (2009) Epigenetic mechanisms in neurological diseases: genes, syndromes, and therapies. Lancet Neurol 8:1056–1072

Valinluck V, Sowers LC (2007) Endogenous cytosine damage products alter the site selectivity of human DNA maintenance methyltransferase DNMT1. Cancer Res 67:946–950

Vallee M, MacCari S, Dellu F, Simon H, Le Moal M, Mayo W (1999) Long-term effects of prenatal stress and postnatal handling on age-related glucocorticoid secretion and cognitive performance: a longitudinal study in the rat. Eur J Neurosci 11:2906–2916

Van den Bergh BR, Mulder EJ, Mennes M, Glover V (2005) Antenatal maternal anxiety and stress and the neurobehavioural development of the fetus and child: links and possible mechanisms. A review. Neurosci Biobehav Rev 29:237–258

Van den Hove DL, Steinbusch HW, Scheepens A, Van deBWD, Kooiman LA, Boosten BJ, Prickaerts J, Blanco CE (2006) Prenatal stress and neonatal rat brain development. Neuroscience 137:145–155

Vaquero A, Loyola A, Reinberg D (2003) The constantly changing face of chromatin. Sci Aging Knowledge Environ 2003:RE4

Waddington CH (1942) The epigenotype. Endeavour 1:18–20

Wadhwa PD, Sandman CA, Porto M, Dunkel-Schetter C, Garite TJ (1993) The association between prenatal stress and infant birth weight and gestational age at birth: a prospective investigation. Am J Obstet Gynecol 169:858–865

Weaver IC, Cervoni N, Champagne FA, D'Alessio AC, Sharma S, Seckl JR, Dymov S, Szyf M, Meaney MJ (2004) Epigenetic programming by maternal behavior. Nat Neurosci 7:847–854

Weaver IC, Champagne FA, Brown SE, Dymov S, Sharma S, Meaney MJ, Szyf M (2005) Reversal of maternal programming of stress responses in adult offspring through methyl supplementation: altering epigenetic marking later in life. J Neurosci 25:11045–11054

Weaver IC, DiAlessio AC, Brown SE, Hellstrom IC, Dymov S, Diorio J et al (2007) The trasncription factor NGFI-A mediates epigenetic programming: Altering epigenetic marking through immediate early genes. J Neurosci 27:1756–1768

Weber M, Hellmann I, Stadler MB, Ramos L, Paabo S, Rebhan M, Schubeler D (2007) Distribution, silencing potential and evolutionary impact of promoter DNA methylation in the human genome. Nat Genet 39:457–466

Weinstock M (2001) Alterations induced by gestational stress in brain morphology and behaviour of the offspring. Prog Neurobiol 65:427–451

Wilkinson MB, Xiao G, Kumar A, LaPlant Q, Renthal W, Sikder D, Kodadek TJ, Nestler EJ (2009) Imipramine treatment and resiliency exhibit similar chromatin regulation in the mouse nucleus accumbens in depression models. J Neurosci 29:7820–7832

Wilson RS, Schneider JA, Boyle PA, Arnold SE, Tang Y, Bennett DA (2007) Chronic distress and incidence of mild cognitive impairment. Neurology 68:2085–2092

Yamada K, Mizuno M, Nabeshima T (2002) Role for brain-derived neurotrophic factor in learning and memory. Life Sci 70:735–744

Zuena AR, Mairesse J, Casolini P, Cinque C, Alema GS, Morley-Fletcher S, Chiodi V, Spagnoli LG, Gradini R, Catalani A, Nicoletti F, Maccari S (2008) Prenatal restraint stress generates two distinct behavioral and neurochemical profiles in male and female rats. PLoS ONE 3:e2170

Chapter 17
Epigenetic Mechanisms of Perinatal Programming: Translational Approaches from Rodent to Human and Back

Patrick O. McGowan

Abstract Perinatal life is a period of enhanced plasticity and susceptibility to environmental effects via the maternal environment or parental care. A variety of studies have indicated that epigenetic mechanisms, which can alter gene function without a change in gene sequence, play a role in setting developmental trajectories that impact health, including mental health. This chapter reviews examples of translational approaches to the study of biological embedding of mental health via differences in parental care.

17.1 Introduction

A prominent feature of parent care effects on mental health is its influence on the hypothalamic–pituitary axis, a major endocrine regulator of the response to psychosocial stress. Laboratory rodent models have been particularly useful in identifying mechanisms of epigenetic regulation in the brain that have then been used to generate hypotheses in humans. At the same time, recent advances in genomics have provided new means to address these questions in large numbers of human subjects in an increasingly comprehensive and powerful manner.

17.1.1 How Mechanisms of Gene Regulation "Above the Genome" Contribute to Interindividual Differences in Behavior

The advent of new high-throughput DNA sequencing technologies, initiated near the turn of the last century, has allowed the elucidation of the genetic sequence identities of humans and a growing list of other species. Where human health is concerned, these technologies were thought to herald a complete understanding of

P. O. McGowan (✉)
Department of Biological Sciences, Centre for Environmental Epigenetics and Development, University of Toronto, Scarborough, Toronto, ON, Canada
e-mail: pmcgowan@utsc.utoronto.ca

M. C. Antonelli (ed.), *Perinatal Programming of Neurodevelopment*, Advances in Neurobiology 10, DOI 10.1007/978-1-4939-1372-5_17,

biological diversity, with strong implications for understanding the ultimate origins of behavior and psychopathology. However, this exuberance turned out to be misplaced. Our current understanding of genome function indicates that the DNA code must be exquisitely programmed by molecular mechanisms "above the genome," termed epigenetic. At a fundamental level, these mechanisms are now known to be crucial for conferring cell-type specificity during development by programming unique patterns of gene expression among the body's 200+ cell types. It is also now known that cell-type-specific patterns of epigenetic modifications are not exclusively genetically determined, and are to some degree responsive to environmental signaling throughout life. As we discuss below with specific examples, there is accumulating evidence indicating that phenotypic variation observed in humans and animal models can result from changes in gene function via epigenetic changes. The dynamic nature of epigenetic signaling contrasts with the static nature of the genetic code, and provides a mechanism of gene × environment interactions that bridge inherited variation with variation originating from environmental and stochastic sources. It should be noted, however, that these sources of variation are not necessarily mutually exclusive, as genetic variation can interact with epigenetic variation. However, in contrast to the static nature of genetic variation, epigenetic variation is potentially amenable to environmental or therapeutic intervention. Thus, we and others have argued that the study origins of health and human disease—including psychiatric disorders—is incomplete without an understanding of genetic, environmental and stochastic contributions to epigenetic signaling (Petronis 2010; Sasaki et al. 2013).

17.1.2 What Is "Epigenetics"?

We have alluded to epigenetic mechanisms as those that change gene function in the absence of a change in DNA sequence. It should be noted that the definition of the term "epigenetics" continues to be a matter of debate. The historical definition emphasized the heritability of changes in gene function without a change in gene sequence. This definition implied that such changes were trans-generational, passing from the maternal environment to the offspring via gametic inheritance. However, it is clear that such a definition leaves out important changes in gene function that do not involve changes in gene sequence, yet are relatively stable and passed from cell to daughter cell during mitosis. It also leaves out mechanisms that program changes in gene function in postmitotic cells such as neurons, heart cells, and other cells that do not replicate and would thus be left out of the epigenetic sphere. It may be equally unwise, however, to characterize all changes of gene function in the absence of a change in gene sequence as "epigenetic." The firing of neurons is a process that initiates a change in the expression of hundreds of genes, yet many of these are distinct from processes that result in the long-term potentiation of neural responses. Clearly, extending the definition of epigenetics to encompass mechanisms altering gene expression outside the realm of the DNA sequence requires careful consideration of the *outcomes* associated with the epigenetic change. We and others have proposed a broad definition of epigenetics that includes long-term changes in gene function that are meiotically or mitotically heritable (Sasaki et al. 2013).

17.1.3 Gene Sequence and Its Relationship to Epigenetic Changes

The pattern of gene expression in each cell type confers its tissue-specific phenotype. As a result, a change in gene function as a result of a sequence alteration could alter the structure of the protein encoded by the gene or its activity by interfering with factors that increase or decrease its activity. Sequence variation has been associated with behavioral pathologies but with the exception of a minority of psychopathology linked to Mendelian disorders, such variation is inevitably linked to specific environmental conditions and thus is very rarely predicted solely on the basis of genetic variance. This has created something of a crisis in psychiatric research and a search by some for intermediary mechanisms that might help explain a missing heritability not directly attributable to genetic variance (Petronis 2010). In fact, it has been known for some time that epigenetic and genetic variation can predict the same biological outcomes.

We propose that understanding epigenetic mechanisms associated with psychopathology is important for at least three reasons: (1) Identifying alterations in epigenetic signaling associated with psychopathology may help explain intermediate mechanisms between the genome and stochastic and environmental factors involved in behavioral pathology; (2) defining these relationships could provide a biological understanding of gene × environment interactions; (3) because epigenetic mechanisms are potentially reversible, we may be able to identify novel therapeutic interventions to prevent or reverse these changes. In what follows, we provide a synopsis of our work addressing stable epigenetic changes associated with early-life environment in humans and rodent models. To do so, we first describe known molecular epigenetic processes that are best understood as mediators of these effects.

17.2 The Epigenome

17.2.1 Chromatin and the Histone Code

The epigenome consists of the chromatin and its modifications as well as a covalent modification by methylation of cytosine rings found at the dinucleotide sequence CG (Razin 1998). The epigenome determines the accessibility of the transcription machinery, which transcribes the genes into messenger RNA, to the DNA. Inaccessible genes are therefore silent whereas accessible genes are transcribed. We therefore distinguish between open and closed configuration of chromatin (Groudine et al. 1983; Marks et al. 1985; Ramain et al. 1986; Grunstein 1997; Varga-Weisz and Becker 2006). Densely packaged chromatin can be visualized microscopically and is termed heterochromatin while open accessible chromatin is termed euchromatin. Another level of epigenetic regulation by small noncoding RNAs termed microRNA (miRNA) has been discovered (Bergmann and Lane 2003). miRNAs regulate gene expression at different levels; silencing of chromatin, degradation of mRNA, and blocking translation. miRNAs have been linked to behavioral pathologies in humans and regulate gene function through a variety of mechanisms, as has been extensively reviewed elsewhere (Vo et al. 2005; Mehler and Mattick 2006, 2007; Qureshi and Mehler 2009). These mechanisms are currently the subject of intense investigation. miRNA expression in adulthood has been linked to early-life stress in rats (Uchida et al. 2010) and almost certainly plays a more important role than is currently understood.

The DNA is wrapped around a protein-based structure termed chromatin. Chromatin is formed by an octamer of histone proteins termed a nucleosome. Variants of histone proteins, H1, H2A, H3B, H3, and H4 (Finch et al. 1977) and other minor variants have specific functions in DNA repair and gene activity (Sarma and Reinberg 2005). The octamer structure of the nucleosome is composed of a H3–H4 tetramer flanked on either side with a H2A–H2B dimer (Finch et al. 1977). The N terminal tails of these histones are extensively modified by methylation (Jenuwein 2001), phosphorylation, acetylation (Wade et al. 1997), and ubiquitination (Shilatifard 2006). The state of modification of these tails plays an important role in defining the accessibility of the DNA. The amino terminal tails of H3 and H4 histones that are positively charged form tight interactions with the negatively charged DNA backbone, thus blocking the interaction of transcription factors with the DNA. Modifications of the tails neutralize the charge on the tails, thus relaxing the tight grip of the histone tails. Different histone variants, which replace the standard isoforms also play a regulatory role and serve to mark active genes in some instances (Henikoff et al. 2004). The specific pattern of histone modifications was proposed to form a "histone code" that delineates the parts of the genome to be expressed at a given point in time in a given cell type (Jenuwein and Allis 2001). A change in histone modifications around a gene will change its level of expression and could convert a gene from an active state to a silent state, resulting in "loss of function" or switch a gene from a silent state to an active state, leading to "gain of function."

17.2.2 Histone-Modifying Enzymes and Chromatin Remodeling

The most-investigated histone-modifying enzymes are histone acetyltransferases (HAT), which acetylate histone H3 at the K9 residue as well as other residues and H4 tails at a number of residues, and histone deacetylases (HDAC), that deacetylate histone tails (Kuo and Allis 1998). Histone acetylation is believed to be a predominant signal for an active chromatin configuration (Perry and Chalkley 1982; Lee et al. 1993). Deacetylated histones signal inactive chromatin, associated with inactive genes. Many repressors and repressor complexes recruit HDACs to genes, thus leading to their inactivation (Wolffe 1996). Histone tail acetylation is believed to enhance the accessibility of a gene to the transcription machinery whereas deacetylated tails are highly charged and believed to be tightly associated with the DNA backbone and thus limiting accessibility of genes to transcription factors (Kuo and Allis 1998).

Histone modification by methylation is catalyzed by different histone methyltransferases. Some specific methylation events are associated with gene silencing and some with gene activation (Lachner et al. 2001). Particular factors recognize histone modifications and further stabilize an inactive state. For example, the heterochromatin-associated protein HP-1 binds H3-histone tails methylated at the K9 residue and precipitates an inactive chromatin structure (Lachner et al. 2001). Recently described histone demethylases remove the methylation mark causing either activation or repression of gene expression (Shi et al. 2004; Tsukada et al. 2006).

Chromatin remodeling complexes, which are ATP dependent, alter the position of nucleosomes around the transcription initiation site and define its accessibility to the transcription machinery (Varga-Weisz and Becker 2006). It is becoming clear now that there is an interrelationship between chromatin modification and chromatin remodeling. For example, active regions of the chromatin are associated with hypomethylated DNA, and hypermethylated DNA is packaged in inactive chromatin (Razin 1998; Razin and Cedar 1977).

17.2.3 Targeting of Chromatin-Modifying Enzymes to Specific Genes

To date, there are relatively few examples from neuroscience research of gene targeting of epigenetic mechanisms, though we review one such example below for maternal care. In our view, this is a fundamental principle of epigenetic regulation of gene expression that will shed important light on neuronal gene regulation. There are, however, many examples of targeting from other areas of research, a few of which we describe below as illustrative examples. Transcription factors and repressors recruit the nonspecific histone-modifying enzymes to specific genomic loci and target-specific genes (Jenuwein and Allis 2001). Transcription factors and repressors recognize specific *cis*-acing sequences in genes, bind to these sequences, and attract specific chromatin-modifying enzymes to genes through protein–protein interactions. Specific transacting factors are responsive to cellular signaling pathways. Signal transduction pathways are activated by cell surface receptors and could thus serve as conduits for epigenetic change linking the environmental trigger at cell surface receptors with gene-specific chromatin alterations and modulation of programming of gene activity. For example, numerous signaling pathways including those triggered by G-protein-coupled cell surface receptors in the brain alter the concentration of cyclic adenosine monophosphate (cAMP) in the cell. One of the transcription factors which respond to increased cAMP is cAMP response element-binding protein (CREB). CREB binds cAMP response elements in certain genes. CREB also recruits CREB-binding protein CBP. CBP is a HAT, which acetylates histones (Ogryzko et al. 1996). Thus, elevation of cAMP levels in response to an extracellular signal would result in a change in the state of histone acetylation in specific genes. Environmental or physiological events that interfere at any point along the signaling pathway may result in chromatin alterations. Below, we discuss an example of such a pathway that leads from maternal behavior to long-term programming of gene expression in the hippocampus (Meaney and Szyf 2005).

17.2.4 DNA Methylation

In addition to chromatin, which is associated with DNA, the DNA molecule itself is chemically modified by methyl residues at the 5′ position of the cytosine rings in the dinucleotide sequence CG in vertebrates (Razin 1998). Other modifications to DNA, including hydroxymethylation (5-hmC) and several other DNA modifications, are attracting increasing interest as potential gene regulatory mechanisms (Labrie et al. 2012). It should be noted that conventional methods used for mapping 5-mC, such as bisulfite sequencing and methylation-sensitive restriction enzyme-based approaches, do not differentiate it from 5-hmC, although it is possible to use enzyme-based glycosylation of 5-hmC followed by restriction enzyme-based detection of 5-hmC and 5-mC or as via modified bisulfite sequencing (Booth et al. 2012). In this chapter, we use the term "DNA methylation" to denote epigenetic changes associated with the DNA itself, though the term "DNA modification" is perhaps more accurate given the current knowledge.

Among different cell types, distinct CG methylation generates cell-type-specific epigenetic patterns. Thus, the DNA methylation pattern confers upon the genome its cell-type identity (Razin 1998). The DNA methylation pattern is established during development and is then maintained throughout life by the maintenance DNA methyltransferases (DNMT; Razin and Riggs 1980). DNA methylation in distinct regulatory regions is believed to generally mark silent genes. Thus, aberrant methylation will silence a gene, resulting in "loss of function," which will have a similar consequence as a loss of function by genetic mechanism such as mutation, deletion, or rearrangement.

17.2.5 DNA Methylation Enzymes

The DNA methylation pattern is not copied by the DNA replication machinery, but by independent enzymatic machinery the DNMT (Razin and Cedar 1977). The DNA methylation machinery in vertebrates has two main roles. First, it establishes new cell-type-specific DNA methylation patterns during development and possibly during adulthood in response to new signals. Second, it maintains these patterns during downstream cell divisions and after DNA repair. The different enzymes and proteins of the DNA methylation machinery must address these different tasks. The methylation of DNA occurs immediately after replication by a transfer of a methyl moiety from the donor *S*-adenosyl-L-methionine (AdoMet) in a reaction catalyzed by DNMTs. The maintenance DNMT1 prefers a hemimethylated substrate (Razin and Riggs 1980). Since hemimethylated sites are generated during DNA replication when a nascent unmethylated C is synthesized across a methylated C in the template parental strand, the DNMT accurately copies the methylation pattern of the template strand. Three distinct phylogenic DNMT were identified in mammals. DNMT1 shows preference for hemimethylated DNA in vitro, which is consistent with its role as a maintenance DNMT, whereas DNMT3a and DNMT3b methylate unmethylated and methylated DNA at an equal rate which is consistent with a de novo DNMT role (Okano et al. 1998). Knockout mouse data indicate that DNMT1 is responsible for a majority of DNA methylation marks in the mouse genome (Li et al. 1992) whereas DNMT3a and DNMT3b are responsible for some but not all de novo methylation during development (Okano et al. 1999).

The answer to the question of whether the DNA methylation is reversible or not has important implications on the possibility that DNA methylation is dynamic and responsive to physiological and environmental signals throughout life. DNMTs are present in neurons (Goto et al. 1994) and there are data suggesting that DNMT levels in neurons change in certain pathological conditions such as schizophrenia (Veldic et al. 2005). The presence of DNMT in neurons suggests that DNA methylation is dynamic in postmitotic tissues and is a balance of methylation and demethylation reactions (Szyf 2001).

17.2.6 DNA Demethylation Enzymes

We and others have proposed that the DNA methylation pattern is a balance of methylation and demethylation reactions that are responsive to physiological and environmental signals and thus serves as a biological manifestation of gene–environment interactions (Sasaki et al. 2013; Szyf et al. 2007; McGowan and Szyf 2010). There are now convincing examples of active, replication-independent DNA demethylation during development as well as in somatic tissues. Active demethylation was reported for the *myosin* gene in differentiating myoblast cells (Lucarelli et al. 2001), Il2 gene upon T cell activation (Bruniquel and Schwartz 2003), the interferon γ gene upon antigen exposure of memory CD8 T cells (Kersh et al. 2006), in the glucocorticoid receptor (GR) gene promoter in adult rat brains upon treatment with the HDAC inhibitor (HDACi) TSA (Weaver et al. 2004), and in neurons as a function of neural activity (Rudenko et al. 2013).

The precise mechanisms governing DNA demethylation are the subject of intense investigation. One proposal has been that a G/T mismatch repair glycosylase also functions as a 5-methylcytosine DNA glycosylase, recognizes methyl cytosines and cleaves the bond between the sugar and the base. The abasic site is then repaired and replaced with a nonmethylated cytosine resulting in demethylation

(Jost 1993). An additional protein with a similar activity was identified as the methylated DNA-binding protein 4 (MBD4; Zhu et al. 2000). Active demethylation early in embryogenesis as well as in somatic cells was also shown to be catalyzed by a nucleotide excision repair mechanism, whereby methylated cytosines are replaced by unmethylated cytosines, which involves the growth arrest and damage response protein Gadd45a and the DNA repair endonuclease XPG (Barreto et al. 2007). It has been proposed that the pathway from methylated to unmethylated DNA involves 5-mC hydroxylation via TET enzymes (Guo et al. 2011). As such, 5-hmC may be an intermediary marker of demethylation. In sum, though a number of biochemical processes were implicated in demethylation, it is unclear how and when these different enzymes participate in shaping and maintaining the overall pattern of methylation and how these activities respond to different environmental exposures.

17.2.7 Targeting DNA Methylation and Demethylation to Specific Genes

A central question regarding gene-specific changes in DNA methylation associated with the environment concerns the targeting of these changes to specific loci in the genome. Because DNA methylating and demethylating enzymes are nonspecific, targeting must be achieved via other mechanisms. There is evidence that chromatin configuration can regulate the accessibility of genes to either DNA methylation or demethylation machineries (Cervoni and Szyf 2001; D'Alessio and Szyf 2006). For example, the HDACi trichostatin A (TSA), which leads to hyperacetylated chromatin, also leads to active DNA demethylation (Cervoni and Szyf 2001). A change in histone acetylation is normally caused by transcription factors that recruit HATs, which may cause histone acetylation and facilitate demethylation. Examples of histone-modifying enzymes shown to interact with DNMT1 are HDAC1, HDAC2, the histone methyltransferases SUV3–9 and enhancer of zeste homolog 2 (EZH2), a member of the multi-protein Polycomb complex PRC2 that methylates H3 histone at the K27 residue (Fuks et al. 2000, 2003; Rountree et al. 2000; Vire et al. 2005). DNMT3a was also shown to interact with EZH2 which targets the DNA methylation-histone modification multi-protein complexes to specific sequences in DNA (Vire et al. 2005).

Sequence-specific transcription factors may target demethylation to specific genes. *Trans*-acting repressors target both histone-modifying enzymes and DNMTs to specific *cis*-acting signals in regulatory regions of particular genes causing gene-specific DNA methylation and chromatin modification. For example, the promyelocytic leukemia PML-RAR fusion protein targets HDAC and DNMTs to its binding sequences and produces de novo DNA methylation of adjacent genes (Di Croce et al. 2002). The intronic kappa chain enhancer and the transcription factor NF-kappaB are required for B-cell-specific demethylation of the kappa immunoglobulin gene (Lichtenstein et al. 1994). As we discuss below, maternal care may be a behavioral mechanism to program gene expression through recruitment of the transcription factor nerve growth factor-induced protein A (NGFI-A) to a promoter region of the GR in the hippocampus (Weaver et al. 2007).

In summary, we and others have proposed that DNA modifications are maintained in an equilibrium between methylation and demethylation as long as this equilibrium of sequence-specific factors engagement of the genes is maintained (Sasaki et al. 2013; McGowan and Szyf 2010). This process is essential for normal development and the process of tissue-specific cellular differentiation. Physiological or environmental signals, which alter the signaling pathways in the cell, may result in altering this balance by activating or suppressing specific *trans* acting factors.

17.2.8 DNA Methylation and Gene Repression

DNA methylation in critical gene regulatory regions can silence gene expression. There are two main mechanisms by which cytosine methylation suppresses gene expression. The first mechanism involves direct interference of the methylation mark with the binding of a transcription factor to its recognition element in the gene. The interaction of transcription factors with genes is required for activation of the gene; lack of binding of a transcription factor would result in silencing of gene expression (Comb and Goodman 1990; Inamdar et al. 1991). This form of inhibition of transcription by methylation requires that the methylation events occur within the recognition sequence for a transcription factor. A second mechanism is indirect. A certain density of DNA methylation moieties in the region of the gene attracts the binding of methylated-DNA-binding proteins such as MBD1, MBD2, MBD3, and MeCP2, which lead to the formation of a "closed" chromatin configuration and silencing of gene expression (Nan et al. 1997; Hendrich and Bird 1998; Ng et al. 1999; Fujita et al. 1999). For example, MeCP2 recruits other proteins such as SIN3A and histone-modifying enzymes, modifying chromatin conformation via this mechanism (Nan et al. 1997).

17.2.9 Cross Talk Between Chromatin Structure and DNA Methylation

As described previously, DNA methylation can define chromatin structure by recruiting chromatin-modifying enzymes. Emerging evidence indicates that targeting may be a result of both the cross talk between DNA methylation and chromatin modifications, and transcription factor recruitment of DNA-modifying proteins to gene regulatory elements. The loss of DNA methylation will result in the "opening" of chromatin configuration due to increased levels of histone acetylation. Thus, chromatin structure and DNA methylation exhibit a substantial cross talk that creates a feedback loop whereby DNA methylation increases chromatin compaction and transcriptional repression. DNA demethylation likewise increases chromatin activation, further enhancing DNA demethylation (Cedar and Bergman 2009).

17.3 Maternal Care, Epigenetics, and the HPA Axis: Laboratory Animal Studies

Research findings by Weaver, Meaney, Szyf, and colleagues in the early 2000s launched epigenetics to the forefront of research on mechanisms leading from the maternal behavior to long-term programming of gene expression the offspring. Earlier work by the Meaney laboratory and others had established that naturally occurring differences in maternal care in the early postnatal period—during the first week of life in the rat—lead to long-term effects on stress and stress-related behavior. The offspring of mothers who naturally exhibit high levels of care show elevated transcript abundance of the GR in the hippocampus, enhanced negative feedback sensitivity, and a more modest response to stressors in adulthood (Liu et al. 1997). Cross-fostering studies showed that this phenotype is directly attributable to maternal behavior rather than factors related to the prenatal environment, as offspring phenotype was shown to match that of the adoptive mother rather than that of the biological mother (Francis et al. 1999). Weaver and colleagues showed that the accompanying change in GR expression was regulated by DNA methylation of the $GR1_7$ splice variant in the hippocampus by inhibiting the binding of NGFI-A, a transcription factor that drives GR expression (Weaver et al. 2004). GR17 is 1 of at least 11 untranslated first exons of the GR gene. Though it is ubiquitously expressed in virtually all cells, levels of expression of GR vary and are controlled in part by tissue-specific expression of GR exon 1 splice variants (this is also true for the human GR exon 1 as is discussed; Turner and Muller 2005; McCormick et al. 2000). In the hippocampus, GR17 was previously shown to vary in expression as a function of the level of maternal care received (McCormick et al. 2000). Interestingly, relatively high levels of DNA methylation were maintained among the offspring of low maternal care mothers, whereas offspring of high maternal care mothers showed demethylation of this promoter during the first week of life, coinciding with emergence of differences in maternal care between the two litter types. The results implied that DNA demethylation (through a yet unknown process) leads to an increased number of GRs and an attenuated response to stress. DNA methylation differences were stable throughout adulthood in these animals, but were reversible by infusion of trichostatin A (TSA), an HDACi, which also leads to increased gene expression in hundreds of other genes (Weaver et al. 2006). Likewise, lower levels of DNA methylation observed among the offspring of high maternal care mothers resembled that of offspring of high maternal care mothers given central infusions of the methyl donor l-methionine, indicating that enzymes responsible for DNA methylation were poised to act in the adult brain in response to methyl donor.

A recent study has challenged the idea that $GR1_7$ transcript is regulated by DNA methylation of the NGFI-A response element in rats exposed to stress paradigms that lead to altered NGFI-A levels, though stress does appear to modulate the methylation of other CG sites within the $GR1_7$ promoter (Witzmann et al. 2012). It is likely that DNA methylation of $GR1_7$ gene expression involves the binding of additional transcription factors and/or is context and brain region specific. It is also likely that the $GR1_7$ is itself part of a response mechanism that involves additional splice variants of GR and perhaps other transcription factors.

We performed a microarray analysis of DNA methylation, H3K9 acetylation, and gene expression in a 7-million base pair region containing the GR gene in the rat hippocampus (McGowan et al. 2011). We found that epigenetic differences in adulthood that were associated with early maternal care occurred in clustered regions of up to 100 kbp but were nonetheless exquisitely patterned, whereby increased transcription occurred in conjunction with hyperacetylation and hypermethylation of exons, and hypomethylation of promoters. We found epigenetic differences in association with altered transcription as a function of maternal care across several GR1 splice variants. Large epigenetic differences were noted in proximity to the transcription start site of GR, within the first coding exon (exon 2) and within GR introns, suggesting there may be additional regulation of GR via yet-to-be-identified noncoding RNAs within the GR locus. These data were the first to link epigenetic changes across both coding and noncoding regions in the mammalian brain, and implicate a nonrandom "epigenetic programming" across large-scale loci in response to differences in early care. Accumulating evidence indicates that additional genes in the neural pathway mediating the stress response are epigenetically regulated by DNA methylation of gene regulatory elements in association with early-life stress, for example, arginine vasopressin in the hypothalamus (Murgatroyd et al. 2009), BDNF in the hippocampus (Roth et al. 2009), and GAD67 in the prefrontal cortex (Zhang et al. 2010).

These postnatal programming effects appear to derive from environmentally induced alterations of maternal–neonatal interactions, involving systems that determine methylation patterns of GR gene promoter sequences and additional loci. It will be important to understand the precise nature of the maternal–neonatal interaction that mediates these changes. For example, there is evidence that artificial stimulation of pups with a paintbrush as a substitute for maternal licking can alter DNA methylation of a promoter region so of the estrogen receptor alpha gene in the preoptic area of the hypothalamus (Kurian et al. 2010). These data have important implications for studies of trans-generational impacts related to maternal care via epigenetic mechanisms, through via *behavioral* mechanism of inheritance rather than gametic inheritance, as maternal behavior tends to correlate with the maternal care provided by offspring to their progeny (Champagne et al. 2003).

17.4 Maternal Care, Epigenetics, and the HPA Axis: Human Studies

In the previous section, we discussed evidence from our studies of widespread but specific epigenetic and transcriptional alterations of the GR gene extending far beyond the GR promoter associated with differences in maternal care (McGowan et al. 2011; Murgatroyd et al. 2009; Roth et al. 2009; Zhang et al. 2010). Thus, there is mounting evidence that epigenetic mechanisms coordinate widespread changes in gene expression in response to differences in early maternal care or adversity.

GR promoter DNA methylation has been associated in peripheral blood with a variety of outcomes related to HPA dysfunction. In one of the earliest reports, DNA methylation of GR promoter in infants' cord blood was found to differ as a function of maternal mood during pregnancy and correlate with infants' cortisol response (Oberlander et al. 2008). These data suggest that GR promoter methylation in the brain and in lymphocytes is under epigenetic control as a function of the pre- and postnatal factors. A more recent study indicated that DNA methylation of GR promoter in placenta was associated with birth weight, implicating GR methylation in placental function and suggesting that environmental factors alter metabolic processes in part via epigenetic changes in GR (Filiberto et al. 2011). Other recent research has identified GR DNA methylation as a predictor of treatment outcome in PTSD patients (Yehuda et al. 2013).

We examined postmortem brain tissue from adults with well-characterized life histories to investigate the influence of early-life adversity on GR DNA methylation in adults with a history of trauma. Our focus was on individuals with a history of severe physical or sexual abuse or neglect during childhood, which is common among suicide victims, and is an important risk factor for suicide (Turecki et al. 2012). We examined the GR1F promoter in the hippocampus of human suicide victims and controls (McGowan et al. 2009). Family dysfunction and childhood adversity are linked to altered HPA stress responses and an increased risk for suicide. The promoter region we examined is upstream of one of several untranslated exon 1 splice variants that are known to regulate tissue-specific expression of GR, akin to the function that the GR exon 1 splice variants serve in the rodent (Turner and Muller 2005). The study included three conditions: (1) suicide completers with a history of childhood abuse or severe neglect, (2) suicide completers without a history of childhood abuse or neglect, and (3) individuals who have neither committed suicide nor had a history of childhood abuse or neglect. A fourth group of nonsuicide victims with a history of abuse or neglect was not available, partly due to the fact that tissues from such a "control" group are exceedingly rare, and were unavailable for our study. In this study, we found that the GR gene was differentially methylated among suicide victims with a history of abuse in childhood, but not among suicide victims with a negative history of childhood abuse, compared to control individuals without a history of suicide. The data suggest that epigenetic processes might mediate the effects of the social environment during childhood on hippocampal gene expression and that stable epigenetic marks such as DNA methylation might then persist into adulthood and influence vulnerability for psychopathology through effects on intermediate levels of function such as activity of the HPA axis that regulates the stress response. However, it is still unclear whether the epigenetic aberrations were present in the germ line, whether they were introduced during embryogenesis, or whether they were truly changes occurring during early childhood. We also do not yet know the extent to which parental factors per se play a role in this phenotype. Despite these important caveats, these data were the first to link the early-life environment to changes in the GR gene in humans. The data parallel that in the rodent study mentioned above, though in a very different context.

We have applied high-throughput approaches to examining DNA methylation, chromatin modifications, and mRNA expression in gene regulatory, coding, intragenic, and intergenic regions in humans in a study that paralleled that described above in rats. We analyzed the GR gene locus by interrogating a 7-Mbp region containing the GR gene in hippocampi of adult suicide victims who were abused early in life compared to controls using high-throughput DNA microarray (Suderman et al. 2012). The GR gene locus shows substantial conservation with the same locus in rodents, with an almost identical order or orthologous genes across the locus. Like the study in the rat (McGowan et al. 2011), methylation levels were nonrandomly distributed across the locus, indicating that stochastic processes are unlikely to account for the range of variation that we observed in this study. Proximal to the GR gene itself, we found a large region hypermethylated in suicide completers relative to controls within the first coding exon of the GR gene and its proximal promoters, extending previous observations of hypermethylation of the GR1F promoter among suicide victims with a history of abuse (McGowan et al. 2009). This analysis also revealed differences in DNA methylation in intragenic regions of the GR gene. At this time, we can only speculate that unrecognized noncoding RNAs may reside within this region and affect GR expression. Other differences were discovered within coding regions and the 3′ UTR of the GR gene. These data suggest that GR is epigenetically labile in response to the early-life social environment in both rodents and humans, though the specific alterations that we observed are not identical in both species (Suderman et al. 2012). Nevertheless, the data indicate that the animal model of parental care may have broad applicability for translational studies aimed at understanding the consequences of epigenetic modification of GR in humans.

Though the most cost-effective means of targeting specific loci for epigenetic analyses remain microarray approaches combined with immunoprecipitation, such studies are not without limitations. Such approaches suffer from a lack of resolution (~200 bp) compared to single-nucleotide resolution sequencing-based analyses. As the cost of sequencing continues to decrease, it is now becoming feasible to employ sequencing-based epigenetic analyses of DNA methylation (via meDIP-Seq or Bisulfite Sequencing [BIS-seq]) and chromatin modifications (via ChIP-seq). These two have important limitations, as have been reviewed elsewhere (e.g., Bock 2012). High-throughput studies such as the ones described above in the rat and human open up a number of questions—undoubtedly more than are answered. It is clear that these technological advances that allow whole-epigenome analysis must be coupled with equally powerful phenotypic screens using appropriate cell types and conditions.

17.5 Conclusions and Prospective

A more complete understanding of the role of epigenetic mechanisms in perinatal programming will be afforded by studies that address several basic questions. First, in what contexts is the epigenome labile in response to early environment? Are there indeed critical time windows for the influence of the environment on epigenetic trajectories? A number of studies have linked early-life events to changes in neuroplasticity that have a lasting impact of endocrine systems mediating the response to stress (McEwen 2012). It is not always clear, however, which cell types are relevant to the question under study. This is particularly problematic for studies in humans, where access to neural tissue is nonexistent or limited. Peripheral cells such as peripheral blood mononuclear cells (PBMCs) offer an avenue to examine the HPA, as PBMCs are sensitive to endocrine modulation of HPA [ref]. Whole blood has also been used, but each tissue type is known to be sensitive to differences in constituent cell numbers, which may bias the results (Lam et al. 2013; Suderman et al. 2013). However, in studying environmental impacts prospectively in children, it is not often possible to obtain blood samples and other tissues must be used. The most commonly used tissues in such epigenetic studies are buccal cells from mouth swabs or saliva. Intriguingly, there is some evidence that such tissue is responsive to early-life adversity, though perhaps not via epigenetic changes in GR per se (Essex et al. 2013). Buccal cells complement studies of adversity in neurons in the sense that they do represent cells with a common embryonic origin. Such studies will provide a valuable means of resampling to examine epigenetic variance over time and with interventions. In animal studies, a goal going forward for translational work will be to identify labile epigenetic regions like the GR that can be assessed in brain and blood in order to generate hypotheses and biomarkers that can be examined in humans. Such research stands to offer critical insights into the manner by which the biological embedding occurs during the perinatal period.

Conflict of Interest The authors declare no conflicts of interest.

References

Barreto G, Schafer A, Marhold J, Stach D, Swaminathan SK, Handa V, Doderlein G, Maltry N, Wu W, Lyko F, Niehrs C (2007) Gadd45a promotes epigenetic gene activation by repair-mediated DNA demethylation. Nature 445:671–675

Bergmann A, Lane ME (2003) HIDden targets of microRNAs for growth control. Trends Biochem Sci 28:461–463

Bock C (2012) Analysing and interpreting DNA methylation data. Nat Rev Genet 13:705–719

Booth MJ, Branco MR, Ficz G, Oxley D, Krueger F, Reik W, Balasubramanian S (2012) Quantitative sequencing of 5-methylcytosine and 5-hydroxymethylcytosine at single-base resolution. Science 336:934–937

Bruniquel D, Schwartz RH (2003) Selective, stable demethylation of the interleukin-2 gene enhances transcription by an active process. Nat Immunol 4:235–240

Cedar H, Bergman Y (2009) Linking DNA methylation and histone modification: patterns and paradigms. Nat Rev Genet 10:295–304

Cervoni N, Szyf M (2001) Demethylase activity is directed by histone acetylation. J Biol Chem 276:40778–40787

Champagne FA, Francis DD, Mar A, Meaney ZJ (2003) Variations in maternal care in the rat as a mediating influence for the effects of environment on development. Physiol Behav 79:359–371

Comb M, Goodman HM (1990) CpG methylation inhibits proenkephalin gene expression and binding of the transcription factor AP-2. Nucleic Acids Res 18:3975–3982

D'Alessio AC, Szyf M (2006) Epigenetic tete-a-tete: the bilateral relationship between chromatin modifications and DNA methylation. Biochem Cell Biol 84:463–476

Di Croce L, Raker VA, Corsaro M, Fazi F, Fanelli M, Faretta M, Fuks F, Lo Coco F, Kouzarides T, Nervi C, Minucci S, Pelicci PG (2002) Methyltransferase recruitment and DNA hypermethylation of target promoters by an oncogenic transcription factor. Science 295:1079–1082

Essex MJ, Boyce WT, Hertzman C, Lam LL, Armstrong JM, Neumann SM, Kobor MS (2013) Epigenetic vestiges of early developmental adversity: childhood stress exposure and DNA methylation in adolescence. Child Dev 84:58–75

Filiberto AC, Maccani MA, Koestler D, Wilhelm-Benartzi C, Avissar-Whiting M, Banister CE, Gagne LA, Marsit CJ (2011) Birthweight is associated with DNA promoter methylation of the glucocorticoid receptor in human placenta. Epigenetics 6:566–572

Finch JT, Lutter LC, Rhodes D, Brown RS, Rushton B, Levitt M, Klug A (1977) Structure of nucleosome core particles of chromatin. Nature 269:29–36

Francis D, Diorio J, Liu D, Meaney MJ (1999) Nongenomic transmission across generations of maternal behavior and stress responses in the rat. Science 286:1155–1158

Fujita N, Takebayashi S, Okumura K, Kudo S, Chiba T, Saya H, Nakao M (1999) Methylation-mediated transcriptional silencing in euchromatin by methyl-CpG binding protein MBD1 isoforms. Mol Cell Biol 19:6415–6426

Fuks F, Burgers WA, Brehm A, Hughes-Davies L, Kouzarides T (2000) DNA methyltransferase Dnmt1 associates with histone deacetylase activity. Nat Genet 24:88–91

Fuks F, Hurd PJ, Wolf D, Nan X, Bird AP, Kouzarides T (2003) The methyl-CpG-binding protein MeCP2 links DNA methylation to histone methylation. J Biol Chem 278:4035–4040

Goto K, Numata M, Komura JI, Ono T, Bestor TH, Kondo H (1994) Expression of DNA methyltransferase gene in mature and immature neurons as well as proliferating cells in mice. Differentiation 56:39–44

Groudine M, Eisenman R, Gelinas R, Weintraub H (1983) Developmental aspects of chromatin structure and gene expression. Prog Clin Biol Res 134:159–182

Grunstein M (1997) Histone acetylation in chromatin structure and transcription. Nature 389:349–352

Guo JU, Su Y, Zhong C, Ming GL, Song H (2011) Hydroxylation of 5-methylcytosine by TET1 promotes active DNA demethylation in the adult brain. Cell 145:423–434

Hendrich B, Bird A (1998) Identification and characterization of a family of mammalian methyl-CpG binding proteins. Mol Cell Biol 18:6538–6547

Henikoff S, McKittrick E, Ahmad K (2004) Epigenetics, histone H3 variants, and the inheritance of chromatin states. Cold Spring Harb Symp Quant Biol 69:235–243

Inamdar NM, Ehrlich KC, Ehrlich M (1991) CpG methylation inhibits binding of several sequence-specific DNA-binding proteins from pea, wheat, soybean and cauliflower. Plant Mol Biol 17:111–123

Jenuwein T (2001) Re-SET-ting heterochromatin by histone methyltransferases. Trends Cell Biol 11:266–273

Jenuwein T, Allis CD (2001) Translating the histone code. Science 293:1074–1080

Jost JP (1993) Nuclear extracts of chicken embryos promote an active demethylation of DNA by excision repair of 5-methyldeoxycytidine. Proc Natl Acad Sci U S A 90:4684–4688

Kersh EN, Fitzpatrick DR, Murali-Krishna K, Shires J, Speck SH, Boss JM, Ahmed R (2006) Rapid demethylation of the IFN-gamma gene occurs in memory but not naive CD8 T cells. J Immunol 176:4083–4093

Kuo MH, Allis CD (1998) Roles of histone acetyltransferases and deacetylases in gene regulation. Bioessays 20:615–626

Kurian JR, Olesen KM, Auger AP (2010) Sex differences in epigenetic regulation of the estrogen receptor-alpha promoter within the developing preoptic area. Endocrinology 151:2297–2305

Labrie V, Pai S, Petronis A (2012) Epigenetics of major psychosis: progress, problems and perspectives. Trends Genet 28:427–435

Lachner M, O'Carroll D, Rea S, Mechtler K, Jenuwein T (2001) Methylation of histone H3 lysine 9 creates a binding site for HP1 proteins. Nature 410:116–120
Lam LL, Emberly E, Fraser HB, Neumann SM, Chen E, Miller GE, Kobor MS (2013) Reply to Suderman et al.: importance of accounting for blood cell composition in epigenetic studies. Proc Natl Acad Sci U S A 110:E1247
Lee DY, Hayes JJ, Pruss D, Wolffe AP (1993) A positive role for histone acetylation in transcription factor access to nucleosomal DNA. Cell 72:73–84
Li E, Bestor TH, Jaenisch R (1992) Targeted mutation of the DNA methyltransferase gene results in embryonic lethality. Cell 69:915–926
Lichtenstein M, Keini G, Cedar H, Bergman Y (1994) B cell-specific demethylation: a novel role for the intronic kappa chain enhancer sequence. Cell 76:913–923
Liu D, Diorio J, Tannenbaum B, Caldji C, Francis D, Freedman A, Sharma S, Pearson D, Plotsky PM, Meaney MJ (1997) Maternal care, hippocampal glucocorticoid receptors, and hypothalamic-pituitary-adrenal responses to stress. Science 277:1659–1662
Lucarelli M, Fuso A, Strom R, Scarpa S (2001) The dynamics of myogenin site-specific demethylation is strongly correlated with its expression and with muscle differentiation. J Biol Chem 276:7500–7506
Marks PA, Sheffery M, Rifkind RA (1985) Modulation of gene expression during terminal cell differentiation. Prog Clin Biol Res 191:185–203
McCormick JA, Lyons V, Jacobson MD, Noble J, Diorio J, Nyirenda M, Weaver S, Ester W, Yau JL, Meaney MJ, Seckl JR, Chapman KE (2000) 5′-heterogeneity of glucocorticoid receptor messenger RNA is tissue specific: differential regulation of variant transcripts by early-life events. Mol Endocrinol 14:506–517
McEwen BS (2012) Brain on stress: how the social environment gets under the skin. Proc Natl Acad Sci U S A 109(suppl 2):17180–17185
McGowan PO, Szyf M (2010) The epigenetics of social adversity in early life: implications for mental health outcomes. Neurobiol Dis 39:66–72
McGowan PO, Sasaki A, D'Alessio AC, Dymov S, Labonte B, Szyf M, Turecki G, Meaney MJ (2009) Epigenetic regulation of the glucocorticoid receptor in human brain associates with childhood abuse. Nat Neurosci 12:342–348
McGowan PO, Suderman M, Sasaki A, Huang TC, Hallett M, Meaney MJ, Szyf M (2011) Broad epigenetic signature of maternal care in the brain of adult rats. PLoS ONE 6:e14739
Meaney MJ, Szyf M (2005) Maternal care as a model for experience-dependent chromatin plasticity? Trends Neurosci 28:456–463
Mehler MF, Mattick JS (2006) Non-coding RNAs in the nervous system. J Physiol 575:333–341
Mehler MF, Mattick JS (2007) Noncoding RNAs and RNA editing in brain development, functional diversification, and neurological disease. Physiol Rev 87:799–823
Murgatroyd C, Patchev AV, Wu Y, Micale V, Bockmuhl Y, Fischer D, Holsboer F, Wotjak CT, Almeida OF, Spengler D (2009) Dynamic DNA methylation programs persistent adverse effects of early-life stress. Nat Neurosci 12:1559–1566
Nan X, Campoy FJ, Bird A (1997) MeCP2 is a transcriptional repressor with abundant binding sites in genomic chromatin. Cell 88:471–481
Ng HH, Zhang Y, Hendrich B, Johnson CA, Turner BM, Erdjument-Bromage H, Tempst P, Reinberg D, Bird A (1999) MBD2 is a transcriptional repressor belonging to the MeCP1 histone deacetylase complex. Nat Genet 23:58–61
Oberlander TF, Weinberg J, Papsdorf M, Grunau R, Misri S, Devlin AM (2008) Prenatal exposure to maternal depression, neonatal methylation of human glucocorticoid receptor gene (NR3C1) and infant cortisol stress responses. Epigenetics 3:97–106
Ogryzko VV, Schiltz RL, Russanova V, Howard BH, Nakatani Y (1996) The transcriptional coactivators p300 and CBP are histone acetyltransferases. Cell 87:953–359
Okano M, Xie S, Li E (1998) Cloning and characterization of a family of novel mammalian DNA (cytosine-5) methyltransferases. Nat Genet 19:219–220
Okano M, Bell DW, Haber DA, Li E (1999) DNA methyltransferases Dnmt3a and Dnmt3b are essential for de novo methylation and mammalian development. Cell 99:247–257

Perry M, Chalkley R (1982) Histone acetylation increases the solubility of chromatin and occurs sequentially over most of the chromatin. A novel model for the biological role of histone acetylation. J Biol Chem 257:7336–7347

Petronis A (2010) Epigenetics as a unifying principle in the aetiology of complex traits and diseases. Nature 465:721–727

Qureshi IA, Mehler MF (2009) Regulation of non-coding RNA networks in the nervous system-what's the REST of the story? Neurosci Lett 466:73–80

Ramain P, Bourouis M, Dretzen G, Richards G, Sobkowiak A, Bellard M (1986) Changes in the chromatin structure of *Drosophila* glue genes accompany developmental cessation of transcription in wild type and transformed strains. Cell 45:545–553

Razin A (1998) CpG methylation, chromatin structure and gene silencing-a three-way connection. EMBO J 17:4905–4908

Razin A, Cedar H (1977) Distribution of 5-methylcytosine in chromatin. Proc Natl Acad Sci U S A 74:2725–2728

Razin A, Riggs AD (1980) DNA methylation and gene function. Science 210:604–610

Roth TL, Lubin FD, Funk AJ, Sweatt JD (2009) Lasting epigenetic influence of early-life adversity on the *BDNF* gene. Biol Psychiatry 65:760–769

Rountree MR, Bachman KE, Baylin SB (2000) DNMT1 binds HDAC2 and a new co-repressor, DMAP1, to form a complex at replication foci. Nat Genet 25:269–277

Rudenko A, Dawlaty MM, Seo J, Cheng AW, Meng J, Le T, Faull KF, Jaenisch R, Tsai LH (2013) Tet1 is critical for neuronal activity-regulated gene expression and memory extinction. Neuron 79:1109–1122

Sarma K, Reinberg D (2005) Histone variants meet their match. Nat Rev Mol Cell Biol 6:139–149

Sasaki A, de Vega WC, McGowan PO (2013) Biological embedding in mental health: an epigenomic perspective. Biochem Cell Biol 91:14–21

Shi Y, Lan F, Matson C, Mulligan P, Whetstine JR, Cole PA, Casero RA, Shi Y (2004) Histone demethylation mediated by the nuclear amine oxidase homolog LSD1. Cell 119:941–953

Shilatifard A (2006) Chromatin modifications by methylation and ubiquitination: implications in the regulation of gene expression. Annu Rev Biochem 75:243–269

Suderman M, McGowan PO, Sasaki A, Huang TC, Hallett MT, Meaney MJ, Turecki G, Szyf M (2012) Conserved epigenetic sensitivity to early life experience in the rat and human hippocampus. Proc Natl Acad Sci U S A 109(suppl 2):17266–17272

Suderman M, Borghol N, Pappas JJ, McArdle W, Racine A, Hallett MT, Pembrey ME, Hertzman C, Power C, Szyf M (2013) Epigenomic socioeconomic studies more similar than different. Proc Natl Acad Sci U S A 110:E1246

Szyf M (2001) Towards a pharmacology of DNA methylation. Trends Pharmacol Sci 22:350–354

Szyf M, McGowan P, Meaney MJ (2007) The social environment and the epigenome. Environ Mol Mutagen 49:46–60

Tsukada Y, Fang J, Erdjument-Bromage H, Warren ME, Borchers CH, Tempst P, Zhang Y (2006) Histone demethylation by a family of JmjC domain-containing proteins. Nature 439:811–816

Turecki G, Ernst C, Jollant F, Labonte B, Mechawar N (2012) The neurodevelopmental origins of suicidal behavior. Trends Neurosci 35:14–23

Turner JD, Muller CP (2005) Structure of the glucocorticoid receptor (NR3C1) gene 5′ untranslated region: identification, and tissue distribution of multiple new human exon 1. J Mol Endocrinol 35:283–292

Uchida S, Hara K, Kobayashi A, Funato H, Hobara T, Otsuki K, Yamagata H, McEwen BS, Watanabe Y (2010) Early life stress enhances behavioral vulnerability to stress through the activation of REST4-mediated gene transcription in the medial prefrontal cortex of rodents. J Neurosci 30:15007–15018

Varga-Weisz PD, Becker PB (2006) Regulation of higher-order chromatin structures by nucleosome-remodelling factors. Curr Opin Genet Dev 16:151–156

Veldic M, Guidotti A, Maloku E, Davis JM, Costa E (2005) In psychosis, cortical interneurons overexpress DNA-methyltransferase 1. Proc Natl Acad Sci U S A 102:2152–2157

Vire E, Brenner C, Deplus R, Blanchon L, Fraga M, Didelot C, Morey L, Van Eynde A, Bernard D, Vanderwinden JM, Bollen M, Esteller M, Di Croce L, de Launoit Y, Fuks F (2005) The polycomb group protein EZH2 directly controls DNA methylation. Nature 439:871–874
Vo N, Klein ME, Varlamova O, Keller DM, Yamamoto T, Goodman RH, Impey S (2005) A cAMP-response element binding protein-induced microRNA regulates neuronal morphogenesis. Proc Natl Acad Sci U S A 102:16426–16431
Wade PA, Pruss D, Wolffe AP (1997) Histone acetylation: chromatin in action. Trends Biochem Sci 22:128–132
Weaver IC, Cervoni N, Champagne FA, D'Alessio AC, Sharma S, Seckl JR, Dymov S, Szyf M, Meaney MJ (2004) Epigenetic programming by maternal behavior. Nat Neurosci 7:847–854
Weaver IC, Meaney MJ, Szyf M (2006) Maternal care effects on the hippocampal transcriptome and anxiety-mediated behaviors in the offspring that are reversible in adulthood. Proc Natl Acad Sci U S A 103:3480–3485
Weaver IC, D'Alessio AC, Brown SE, Hellstrom IC, Dymov S, Sharma S, Szyf M, Meaney MJ (2007) The transcription factor nerve growth factor-inducible protein a mediates epigenetic programming: altering epigenetic marks by immediate-early genes. J Neurosci 27:1756–1768
Witzmann SR, Turner JD, Meriaux SB, Meijer OC, Muller CP (2012) Epigenetic regulation of the glucocorticoid receptor promoter 1(7) in adult rats. Epigenetics 7:1290–1301
Wolffe AP (1996) Histone deacetylase: a regulator of transcription. Science 272:371–372
Yehuda R, Daskalakis NP, Desarnaud F, Makotkine I, Lehrner AL, Koch E, Flory JD, Buxbaum JD, Meaney MJ, Bierer LM (2013) Epigenetic biomarkers as predictors and correlates of symptom improvement following psychotherapy in combat veterans with PTSD. Front Psychiatry 4:118
Zhang TY, Hellstrom IC, Bagot RC, Wen X, Diorio J, Meaney MJ (2010) Maternal care and DNA methylation of a glutamic acid decarboxylase 1 promoter in rat hippocampus. J Neurosci 30:13130–13137
Zhu B, Zheng Y, Hess D, Angliker H, Schwarz S, Siegmann M, Thiry S, Jost JP (2000) 5-methylcytosine-DNA glycosylase activity is present in a cloned G/T mismatch DNA glycosylase associated with the chicken embryo DNA demethylation complex. Proc Natl Acad Sci U S A 97:5135–5139

Chapter 18
Perinatal Administration of Aromatase Inhibitors in Rodents as Animal Models of Human Male Homosexuality: Similarities and Differences

Sandra Olvera-Hernández and Alonso Fernández-Guasti

Abstract In this chapter we briefly review the evidence supporting the existence of biological influences on sexual orientation. We focus on basic research studies that have affected the estrogen synthesis during the critical periods of brain sexual differentiation in male rat offspring with the use of aromatase inhibitors, such as 1,4,6-androstatriene-3,17 (ATD) and letrozole. The results after prenatal and/or postnatal treatment with ATD reveal that these animals, when adults, show female sexual responses, such as lordosis or proceptive behaviors, but retain their ability to display male sexual activity with a receptive female. Interestingly, the preference and sexual behavior of these rats vary depending upon the circadian rhythm.

Recently, we have established that the treatment with low doses of letrozole during the second half of pregnancy produces male rat offspring, that when adults spend more time in the company of a sexually active male than with a receptive female in a preference test. In addition, they display female sexual behavior when forced to interact with a sexually experienced male and some typical male sexual behavior when faced with a sexually receptive female. Interestingly, these males displayed both sexual behavior patterns spontaneously, i.e., in absence of exogenous steroid hormone treatment. Most of these features correspond with those found in human male homosexuals; however, the "bisexual" behavior shown by the letrozole-treated rats may be related to a particular human population. All these data, taken together, permit to propose letrozole prenatal treatment as a suitable animal model to study human male homosexuality and reinforce the hypothesis that human sexual orientation is underlied by changes in the endocrine milieu during early development.

A. Fernández-Guasti (✉) · S. Olvera-Hernández
Department of Pharmacobiology, Center for Research and Advanced Studies (CINVESTAV), Calzada de los Tenorios 235 Colonia Granjas Coapa, 14330 México, D.F., Mexico
e-mail: jfernand@cinvestav.mx

M. C. Antonelli (ed.), *Perinatal Programming of Neurodevelopment*,
Advances in Neurobiology 10, DOI 10.1007/978-1-4939-1372-5_18,

18.1 Sexual Orientation in Humans

Human sexuality is a multidimensional phenomenon that includes several aspects of behavior and personality. Four dimensions can be distinguished: (1) the type of specific action patterns that are produced by the individual (performance) and the motivation underlying the expressions of these behaviors, (2) the orientation of the behavior and of the sexual fantasies associated with it (sexual orientation), (3) the sexual identity that the person believes he/she has, and (4) the sexual role that the individual plays into the society (Balthazart 2012).

In 1948, Kinsey classified the sexual orientation in a scale where 0 is the indicator for exclusively heterosexual behavior and desire and 6 is the indicator for exclusively homosexual behavior and desire, with 1–5 for degrees in between. This classification takes into account the individual's sexual attraction, sexual fantasies, sexual behavior, and self-identification (Kinsey et al. 1948). Accordingly, the majority of people are sexually attracted to individuals of the opposite sex—heterosexuals—however, there is a low percentage (2–10%) of men and women who are exclusively attracted to individuals of their own sex—homosexuals (Bagley and Tremblay 1998; Diamond 1993; Sell et al. 1995). The variation in the incidence of homosexuality or bisexuality primarily depends on self-acceptance. In this regard, the self-definition of heterosexual, homosexual, or bisexual may take some sexual experimentation. It is likely that the prevailing community context with its social, legal, and religious norms, and their internalization by subjects play a significant role in the sometimes long searching process of one's true sexual orientation. A societal context with a hostile attitude toward homosexuality may be so intimidating to individuals that it leads to denial or reluctance to (self)-admit one's sexual orientation (Gooren 2006). The prevalence of homosexual or bisexual behavior, defined as sexual activity with same-sex subjects, is largely practiced in nature and can be seen in species varying from worms to primates (Poiani 2010), an observation that suggests a biological role in its establishment.

The causes of homosexual orientation have been subject of controversy. There is a variety of theories ranging from Freudian psychoanalysis to social constructivism, where homosexual orientation has been considered as the result of social experiences during early childhood. In the 1970s, John Money applied this concept to a male baby: the John–Joan–John case. Unfortunately, a boy lost his penis accidentally during a phimosis repair. His parents were consulted and they decided to raise their son as a girl. Money (1975) published that the treatment was successful because the boy accepted the new identity as a girl. However, several years later, during puberty, Joan became acquainted with the history and decided to switch sex and to live as a male: He turned once more to John, married a woman, and adopted a child (Diamond and Sigmundson 1997). After various difficult years, he committed suicide. This case suggests that the development of sexual orientation in humans is primarily determined during pregnancy and that social experiences only possess a modulatory role.

Table 18.1 Biological causative factors of male homosexuality

Evidence	References
Genetic: polymorphic X-linked genes Xq28, 7q36	Bailey et al. 1999; Camperio-Ciani et al. 2004, 2012; Hamer et al. 1993; Mustanski et al. 2005
Autoimmune: fraternal birth order effect	Blanchard and Bogaert 1996a; Blanchard and Klassen 1997; Bogaert and Skorska 2011
Maternal stress	Dörner et al. 1980, 1983; Ellis et al. 1988; Ellis and Cole-Harding 2001
Hormonal influences	Bogaert et al. 2002; Bogaert and Blanchard 1996; Meyer-Bahlburg 1977; Rahman 2005; Rahman and Wilson 2003; Robinson and Manning 2000

18.2 Biological Basis of Homosexuality

Several data suggest that biological factors are important regulators of both heterosexual and homosexual orientation. The scientific study of human homosexual behavior is relatively recent and primarily centered around male homosexuality possibly because it threatens society more deeply than its female counterpart. The present chapter is primarily devoted to analyze male homosexuality, although we are aware of the increasing literature emerging in relation with the biological bases of female homosexuality (Mustanski et al. 2002; Peplau and Huppin 2008; Veniegas and Conley 2000), including hormonal factors (Meyer-Bahlburg 1979). The main evidences of genetic, autoimmune, and neurohormonal factors possibly causing male homosexuality are summarized in Table 18.1. In addition to these internal factors, others have proposed that maternal stress could favor the expression of male homosexuality. Regardless of the sources of male homosexuality, there is a growing body of evidence indicating anatomical and functional differences in the central nervous system (CNS) between homosexual and heterosexual men. In this chapter, we will give a general view of these features with special focus on the endocrine factors.

18.2.1 Genetic Influences

Family and twin studies have provided evidence for a genetic component in male homosexuality. Family studies have found a higher rate of homosexuality in the male siblings and the maternal uncles of homosexual men. Brothers of homosexual men had a median rate for homosexuality of 9 %, which is above the expected frequency (Pillard and Bailey 1995). Also, male homosexuality appears to be inherited more frequently from the maternal line (Bailey et al. 1999; Pillard et al. 1981, 1982), suggesting the existence of polymorphic and heritable maternal genes favoring male homosexual orientation. Dean Hamer's group proposed a gene linked to male homosexuality found in the distal region of the long arm of the X-chromosome, named Xq28 (Hamer et al. 1993). This result was replicated in a subsequent study (Hu et al. 1995) using a different sample of male homosexuals.

In this line, a recent study proposed new regions of genetic interest possibly involved in sexual orientation determination, for instance 7q36, which contain important candidate genes such as the vasoactive intestinal peptide receptor (Mustanski et al. 2005). This receptor is necessary for maintaining synchronous rhythms among neurons within the suprachiasmatic nucleus (Welsh et al. 2010) that has been proposed to be involved in homosexual orientation (*vide infra*). Interestingly, a group has suggested that the biological advantage of preserving such genes in the population is related with fertility. Thus, it has been reported that women related to homosexual men in the maternal line are more fertile (Camperio-Ciani et al. 2004, 2012; Lemmola and Camperio-Ciani 2009). Accordingly, other studies have confirmed that homosexual men have a high number of relatives from the maternal line (King et al. 2005; Rahman et al. 2008). Even the genetic causes of homosexuality is a promising field; no contributions have been made regarding the functional role of the proteins that may differ between homosexual and heterosexual men. In addition, some homosexual individuals do not share such genetic sequences with other homosexual men suggesting that other causative factors might be involved.

18.2.2 Fraternal Birth Order Effect

Recently, it was posed that there is a relationship between birth order and the incidence of male homosexuality. The occurrence of male homosexuality is positively correlated with the number of older brothers but not with the number of older sisters. This is referred to as the "fraternal birth order effect," in which each older brother increases the odds that a male will be homosexual by 33 % (Blanchard and Bogaert 1996a). This hypothesis proposes that this effect reflects the progressive immunization of some mothers to male-specific antigens by each succeeding male fetus. Male H–Y antigens are foreign to a woman's body, and thereby the female's immune system becomes more efficient to produce antibodies to attack a male fetus with each successive pregnancy (Blanchard and Klassen 1997; Bogaert and Skorska 2011).

18.2.3 Maternal Stress

Physical and emotional stress produces profound effects on the body. The stress of the mother during pregnancy alters the course of fetal development (Ferreira et al. present book) and several researchers have explored the effects of prenatal maternal stress on human sexual orientation, leading to controversial results. The first suggestion was made by Dörner, who surveyed a group of homosexuals born in Germany between 1934 and 1953. He found that a high number of homosexual men were born during and just after World War II, presumably during a stressful time in Germany. He assumed that women who were pregnant during and immediately after the war were more likely to experience stress (Dörner et al. 1980, 1983). Another group found that mothers exposed to stress during the second trimester of pregnancy were more prone to give birth to homosexual men (Ellis et al. 1988; Ellis and Cole-Harding 2001), while others failed to find an association between male homosexuality and prenatal stress (Bailey et al. 1991; Schmidt and Clement 1990). Even controversial, maternal stress could contribute to disrupt the endocrine milieu during pregnancy, thus affecting developmental processes involved in sexual orientation. The role of hormones during development in the establishment of sexual preference is largely discussed in an independent section (*vide infra*).

18.2.4 Anatomical and Functional Changes in CNS in Homosexual Men

As aforementioned, independently of the causes underlying homosexuality, a vast series of results show anatomical and functional changes in the CNS that appear to be related with a particular sexual orientation. Interestingly, most of these studies include Kinsey 6 "purely" homosexual subjects, thus intending to associate a particular feature with this particular sexual orientation. In this regard, it is worth mentioning that practically no research has been done in bisexual subjects, a category which has been even discussed by some authors (Dworkin 2001; Eliason 1997, 2001; Reiger et al. 2005). In 1990, Dick Swaab was the first to describe anatomical brain differences between homosexual and heterosexual men. He found that the suprachiasmatic nucleus is twice the size and contains a great number of the arginine vasopressin neurons in homosexual men in comparison with heterosexuals (Swaab and Hoffman 1990). Following this pioneer study, Simon LeVay reported that the third interstitial nucleus of the anterior hypothalamus (INAH-3) is larger in heterosexual men than in homosexuals (LeVay 1991). Another example was given by Laura Allen and Roger Gorski who showed that the midsagittal area of the anterior commissure is larger in homosexual men (Allen and Gorski 1992). Interestingly, this area plays a role in the inter-hemispheric integration of sensory information (Risse et al. 1978), suggesting that some behavioral traits that differ between sexual orientations have biological bases. In line, the size of the isthmus area of the corpus callosum is increased in homosexual men (Witelson et al. 2008). This region contains axons that connect right and left parietotemporal cortical areas and is involved in language and spatial cognition (Gazzaniga 2000). Remarkably, some of these anatomical brain differences are susceptible to be modified by sex hormones during development (Swaab et al. 1987, 1990), suggesting this as a contributory factor.

With respect to functional differences between homosexual and heterosexual men, it is necessary to mention the dissimilar brain activity in response to the antidepressant, fluoxetine (Kinnunen et al. 2004). This important finding suggests sexual-orientation differences in the activity of the serotoninergic system. In addition, recent studies have reported a divergent activation of the anterior hypothalamus in response to pheromones between homosexual and heterosexual men (Savic et al. 2005). The same group found sex-atypical cerebral asymmetry and functional connections in homosexual men in comparison with heterosexuals (Savic and Lindström 2008). Finally, numerous studies have found that homosexual men performed less well than their heterosexual counterpart in some spatial ability tests (McCormick and Witelson 1991; Neave et al. 1999; Rahman and Koerting 2008;

Rahman et al. 2003). All these studies indicate that the neural pathways differ between homosexual and heterosexual men. That is, sexual orientation finds its biological foundation in the wiring between specific neurons.

18.2.5 *Nonright-handedness*

Various studies have indicated that the prevalence of left-handedness is sexually dimorphic with 13 % in males and 11 % in females (Gilbert and Wysocki 1992; Peters et al. 2006). Geschwind and Galaburda (1985) proposed that an excess of prenatal testosterone disrupts the development of the left brain hemisphere so that corresponding regions on the right one develop more rapidly, resulting in nonright-handedness dominance. Some studies have associated male homosexuality and left-handedness because homosexuals show higher rates of nonright-handedness in comparison to heterosexuals (Lalumière et al. 2000; Lippa 2003; McCormick et al. 1990). These data suggest that high levels of testosterone, present during gestation, could be responsible for both processes.

18.3 Hormonal Influences

The role that sex hormones play in homosexual orientation has been widely researched. After the initial discovery that sex hormones, particularly testosterone, were able to induce male sexual behavior in animals, it was proposed that the homosexual orientation was due to low levels of serum testosterone (Meyer-Bahlburg 1977). In support, some studies show that homosexual men had reduced plasma testosterone levels and impaired spermatogenesis (Kolodny et al. 1971, 1972). However, more recent studies have consistently demonstrated a lack of difference in systemic hormone levels between heterosexual and homosexual men (Pillard et al. 1974; Savic et al. 2005). Furthermore, the male homosexuals' gonadotropin response is similar to that of heterosexuals after a challenge with exogenous steroids (Gooren 1986; Hendricks et al. 1989). In this line, there is also a lack of relationship between the density, hypothalamic distribution, and sequence of the androgen receptor and the sexual orientation (Kruijver et al. 2001; Macke et al. 1993). These data and others reinforce the idea that male homosexuality does not depend on changes in the effects of testosterone in adulthood. That is, there is no experimental support sustaining the idea that male homosexuality is due to endocrine alterations in adulthood.

However, we now know that hormones, particularly sex steroids, have a paramount role in the organization of the brain during development (Bao and Swaab 2011; Simerly 2002). Thus, the endocrine changes that underlie sexual orientation may have occurred in early stages. Some anthropometric characteristics have been explored in relation with sexual orientation, as the result of nonspecific effects of hormones during development, for example, the sexual dimorphic 2D:4D ratio (the ratio between the index finger and the ring finger, respectively) that could be used as a window to infer the prenatal hormone milieu. Usually, males show a reduced 2D:4D ratio because they have a 2D shorter than 4D, whereas females have a greater ratio due to the fact that both finger lengths are similar. It has been proposed that such ratio is a marker of prenatal exposure to androgens (Manning et al. 1998). Interestingly, homosexual men show a lower 2D:4D rations in comparison with heterosexuals (Rahman 2005; Rahman and Wilson 2003; Robinson and Manning 2000). In this line of thought, men and women have sex differences in height, weight, and onset of puberty. Thus, men, on average, are taller, heavier, and present a later onset of puberty as compared with women (Lee 1980; Tanner 1986). In contrast, homosexual men are significantly shorter, lighter (Blanchard and Bogaert 1996b; Bogaert 2010; Bogaert and Blanchard 1996), and show an earlier pubertal onset than their heterosexual counterpart (Bogaert and Blanchard 1996; Bogaert et al. 2002).

Another series of results that further support the role of the hormonal prenatal environment in the establishment of sexual orientation are those that used men and women with unusual sex hormone exposure during early development (Ellis and Ames 1987), for example, the case of daughters of women with congenital adrenal hyperplasia (CAH), characterized by abnormally high amounts of adrenal androgens due to a deficiency of the enzyme steroid 21 hydroxylase, required for cortisol biosynthesis. In this condition, the daughters of such women are exposed to high testosterone levels prenatally, which resemble those found in healthy male fetuses. Consequently, they are born with partially masculinized external genitalia and brains. Although the girls with CAH are treated to correct the hormone abnormality, they show an increased bisexuality and homosexuality and male-typical behavior across their life span (Meyer-Bahlburg et al. 2008; Zucker et al. 1996).

The direct analysis of in-uterus steroid hormone levels and sexual orientation of these subjects in adulthood has not been made. Even more difficult seems the relation between functional steroid action in uterus (e.g., brain androgen or estrogen receptor distribution) and later sexual orientation; indeed, present technique impediments preclude such putative associations. Nevertheless, the use of animal models permits to experimentally analyze this approach. However, the comparison of sexual behavior and orientation between animals and humans has to be cautiously considered primarily because animal studies are incapable to mimic the complex biological, psychological, and social factor interaction that underlies human sexuality (Pfaus et al. 2003), but we and others believe that the basic mechanisms underlying the process of sexual orientation/preference are conserved across the species.

18.4 Basic Research: Organization Effects of Sex Hormones

In mammals, sexual differentiation fundamentally results after the expression of the sex-divergent chromosomal charge, males possess an XY and females an XX. Genetic sex is determined at the time of fertilization by the entry of an X or a Y chromosome from the spermatozoon into the ovule. The main role of the sex chromosome is to differentiate the gonads as testes or ovaries. In males, the differentiation of the genital ridge into testis, with the formation of Sertoli cells, only occurs in the presence of a Y chromosome. Androgens are essential for normal male sex differentiation: During fetal development, testosterone promotes virilization of the urogenital tract and it stimulates the Wolffian ducts to develop and differentiate into the epididymides, seminal vesicles, and vasa deferentia. In the external genitalia, testosterone is rapidly transformed into 5α-dihydrotestosterone (DHT), by the enzyme 5α-reductase, to induce the development of the male urethra, prostate, penis, and scrotum. The development of the female sexual organs in the womb is primarily, but not exclusively, based upon the absence of androgens (Fleming and Vilain 2005).

The concept of hormonal regulation of sexual differentiation of the mammalian reproductive system was established by Jost's classic experiments in which the testes were removed from fetal male rabbits, inducing a female phenotype, and the transplantation of testis into female embryos producing a male phenotype at birth (Jost et al. 1973). Also classic is the organizational–activational hypothesis that comes from the study done by Phoenix and coworkers. They showed that the administration of testosterone to pregnant guinea pigs produced female offspring that showed increased capacity in adulthood for male-typical sexual behavior and decreased competence for female-typical sexual responses. They contrasted these early and permanent effects of testosterone—which is called organizational due to their changes in the organization of the neural systems—with the later and transient effects of hormones during and after puberty, named activational because they reflect transient activation of the previously organized systems (Phoenix et al. 1959).

A large body of evidence supports the notion that the sexual differentiation of the neural tissue destined to mediate sexual preference and behavior in the rat occurs as a result of steroid hormone action during development, specifically by estradiol which is synthesized via local aromatization of testosterone secreted by the testes (Lephart 1996). It should be taken into account that in primates, including men, the brain sexual differentiation process does not seem to depend exclusively upon the conversion of testosterone to estradiol. That is, testosterone and possibly other androgens exert an important role in this process (Amateau et al. 2004; Lenz et al. 2012; McCarthy 2008). In rodents, exposure to androgens during this period results in the organization of masculine behavior—masculinization (increased mount, intromission, and ejaculatory behaviors, for definition see Larsson 1979)—in the presence of a sexually receptive female and in a decrease in the probability to display feminine behavior in presence of a male—defeminization (lordosis, dorsiflexion reflex posture with the consequent pelvic elevation that allows the penis entrance into the vagina; Beach 1976).

18.5 Animal Models

According to the criteria proposed by Willner (Willner and Mitchell 2002), an ideal animal model requires to fulfill various validity criteria such as: face validity, that is, phenomenological similarity between the model and the disorder modeled (symptomatology), and construct validity, that is, the model either relies on or elucidates the same underlying mechanism responsible for the human alteration (causative factors; Willner 1984). Even these criteria are vastly discussed for animal models of psychiatric diseases and homosexuality was deleted as mental illness from the *Diagnostic and Statistical Manual* (*DSM*) in 1973; however, the appliance of the face and construct validity criteria to establish a good animal model of homosexuality remains actual. In the following section, we review some animal models of homosexuality that could be useful to understand the causes and the biological features of human homosexuality.

Perkins and Fitzgerald (1992) described that in domestic rams there is a small subpopulation that exclusively mount other rams even in the presence of estrous ewes. They named these animals male-oriented rams. The frequency of this phenotype (approximately 8%) is similar to that observed in humans. In ovine, as in other species, there is a sexually dimorphic cluster of neurons in the preoptic area that expresses high levels of aromatase mRNA; this area was named ovine sexual dimorphic nucleus (oSDN) and is larger in female-oriented rams than in male-oriented ones (Roselli et al. 2004). The size of this nucleus is controlled directly by the amount of androgen present during perinatal development. This observation suggests that during development, the brain of male-oriented rams was exposed to low levels of estradiol that may be critical for the establishment of same-sex preference in adulthood in this species (Roselli et al. 2011). The use of male-oriented rams is probably the best animal model of human homosexuality since it clearly fulfills the face validity criterion. Unfortunately, the use of rams as experimental subjects requires ad hoc expensive settings.

In rodents, several animal models of human homosexuality have been proposed (Baum 2006; Fausto-Sterling 1995). Most of them have used changes in the endocrine milieu (Bakker et al. 1993a), stress (Meek et al. 2006; Popova et al. 2011; Wang et al. 2006), and cohabitation with same-sex partners (Cibrian-Llanderal et al. 2012; Triana-Del Rio et al. 2011). Following the initial ideas proposed by Phoenix in the 1950s (*vide supra*), various research groups have administered testosterone or estradiol to pregnant dams and/or to neonatal pups and analyzed their preference and sexual behavior in adulthood (Henley et al. 2010; Manning and McGill 1974). Most studies found that the female offspring displayed high levels of male sexual orientation and behavior (Beach 1942; Clemens et al. 1970; de Jonge et al 1988; Whalen and Edwards 1967). Conversely, some studies analyzed the effect of the absence of androgens or estrogens in male offspring (usually by neonatal castration) and found that such animals, as adults, showed female sexual responses after treatment with estradiol plus progesterone (Brand and Slob 1991; Feder and Whalen 1965; Hendricks 1969; Merkx 1984). The important finding revealing that testosterone exerts its virilizing action through its aromatization to estradiol in the rat invited various authors to analyze the role of aromatase inhibitors on sexual preference and behavior (see Table 18.2).

Table 18.2 shows that the male rats that were prenatally and/or neonatally treated with various aromatase inhibitors showed female sexual behavior and orientation when adults. Within them, the Slob's group reported that the neonatal administration of 1,4,6-androstatriene-3,17 (ATD) induces bisexual behavior when males were

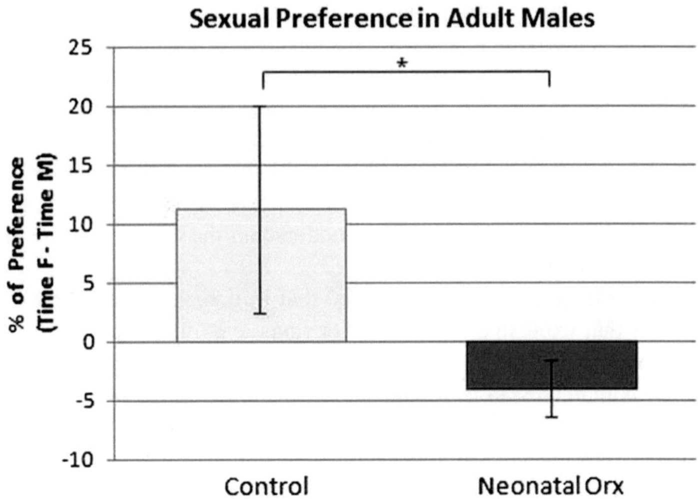

Fig. 18.1 The figure shows the mean ± S.E of the percentage of preference calculated as the difference in time with the female minus time with the male. Mann–Whitney U test, $*p = 0.06$

tested in a three-compartment box where they could sexually interact with both, a sexually active male or a receptive female. Interestingly, with the receptive female the ATD males displayed mount and intromissions, whereas with the male they exhibited proceptive (female sexual inviting behaviors such as hop and darting) and lordosis behaviors (Brand et al. 1991; Swaab et al. 1995). Interestingly, this partner-preference behavior shows a nocturnal rhythmicity: When tested late in the dark phase, ATD males show a clear preference for the female, while when tested early in the dark phase they showed a lesser preference for the female, or no preference at all (Bakker et al. 1993b, 1995). The nature of this circadian shift remains to be studied. Interestingly, the ATD male rats possess a higher number of vasopressin neurons in the suprachiasmatic nucleus (Swaab et al. 1995). This difference is similar to that found in homosexual men (*vide supra*; Swaab and Hoffman 1990), reinforcing the idea that the suprachiasmatic nucleus, particularly the neurons using vasopressin as neurotransmitter, participates in sexual orientation and suggesting that the modification of the endocrine milieu in these neurons during the sexual differentiation process participates in the establishment of male–male sexual orientation.

18.5.1 Neonatal Castration

In our laboratory, the first approximation to study the relationship between male-sex preference in rats and postnatal endocrine milieu was castrating male rats before the 5th day postnatally. Figure 18.1 shows the relevance of testicular hormones in the preference of a male for a receptive female in behavioral estrous using a Y maze. Intact control males showed a clear preference for an estrous female while

Table 18.2 Studies in rat preference and sexual behavior using aromatase inhibitors

Strain	Drug	Treatment	Sexual preference	Sexual behavior	References
CD	1,4,6-androstatri-ene-3,17, dione (ATD)	Pups received implant on 2nd postnatal day and it was removed on day 10 postnatally		Castrated in adulthood and under treatment with estradiol and progesterone: 89 % displayed lordosis and the majority showed ear wiggling and some darts	McEwen et al. 1977
Long–Evans	1,4,6-androstatri-ene-3,17, dione (ATD)	Pregnant females received 1 mg/day or 5 mg/day from day 10 to 22 of gestation		Castrated in adulthood and treated with several doses of estradiol plus progesterone. Dose-response effect on lordosis. High receptivity after administration of estradiol plus progesterone	Clemens and Gladue 1978
Sprague Dawley	Androst-4-ene-3,6,17-trione (ADT)	Pups were castrated within 24 h of birth and received 500 μg during the first 5 days of life		Estradiol alone induced lordosis Cyclic gonadotropin secretion in males Diminution of male sexual behavior after testosterone propionate	Booth 1978
CD	1,4,6-androstatri-ene-3,17, dione (ATD)	Pups received implant on 2nd postnatal day and it was removed on day 10 postnatally	Gonadally intact; more time with the female than with the male; similar to controls	Gonadally intact: normal male sexual behavior, 50 % displayed lordosis behavior with male Castrated under treatment with estradiol and progesterone: 100 % displayed lordosis and the78 % showed solicitation behavior	Davis et al. 1979
Long–Evans	1,4,6-androstatri-ene-3,17, dione (ATD)	Pregnant females received 5 mg/day from day of gestation 10 to 21 or 10 to 22 or 10 to 23 or 20 to 23		Castrated in adulthood and treated with several doses of estradiol plus progesterone. Some displayed lordosis	Whalen and Olsen 1981
Sprague Dawley	1,4,6-androstatri-ene-3,17, dione (ATD)	Pregnant females received 5 mg/day from day of gestation 10 to 22 or 10 to 24 or 20 to 22 or 20 to 24 of gestation		Castrated in adulthood and with treatment with several doses of estradiol plus progesterone displayed lordosis, but lower in comparison with Long–Evans	

Table 18.2 (continued)

Strain	Drug	Treatment	Sexual preference	Sexual behavior	References
Wistar	1,4,6-androstatriene-3,17, dione (ATD)	1) Prenatal: pregnant females with silicone implant from 11th day of gestation until parturition 2) Prenatal and neonatal: pregnant females with silicone implant from 11th day of gestation until parturition; pups received an implant within 9 h after birth until 10 days of age 3) Neonatal: pups received an implant within 9 h after birth until 21 days of age	Gonadally intact; Prenatal: preference for a receptive female Prenatal and neonatal: lower preference for the estrous female and high preference for active male	High lordosis after administration of progesterone Prenatal and neonatal: decrease ejaculation with female. High lordosis when mounted by a male, there is an increase after administration of progesterone Neonatal: reduction in ejaculation with receptive female. High lordosis when mounted by male and increase after administration of progesterone	Brand et al. 1991
Wistar	1,4,6-androstatriene-3,17, dione (ATD)	Pups received an implant within 9 h after birth until 21 days of age	Gonadally intact showed a lower preference for receptive female Castrated in adulthood display lower preference for receptive female Castrated in adulthood and tested between 63 and 84 days of age with estradiol for 3 weeks, the rats displayed preference for a sexually active male Castrated in adulthood tested between 63 and 84 days of age and with estradiol plus dihydrotestosterone for 5 weeks. Animals maintain the lower preference for a receptive female	Gonadally intact lower number of mounts and intromissions with receptive female Castrated in adulthood showed a reduction in the number of mounts and intromissions with receptive female Castrated in adulthood and tested between 63 and 84 days of age with estradiol for 3 weeks showed a reduction in the number of mounts and intromissions with receptive female Castrated in adulthood tested between 63 and 84 days of age and with estradiol plus dihydrotestosterone for 5 weeks. Increase in the number of mounts and intromissions with receptive female	Bakker et al. 1993a

Table 18.2 (continued)

Strain	Drug	Treatment	Sexual preference	Sexual behavior	References
Wistar	1,4,6-androstatri-ene-3,17, dione (ATD)	1) Pups received an implant within 2–4 h after birth until 14 days of age 2) Pups received an implant of day 2 after birth until 14 days of age 3) Pups received an implant of day 5 after birth until 14 days of age	Gonadally intact showed preference for receptive female in the late dark phase and in the early dark phase preferred the sexually active male		Bakker et al. 1993b
Wistar	1,4,6-androstatri-ene-3,17, dione (ATD)	Pups received an implant within 2–4 h after birth until 21 days of age	Gonadally intact showed preference for receptive female in the late dark phase and in the early dark phase preferred the sexually active male Castrated at 7 months and tested with testosterone implant, the preference for the active male increased over test Gonadally intact and castrated showed proceptive and receptive behaviors in the early part of the dark period		Bakker et al. 1995

Table 18.2 (continued)

Strain	Drug	Treatment	Sexual preference	Sexual behavior	References
Wistar	1,4,6-androstatriene-3,17, dione (ATD)	1) Prenatal: pregnant females received 5 mg/day from day 10 to 22 of gestation 2) Prenatal and neonatal: pregnant females received 5 mg/day from day 10 to 22 of gestation and pups received an implant within 9 h after birth until 21 days of age	Gonally intact; Prenatal and neonatal group had lower preference for the estrous female	Prenatal and neonatal group had lower male sexual behavior	Houtsmuller et al. 1994
Wistar	1,4,6-androstatriene-3,17, dione (ATD)	1) Prenatal: pregnant females received 5 mg/day from day 10 to 22 of gestation 2) Prenatal and neonatal: pregnant females received 5 mg/day from day 10 until parturition and pups received an implant between 3 and 9 h after birth until 21days of age	Gonally intact; Prenatal and neonatal group had lower preference for the estrous female	During the early dark phase, prenatal rats displayed the typical male behavior with a estrous female Prenatal and neonatal: lordosis when tested with a sexually active male	Swaab et al. 1995
Wistar	Letrozole	Pregnant females received 1 mg/kg/day on days 21 and 22 of gestation		Castrated in adulthood and under treatment with testosterone propionate 24 h before the test, less than 50 % showed a normal male sexual behavior in presence of receptive female Castrated in adulthood and under treatment with estradiol benzoate 24 hour before the test, the 25 % showed female sexual behavior	Gerardin and Pereira 2002

neonatally castrated males displayed a male preference. It has to be mentioned that both groups were tested for sexual preference without any adult manipulation and without previous sexual behavior training tests. This study, however, does not analyze if the essential role of these hormones occurs during early development or in adulthood because, as aforementioned, the neonatally orchidectomized males were untreated as adults. In line, Brand and Slob (1991) explored the role of neonatal castration on sexual preference in the three-compartment box. Clearly, these rats without treatment in adulthood had a lower female preference than controls, but after 3 weeks of testosterone treatment they chose to spend more time with an estrous female than with a male. These data reveal that the low preference of a neonatally castrated male for a receptive female is more likely due to the lack of an activational effect of androgens.

In addition, the castration approach is far from considering the causative factors underlying human homosexuality. That is, such sexual preference in humans is not the result of gonadal loss during development or later in life.

18.5.2 Prenatal Administration of Letrozole

Following the idea that homosexuality is caused by atypical sex steroid levels in uterus, our next approximation consisted in altering the levels of estradiol with an aromatase inhibitor, letrozole. Female rats in proestrus were timed mated (the day of mating was considered as day 0 of pregnancy, and the presence of spermatozoa in a vaginal smear confirmed pregnancy). These dams received a daily subcutaneous (s.c.) injection of letrozole at doses of 0.31 and 0.56 µg/kg in 0.1 ml (dissolved in corn oil) or vehicle from day 10 of pregnancy until parturition. This procedure was similar to that used by Houtsmuller et al. (1994). On the day of birth, the pups were culled to five males and five females and they were weaned at 21 days of age. The animals were tested for partner preference and sexual behavior when they were 3–4 months old. The testing apparatus for sexual preference consisted in a three-compartment box similar to that used by Brand et al. (1991). The lateral compartments contained the stimulus animals: These incentives were a sexually active male and an estrous female (induced by the sequential s.c. injections of estradiol benzoate, 8 µg/rat, followed by progesterone, 2 mg/rat, 24 and 4 h before the test, respectively).

In line with previous reports (Brand et al. 1991; Bakker et al. 1995), control subjects spent more time in the female chamber in the preference test (see Fig. 18.2). The lower dose of letrozole (0.31 µg/kg) produced male rats with female preference and only one subject displayed a male-preference behavior, accompanied by proceptive behaviors. The amount of time spent with a male increased after raising the letrozole dose (0.56 µg/kg). Thus, 8 out of 27 males displayed a clear preference for the sexually active male. These subjects formed the group termed 0.56 µg/kg MP. Such increased preference for the male chamber was unaccompanied by changes in the time in the middle compartment suggesting specific actions. In general, only one male of the litter (of mothers treated with letrozole at 0.56 µg/kg) showed a

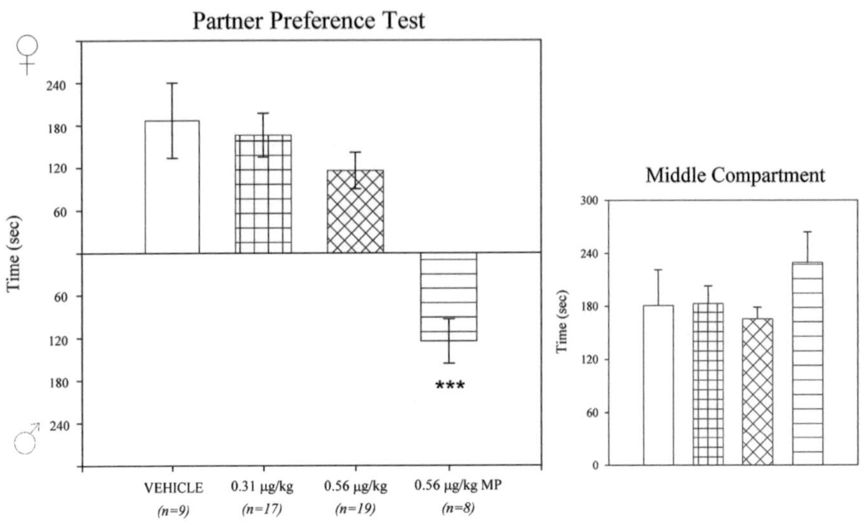

Fig. 18.2 The figure shows mean ± S.E of difference in the time spent in the estrus female compartment minus the time spent in the male compartment. Kruskal–Wallis ANOVA (H=20.56) followed by Dunn's test: differences versus others groups, ***p 0.56) followed by Dunn's test

Table 18.3 Sexual behavior in the presence of a receptive female or a sexually active male

	Female			Male			
Treatment	Mounts	Intromissions	Ejaculations	Mounts to male	Mounts from male	Lordosis	Lordosis quotient
Vehicle (7/9)	100% (7/7)	100% (7/7)	71% (5/7)	67% (6/9)	100% (9/9)	0% (0/9)	–
Let 0.31 µg/kg (11/17)	100% (11/11)	91% (10/11)	18% (2/11)*	53% (9/17)	53% (9/17)	12% (2/17)	82%
Let 0.56 µg/kg (16/19)	100% (16/16)	81% (13/16)	13% (1/16)**	58% (11/19)	53% (10/19)	16% (3/19)	90%
Let 0.56 µg/kg MP (6/8)	100% (6/6)	100% (6/6)	67% (4/6)	100% (8/8)	100% (8/8)	63% (5/8)**	96%

MP male preference
Fisher test *p<0.05, **p <0.01 versus vehicle

male-oriented preference. The reason for such proportion and the causes underlying why only one subject was altered remains to be studied.

Using a cylindrical cage, these subjects were forced to interact with a receptive female and with a sexually active male during 30 min. The sexual behavior of the animals treated with the high letrozole dose (0.56 µg/kg) was analyzed dependently for the preference for the receptive female or the sexually active male. The results of this experiment are summarized in Table 18.3.

In the presence of a sexually receptive female, most of these males, regardless of the prenatal treatment, showed mounting and intromitting behaviors with a similar median number of mounts (data not shown). However, most control males (71 %) showed ejaculation in this period, while much less percentages (13 and 18 %) were found in the letrozole-treated animals with a female preference. That is, most letrozole-treated animals were sluggish. Interestingly, the subjects that had a male preference displayed a similar percentage of ejaculatory behavior than controls. When faced with a sexually active male, 67 % of the control animals mounted the stimulus male and all of the control animals received mounts from the active male. However, no control subject displayed lordosis after being mounted. Around half of the males prenatally treated with letrozole that showed a female preference were mounted or mounted the stimulus male; in those groups only 2 out of 17 animals treated with the low letrozole dose (0.31 μg/kg) and 3 out of 19 with the high dose (0.56 μg/kg) performed lordosis behavior consistently (with high lordosis quotients established by dividing the number of lordosis between the numbers of mounts). Interestingly, all the males prenatally treated with the high dose of letrozole (0.56 μg/kg MP) that preferred the male showed mounts toward and from the stimulus male and most of them (63 %) displayed lordosis behavior.

In basic research, most studies analyzing the adult effects of altering the hormonal milieu during gestation have studied performance aspects of masculine or feminine sexual behavior (see Table 18.2). However, an important aspect of precopulatory behavior is the preference for a potential sexual partner. When given the choice for mating partners, most of the individuals usually choose a partner of the opposite sex. The prenatal inhibition in estradiol synthesis, by the aromatase inhibitor letrozole, clearly shifted the sexual preference of some of the male offspring. This approach is in line with the theory of Ellis and Ames (1987) stating that alterations in the endocrine milieu underlie male homosexuality, at least partly fulfilling the construct validity criterion of animal models. In addition, the preference and sexual behavior displayed by these males resemble human homosexual orientation primarily because the animal is free to choose between a receptive female and a sexually active male, satisfying the face validity condition (*vide supra*).

In female rodents, typical sexual behavior is the result of the sequential hormonal priming with estrogen and progesterone which promote the display of receptivity and proceptivity (Beach 1976). Receptivity is usually determined by the lordosis quotient. It is known that estrogen alone (usually at high doses or after repeated treatment) is sufficient to elicit full lordosis, but progesterone facilitates this behavior (Beach 1976; Whalen 1974). Proceptivity includes various behaviors (ear wiggling, hopping, and darting) that the female exhibits to stimulate and increase the male's attention (de Jonge et al. 1986; Erskine 1989). The expression of these behaviors requires both estrogen and progesterone (Albert et al. 1991; Beach 1976; Clark et al. 2004), and may be fully presented by neonatally castrated males treated with these hormones (Brand and Slob 1991; Vega-Matuszczyk et al. 1988). Furthermore, it is also known that estradiol and testosterone participate in the female's preference for a sexually active male (de Jonge et al. 1986; Vega-Matuszczyk et al. 1988). All these data indicate that these parameters are highly influenced by the adult hormonal profile. Thereby, it is of crucial importance to know the adult serum levels of testosterone, estradiol, and progesterone of prenatally letrozole-treated males in order to establish if the changes in sexual behavior and preference are related with adult hormonal variations. This information is also important to establish if this is a good animal model of human homosexuality, since, as aforementioned, homosexual men do not have an altered endocrine profile.

It is remarkable that various males prenatally treated with letrozole have a male sexual preference and—when forced to interact—displayed all components of female sexual behavior. Conversely, few other letrozole-treated males showed preference for the female but when faced to the male displayed lordosis and soliciting behaviors. Finally, a third population, mostly formed by control and low-dose letrozole-treated subjects, had a clear preference for the female but did not show female sexual behavior. The two former groups could represent the biological behavioral diversity that characterizes human homosexuality (LeVay 2011). In addition—similar to controls—most of the males treated with letrozole showed typical masculine sexual behavior, suggesting that the masculinization process does not depend upon the androgen aromatization in late pregnancy or that in this species "pure" homosexual subjects cannot be experimentally found. In favor of the latter, no experimental manipulation, besides perinatal castration, leads males with only feminine sexual behavior (Fausto-Sterling 1995).

It is worth mentioning that the treatment with the aromatase inhibitor ATD produces males that develop male preference after several sexual behavior trainings (Bakker et al. 1993b, 1995). That is, experience seems to account for the sexual preference development. An advantage of letrozole is that some of the animals prenatally treated showed such preference and behavior in the first test. These results invited to study the partner preference along several assays in letrozole-treated males. At present we are analyzing their preference repeatdly each 15 days in males from 45 to 100 days. Another future research line is to analyze if the circadian-dependent changes in sexual preference in perinatally treated ATD males (Bakker et al. 1993b, 1995) is also present in letrozole-treated animals. Finally, it is worth studying if the males sexually oriented to other males possess higher number of vasopressinergic neurons in the suprachiasmatic nucleus, as do ATD-treated rats (Swaab et al. 1995) and homosexual men (Swaab and Hoffman 1990).

18.5.3 Prenatal Administration of Diethylstilbestrol

As previously stated, it has been proposed that an excess of steroids (androgens or estrogens) during development may also account for men homosexuality (*vide supra*). That is, it is suggested that either excess or a deficiency of androgens may produce a homosexual orientation. Diethylstilbestrol (DES) is a potent synthetic non-steroidal estrogen that was administered to pregnant women (from 1947 to 1971) in an effort to preserve pregnancy (Ehrhardt and Meyer-Bahlburg 1981; Noller and Fish 1974). In addition, in 1985 there was a study revealing that the women exposed to high levels of this compound during pregnancy had an increased prevalence in

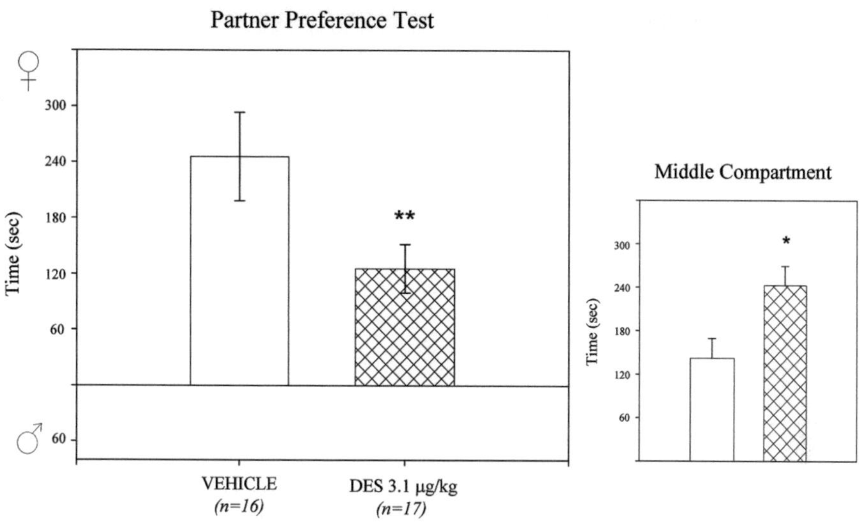

Fig. 18.3 The figure shows mean ± S.E of difference in the time spent in the estrus female compartment minus the time spent in the male compartment. Mann–Whitney test $*p<0.05$, $**p<0.01$

their daughters to show bisexuality or homosexuality (Ehrhardt et al. 1985; Meyer-Bahlburg et al. 1995). These data reinforce the notion that sexual orientation in humans is endocrinologically influenced during development. On the bases of these results, we performed the following experiment: DES was daily injected to pregnant females from the 7th gestation day until delivery at the dose of 3.1 µg/kg/day. This schedule was similar to that done by Kobayasy et al. (2009) to disrupt the steroidogenesis and spermatogenesis. In this experiment, we followed the same methodology as previously described.

Figure 18.3 shows the preference of male rats prenatally treated with DES. Clearly, in these animals there is a reduction in the time they spent in the sexually receptive female compartment. Such reduction, however, was not accompanied by an increase in the male area but in the neutral middle compartment. This finding suggests that prenatal DES treatment renders unspecific actions on sexual preference or produces less sociable rats.

Table 18.4 summarizes the sexual behavior shown by control and prenatally treated DES males when forced to interact with a receptive female or with an active male. When both the control and the DES-treated subjects interacted with a receptive female, all displayed typical male sexual behavior at comparable levels. In the presence of a sexually active male, around of 28 % of the subjects prenatally treated with DES displayed lordosis, while no control male showed this behavior, suggesting that DES partly feminized these subjects.

The reduced preference for the female seen in DES-treated rats is most likely due to the low serum testosterone and estradiol levels found in these subjects as adults (data not shown). This prenatal manipulation increased estrogenic activity during development, but failed to shift the sexual preference. Various reasons may account

Table 18.4 Sexual behavior of subjects prenatally treated with DES

	Female			Male			
Treatment	Mounts	Intromissions	Ejaculations	Mounts to sexual male	Mounts from sexual male	Lordosis	Lordosis quotient
Vehicle (11/16)	100% (11/11)	100% (11/11)	71% (7/11)	75% (12/16)	75% (12/16)	0% (0/16)	–
DES 3.1 μg/kg (16/18)	100% (16/16)	100% (16/16)	92% (16/11)	83% (15/18)	61% (11/18)	28% (5/18)*	43%

Fisher test: $^{*}p<0.05$

for this finding including the DES dose and time of treatment. Thus, it has been suggested that there is a window during gestation that is crucial for the development of partner preference (Vega Matuszczyk and Larsson 1995; Vega Matuszczyk et al. 1988). Additionally, it has been suggested that an excess of prenatal androgens, rather than estrogens, accounts for left-handedness and male homosexuality (Holtzen 1994; Lindesay 1987).

In comparison with letrozole-treated animals, this manipulation provoked a lower proportion of animals showing lordosis with a lower lordosis quotient and failed to alter sexual preference. These results, taken together, suggest that letrozole administered prenatally is a better method than DES to investigate changes in sexual preference and behavior.

In closing, we would like to remark that from the earliest stages of fetal brain development several areas have steroid receptors (Kühnemann et al. 1995; Vito and Fox 1981). These hormones are necessary for the development of brain structures and circuits that will be maintained during the rest of life. Consequently, factors that disrupt the interaction between hormones and the developing brain would permanently influence behavior (Bao and Swaab 2011; Meyer-Bahlburg et al. 1996). The rising in hormone levels during puberty activates circuits that were built during fetal development, and thereby stimulates the expression of behavioral patterns created in gestation (Schulz et al. 2009; Vigil et al. 2011). The series of present data support the idea that the interaction of sex hormones with different brain structures along development determines sexual preference. Finally, the use of animal models of human sexual orientation would permit important research on the biological causes determining sexual preference and on the influence of biological factors on the development of associated features such as cognitive, psychiatric alteration, and even psychoactive drug-response peculiarities.

Acknowledgments Authors wish to thank Rebeca Reyes (M.Sc.), Martha Luna (M.Sc.), and Héctor Vázquez Espinosa for technical support and Conacyt for fellowship (219853) to S.O.H.

Conflicts of Interest The authors declare no conflicts of interest.

References

Albert DJ, Jonik RH, Gorzalka BB et al (1991) Serum estradiol concentration required to maintain body weight, attractivity, proceptivity, and receptivity in the ovariectomized female rat. Physiol Behav 49:225–231.

Allen LS, Gorski RA (1992) Sexual orientation and the size of the anterior commissure in the human brain. Proc Natl Acad Sci 89:7199–7202

Amateau SK, Alt JJ, Stamps CL et al (2004) Brain estradiol content in newborn rats: sex differences, regional heterogeneity, and possible de novo synthesis by the female telencephalon. Endocrinology 145:2906–2917

Bagley C, Tremblay P (1998) On the prevalence of homosexuality and bisexuality, in a random community survey of 750 men aged 18 to 27. J Homosex 36:1–18

Bailey JM, Willerman L, Parks C (1991) A test of the maternal stress theory of human male homosexuality. Arch Sex Behav 20:277–293

Bailey JM, Pillard RC, Dawood K et al (1999) A family history study of male sexual orientation using three independent samples. Behav Genet 29:79–86

Bakker J, Brand T, van Ophemert J et al (1993a) Hormonal regulation of adult partner preference behavior in neonatally ATD-treated male rats. Behav Neurosci 107:480–487

Bakker J, van Ophemert J, Slob AK (1993b) Organization of partner preference and sexual behavior and its nocturnal rhythmicity in male rats. Behav Neurosci 107:1049–1058

Bakker J, van Ophemert J, Timmerman MA et al (1995) Endogenous reproductive hormones and nocturnal rhythms in partner preference and sexual behavior of ATD-treated male rats. Neuroendocrinology 62:396–405.

Balthazart J (ed) (2012) The biology of homosexuality. Oxford University Press, New York

Bao AM, Swaab DF (2011) Sexual differentiation of the human brain: relation to gender identity, sexual orientation and neuropsychiatric disorders. Front Neuroendocrinol 32:214–226

Baum MJ (2006) Mammalian animal models of psychosexual differentiation: when is 'translation' to the human situation possible? Horm Behav 50:579–588

Beach FA (1942) Male and female mating behavior in prepuberally castrated female rats treated with androgens. Endocrinology 31: 373–378

Beach FA (1976) Sexual attractivity, proceptivity, and receptivity in female mammals. Horm Behav 7:105–138

Blanchard R, Bogaert AF (1996a). Homosexuality in men and number of older brothers. Am J Psychiatry 153:27–31

Blanchard R, Bogaert AF (1996b) Biodemographic comparisons of homosexual and heterosexual men in the Kinsey Interview Data. Arch Sex Behav 25:551–579

Blanchard R, Klassen P (1997) H-Y antigen and homosexuality in men. J Theor Biol 185:373–378

Bogaert AF (2010) Physical development and sexual orientation in men and women: an analysis of NATSAL-2000. Arch Sex Behav 39:110–116

Bogaert AF, Blanchard R (1996) Physical development and sexual orientation in men: height, weight, and age of puberty differences. Pers Individ Dif 21:77–84

Bogaert AF, Friesen C, Klentrou P (2002) Age of puberty and sexual orientation in a national probability sample. Arch Sex Behav 31:73–81

Bogaert AF, Skorska M (2011) Sexual orientation, fraternal birth order, and the maternal immune hypothesis: a review. Front Neuroendocrinol 32:247–254

Booth JE (1978) Effects of the aromatization inhibitor androst-4-ene-3,6,17-trione on sexual differentiation induced by testosterone in the neonatally castrated rat. J Endocrinol 79:69–76

Brand T, Slob AK (1991) Neonatal organization of adult partner preference behavior in male rats. Physiol Behav 49:107–111

Brand T, Kroonen J, Mos J et al (1991) Adult partner preference and sexual behavior of male rats affected by perinatal endocrine manipulations. Horm Behav 25:323–341

Camperio-Ciani A, Corna F, Capiluppi C (2004) Evidence for maternally inherited factors favouring male homosexuality and promoting female fecundity. Proc Biol Sci 271:2217–2221

Camperio-Ciani A, Fontanesi L, Lemmola F et al (2012) Factors associated with higher fecundity in female maternal relatives of homosexual men. J Sex Med 9:2878–2887
Cibrian-Llanderal T, Rosas-Aguilar V, Triana-Del RR et al (2012) Enhaced D2-type receptor activity facilitates the development of conditioned same-sex partner preference in male rats. Pharmacol Biochem Behav 102:177–183
Clark AS, Kelton MC, Guarraci FA et al (2004) Hormonal status and test condition, but not sexual experience, modulate partner preference in female rats. Horm Behav 45:314–323.
Clemens LG, Gladue BA (1978) Feminine sexual behavior in rats enhanced by prenatal inhibition of androgen aromatization. Horm Behav 11:190–201
Clemens LG, Shryne J, Gorski RA (1970) Androgen and development of progesterone responsiveness in male and female rats. Physiol Behav 5:673–678
Davis PG, Chaptal CV, McEwen BS (1979) Independence of the differentiation of masculine and feminine sexual behavior in rats. Horm Behav 12:12–19
de Jonge FH, Eerland EM, van de Poll NE (1986) The influence of estrogen, testosterone and progesterone on partner preference, receptivity and proceptivity. Physiol Behav 37:885–891
de Jonge FH, Muntjewerff, Louwerse AL et al (1988) Sexual behavior and sexual orientation of the female rat after hormonal treatment during various stages of development. Horm Behav 22:100–115
Diamond M (1993) Homosexuality and bisexuality in different populations. Arch Sex Behav 22:291–310
Diamond M, Sigmundson HK (1997) Sex reassignment at birth: long-term review and clinical implications. Arch Pediatr Adolesc Med Mar 151:298–304
Dörner G, Geier T, Ahrens L et al (1980) Prenatal stress as possible aetiogenetic factor of homosexuality in human males. Endokrinologie 75:365–368
Dörner G, Schenk B, Schmiedel B et al (1983) Stressful events in prenatal life of bi- and homosexual men. Exp Clin Endocrinol 81:83–87
Dworkin SH (2001) Treating the bisexual client. J Clin Psychol 57:671–680
Ehrhardt AA, Meyer-Bahlburg HF (1981) Effects of prenatal hormones on gender-related behavior. Science 211:1312–1318
Ehrhardt AA, Meyer-Bahlburg HF, Rosen LR et al (1985) Sexual orientation after prenatal exposure to exogenous estrogen. Arch Sex Behav 14:57–77
Eliason M (1997) The prevalence and nature of biphobia in heterosexual undergraduate students. Arch Sex Behav 26:317–326
Eliason M (2001) Bi-negativity: the stigma facing bisexual men. In Brett B, Steinman E (eds) Bisexuality in the lives of men: facts and fictions, 1st edn. Haworth, Binghamton, pp 137–154
Ellis L, Ames MA (1987) Neurohormonal functioning and sexual orientation: a theory of homosexuality-heterosexuality. Psychol Bull 101:233–258
Ellis L, Cole-Harding S (2001) The effects of prenatal stress, and of prenatal alcohol and nicotine exposure, on human sexual orientation. Physiol Behav 74:213–226
Ellis L, Ames MA, Peckham W et al (1988) Sexual orientation of human offspring may be altered by severe maternal stress during pregnancy. J Sex Res 25:152–157
Erskine MS (1989) Solicitation behavior in the estrous female rat: a review. Horm Behav 23:473–502
Fausto-Sterling A (1995) Animals models for the development of human sexuality: a critical evaluation. J Homosex 28:217–236
Feder HH, Whalen RE (1965) Feminine behavior in neonatally castrated and estrogen-treated male rats. Science 147:306–307
Fleming A, Vilain E (2005) The endless quest for sex determination genes. Clin Genet 67:15–25.
Gazzaniga MS (2000) Cerebral specialization and interhemispheric communication: does the corpus callosum enable the human condition? Brain 123:1293–1326
Gerardin DC, Pereira OC (2002) Reproductive changes in male rats treated perinatally with an aromatase inhibitor. Pharmacol Biochem Behav 71:301–315
Geschwind N, Galaburda AM (1985) Cerebral lateralization: biological mechanisms, associations, and pathology: I. A hypothesis and a program for research. Arch Neurol 42:428–459

Gilbert AN, Wysocki CJ (1992) Hand preference and age in the United States. Neuropsychologia 30:601–608
Gooren L (1986) The neuroendocrine response of luteinizing hormone to estrogen administration in heterosexual, homosexual, and transsexual subjects. J Clin Endocrinol Metab 63:583–588
Gooren L (2006) The biology of human psychosexual differentiation. Horm Behav 50:589–601.
Hamer DH, Hu S, Magnuson VL et al (1993) A linkage between DNA markers on the X chromosome and male sexual orientation. Science 261:321–327
Hendricks SE (1969) Influence of neonatally administered hormones and early gonadectomy on rats' sexual behavior. J Comp Physiol Psychol 69: 408–413.
Hendricks SE, Graber B, Rodriguez-Sierra JF (1989) Neuroendocrine responses to exogenous estrogen: no differences between heterosexual and homosexual men Psychoneuroendocrinology 14:177–185
Henley CL, Nunez AA, Clemens LG (2010) Exogenous androgen during development alters adult partner preference and mating behavior in gonadally intact male rats Horm Behav 57:488–495
Holtzen DW (1994) Handedness and sexual orientation. J Clin Exp Neuropsychol 16:702–712
Houtsmuller EJ, Brand T, de Jonge FH et al (1994) SDN-POA volume, sexual behavior, and partner preference of male rats affected by perinatal treatment with ATD. Physiol Behav 56:535–541.
Hu S, Pattatucci AM, Patterson C et al (1995) Linkage between sexual orientation and chromosome Xq28 in males but not in females. Nat Genet 11:248–256
Jost A, Vigier B, Prépin J et al (1973) Studies on sex differentiation in mammals. Recent Prog Horm Res 29:1–41
King M, Green J, Osborn DP et al (2005) Family size in white gay and heterosexual men. Arch Sex Behav 34:117–122
Kinnunen LH, Moltz H, Metz J et al (2004) Differential brain activation in exclusively homosexual and heterosexual men produced by the selective serotonin reuptake inhibitor, fluoxetine. Brain Res 1024:251–254
Kinsey AC, Pomery WB, Martin CE (eds) (1948) Sexual behavior in the human male. Saunders, Philadelphia
Kobayasy T, Shirai M, Sakaue M et al (2009) Effects of maternal exposure to low doses of DES on testicular steroidogenesis and spermatogenesis in male rat offspring. J Reprod Dev 55:629—637
Kolodny RC, Masters WH, Hendryx J et al (1971) Plasma testosterone and semen analysis in male homosexuals. N Engl J Med 285:1170–1174
Kolodny RC, Jacobs LS, Masters WH et al (1972) Plasma gonadotrophins and prolactin in male homosexuals. Lancet 2:18–20
Kruijver FP, Fernández-Guasti A, Fodor M et al (2001) Sex differences in androgen receptors of the human mamillary bodies are related to endocrine status rather than to sexual orientation or transsexuality. J Clin Endocrinol Metab 86:818–827
Kühnemann S, Brown TJ, Hochberg RB et al (1995) Sexual differentiation of estrogen receptor concentrations in the rat brain: effects of neonatal testosterone exposure. Brain Res 691:229–234
Lalumière ML, Blanchard R, Zucker KJ (2000) Sexual orientation and handedness in men and women: a meta-analysis. Psychol Bull 126:575–592
Larsson K (1979) Features of the neuroendocrine regulation of masculine sexual behavior. In Beyer C (ed) Endocrine control of sexual behavior, 1st edn. Raven, New York, pp 77–163
Lee PA (1980) Normal ages of pubertal events among American males and females. J Adolesc Health Care 1:26–29
Lemmola F, Camperio-Ciani A (2009) New evidence of genetic factors influencing sexual orientation in men: female fecundity increase in the maternal line. Arch Sex Behav 38:393–399
Lenz KM, Nugent BM, McCarthy MM (2012) Sexual differentiation of the rodent brain: dogma and beyond. Front Neurosci 6:1–13
Lephart ED (1996) A review of brain aromatase cytochrome P450. Brain Res Brain Res Rev 22:1–26

LeVay S (1991) A difference in hypothalamic structure between heterosexual and homosexual men. Science 253:1034–1037
LeVay S (ed) (2011) Gay straight, and the reason why: the science of sexual orientation. Oxford University Press, New York
Lindesay J (1987) Laterality shift in homosexual men. Neuropsychologia 25:965–969
Lippa RA (2003) Handedness, sexual orientation, and gender-related personality traits in men and women. Arch Sex Behav 32:103–114
Macke JP, Hu N, Hu S et al (1993) Sequence variation in the androgen receptor gene is not a common determinant of male sexual orientation. Am J Hum Genet 53:844–852
Manning A, McGill TE (1974) Neonatal androgen and sexual behavior in female house mice. Horm Behav 5:19–31.
Manning JT, Scutt D, Wilson J et al (1998) The ratio of 2nd to 4th digit length: a predictor of sperm numbers and concentrations of testosterone, luteinizing hormone and oestrogen. Hum Reprod 13:3000–3004
McCarthy MM (2008) Estradiol and the developing brain. Physiol Rev 88:91–124
McCormick CM, Witelson SF (1991) A cognitive profile of homosexual men compared to heterosexual men and women. Psychoneuroendocrinology 16:459–473
McCormick CM, Witelson SF, Kingstone E (1990) Left-handedness in homosexual men and women: neuroendocrine implications. Psychoneuroendocrinology 15:69–76
McEwen BS Lieberburg I, Chaptal C et al (1977) Aromatization: important for sexual differentiation of the neonatal rat brain. Horm Behav 9:249–263
Meek LR, Schulz KM, Keith CA (2006) Effects of prenatal stress on sexual partner preference in mice. Physiol Behav 89:133–138
Merkx J (1984) Effect of castration and subsequent substitution with testosterone, dihydrotestosterone and oestradiol on sexual preference behaviour in the male rat. Behav Brain Res 11:59–65
Meyer-Bahlburg HF (1977) Sex hormones and male homosexuality in comparative perspective. Arch Sex Behav 6:297–325
Meyer-Bahlburg HF (1979) Sex hormones and female homosexuality: a critical examination. Arch Sex Behav 8:101–119
Meyer-Bahlburg HF, Ehrhardt AA, Rosen LR et al (1995) Prenatal estrogens and the development of homosexual orientation. Develop Psyc 31:12–21
Meyer-Bahlburg HF, Gruen RS, New MI et al (1996) Gender change from female to male in classical congenital adrenal hyperplasia. Horm Behav 30:319–332
Meyer-Bahlburg HF, Dolezal C, Baker SW et al (2008) Sexual orientation in women with classical or non-classical congenital adrenal hyperplasia as a function of degree of prenatal androgen excess. Arch Sex Behav 37:85–99
Money J (1975) Ablatio penis: normal male infant sex-reassigned as a girl. Arch Sex Behav 4:65–71
Mustanski BS, Chivers ML, Bailey JM (2002) A critical review of recent biological research on human sexual orientation. Annu Rev Sex Res 13:89–140
Mustanski BS, Dupree MG, Nievergelt CM et al (2005) A genomewide scan of male sexual orientation. Hum Genet 116:272–278
Neave N, Menaged M, Weightman DR (1999) Sex differences in cognition: the role of testosterone and sexual orientation. Brain Cogn 41:245–262
Noller KL, Fish CR (1974) Diethylstilbestrol usage: its interesting past, important present and questionable future. Med Clin North Am 58:793–810
Peplau LA, Huppin M (2008) Masculinity, femininity and the development of sexual orientation in women. In Mathy R (ed) Childhood gender nonconformity and the development of adult homosexuality, 1st edn. Haworth, Binghamton, pp 147–167
Perkins A, Fitzgerald JA (1992) Luteinizing hormone, testosterone, and behavioral response of male-oriented rams to estrous ewes and rams. J Anim Sci 70:1787–1794
Peters M, Reimers S, Manning JT (2006) Hand preference for writing and associations with selected demographic and behavioral variables in 255,100 subjects: the BBC internet study. Brain Cogn 62:177–189

Pfaus JG, Kippin TE, Coria-Avila G (2003) What can animal models tell us about human sexual response? Annu Rev Sex Res 14:1–63
Phoenix CH, Goy RW, Gerall AA (1959) Organizing action of prenatally administered testosterone propionate on the tissues mediating mating behavior in the female guinea pig. Endocrinology 65:369–382
Pillard RC, Bailey JM (1995) A biologic perspective on sexual orientation. Psychiatr Clin North Am 18(1):1871–1884
Pillard RC, Rose RM, Sherwood M (1974) Plasma testosterone levels in homosexual men. Arch Sex Behav 3:453–458
Pillard RC, Poumadere J, Carretta RA (1981) Is homosexuality familial? A review, some data, and a suggestion. Arch Sex Behav 10:465–475
Pillard RC, Poumadere J, Carretta RA (1982) A family study of sexual orientation. Arch Sex Behav 11:511–520
Poiani A (ed) (2010) Animal Homosexuality: A Biosocial Perspective. Cambridge University Press, New York
Popova NK, Morozova MV, Amstislavskaya TG (2011) Prenatal stress and ethanol exposure produces inversion of sexual partner preference in mice. Neurosci Lett 489:48–52
Rahman Q (2005) Fluctuating asymmetry, second to fourth finger length ratios and human sexual orientation. Psychoneuroendocrinology 30:382–391
Rahman Q, Koerting J (2008) Sexual orientation-related differences in allocentric spatial memory tasks. Hippocampus 18:55–63
Rahman Q, Wilson GD (2003) Sexual orientation and the 2nd to 4th finger length ratio: evidence for organising effects of sex hormones or developmental instability? Psychoneuroendocrinology 28:288–303
Rahman Q, Wilson GD, Abrahams S (2003) Sexual orientation related differences in spatial memory. J Int Neuropsychol Soc 9:376–383
Rahman Q, Collins A, Morrison M et al (2008) Maternal inheritance and familial fecundity factors in male homosexuality. Arch Sex Behav 37:962–969
Rieger G, Chivers ML, Bailey JM (2005) Sexual arousal patterns of bisexual men. Psychol Sci 16:579–584
Risse GL, LeDoux J, Springer SP et al (1978) The anterior commissure in man: functional variation in a multisensory system. Neuropsychologia 16:23–31
Robinson SJ, Manning JT (2000) The ratio of 2nd to 4th digit length and male homosexuality. Evol Hum Behav 21:333–345
Roselli CE, Larkin K, Resko JA et al (2004) The volume of a sexually dimorphic nucleus in the ovine medial preoptic area/anterior hypothalamus varies with sexual partner preference. Endocrinology 145:478–483
Roselli CE, Reddy RC, Kaufman KR (2011) The development of male-oriented behavior in rams. Front Neuroendocrinol 32:164–169
Savic I, Lindström P (2008) PET and MRI show differences in cerebral asymmetry and functional connectivity between homo- and heterosexual subjects. Proc Natl Acad Sci U S A 105:9403–9408
Savic I, Berglund H, Lindström P (2005) Brain response to putative pheromones in homosexual men. Proc Natl Acad Sci U S A 102:7356–7361
Schmidt G, Clement U (1990) Does peace prevent homosexuality? Arch Sex Behav 19:183–187
Schulz KM, Zehr JL, Salas-Ramirez KY et al (2009) Testosterone programs adult social behavior before and during, but not after, adolescence. Endocrinology 150:3690–3698
Sell RL, Wells JA, Wypij D (1995) The prevalence of homosexual behavior and attraction in the United States, the United Kingdom and France: results of national population-based samples. Arch Sex Behav 24:235–248
Simerly RB (2002) Wired for reproduction: organization and development of sexually dimorphic circuits in the mammalian forebrain. Ann Rev Neurosci 25:507–536
Swaab DF, Hoffman MA (1990) An enlarged suprachiasmatic nucleus in homosexual men. Brain Res 537:141–148

Swaab DF, Roozendaal B, Ravid R (1987) Suprachiasmatic nucleus in aging, Alzheimer's disease, transsexuality and Prader-Willi syndrome. In de Kloet ER, Wiegant VM, de Wied D (eds), Neuropeptides and brain function, Progress in brain research, vol. 72, 2nd edn. Elsevier, Amsterdam, pp 301–311
Swaab DF, Hoffman MA, Honnebier MB (1990) Development of vasopressin neurons in the human suprachiasmatic nucleus in relation to birth. Brain Res Dev Brain Res 52:289–293
Swaab DF, Slob AK, Houtsmuller EJ et al (1995) Increased number of vasopressin neurons in the suprachiasmatic nucleus (SCN) of 'bisexual' adult male rats following perinatal treatment with the aromatase blocker ATD. Brain Res Dev Brain Res 85:273–279
Tanner JM (1986) Normal growth and techniques of growth assessment. Clin Endocrinol Metab 15:411–451
Triana-Del Rio R, Montero-Domínguez F, Cibrian-Llanderal T et al (2011) Same-sex cohabitation under the effects of quinpirole induces a conditioned socio-sexual partner preference in males, but not in female rats. Pharmacol Biochem Behav 99:604–613
Vega Matuszczyk J, Larsson K (1995) Sexual preference and feminine and masculine sexual behavior of male rats prenatally exposed to antiandrogen or antiestrogen. Horm Behav 29:191–206
Vega Matuszczyk J, Fernandez-Guasti A, Larsson K (1988) Sexual orientation, proceptivity, and receptivity in the male rat as a function of neonatal hormonal manipulation. Horm Behav 22:362–378.
Veniegas, RC, Conley TD (2000) Biological research on women's sexual orientations: evaluating the scientific evidence. J Soc Issues 56:267–282
Vigil P, Orellana RF, Cortés ME et al (2011) Endocrine modulation of the adolescent brain: a review. J Pediatr Adolesc Gynecol 24:330–337
Vito CC, Fox TO (1981) Androgen and estrogen receptors in embryonic and neonatal rat brain. Brain Res 254:97–110
Wang CT, Shui HA, Huang RL et al (2006) Sexual motivation is demasculinized, but not feminized, in prenatally stressed male rats. Neuroscience 138:357–364
Welsh DK, Takahashi JS, Kay SA (2010) Suprachiasmatic nucleus: cell autonomy and network properties. Annu Rev Physiol 72:551–577
Whalen RE (1974) Estrogen-progesterone induction of mating in female rats. Horm Behav 5:157–162
Whalen RE, Edwards DA (1967) Hormonal determinants of the development of masculine and feminine behavior in male and female rats Anat Rec 157:173–180.
Whalen RE, Olsen KL (1981) Role of aromatization in sexual differentiation: effects of prenatal ATD treatment and neonatal castration. Horm Behav 15:107–122
Willner P (1984) The validity of animal models of depression. Psychopharmacology (Berl) 83:1–16
Willner P, Mitchell PJ (2002) The validity of animal models of predisposition to depression. Behav Pharmacol 13:169–188
Witelson SF, Kigar DL, Scamvougeras A (2008) Corpus callosum anatomy in right-handed homosexual and heterosexual men. Arch Sex Behav 37:857–863
Zucker KJ, Bradley SJ, Oliver G (1996) Psychosexual development of women with congenital adrenal hyperplasia. Horm Behav 30:300–318

Part IV
Prevention Programs

Chapter 19
Impact of the Perinatal Environment on the Child's Development: Implications for Prevention Policies

Françoise Molenat

Prevention is a bargain compared to the current cost of our failures

—Lisbeth B. Schorr 1988

Abstract *Basic emotional security* is central to the construction of the child and has an impact on the brain's organisation, the personal autonomy and the capacity to explore the world. The key concept of the attachment theory is supported by recent neuroimaging findings of brain development and the structuring of the hypothalamic–pituitary–adrenal axonal regulatory systems.

In addition to the child's potential, the essential variable lies in the *quality of the environment's responses,* and consequently in the quality of the maternal security, from the very early intrauterine life. The understanding of the effects of parental stress during the early developmental stages is advancing. In France, the *emotional security* of pregnant women and future parents has become a major stake of perinatal policies for the prevention of developmental disorders.

Specific strategies are being developed to improve both the maternal and the infant well-being. These are not restricted only to mental health specialists but rather involve every health-care professional of the perinatal period. The mechanisms of change for vulnerable parents emerge from the prospective analysis of support methods. Continuity and coherence of such care serve as a holding function, which enables the restructuring of previous emotional traumas.

A new interdisciplinary perinatal medicine is emerging, structured rigorously around a well-coordinated obstetrical and paediatric follow-up. Considering the future of children, teenagers and adults, the stakes are enormous.

F. Molenat (✉)
Association for the Education and the Research on the Child and its Environment (AFREE),
BP 64164 34092 Montpellier Cedex 05, France
e-mail: afree@afree.asso.fr; fmolenat@yahoo.fr

M. C. Antonelli (ed.), *Perinatal Programming of Neurodevelopment,*
Advances in Neurobiology 10, DOI 10.1007/978-1-4939-1372-5_19,

19.1 Introduction

Many societies are facing crucial problems regarding the education of children, behaviour troubles and mental disorders during adolescence and later stages of human life. Advancement in imaging techniques has resulted in considerable progress in the field of maternal and infant mortality. However, other forms of suffering prevail and prevent children, the futute adults, to reach their maximum potential. Material and social living conditions can heavily increase the insecurity of parents, let alone children. Parenthood is from now on identified as a progressive development, which is not always straightforward, depending on the present or past living conditions/experiences. Poorly built foundations can have major consequences from both a psychosocial and a financial point of view. Taking care of any mental health issue is lengthy and costly procedure that often implies a sort of helpnessness. Logically, any necessary and preventive strategy should be implemented as early as possible for both future parents and their child. Traumatims and negative emotions triggered by medical complications have been taken into account during the last years. However, psychosocial vulnerability situations still remain difficult to be addressed in an appropriate way and time. One of the difficulties comes from the fact that health professionals who meet a family after a birth are not those who will assess and treat disorders at a later stage.

Unfortunately, these well-known observations have not led to sufficient and efficient prevention strategies. Once the complications have occurred, family distress affects in return parents who were already vulnerable, and the loss of self-confidence increases depression and mental health risks. In this context, distrust in the health care system that failed to detect or repair any kind of distress or disorder can emerge. Therefore, it's essential that efforts should be made in time before parents lose their faith and courage and before troubles crystallize.

In spite of previous international recommendations and local good practices, it is still difficult to define a common framework. In this framework, every implicated specialist would contribute to anticipate and provide the required preventive strategies.

The worldwide diversity and dispersion of health care organizations, settings and systems make the assessment and the dissemination of knowledge and action policies really difficult as well as the construction of a common paradigm. Old's studies (1988) clearly showed that financial recovery from an intensive intervention for high risk populations occurs within four years and in a cost-effective manner. However, these conclusions failed to be converted into health policies, while the rhythm may vary substantially from one country to another.

The care of a child's development in its integrative/ holistic nature (physical, psychological and social aspects) is undermined by the phenomenon of our era: over-specialized professionals with tightly defined disciplines BUT WITH poor communication between them. It is much easier for any health professional to treat a disorder relevant to his expertise partially and independently from the rest of the follow-up rather than work together with other colleagues in a multidisciplinary network. However, the developing interest of clinicians and researchers in the perinatal period leads to high expectations. The part Neurobiology has largely contributed to this: The concept of neuronal plasticity and the role of the environment in the development of the foetal and neonatal brain offered the necessary objective basis and proof to the practicians' previous intuitions concerning the value and necessity of an early intervention. The confirmation of early skills of infant newly born, their role in the strengthening of human attachment, the restructure of any past traumatic experiences of the adult, the better knowledge of the mechanisms between biology/emotion/thought opened up/ created excting intervention and research perspectives. It might not everything be defined at birth, but still the hormonal regulation system is organized from the very early stages of life while being in a permanent interaction with the human environment. Doing everything to preserve and improve this environment would be the obvious *ecological* approach. However, the complexity of procedures at stake demands a reorganization of disciplines concerned and new training methods.

The gestational period appears as a special occasion and time. (FOR WHAT?) Becoming a parent especially for the first time demands particular skills to adapt and adjust to new, unknown sometimes even terrifying situations. This transitional period is characterised by a special permeability to the environment which permits to experience situations and relationships with the professionals in new, different ways. Being pregnant implicates the body which leads to meeting with the health care system and experience new forms of a secure, non-threatening proximity, of being protected and safe. Securing the mother's well-being is securing that one of the child, too, and vice versa. For instance, a woman having addiction troubles can accept to receive help for the child's sake, while she wouldn't do so or haven't so far requested any for herself. Expecting a child is a positive stage, which is generally valued by society. Parents hope to get the best for their child, especially if they did not benefit from optimal conditions in their own lives. The quality of care, the availability of a professional mobilise. Any negative experiences not previously verbalised. The future father equally benefits from a framework where his active role is encouraged and recognized, whatever his personal story and difficulties are. Trusting someone (=a professional) can be a new adventure for adults who had to grow up in a premature self-reliance mode. During pregnancy, parents and professionals share the same goal: to give every chance to the child. The alliance is easier to establish than when back at home while facing recognized difficulties. The confidence established at the start will facilitate the work of those who will intervene later, in the conditions that we will develop further on. The reception of anxiety by a securized professional reduces its intensity. The list would be long if we wanted to present the whole range of opportunities on offer, when expecting a child is accompanied in an adequate manner. Finally, the perinatal period is a unique occasion for the future mother as she is taken care of both physically and psychologically with all the corrective benefits mentioned above. Considering the risks of repeating early disruptions in the parents' life, we can imagine the prodigious potential of reconstruction. The conditions of training and assessment organization remain to be defined in order to make such a paradigm's and culture's change possible in all the disciplines concerned.

19.2 Scientific Arguments for a Perinatal Prevention Policy

Since the end of the 1960s, many studies have enlightened the early stages of child development. In the present review, we discuss the data originated from interdisciplinary studies. while this collection is debatable, our experience in a local environment lead us to consider these data as empowering for the development of of a multidisciplinary culture.

19.2.1 Secure Base, Attachment, Self-feeling

Social and emotional development in early infancy is recognized as essential for all aspects of functioning throughout the lifespan. The notion of *secure base* has first emerged from ethological studies, then from neurophysiologic exploration in humans. The secure base builds up in a succession of interactions with a human environment in specific conditions of adjustment. The progressive discovery of one's individuality while being related with the Other leads to the self construction. In an appropriate environment, the child becomes able to relate to the social world without fear. It progressively acquires an understanding of the human environment and of itself, which will allow it to face the unexpected.

The attachment theoriticians (Bowlby 1969–1982; Cassidy and Shaver 1999; Grossmann 2005, 2006) have highlighted the major role of the caregiver; generally the mother due to foetal preparation and breastfeeding, and also the father or any adult acting as a replacement. In a childbearing woman, pregnancy activates her own early attachment experiences, which she had internalised during infancy. These mental models (*internal working models*) were revised at different stages in life and work as point of reference or a filter through which the new interpersonal relations are formed, shaped. They are particularly active before a birth. In this way different factors may interfere with the quality of the child's emotional and relational development. These factors may either be child-specific (prematurity, genetic illnesses, etc.) or derive from the mother (insecurity elements) and the family environment (Murray et al. 1999).

Pregnancy and childbirth are powerful stress activators to an unknown situation, especially for the first child. Several other authors have described the interaction methods and their disruptions. Guedeney (2011) has synthesised the concept of synchrony: "*Within attachment theory, synchronization plays a major role with the concept of sensitivity of the caregiver response to infant stress*." This observation indicates the maternal security's major role which can be affected by numerous factors including her psychosocial living conditions and experiences. Guedeney refers to the works of Hesse and Main (2006) that describe the impact of a trauma experienced by the mother: the elements are not integrated as a whole but "*rather stored as isolated fragments of sensory perceptions of affective states that can be abruptly and easily activated by stimuli associated with the traumatic event*". This dissociative mechanism has been analysed by Fraiberg (1981) as underlying the "*ghosts in the nursery phenomenon, resulting from unresolved traumatic attachement experiences of the caregiver*". A discordance between the mother's own needs and those of the child prevents the caregiver from intervening effectively. The mother who feels insecure needs a transitional protective relationship for herself. This situation can generate a major anxiety, which can be found at the origin of the post-partum depression (PPD) or in the child's functional patterns.

The transgenerational transmission of the attachment methods has been particularly examined. It determines the risks of disorder repetition from one generation to the next: abuse, deficiency and psychopathology. Fonagy and his colleagues (1997, 2002, 2005) have particularly examined the role of reflective functioning, which allows the mentalisation of the experience and its psychological integration. In the secure attachment relationship, the child could freely explore the adults' mental states and understand its own. This ability would help the child to better regulate his emotions and to develop harmonious connections with others. If the adult has not been able to think about, understand and transformate his/her early traumatic experiences, new restauring opportunities occur with reassuring significant others (???) within secure contexts such as a partner, a psychotherapeutic relationship, And the alliance with professionals of the perinatal period. RACCOURCI Den katalavaino poli kala ti thelei na pei, an katalavaino kala tote: However, if the adult has not been able to overcome early traumatic experiences and failed to develop a healthy mental state, an intimqte relqtionship such as a partner, a psychotherapist and/or the perinatal professionals can offer a secure and reassuring environment and new corrective opportunities. Such a close relationship involving body, emotions and their verbalisation offers a unique occasion to both parents to revisit past experiences through the feelings and situations experienced at the present. The father is a vital interlocutator of the child that shouldn't be overlooked. Moreover, by actively participating in the perinatal follow-up he can benefit from such a "corrective attachment experience" himself too.

19.2.2 Negative Parental Experiences and Memory Reorganisation

The studies of Stern (2004), in general, recent studies on implicit memory functioning, enlighten the conditions of reactivation of the parent's early childhood experiences, a period when the neurological equipment did not allow mentalisation. The conscience is unmarked but the neuronal inscription persists (Damasio 1999). The emotion or the sensation can be reactivated if the situation displays analogies with the past experience. The expectance and the arrival of a baby can reactivate moments of the parents' early childhood. Within a new human context where the processes of dependence and autonomy are reactivated, any past traumatism can be reorganised especially based on an intimate, protective and trustful relationship. In this way, new possibilities to avoid transgenerational repetion are offered. We shouldn't forget that many adults don't consult a psychotherapist spontaneously; either because they are unaware of their old traumas or they are afraid of having them arisen after having them costly pushed away by building fragile protections. In addition, psychotherapists underline the importance of the timing and conditions: if one person is not appropriately surrounded, forcing the trauma's verbalisation can lead to the opposity results and aggravate it. Through a trust-based consistent and reliable relationship, new positive feelings can counterbalance the resurgence of the trauma's stressful feelings. Thus, "the present can rewrite the past" (Stern, 2004) within a close and empathetic relationship.

19.2.3 Stress: Transmission, Effects, Regulation

More recent studies have focused on stress in the perinatal period. The organization of the stress regulation system and especially the epigenetic mechanisms of foetal development or reprogramming were examined on mammals. At the same time as fundamental research for a better understanding of mother to foetus and neonatal transmission, several epidemiologic studies have found a connection between amajor anxiety disorders and high levels of stress during pregnancy and the occurence of pathology in adulthood. An exhaustive review of these models was published by Fumagalli et al. (2007). Such theories are in accordance with the clinical experience in perinatal psychology and encourage greater attention to the emotions of pregnant women. Moreover, the comprehensive studies that Van den Bergh has perfectly synthesised (2011) describe the effects of a certain level of stress in pregnant women on the short- and long-term development of the child. Many prospective studies have revealed a link between prenatal exposition to maternal stress, anxiety and depression and some neuro-developmental changes. Socio-emotional, cognitive and behavioural problems could appear at birth or up to the age of 20. According to Meaney (2010), quoted by Van den Bergh, these epigenetic processes would bring physical support to the influence of prenatal environment on the future of a person. Postnatal care could have a moderating effect. The reduction of the effects of prenatal stress justifies the implementation of targeted pre- and postnatal intervention strategies. Preventive interventions could significanty affect the child's well-being.

Prenatal stress has also been identified long time ago as a contributing factor to premature childbirth and low birth weight. In spite of the fact that all these have been taken into account and efforts have been made, neither specific tools have been developed nor the rates of preterm births have been significantly reduced in most countries.

All these results are to be handled with caution, without forgetting the positive function of adaptation to stress in adults and in children. Identifying the stress factors is more beneficial as it allows the assessment of severity and persistence of symptoms as well as to define an indivualised treatment. The investigation of any stressful life events before or during pregnancy sold be part of any monitoring and follow-up procedure. It's well known that the simple fact of sharing them with a specialized, empathetic professional can reduce their negative impact.

19.2.4 Parental Depression: Repercussions, Impact on the Child

Studies on perinatal depression have accumulated during the past few decades. Depression is one of the most incapacitating illnesses among mental health disorders, and according to the WHO's forecast (WHO 2001), it will be the second most serious health issue in 2020 (Alderdice et al. 2012). The procreation period is recognized as a high-risk zone of depression. The specificity of PPD has progressively emerged due its consequences the woman, the child and the family as a whole. It is estimated that one woman out of seven suffers from PPD. In the last decade, several indicators have been identified that helped to isolate the entity of the antenatal depression and enlight the links between the pre- and the post-natal depression. The limited knowledge on this issue is atrributed to two main reasons: Firstly, detection of clear/specific symptoms is mostly possible quite late after childbirth and once at a distance from the regular obstetrical and pediatric follow-up. Secondly, the dispersion of literature itself has restricted the interdisciplinary dissemination of knowledge and communication. In the last decade, several indicators have been identified that helped to isolate the entity of the antenatal depression and enlight the links between the pre- and the post-natal depression.

The limited knowledge on this issue is atrributed to two main reasons: Firstly, detection of clear/specific symptoms is mostly possible quite late after childbirth and once at a distance from the regular obstetrical and pediatric follow-up. Secondly, the dispersion of literature itself has restricted the interdisciplinary dissemination of knowledge and communication. The actors of a potential prevention are not those of the treatment. Since a long time, psychiatry reviews have been reporting the epidemiology, evolution and treatment of PPD. In the past 10 years, the disciplines concerned (obstetrics, paediatrics, general practice, psychiatry) have been coming closer. This permits a better understanding of semiology and thus a better screening even if a differential diagnosis remains essential between gestation-related emotional changes and precursors symptoms of PPD. It's identified that ineffectively-resolved conflicts lie at the core of the perinatal depression that are reactivated by parenthood (although not necessarily by the first birth), while other external factors may interfere. Maternal depression can have a negative impact on the mother-infant attachment due to the emotional and sometimes even physical unavailability of the mother.

Internationally, health systems aware of the need for change in practices and policies are trying in disparate enough ways and speed to proceed to the necessary reforms. The growing nuber of studies published outlining the risks of a chronic depression allowed women to break their silence that they had previously chosen either by the fear of stigma or by ignorance. Self-blame, lack of knowledge of the family environment and/or the practicioner have been for a long time preventing screening and treatment. Even if predictive signs during pregnancy are known (anxiety, unease, tiredness, sleep disorders, consultations at emergency rooms, poor psychosocial conditions, etc.), pregnant women don't easily confess any feelings of unease or previous episodes potentially stigmatising. Either by fear of being judged or by ignorance that these symptoms deserve to be treated, they don't confide in. More rarely, the pregnancy goes well and the depression occurs in the first few weeks or months after returning home. In this case, the woman may have risk factors in her past, not previously identified, even if their negative impact on her well-being is obvious.

The risk of paternal depression, estimated at 10% is even more unknown in spite of its impact on family relations. It is expressed by somatic or behavioural manifestations: tiredness, sleep disorders, alcohol consumption, avoiding behaviour or hyperactivity. Likewise for women, these disorders can easily be interpreted by the family as unwillingness or indifference making worse the misunderstnding and the emotional mess.

Goodman (2008) described the effects of maternal depression on the child from intrauterine life to adulthood. These correlations alert psychiatrists treating adults about the importance of perinatal care to prevent later psychopathology. Several mechanisms are at stake:

- Increase of maternal cortisol in severe states of depression
- Impact of maternal cortisol on the activity of the foetus' HPA axis
- Consequences on the psychomotor and cognitive development

Concerning to the impact on the mother–child relationship, considerable literature shows the long-term effects of chronic maternal PPD. The emotional availability of the mother and her resposiveness to the child's needs are reduced or impoverished as well as the mother-infant interaction itself. Attachment disorders are well documented. Unfortunately, the quality of the environmental context is rarely taken into account in studies, thus, it is an important variable. Family can sometimes compensate the maternal unavailability. Besides, the depressive state does not systematically affect the *holding* quality; according to the studies, only a limited number of children will suffer from its negative effects.

19.3 Innovative Prevention Strategies

19.3.1 2004 French Report

In France, the Health Ministry has requested a report on perinatal practices in order to prevent the emotional complications of gestational period and their effects on child development. The emphasis was put in particular on the methods of collaboration between the domains of medical health care and mental health care. A study group gathered representatives of all the disciplines concerned: obstetricians, paediatricians, midwifes, nurses, general practicioners, social workers, psychologists, cchild and adult psychiatrists during a year. Many professionals were interviewed to analyse the difficulties in the field.

The basic principles derived from:

1. The data from literature on the early needs of children for an optimal development
2. The experience acquired in certain areas through the successful cooperation of pregnancy and early childhood practicioners on the one hand, and social workers on the other.

3. Prospective clinical studies on the impact of professional attitudes on parents–infant bonding (Molenat 2001)
4. The European collaborations established by Association for the Education and the Research on the Child and its Environment since 1990 which have helped examining the reproducibility of interdisciplinary training and clinical research in various countries
5. A close collaboration with Sainte-Justine Hospital in Montreal (Canada) through studies about infant development and strategies of prevention

The main theories were based on the following assessment:

- A child's basic emotional security is organised through the global security of their mother's security which depends on the quality not only of the family but also of the professionals involved from the start of the pregnancy till early childhood.
- The notion of security implies a minimum of continuity and consistency for the child but also for the pregnant woman.
- Professionals ensuring perinatal health care also need to be secured themselves to face family situations that make the communication difficult and generate strong emotions.

To reach these objectives, it was necessary to have a good knowledge of the specific needs of pregnant woman (and future father). However, childbirth is not only restricted to biological mechanisms: emotional and scial elements that cannot be easily expressed during a traditional medical consutation interfere too.

The organisation of health care in France, like in other countries, did not promote the feeling of continuity and consistency. Childbearing women often deal with different even contrary opinions. They sometimes meet different professionals even in the traditional follow-up of a pregnancy without complications (obstetricians, family doctors, midwifes). patient's file helps to pass on the medical data but the quality of the meetings is not ensured and the course of the women's emotions could break. If there are some risk factors, they will have to be submitted to numerous interventions involving professionals who are isolated from each other and belong to different disciplines. A cross-cultural combination of the somatic and the psychosocial approach was—and still is—insufficient.

An official report entitled "2005–2007 Perinatalogy Plan"[1] set the framework for a policy aiming at modernizing the pregnancy and childbirth environment. It proposed a series of measures aiming at improving health-care security and quality while developing a more human and individualised context. It also aimed at improving the mutual knowledge of health actors and to make their organisation more clearer. One of the main issues consisted of improving the psychological and social environment of parents and children. Three essential measures were implemented:

[1] Plan « Perinatology ». Humanity, proximity, security, quality, 2005–2007. Official Report no 2006-4. http://www.sante.gouv.fr/htm/actu/perinatalite04/Circular DHOS/DGS/O2/6 C n° 2005–300 of 4th July 2005 relative to the promotion of the medical and psychological collaboration in perinatal mental health: http://www.sante.gouv.fr/adm/dagpb/bo/2005/05-08/a0080026.htm

- A thorough personal interview, called early prenatal interview conducted by a mid-wife or a doctor around the 4th month of gestation;
- The improvement of the psychological environment through the creation of psychologist positions in maternity units
- The development of interdisciplinary methods of training

This plan reminds us of the importance of the emotional aspect in the birth process, which will determine the good development of the physiologic evolution as well as the harmonious construction of family ties. A birth is indeed a bio-psychosocial event. A strictly technical approach proves to be deleterious just as an accumulation of uncoordinated interventions. *Early and quality support to parenthood from the pre- and the early post-partum, are considered as preventive factors against child violence and abuse as well as psychopathology. In these terms, such a support consists a matter of public health.*

Based on some revolutionary and individual good practices, the articulation of psychiatry and perinatal medicine were prioritized. Every time, the forces of both disciplines were united, this did helped to respond to difficult situations such as maternal psychopathology, addiction troubles, medical complications and risks (foetal pathology, severe prematurity, etc.). All these situations can be the trigger for major affective disorders for both parents and they should be taken into account in the follow-up. One of the specific objectives of the plan was to adress the issue of this vital interdisciplinary collaboration in particular for families facing multiple risks and vulnerability factors.

The primary objective was to identify as early as possible the needs of the pregnant woman, her family and the professionals concerned. In order to do so, the global aim of continuity and coherence was set in the centre of efforts demanding collaborations between the different health actor concerned in a coordinated team work (who often had never talked before to each other). Parents' security, which is necessary to the child, is particularly related to the security of the professionals who surround them. Interdisciplinary training methods will be an essential means for all health-care actors to develop a mutual trust in order to achieve better collaboration and consistency.

In the plan, *both concepts of continuity and consistency* are considered as the milestones for perinatal medical–psychological care. This principle leads to consider the health-care system in terms of integration, articulation and take-over. It is about offering customized care whatever the specific needs of each childbearing woman and family are. This implies a reduction in the separation and isolation between the different domains of practice concerned: medical, social, psychological, public or private.

Three new essential tools have been proposed and tested:

The early prenatal interview (around the 4th month of gestation and usually conducted by a midwife) It is a question of establishing a reassuring alliance between the pregnant woman, her partner and a first professional. offers the opportunity for the future mother to express her fears, desires and any difficulties at home or at work. Simultaneously, her local network and her social support capital (family and friends) are explored. The differentiating point comparing to the traditional practices is the anticipation of the stages later on, any risk factors that could potentially weaken the parents. Some key points of the stages to anticipate are: the pregnancy follow-up, reactions to the echography, once at the labour room, returning home, community care and support if necessary. Experience showed that a vulnerable woman talks more easily with a midwife rather than a doctor, due to the fear to disturb, short consultation time, lack in interprofessional communication.

If the eraly prenatal interview helps to identify any particular riks concerning the woman or the couple, the coordination function appears as a way to reduce these risks accordingly to the recent literature. The orientation towards a mental health professional or a social institution should be handled with caution as self-image is directly concerned and can lead to a fear of being abandoned by the first professional or judged.

The early prenatal interview is therefore based on respect and trust, the two indispensable components so that parents would open up and express any negative feelings and situations. The quality of this first consultation influences what will follow in a profound way. The professional himself/herself must feel secure in his/her practices and have a good knowledge of the health care actors and system in order to be able to suggest any appropriate referrals and liaisons in response to the issues arisen.

Rigorous referrals aim at preserving confidentiality and the pregnant woman's feeling of control over her follow-up. The points to transmit must be defined with the woman and the couple especially when they concern negative emotions or behaviors- such as addiction troubles, and they are susceptible to affect the self-image. Some guidelines for this first interview have defined and need to become even clearer as the interview's practice expands:

- To have the necessary training in order to have a thorough knowledge of the health care system and its actors
- To be able to organise and coordinate a variety of interventions
- To benefit not only from basic psychological training but also from the appropriate supervision for complex cases

In this way, the parent can experience how is it confiding in without being judged or stigmatised, having his remarks taken into account in order to adjust the follow-up, being supported by a professional to handle any bargain, being surrounded by a team of health care actors that work together and respect each other.

Starting from this first interview, the midwife will assess the specific needs and will adjust her role according to the family and interprofessional context.

Experience shows that a good coordination between the different actors involved is really reassuring parents who don't usually expect such a collective and well-coordinated attention to their needs. The most vulnerable parents talk about a new experience that they had never before: being the center of a caring, securing group of professionals whose roles are well-defined and complementary, while adjusting to their owns needs expresses. (*here, we note the corrective experience of the attachment theory*). This experience is the first and necessary occasion for parenting education. Any professional advice at a later stage when the parents would have experience the childbirth without any professional support, would be most likely ignored.

Why a midwife has been chosen for this interview (sometimes a medical doctor)? In fact, the pregnancy-related medical follow-up is understandable, not stigmatizing, and protective, while broadening the attention to the whole of the needs does not sort out of the global health field. It is known that the most vulnerable women are afraid of being judged, meeting new professionals, confiding in and/or being abandoned. Once the trust is established, restored, the medical health care is a source of reward and also security. It will help to take into account the negative elements, that the fisrt caregiver listened to/aknowledged in a positive way.

The creation of perinatal psychologists' positions Some psychologists and psychiatrists had to develop and change their culture in order to take part in the birth follow-up. The support of first-line actors is a very important mission: in case of complicated situations, without this support and supervision, professionals can have difficulties in establishing an intersubjective relationship with the patient and even more in making any difficults, though necessary, referrals to specialized professionals. This function of the perinatal psychologist called "indirect " or "second line" must be recognized from an administrative and scientific point of view.

Psychologists and psychiatrists' training should allow them to look in understand the psychological impact of caregivers and to better guide direct interventions. "Psychological support" is before all a human support, the specialist having thus the mental health professional should act in a direct way only for complex cases that need his/her expertise.

In France, plenty public resources were given for psychiatric system, even if nowadays it is saturated running out of capacity. It is clear that the isolation of the professionals has not facilitated the treatment of new forms of developmental or mental pathologies, which fall within interdisciplinarity. The current effort to coordinate actions should help to improve their efficiency and to avoid superimposition overlays.

The discovery by psyche experts of the "caregiver's relational skills" in practice conditions allowing them to blossom has become the driving force of interdisciplinary training but has not yet entered academic teaching.

An Innovative Methodology for Interdisciplinary Training training sessions gather professionals from all disciplines (medical, social, psychological), focusing on concrete situations analyze real situations covering the whole perinatal period (from the start of the pregnancy till the early childhood). They help to apprehend the role of the different actors, their specificities, their complementarity, in the spirit of consistency and continuity. The presentation of a case is done by those who have intervened in chronological order, and in a prospective manner. Some breaks are proposed at "key moments" and the pluri-disciplinary audience is invited to share their theories and suggestions as if they were acting in the place of the speaker (obstetrician, midwife, family doctor). The group learns to analyse the situation collectively and to establish a health-care plan by anticipating the potential risks for family structuring on one hand, and on the other hand professionals' needs to guarantee overall security. The participants listen to each other and discover the gap between each person's own representations of a single situation according to their working habits, their usual role, their training. These discussions always act as a revelation and are described as very rewarding. Especially given that a woman or a couple with numerous insecurity factors do not usually share the same things with every professional, rather than confide them in partially. A longitudinal report helps to put together the pieces of a puzzle whose entirety is often missing in everyday, conventional practice.

Such a prospective and longitudinal approach from pregnancy to early childhood helps to compare the professional responses to the family's evolution from a medical and affective point of view. The so called semiology of the interdisciplinary work gets more precise. Anticipation, referrals and orientations have become the new tools of an active prevention putting forces together while maintaining the difference between the respective roles. A special vigilance is needed to assure that parents remain at the centre of the decision making processes.

The use of this methodology allowed to confirm the surprising effects of a collective consistency which does not exclude the privileged role of a coordinating midwife for example. Ante- and postnatal continuity is an essential objective either the child is transferred to neonatal paediatric care or going home directly. The prenatal paeditrician visit has been proven useful in all situations of addiction, substance abuse, previous history of complicated births or loss of the previous baby. The established trust can later be transferred to the local network after the baby and his mother return home. Some parents give a testimony for educational purposes: they explain how much this "holding" by a human group of multiple professionals has allowed them to be restructured.

However, this type of training's requirements are sometimes difficult to meet: to choose a case with hindsight, gather the concerned professionals, use an experienced team coordinator with a good knowledge of each discipline and capable of controlling the group. So it is neither about a synthesis nor an assessment or a Balint group but it is an opportunity of screening the working methods and the consistency of all the professionals around a high-risk family, over time and beyond the discontinuities caused by professionals' changes.

19.3.2 2012–2013 Qualitative Assessment

The efficiency of these three measures is currently being assessed. The implementation of the 4th month interview is slow in spite of its funding. There are multiple obstacles. First of all, the inherent change in professional roles and practices puts doctors through a difficult test: resign from their solitary practice and share the responsibility with a midwife. In addition, the culture of multidisciplinary collaborations with every concerned professional of and throughout the perinatal period hasn't sufficiently developed. Network training meets this need and it has started spreading. It covers in France today more than 50 cities or regions. The administrative organisation of the "Regional perinatal networks" promotes interconnections between community professionals and hospitals. The paradigm has spread to several European countries that are implementing the 4th month interview and interdisciplinary training.

What can a new transversal and bio-psychosocial discipline, is emerging, trying to accompany individuals' and families' needs in a global, integrative way. The culture change is slow and difficult, but it gradually happening.

On the other hand, some new tools are being developed. The role of the coordinating midwifes is establishing itself and new positions are being created to deal with difficult situations. Body-centred childbirth preparation methods are starting to be recognized for their impact on both women and babies according to Field's studies (2006).

International literature through well-conducted has started taking into account the new functions of professionals who have been out of the public sight for a long time. Alderdice et al. (2012) publishes the meta-analysis of 32 systematic journals regarding interventions aiming at improving the maternal well-being. A total of 2497 abstracts have been sidentified as relevant to the area of maternal mental health and well-being from the antenatal to the postnatal period. Methodologies have been carefully examined and a small number of studies was selected. The only validated approach which needs further research has been the close presence of a midwife at home once the mother returns. This requires that the risks have been identified before (hence the importance of the early prenatal interview). The authors highlight the fact that the role of the midwife is not taken into account in most studies whereas she has regularly a close role and relationship with the pregnant woman. In fact, the midwife takes up the major role of tailoring or coordination in order to guide the pregnant woman through the maze of the health system.

A year later, was published in the Cochrane Collaboration (Sandall 2013) a review of midwifery-based models of care compared to other models during pregnancy and birth. The results are significant: *midwife-led continuity of care was associated with several benefits for mothers and babies, and had no identified adverse effects compared with the models of medical-led care and shared care. The main benefits were a reduction in the use of epidurals, with fewer episiotomies or instrumental births. These women were less likely to experience preterm birth, or lose their baby before 24 weeks "gestation", although there were no differences after 24 weeks.*

These analyses confirm the theories, which organise the current changes in perinatal practices.

19.4 Conclusion

From the existing knowledge and literature on early development, we have based our reflection on the mother's well-being, the first condition for a good organisation of the baby's regulation systems. Promoting the parental role is an essential component in the future of the child if the parents receive rigorous help starting from the pregnancy, and for the necessary period, each time they express the particular need to strengthen their parenthood.

We need to point out that all these measures come from common sense. The energy involved in them does not prove to be "profitable" in terms of publications and listed acts. The efficiency depends on a collective action and not on isolated efforts.

We could be discouraged, but the collective conscience has made some progress. The somatic world is now convinced for the major role of emotions in general, somatic and mental health. The medical preoccupation is getting wider by integrating the psycho-emotional aspect of the mother as well as of the baby. The role of doctors, midwifes and nurses is essential as it provides security and protection to the future family. Psychology professionals have a lot to gain from using the support of the highly skilled teams which join forces to provide a quality environment to the mother and child. The task is not always easy when dealing with men and women who have suffered in their lives and are afraid of the health-care system. This daily work needs to be highlighted and the recent publications are doing so. The scientific stakes of an interdisciplinary understanding of development are fantastic and should change the landscape of later sufferings which affect a certain number of children and adolescents.

Conflict of Interest The authors declare no conflicts of interest.

References

Alderdice F et al (2012) A systematic review of systematic reviews of interventions to improve maternal mental health and well-being. Midwifery. http://dx.doi.org/10.1016.j.midw.2012.05.010

Bowlby J (1969–1982) Attachment and loss. Basic Books, New York

Cassidy J, Shaver PR (1999) Handbook of attachment: theory, research and clinical applications. The Guilford Press, New York

Damasio AR (1999) The feeling of what happens: body and emotion in the making of consciousness. Harcourt Brace & Company, New York

Fields T et al (2006) New-borns of depressed mothers who received moderate versus light pressure massage during pregnancy. Infant Behav Develop 29:34–38

Fonagy P, Target M (1997) Attachment and reflective function: Their role in self-organization. Develop Psychopathol 9:679–700

Fonagy P, Gergely G, Jurist EL, Target, M (2002) Affect regulation, mentalization and the development of the self. New York: Other Press

Fonagy P, Target M (2005) Bridging the transmission gap: An end to an important mystery of attachment research?

Fraiberg S (1981) Clinical studies in infant mental health: the first year of life. Basic Books, New York

Fumagalli F, Molteni R, Racagni G, Riva MA (2007) Stress during development: impact of neuroplasticity and relevance to psychopathology. Prog Neurobiol 81:197–217

Goodman SHB (2008) Parental psychopathology and its relation to child psychopathology. In: Handbook of clinical psychopathology. MHAM Gross, Melbourne

Grossmann K et al (2005) Early care and the roots of attachment and partnership representations. In Grossmann KE, Grossmann K, Waters E (eds) Attachment from infancy to adulthood: the major longitudinal studies. The Guilford Press, New York, pp 98–136

Grossmann KE, Grossmann K, Waters E (eds). (2006) Attachment from infancy to adulthood: The major longitudinal studies. Guilford Press

Guedeney A et al (2011) Infantrhythms versus parental time: promoting of parent-infant synchrony. J Psysiol Paris 105:195–200
Hesse E, Main M (2006) Frightening, threatening and dissociative behavior in low risk samples: description, discussion and interpretations. Dev Psychopathol 18:300–343
Meaney MJ (2010) Eigenetics ant the biological definition of gene x environment interactions. Child Dev 81:41–79
Molenat F (2001) For an ethical prevention. Editions érès, Toulouse
Murray L, Creswell C, Cooper PJ (1999) The development of anxiety disorders in childhood: an integrative review. Psychol Med 39(9):1413–1423
Olds D, Henderson J, Cole CR, Echenrode R, Kitzman J, Luckey H, et al (1998) Long term effects of home visitation on children's criminal and antisocial behaviour: 15-year follow-up of a randomized controlled trial. JAMA 280(14):1238–1244
Sandall J, Soltani H, Gates S, Shennan A, Devane D (2013) Midwife-led continuity models versus other models of care for childbearing women. Cochrane Database Syst Rev 8(8)
Stern D (2004) The present moment in psychotherapy and everyday life. Norton, New York
Van den Bergh BRH (2011) Prenatal exposure to maternal stress associated to neurodevelopmental alterations? In Roegiers L, Molenat F (eds) Stress et grossesse: quelle prevention pour quel risque? Editions érès, Toulouse
WHO (2001) The World Health Report, Mental Health: new understanding, new hope. WHO, Geneva

Chapter 20
Perinatal Programming Prevention Measures

A. Miguel Larguía, María Aurelia González, Néstor Alejandro Dinerstein and Constanza Soto Conti

Abstract Over the past 10 years, there has been outstanding scientific progress related to perinatal programming and its epigenetic effects in health, and we can anticipate this trend will continue in the near future. We need to make use and apply these achievements to human neurodevelopment via prevention interventions. Based on the concept of the interaction between genome and ambiome, this chapter proposes low-cost easy-implementation preventive strategies for maternal and infant health institutions.

Breastfeeding and human milk administration are the first preventive measures, as has been reviewed in the *policy statement* of the *American Academy of Pediatrics*. Another strategy is the *Safe and Family-Centered Maternity Hospitals* initiative that promotes and empowers the inclusion of the families and the respect for their rights, especially during pregnancy and birth. (This change of paradigm was approved and is recommended by both United Nations Children's Fund, UNICEF, and Pan American Health Organization, PAHO.) Then, there is also an important emphasis given *to the sacred hour*—which highlights the impact of bonding, attachment, and breastfeeding during the first hour of life—the pain prevention and treatment in newborns, the control of the "new morbidity" represented by late preterm infants, and finally, the importance of avoiding intrauterine and extrauterine growth restriction. (However, there are not yet clear recommendations about nutritional interventions in order to diminish the potential metabolic syndrome consequence in the adult.)

A. M. Larguía (✉) · N. A. Dinerstein · C. S. Conti
Division Neonatology, Hospital Materno Infantil R. Sardá, Esteban De Luca 2151, Ciudad Autónoma de Buenos Aires, Argentina
e-mail: miguellarguia@gmail.com; secretaria@funlarguia.org.ar

N. A. Dinerstein
e-mail: alejandro.dinerstein@gmail.com

C. S. Conti
e-mail: cleosc@yahoo.com

M. A. González
Mental Health Section, Hospital Materno Infantil R. Sardá, Esteban De Luca 2151, Ciudad Autónoma de Buenos Aires, Argentina
e-mail: maureliagonzalez@gmail.com

M. C. Antonelli (ed.), *Perinatal Programming of Neurodevelopment*, Advances in Neurobiology 10, DOI 10.1007/978-1-4939-1372-5_20,

Abbreviations

AAP	American Academy of Pediatric
ABM	Academy of Breastfeeding Medicine
AHA	American Heart Association
EUGR	Extrauterine growth restriction
GE	Gestational age
IVH	Intraventricular hemorrhage
NICU	Neonatal intensive care unit
NRP	Neonatal Resuscitation Program
PAHO	Pan American Health Organization
PN	Preterm newborn
PVL	Periventricular leukomalacia
SFCF	Safe and Family Centered Maternity Hospital
SSC	Skin-to-skin contact
WHO	World Health Organization

20.1 Introduction

20.1.1 The Genome–Ambiome Dialogue

Ambiome is defined as the numerous environmental factors that may affect or modify the diverse epigenomic events that will set the final phenotype. Although this interaction can occur in almost all cells in any moment of life, the genome–ambiome interaction has a dramatic importance in the preterm, post-term, and educational life of children.

It is true that this division or dichotomy has been useful to study the development of cells and the entire organism. However, the border between what is internal and external has become more and more uncertain and the interactions between the internal and external factors of the organism make it impossible to identify parts of this dichotomy. The information required for cell-type definition is generated during the same differentiation process with an active role of the cells, in the purest epigeneticist style, and the interactions between the internal and external factors of the cell.

For a better understanding of this process and many other processes of development, it is necessary to resign the extreme *genocentric* position about development and the deeply rooted intrinsic–extrinsic, innate–acquired, constant–variable, or genetic–environmental dichotomies. These divisions portray the way we see the world; it seems that for some reason, if we organize it into categories or divisions it is easier for us to understand. However, these categories do not necessarily exist. During development, cell-type determination is not set by intra-cells or extra cells factors but by a complex combination of both. Likewise, living beings possess almost no completely genetic, environmental, innate, or acquired characteristics.

> Living beings are much more than a gene catalogue and it becomes clearer when we notice that knowing the genetic content of an organism is not enough to understand how it works. Now we know that when cells divide they transmit their DNA sequence (genome) to their offspring along with a series of superimposed molecular markers (epigenome) that will determine which genes are active and which are not. Genetics has proved that the genome remains almost identical in all the cells of the organism for all its life. Epigenetics gives a step forward and states that the epigenome is dynamic and that it changes from a cell type to another and from a moment of life to another in many cases responding to environmental signals. The gene–environment interaction defines a large part of what we are. (Benitez Keinrad 2011; Roldan Arjona 2011)
> The identity of an individual and many characteristics of an organism are not static, they change all the time and they arise from a combination of processes and mechanisms at different scales: genetics, cellular, ecological, geological, cultural and social. (Benitez Keinrad 2011)

The hypothalamo–hypophyseal–adrenal axis remains highly active and labile in the early infancy and begins to organize itself between 2 and 6 months of age through the interactions between the child and his carers. (Huang 2011)

The quality of attention that the child receives during the early development of his life predicts the emergency of his skills for future self-regulation. The primary continent and sensitive care is associated with an improved self-regulation capacity and with an optimal functioning of the child's hypothalamic–pituitary–adrenal (HPA) system.

20.2 Perinatal Interventions

Even though there are many different interventions more or less related to perinatal programming, the following interventions are the ones we chose as *preventive measures*.

20.2.1 Maternal Breastfeeding and Programming

Human breast milk represents the gold standard in the newborn feeding. In a wider point of view, it could be said that breastfeeding is the way humans adapt to extrauterine life. To that purpose, mentioning the following qualities breastfeeding possesses will be hardly surprising.

From a nutritional point of view, it has the highest biological value proteins (lactalbumin and lactoglobulin). It does not only offer to the newborn the essential fatty acids (linoleic and linolenic acids) but also those which are vital for him (arachidonic and docosahexanoic acids). Human milk is privileged by its unique carbohydrate, the lactose disaccharide. Micronutrients are all present in accurate quantities with absorption and bioavailability synergistic mechanisms. For survival purposes, human milk provides immunologically active cells (real transplant with no rejection) as well as anti-infection, anti-inflammatory, and immunomodulators, for example, the secretory immunoglobulin A (IgA) with a broad spectrum of antibodies, muramidase (lysozyme), iron-blocking agents, antioxidants, and pro- and anti-inflammatory cytokines. It also includes growth factors and trophic hormones. Likewise, human milk decreases the incidence of sepsis/meningitis in the infant, necrotizing enterocolitis in the preterm newborn and the incidence of enteropathogens-caused gastroenteritis especially under adverse sanitary conditions. High quantities of oligosaccharides act as prebiotics and promote the growth of bifido bacterias.

In order to reinforce this summary, we cite a number of breastfeeding definitions from the American Academy of Pediatrics Policy Statement (Eidelman and Schandler 2012; Breastfeeding and the use of human milk. Pediatrics 2012, 129, e 827).

> Exclusive breastfeeding for six months and weaning after one year is the most effective intervention with the potential of preventing more than one million infant deaths per year, equal to preventing 13 % of the world's childhood mortality. (Jones et al. 2003)

Besides the above mentioned *immediate* benefits of breastfeeding, there are also long-term effects. "There is a reduction of 52 % in the risk of developing celiac disease in infants who were breastfed at the the time of the gluten exposure" (Akomberg et al. 2006). "Breastfeeding is associated with a 31 % reduction in the risk of childhood inflammatory bowel disease" (Barclay et al. 2009). The protective effect is hypothesized to result from the interaction of the immunomodulating effect of human milk and the underlying genetic susceptibility of the infant. Different patterns of intestinal colonization in breast-fed versus commercial-infant-formula-fed infants (microbiota) may add to the preventing effect of human milk.

> There is a 15–30 % reduction in adolescent and adult obesity rates if any breastfeeding occurred in infancy compared with no breastfeeding. The duration of breastfeeding also is inversely related to the risk overweigh. (Parikh et al. 2009)

Up to 30 % reduction in the incidence of type 1 diabetes mellitus is reported. It has been postulated that the putative mechanism is the infant exposure to cow milk β lactoglobulin, which stimulates an immune immediate process cross-reacting with pancreatic β cells (Ip et al. 2009). Breastfeeding also reduces the risk of acute lymphocytic leukemia and acute myeloid leukemia in infants.

Actually, significant positive effects of human milk feeding on long term neurodevelopment are observed in preterm infants. These data remain significant after adjustment for confounding factors, such as maternal age, education, marital status, race, and infant morbidities. Long-term studies of preterm infants also suggest that human milk feeding is associated with lower rates of metabolic syndrome, and in adolescents it is associated with lower blood pressures and low-density protein concentrations and improved leptin and insulin metabolism.

Moreover, there are also adverse effects caused by the lack of breastfeeding. "Prospective cohorts studies have noted and increased in post partum depression in mothers who do not breastfeed or wean early" (Henderson et al. 2003).

Furthermore, long-term breastfeeding significantly reduces the incidence of hypertension, hyperlipidemia, cardiovascular diseases, and diabetes in mothers who breast-feed (Schwarz et al. 2009). In addition to these benefits, breastfeeding mothers have a lower incidence of breast and ovarian cancer.

Considering the vast list of benefits, it is possible to state that breastfeeding is not only a lifestyle but also a health strategy of utmost importance. Therefore, we must emphasize the importance of the *Mother-Baby Friendly Hospital* initiative to promote and protect breastfeeding launched by United Nations Children's Fund (UNICEF)/World Health Organization (WHO) and step 9 of the *Safe and Family Centered Maternity Program* (Larguia et al. 2012).

20.2.2 Safe and Family-Centered Maternity/Family Inclusion and Personal Rights

Safe and Family-Centered Maternity (SFCM) promotes an organizational culture that concedes parents and families the leading role in their perinatal care. It poses a paradigm shift by praising and ensuring infants and women rights during pregnancy and puerperium. Under the concept of interculturality, SFCM fosters the active role of the family group and the community. It implements safe practices of proven effectiveness and reinforces other current initiatives such as the Mother-Baby Friendly Hospital (UNICEF/WHO), family planning, and social inclusion programs. It consists of ten steps which comprise reproductive health care to the imitation of the model in other health-care centers (Larguia et al. 2012). SFCM has been recognized as an example of good practices by an international health organization (PAHO 2008).

In order to provide examples of what SFCM aims and because of the importance of maternal stress effects in early human development (Fumagalli 2007), we could highlight step 5. It establishes 24-h free and unrestricted access of parents to the neonatology service. It proposes continuous permanence and active participation of parents in the care of their infants, such as providing them training on basic techniques like gavage feeding. It also recognizes the right of parents to skin-to-skin contact (SSC) with their infants (see corresponding text). Likewise, in cooperation with the mental health service, it facilitates the inclusion of the rest of the family under the Grandparents and Siblings Access Program. Health education teams also receive specific training on reproductive health and how to prevent acute lower respiratory infection. With a view to favoring the continuous permanence of parents, SFCM offers (step 6) a Home for Mothers of preterm and sick newborns. The Home for Mothers (step 7) is under the responsibility of the voluntary workers of the community. Breastfeeding promotion (step 9) boosts the discharge of very low birth-weight infants with a high percentage of breastfeeding, which represents a significant aspect for the follow-up monitoring after discharge (step 8). In conclusion, SFCM refers to this set of connected interventions as "the virtuous circle" (Larguia et al. 2006).

20.2.3 The Late Preterm Problem

As a result of obstetric interventions, in many cases arguable interventions, in the past years a new neonatal morbidity has arisen, the *late preterm*. It comprises those cases in which gestation is interrupted between 34 and 36 weeks. These newborns have a greater incidence of respiratory distress syndrome, hyperbilirubinemia, feeding difficulties, and longer hospitalization.

Preterm birth is a major public health problem, principally because of the high risk for adverse medical and developmental outcomes in survivors. In addition, there is a growing literature reporting increased rates of psychiatric disorders in preterm children, specifically anxiety, depression, attention-deficit/hyperactivity disorder (ADHD), particularly the inattentive type, and autism (Johnson et al. 2010). The majority of research, however, has focused on children born before 30 weeks of gestation. Yet, late preterm births, typically between 34 and 36 weeks gestation, are the largest cohort of premature children comprising 70% of preterm births in 2009. Recent evidence supports an increased rate of psychiatric symptoms in late preterm children as well, particularly ADHD, "emotional," and anxiety symptoms (Westrupp et al. 2012).

> Late preterm children (34–36 weeks of gestation) were at increased risk for anxiety disorders at preschool age. A history of maternal depression mediated this association. Findings confirm the extension of the risk of psychiatric disorders associated with prematurity to the late preterm group, and suggest that maternal depression may play a key role in this risk trajectory. (Rogers et al. 2013; J. Am. Acad. Child Adolesc. Psychiatry; 2013; 52(3):309–318)

Another important factor is the potential impact of maternal history of depression on the risk for psychiatric disorders among late preterm and early term children. Postpartum depressive symptoms are more among mothers of preterm infants. Maternal depression is strongly related to increased risk for psychiatric symptoms and disorders in child offspring. This risk is conferred not only by independent but also interactive genetic and environmental mechanisms (Goodman 2007).

Thus, preterm infants of mothers with a history of depression may be at increased risk for childhood psychopathology because of the interaction of their genetic and biologic risk (poor self-regulation of secondary to altered neurodevelopment), with their exposure to maladaptive parenting behaviors associated with depression. Similarly, mothers of preterm infants have elevated anxiety (Brandon et al. 2011) also known to increase the risk of psychopathology in child offspring. Thus, maternal depression or anxiety may play an important role in the risk for mental disorders in preterm infants, a mechanism not previously well investigated (Schreier et al. 2008).

Depressed mothers may perceive their preterm children as more vulnerable, resulting in "vulnerable child syndrome" and associated increased overprotective parenting. This pattern has been shown to increase anxiety and internalized symptoms in the child.

The results point to the importance of screening and treating mothers of preterm children, even late preterm children, for perinatal depression. Identifying these mothers could both potentially lessen the impact of having a mother with depression and determine which preterm children are at greater risk for psychiatric disorders, to best target prevention efforts.

20.2.4 The Sacred Hour

Uninterrupted SSC in the First Hour after Birth

Developing a care strategy based on reinforcing the bonding and attachment between the mother and her newborn is so important for a better future that it should be considered as standard care.

Concerning this concept, the WHO, the American Academy of Pediatrics (AAP), the Academy of Breastfeeding Medicine (ABM), the American Heart Association (AHA), and the Neonatal Resuscitation Program (NRP) expressed their support.

An interrupted skin-to-skin hour after birth, for the stable baby and mother, is safer for both and has multiple short and long benefits. This "sacred hour" must be honored and protected.

In the new born, brain development in areas of amygdala are in a critical period of maturation in the first 2 months of life. SSC activates the amygdala via the prefrontal–orbital pathway. (Schore 2001). Early sensory experiences like these are disproportionately processed and stored in the infant's right hemisphere. Attachment experiences facilitate its maturation and facilitates maturation of the right brain. Attachment relationships promote the development of the new born brain's self-regulatory mechanism. Infants who spent 1–2 h in SSC had better self-regulation 1 year later (Bystrova et al. 2009).

The benefits of early postpartum SSC: improves physiologic stability for mother and baby, increases maternal attachment behaviors, protects baby from negative effects of separation, supports optimal infant brain development, and increases breastfeeding rates and duration.

Attachment is biologically primed. Maternal caregiving is increased by SSC (endogenous opioid peptides, estrogen and progesterone, prolactin, vasopressin, dopamine, and oxytocin). More specifically, oxytocin has the following actions: contracts uterus, releases mother's milk—colostrum—and increases maternal caregiving behaviors, facial recognition, and attraction. In summary, it increases maternal bonding.

"If the attachment relationship is indeed a major organizer in brain development…then the determinants of attachment relationships are important beyond the provision of a fundamental sense of safety or security" (Fonagy 2005). There is a quality improvement project for changing the practice of SSC after the first hour after birth, from Loma Linda University Medical Center, by Raylene M. Phillips, MD, Assistant Professor of Pediatrics (Philips 2012).

The importance of mother–infant early SSC was demonstrated by a recent randomized trial in which 176 mother–infant pairs were studied. They were divided into two groups; one was the group of mothers who practiced early SSC with the infant in the delivery room and the other was the group of mothers who were separated from their infants. *The mother–infant interaction was videotaped according to the Parent–Child Early Relational Assessment (PCERA) 1 year after birth. The practice of SSC, early suckling, or both during the first 2 h after birth when compared with separation between the mothers and their infants positively affected the PCERA variables on maternal sensitivity infant's self-regulation, and dyadic mutuality and reciprocity at 1 year after birth. The negative effect of a 2-h separation after birth was not compensated for by the practice of rooming in. These findings support the presence of a period after birth (the early "sensitive period") during which close contact between mother and infant may induce long-term positive effect on mother–infant interaction (Bystrova et al.* 2009*)*.

20.2.5 Consequences of Pain in Preterm Newborns

Every year, more than 12 million preterm newborns (PN) are born in the world. Along with an increase in the survival rate of this population, there is also a high prevalence of cognitive impairment, learning difficulties, and abnormal behavior in early infancy and primary school (McCormick et al. 1990). Follow-up studies of ex-preterm infants revealed neurodevelopmental disorders with special assistance needs (McCormick et al. 1992) and a growth in the health and society burden (Slonim et al. 2000).

Preterm newborns experience pain. They must undergo the abrupt change from the safe intrauterine environment to the neonate intensive care unit (NICU). Their survival depends upon highly sophisticated care related to a great number of painful proceedings and stressful situations such as being separated from their mothers during hospitalization. In 3 months, a PN in NICU is likely to undergo about 300 painful proceedings (Grunau et al. 2007).

For human brain development, the moments before and after birth are crucial mainly because there is a growth peak, exuberant synaptogenesis (Rakic 1998), and specific reception group expression (Anand 2008). Preterm newborns are particularly vulnerable to pain and stress effects because of the immaturity of their nervous system (Beggs and Fitzgerald 2007).

While the biological and anatomical mechanisms to transmit pain are already functional in the fetus and then newborn, the physiological systems of *protection* are not yet mature. This preponderant unbalance between pain transmission and systems of protections make the newborn hyperalgesic, with a low perception threshold and an intense and diffuse spatial nociception (Dinerstein and Brundi 1998).

In a PN, acute pain produces short-term adverse changes whereas repeated or chronic pain produces long-term changes related to the processing of pain, the stress response system, and the neurodevelopment.

Prolonged or repeated pain produces changes in pain processing systems (Slater et al. 2010). These changes remain until adolescence and they consist of a greater and prolonged response to future pain stimuli (hyperalgesia), to those stimuli not commonly perceived as painful (allodynia), and there is also greater nociception in the tissues that surround the stimulus (Grunau and Tu 2007).

Even more consequences of continuous adaptation on repeated pain are noticed: permanent neuroanatomic alterations, emotional and behavior disorders, learning difficulties, and functioning disorder of HPA axis in the processing of stress which causes an impact on these infants' health and neurodevelopment (Sullivan et al 2008).

Summary of short and long term consequences of pain:

- Increased heart and respiratory rate
- Increased blood pressure
- Decreased arterial oxygen saturation: hypoxic and hyperoxia episodes
- Cerebrovascular blood flow alterations
- Decreased vagal tone
- Decreased peripheral blood flow
- Intraventricular hemorrhage and periventricular leukomalacia
- Behavioral and physiological disorganization
- Palmar sweating
- Low energy storage used for growth

Long-Term Consequences

- Decreased volume of cortical and subcortical regions caused by excitotoxicity and apoptosis
- Degenerative changes with impact on vulnerable cortical neurons survival
- Permanent alterations in pain processing at spinal and peripheral levels
- Higher hypothalamic hypophyseal axis response to stress
- Neurodevelopmental alterations
- Cognitive and learning development alteration
- Attention deficit disorder
- Inability to face new situations
- Poor adaptive behavior
- Defensive behavior
- Prolonged hypersensitivity

In order to explain cell death and abnormal synaptogenesis in the PN immature nervous system after pain episodes, that induces or generates long-term consequences, it is important to mention two mechanisms, excitotoxicity and apoptosis:

- Neuronal excitotoxicity is mediated by N-methyl-D-aspartate (NMDA) receptors. These receptors trigger Ca^{2+} overload and the latter excitotoxic cell death.
- Apoptosis is mediated by inflammatory cytokines such as tumor necrosis factors (TNF)-α receptors through the activation of a caspase cascade or mitochondrial insult. This cell death predominates in the brain stem and cortex, hippocampus, thalamus, and other subcortical regions (Spreafico et al. 1995).

Although neonatal intensive care provides aggressive treatment for hypoxia, hypoglycemia, or sepsis, it has only recently included treatment for other factors that contribute to neuronal damage: repeated pain and mother–infant separation (Anand 2000).

Recent experimental and clinical observations suggest that pain caused by invasive procedures and mother–infant separation, which leads to a lack of social stimulation (tactile, kinesic and verbal stimulation), may have effects on the immature neuron vulnerability development.

To summarize, PN are pain sensitive and therefore vulnerable to its long- and short-term effects. While their immature nervous system develops, they are prone to excitotoxicity damage and apoptosis leading to anatomical lesions. As a result, the decreased brain volume, at the same time, has an impact on their neurodevelopment and therefore affects their health and entire social group they live in. To that end, systematic approach to pain, its prevention, and adequate treatment have vast consequences on the life of the PN and on the public health in general.

20.2.6 Developmental Interventions in NICU

Several developmental interventions have been implemented in the NICU with a goal of enhancing the neurodevelopmental outcomes of hospitalized preterm infants. The following are the most promising interventions to improve the mental developmental index and psychomotor developmental index scores on the Bayleys Scale (Hussey-Gardner and Famuyide 2009):

a. Application of human touch
b. Intermittent Skin-to-skin care
c. Nested prone and supine positioning
d. Protection against light by incubator covers
e. Exposure to contingent music

To our judgment, from the above mentioned proposals, the most relevant is the intermittent SSC (Moore et al. 2007).

Skin-to-Skin Contact Placing the new born out of the incubator to experience SSC with their parents is one of NICU's main care strategies. This practice is not only possible in extremely low birth weight (ELBW) infants (< 1000 g) but also in patients that require continuous airway pressure with nasal prongs (CPAP), or intermittent mandatory ventilation weaning of mechanical ventilation.

Mother/father–PN SSC contributes to reduce the stress both parents undergo the moment they assume a key role in their infant's life. SSC improves self-esteem and humanizes NICU invasive interventions.

SSC duration depends only on the infant's clinical status and his/her parents' desire and confidence; nevertheless, it requires appropriate monitoring (pulse oxygen, heart rate, respiratory rate, temperature, etc.).

SCC offers multiple benefits for parents. Mothers experience less postpartum depression, they feel more self-confident, and they even embrace, caress, comfort, and smile more at their children. In consequence, the emotional basis of the bond between mother and child is promoted (Brundi 2006).

Regarding breastfeeding, on the one hand prolactin stimulates breastfeeding initiation and continuation since it enhances milk production in lactating mothers. On the other hand, since the first attempt of breastfeeding, infants explore, smell, search for and finally find the nipple and therefore learn to suck earlier. It is essential to bear in mind that mother's breast milk is the best option to feed the infant because it not only nourishes but also provides protection against infections. It is in constant change to adapt itself to the infant's growth needs.

Benefits for the newborn: Through SSC on the mother's (or father's) chest, the PN experiences tactile, auditory, and proprioceptive stimuli. It improves all physiological parameters and contributes to stabilize heart and respiratory rates. A lower need of oxygen therapy, more appropriate weight gain, improved body temperature control, and significant decrease of apnea episodes are observed. It is demonstrated that SSC fosters infant's development, contributes to lower cortisol (stress-related hormone) levels, and therefore stimulates the infant's ability to calm down by himself. The auditory (hearing his mother's voice), tactile, olfactory, and visual stimuli the newborn receives and the fact that keeping direct contact with his mother makes the infant feel safe lead to less crying episodes and better rest and sleep.

SSC between mother and son also leads to increasing deep sleep duration, decreasing motor and muscular activity and lessening of infant crying or stress and his mother's anguish (Larguia et al. 2008).

20.2.7 Nutritional Interventions

The pandemic of metabolic diseases has severe consequences in terms of mortality, morbidity, and life quality. The abrupt population shift towards an obese phenotype suggests that it may have an environmental origin (Yajnik and Deshmukh 2012).

In a recent publication, Glukmann suggests that the fetus adapts to the intrauterine characteristics (Gluckman et al. 2008). These adaptations are caused by intrauterine signals and the changes are evident not only immediately but also later in his life since those signals are postnatal environment predictors. Therefore, if the prediction is adequate it provides the fetus an encouraging prospect of survival; otherwise, the fetus life may be at risk. It has been estimated that more than 60% of weight variation among newborn infants has its origin in the intrauterine environment. One genotype can result in different physiologic or morphologic characteristics in response to different environmental conditions. However, energy and protein supply restriction brings about growth restriction in both prenatal and postnatal life (Thureen 2007). In the long run, this may lead the infant to be prone to health problems (Ehrenkranz et al. 2011; Martin et al. 2009; Poindexter et al. 2006).

A general consensus insists on exclusive breastfeeding up to the first 6 months of life in full-term new borns because of the above-mentioned benefits. However, feeding recommendations become more arduous in PN (less than 37 weeks' gestation) and even more in those PN with less than 32 weeks' gestation or in ELBW (< 1000 g) but a growing survival prognosis. [ENERGY AND PROTEIN DEFICITS GRAPHICS]

Graphic N°2: reflects decrease in energy and protein-accumulated deficits before and after the implementation of an early-intensive-enteral-parenteral feeding protocol during the first 28 days of life (Dinerstein et al. 2006; Figs. 20.1 and 20.2).

There are numerous difficulties for the optimal supply of nutrients in these groups. There is a high incidence of digestive intolerance (higher-to-lower gestational age, GE) and severe complications such as necrotizing enterocolitis. Even

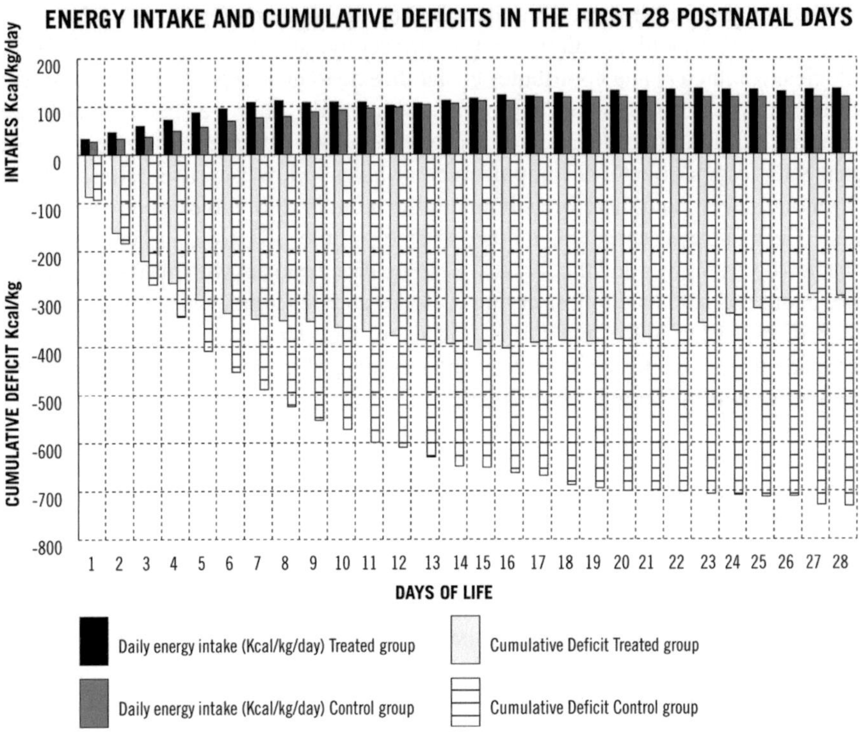

Fig. 20.1 Energy intakes and cumulative deficits in the first 28 postnatal days. Energy intake in aggressive group was higher ($p < 0.001$) and deficit was lower ($p < 0.001$) from birth to day 28

though it is possible to reduce postnatal undernourishment (weight lower than 10th percentile at 40 weeks) with aggressive nutrition interventions (early and progressive enteral feeding with breast milk and parenteral feeding with amino acids and lipids), they do not totally prevent it.

Poor postnatal growth is associated with (Dinerstein et al. 2005) low cognitive scores and developmental compromise (Latal Hajnal et al. 2003). In preterm newborns, growth is a complex process. There are severe consequences for long-term neurological development and health in general not only because of an extrauterine growth restriction (EUGR; neurodevelopmental compromise) but also because nutritional recovery determines a weight gain and abnormal distribution of the adipose tissue (metabolic syndrome risk).

Currently, in preterm newborns the recommendation is to imitate the growth quality and type of a fetus in the same GE (Uhing and Das 2009). Whether this is a desirable or possible-to-achieve goal or not, that is a question that has not been answered yet.

Nowadays, newborns are discharged from NICU with different levels of EUGR; however, they show a perivisceral fat increase. While which is the most adequate growth is still under debate, the growth quality dilemma must also be faced. Should all efforts be made to improve neurological development results or to reduce the

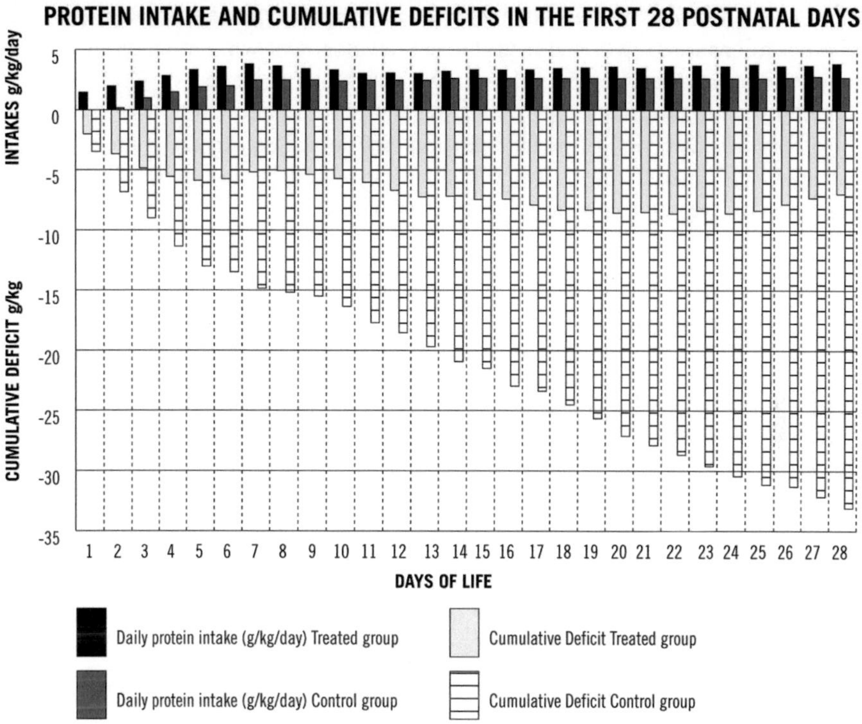

Fig. 20.2 Protein intake and cumulative deficit in the first 28 postnatal days. Protein intake was higher in aggressive group ($p < 0.001$) and deficit was lower ($p < 0.001$)

risk of metabolic syndrome in later stages of life? Currently, the correct strategy to be adopted is still unclear.

Preterm birth continues to contribute disproportionately to neonatal morbidity and subsequent physical and neurodevelopmental disabilities. Epidemiologic studies have described additional long-term health consequences of preterm birth such as an increased risk of hypertension and insulin resistance in adult life. It is not known whether the influence of infant and childhood growth rates and early nutrition on long-term outcomes is the same or different among preterm infants and neonates with IUGR metabolic outcomes.

Present evidence suggests that even brief periods of relative undernutrition during a sensitive period of development have significant adverse effects on later development. Evidences suggest that growth between birth and expected term and 12–18 months post-term has no significant effect on later blood pressure and metabolic syndrome, whereas reduced growth during hospitalization significantly impacts later neurodevelopment. In contrast, growth during late infancy and childhood appears to be a major determinant of later metabolic and cardiovascular well-being, which suggests that nutritional interventions during this period are worthy of more study (Lapillonne and Griffin 2013).

New nutritional strategies will have to be developed to prevent IUGR and EUGR in order to ensure growth and neurological development without increasing metabolic syndrome risk. Therefore, developing nutritional strategies should be a priority followed by the inclusion of these patients in long-term follow-up programs to foster early detection of metabolic syndrome indicators. In the context of Safe and Family-Centered Maternity programs that would be an alternative until the definition of optimal and opportune growth is clear.

20.3 Conclusion

Over the past 10 years, there has been outstanding scientific progress related to perinatal programming and its epigenetic effects in health, and we can anticipate this trend will continue in the near future. We need to make use and apply these achievements to human neurodevelopment via prevention interventions. Based on the concept of the interaction between genome and ambiome, this chapter proposes low-cost easy-implementation preventive strategies for maternal and infant health institutions.

Breastfeeding and human milk administration are the first preventive measures, as has been reviewed in the *policy statement* of the *American Academy of Pediatrics*. Another strategy is the *Safe and Family-Centered Maternity Hospitals* initiative that promotes and empowers the inclusion of the families and the respect for their rights, especially during pregnancy and birth. (This change of paradigm was approved and is recommended by both UNICEF and PAHO.) Then, there is also an important emphasis given to the sacred hour—which highlights the impact in bonding, attachment, and breastfeeding during the first hour of life—the pain prevention and treatment in newborns, the control of the "new morbidity" represented by late preterm infants, and finally, the importance of avoiding IUGR and EUGR. (However, there are not yet clear recommendations about nutritional interventions in order to diminish the potential metabolic syndrome consequence in the adult.)

Conflict of Interest The authors declare no conflicts of interest.

References

Akomberg AK, Ramanan AV, Buchan I (2006) Effect of breast on risk of coeliac disease. a systematic revue and metaanalisis of observational studies. Arch Dis Child 91:39–43

Anand KJS (2000) Pain, plasticity, and premature birth: a prescription for permanent suffering? Nature Med 6:971–973

Anand KJS (2008) From the gate-control theory to brain programs for neonatal pain. In: Buonocuore G, Bellieni C (eds) Neonatal pain. Suffering, pain and risk of brain damage in the fetus and the newborn, 1st edn. Springer, Italy, pp 141–147

Baier C et al (2012) Gestational restraint stress and the developing dopaminergic system: an overview. Neurotox Res 22:16–32

Barclay AR, Russell RK, Wilson ML, Gilmour WH, Satsangi J, Wilson DC (2009) Systematic review: the role of breastfeeding in the development of pediatric inflammatory bowel disease. J Pediatr 155(3):421–426

Beggs S, Fitzgerald M (2007) Development of peripheral and spinal nociceptive systems. In: Anand KJS, Stevens BJ, McGrath PJ (eds) Pain in neonates and infants: pain research and clinical management, 3rd edn. Elsevier, Toronto, pp 11–24

Benítez Keinrad M (2011) DESARROLLO: La odisea del organismo. Centro de Ciencias de la Complejidad. Universidad Nacional Autonóma de México, Mexico

Bond DM, Finnegan EJ (2007) Trends passing the message on: inheritance of epigenetic traits. Plant 12:211–216

Brandon DH, Tully KP, Silva SG et al (2011) Emotional responses of mothers of late-preterm and term infants. J Obstet Gynecol Neonatal Nurs 40:719–731

Brundi M (2006) In Larguia AM, Lomuto C, Gonzalez MA, Guia para la transformación de maternidades convencionales en maternidades centradas en la família. Funlarguia y Ministerio de Salud de la Nación. Rep. Argentina (http://www.funlarguia.org.ar/Herramientas/Guia-para-transformar-maternidades-tradicionales-en-maternidades-centradas-en-la-familia)

Bystrova K, Ivanova V, Edhborg M et al (2009) Early contact versus separation: effects on mother–infant interaction one year later. Birth 36(2):97–109

Das UN (2007) Breastfeeding prevents type 2 diabetes mellitus: but, how and why? Am J Clin Nutr 85(5):1436–1437

Dewey KG, Lönnerdal B (1986) Infant self-regulation of breast milk intake. Acta Paediatr Scand 75(6):893–898

Dinerstein NA, Brundi M (1998) El dolor en el Recién Nacido Prematuro. Rev Hosp Mat Inf Ramón Sardá 17:97–192

Dinerstein A, Schapira I, Nieto R, Solana CL, Aspres N, Perez G (2005) Growth and neurodevelopment in preterm infants receiving early and intensive enteral and parenteral nutrition. E PAS 57:214805

Dinerstein NA, Nieto RM, Solana CL, perez GP, Otheguy LE, Larguía AM (2006) Early and aggressive nutritional strategy (parenteral and enteral) decreases postnatal growth failure in very low birth weight infants. J Perinatol 26:436–442

Ehrenkranz RA, Das A, Wrage LA, Poindexter BB, Higgins RD, Stoll BJ, OH W (2011) Early nutrition mediates the influence of severity of illness on extremely LBW infants. Pediatr Res 69:522–529

Eidelman AL, Schandler RJ (2012) American Academy of Pediatrics, sector on breastfeeding and the use of human milk Pediatrics 129(3) e827–e841

Fonagy P, Target M (2005) Bridging the transmission gap: an end to the important mystery of attachment research? Attach Hum Dev 7:333–343

Fumagalli F, Molteni R, Racagni G et al (2007) Stress during development: impact on neuroplasticity and relevance to psychopathology. Prog Neurobiol 81:197–217

Gluckman PD, Hanson MA, Beedle AS (2007) Early life eventsand their consequences for later disease: a life history and evolutionary perspective. Am J Hum Biol 19:1–19

Gluckman PD, Hanson MA, Cyrus Cooper C, Thornburg KL (2008) Effect of in utero and early-life conditions on adult health and disease. The N Engl J Med 359:61–73

Goodman SH (2007) Depression in mothers. Annu Rev Clin Psychol 3:107–135

Greer FR, Sicherer SH, Burks AW (2008) Effects of early nutritional interventions on the development of atopic disease in infants and children: the role of maternal dietary restriction, breastfeeding, timing of introduction of complementary foods, and hydrolyzed formulas. Pediatrics 121(1):183–191 (American Academy of Pediatrics Committee on Nutrition; American Academy of Pediatrics Section on Allergy and Immunology)

Grunau RE, Tu MT (2007) Long-term consequences of pain in human neonates. In: Anand KJS, Stevens BJ, McGrath PJ (eds) Pain in neonates and infants: pain research and clinical management, 3rd edn. Elsevier, Toronto, pp 45–55

Grunau RE, Haley DW, Whitfield MF et al (2007) Altered basal cortisol levels at 3, 6, 8 and 18 months in preterm infants born at extremely low gestational age. J Pediatr 150:151–156

Henderson JJ, Evans SF, Straton JA, Priest SR, Hagan R (2003) Impact of postnatal depression on breastfeeding duration. Birth 30(3):175–180
Larguia AM, Gonzalez MA, Solana C et al (2012) Maternidad Segura y Centrada en la Familia, con enfoque intercultural 2nd edn. UNICEF. www.funlarguia.org.arhttp://www.funlarguia.org.ar/Herramientas/Manual-de-Maternidades-Seguras-y-Centradas-en-la-Familia
Latal Hajnal B, Von Siebenthal K, Kovari H, Bucher HU, Largo RH (2003) Postnatal Growth in VLBW infants: significant association with neurodevelopment outcomes. J pediatr.143:163–170.
Huang LT (2011) The link between perinatal glucorticoids exposure and psychiatric disorders. Pediatr Res 69(5):2 (University College of Medicine, Taiwan)
Huang LT (2011) Perinatal programming of neuropsychiatric disorders. J Formos Med Assoc 110(6):347–349
Hussey-Gardner B, Famuyide M (2009) Developmental interventions in the NICU: what are the developmental benefis? NeoReviews 10:113–120
Ip S, Chung M, Raman G, Trikalinos TA, Lau J (2009) A summary of the Agency for Healthcare Research and Quality's evidence report on breastfeeding in developed countries. Breastfeed Med 4(suppl 1):S17–S30
Isaacs EB, Fischl BR, Quinn BT, Chong WK, Gadian DG, Lucas A (2010) Impact of breast milk on intelligence quotient, brain size, and white matter development. Pediatr Res 67(4):357–362
Jablonka E, Lamb MJ (2007) Evolution in four dimensions. Behav Brain Sci 30:353–365
Johnson S, Hollis C, Kochhar P et al (2010) Psychiatric disorders in extremely preterm children: longitudinal in dingatage 11 years in the E P I Cure study. J Am Acad Child Adolesc Psychiatry 49:453–463
Jones G, Steketee RW, Black RE, Bhutta ZA, Morris SS (2003) Bellagio child survival study group. How many child deaths can we prevent this year? Lancet 362(9377):65–71
Kramer MS, Chalmers B, Hodnett ED et al (2001) Promotion of Breastfeeding Intervention Trial (PROBIT): a randomized trial in the Republic of Belarus. JAMA 285(4):413–420
Lapillonne A, Griffin A (2013) Feeding preterm infants today for later metabolic and cardiovascular outcomes. J Pediatr 162(3 Suppl):S7–S16
Larguia AM, Lomuto C, Gonzalez MA (2006) "Guia para la transformación de maternidades convencionales en maternidades centradas en la familia". Funlarguia y Ministerio de Salud de la Nación. www.funlarguia.org.ar/Herramientas/Guia-para-transformar-maternidades-tradicionales-en-maternidades-centradas-en-la-familia
Larguia AM, Schapira I, Aspres N et al (2008) Los padres y la familia en los servicios de Neoantología. Guia para padres de prematuros Neonatología, FUNLARGUIA. http://www.funlarguia.org.ar/Herramientas/Lectura-para-padres
Martin CR, Brown YF, Ehrenkranz RA, O´Shea TM, Allred EN, Belfort MB, Mc Cormick M, Levinton A (2009) Nutritional practices and growth velocity in the first month of life in extremely premature infants. Pediatrics 124:649–657
McCormick MC, Gortmarker SL, Sobol AM (1990) Very low birth weight children: behavior problems and school difficult in a national sample. J Pediatr 117:687–693
McCormick MC, Brooks-Gunn J, Workman-Daniels K et al (1992) The health and developmental status of very low-birth-weight children at school age. JAMA 267:2204–2208
Moore ER, Anderson GC, Bergman N (2007) Early skin-to-skin contact for mothers and their healthy newborn infants. Cochrane Database of Systematic Reviews, Issue 3. Art. No.: CD003519. doi:10.1002/14651858CD003519.pub2
PAHO (2008) Ejemplos de Buenas Prácticas: Maternidades Centradas en la Familia.Experiencia del HMIR Sardá, Buenos Aires. Argentina 63. http://publicaciones.ops.org.ar/publicaciones/pubOPS_ARG/pub63.pdf
Parikh NI, Hwang SJ, Ingelsson E et al (2009) Breastfeeding in infancy and adult cardiovascular disease risk factors. Am J Med 122(7):656–663, e1
Phillips R (2012) The sacred hour. Conference at EIP, Madrid
Poindexter BB, Langer JC, Dusick AM, Ehrenkranz RA (2006) Early provision of parenteral amino acids in extremely low birth weight infants: relations to growth and neurodevelopment outcome. J Pediatr 148:300–305

Rakic P (1998) Images in neuroscience. Brain development, VI: radial migration and cortical evolution. Am J Psychiatr 155:1150–1151

Rogers CE, Shannon NL, Joan LL, (2013) Late preterm birth, maternal depression, and risk of preschool psychiatric disorders. J Am Acad Child Adolesc Psychiatry 52(3):309–318

Roldan Arjona T (2011) Conference. Epigenética: una nueva frontera en Biología, (Departamento de Genética, Universidad de Córdoba, España

Schore A (2001). Effects of a secure attachment relationship on right brain development, affect regulation, and infant mental health. Infant Ment Health J 22(1-2):7–66.

Schreier A, Wittchen H-U, Hofler M, Lieb R (2008) Anxiety disorders in mothers and their children: prospective longitudinal community study. Br J Psychiatry 192:308–309

Schwarz EB, Ray RM, Stuebe AM et al (2009) Duration of lactation and risk factors for maternal cardiovascular disease. Obstet Gynecol 113(5):974–982

Slater R, Fabrizi L, Worley A et al (2010) Premature infants display increased noxious-evoked neuronal activity in the brain compared to health age-matched term-born infants. NeuroImage 52(2):583–589

Slonim AD, Patel KM, Ruttimann UE et al (2000) The impact of prematurity: a perspective of pediatric intensive care units. Crit Care Med 28:848–853

Spreafico R, Frassoni C, Arcelli P et al (1995) In situ labeling of apoptotic cell death in the cerebral cortex and thalamus of rats during development. J Comp Neurol 363:281–295

Strathearn L, Mamun AA, Najman JM, O'Callaghan MJ (2009) Does breastfeeding protect against substantiated child abuse and neglect? A 15-year cohort study. Pediatrics 123(2):483–493

Sullivan MC, Hawes K, Winchester SB et al (2008) Developmental origins theory from prematurity to adult disease. J Obstet Gynecol Neonatal Nurs 37:158–164

Sullivan S, Schanler RJ, Kim JH et al (2010) An exclusively human milk-based diet is associated with a lower rate of necrotizing enterocolitis than a diet of human milk and bovine milk-based products. J Pediatr 156(4):562–567

U.S. Department of Health and Human Services (2011) The Surgeon General's call to action to support breastfeeding. www.surgeongeneral.gov/topics/breastfeeding/. Accessed 28 March 2011

Uhing MR, Das UG (2009) Optimizing growth in the preterm infant. Clin Perinatol 36:165–176

Westrupp EM, Mensah FK, Giallo R, Cooklin A, Nicholson JM (2012) Mental health in low to moderate risk preterm, low birth weight, and small for gestational age children at 4 to 5 years: the role of early maternal parenting. J Am Acad Child Adolesc Psychiatry 51:313–323

Yajnik C, Deshmukh U (2012) Early life origins of diabetes and obesity: general aspects and the thin-fat baby paradigm. In: Plagerman A (ed) Perinatal programming: the state of the art. Walter de Gruyter GmbH & Co. KG, Berlin, pp 69–79

Chapter 21
Pregnancy Outcomes After a Maternity Intervention for Stressful Emotions (PROMISES): A Randomised Controlled Trial

Huibert Burger, Claudi L. H. Bockting, Chantal Beijers, Tjitte Verbeek, A. Dennis Stant, Johan Ormel, Ronald P. Stolk, Peter de Jonge, Mariëlle G. van Pampus and Judith Meijer

Abstract There is ample evidence from observational prospective studies that maternal depression or anxiety during pregnancy is a risk factor for adverse psychosocial outcomes in the offspring. However, to date no previous study has demonstrated that treatment of depressive or anxious symptoms in pregnancy actually could prevent psychosocial problems in children. Preventing psychosocial problems in children will eventually bring down the huge public health burden of mental disease. The main objective of this study is to assess the effects of cognitive behavioural therapy in pregnant women with symptoms of anxiety or depression on the child's development as well as behavioural and emotional problems. In addition, we aim to study its effects on the child's development, maternal mental health,

This work is based on a previously published journal article: Meijer JL, Bockting CL, Beijers C, Verbeek T, Stant AD, Ormel J, Stolk RP, de Jonge P, van Pampus MG, Burger H. PRegnancy Outcomes after a Maternity Intervention for Stressful EmotionS (PROMISES): Study protocol for a randomized controlled trial. Trials. 2011 Jun 20;12:157. doi: 10.1186/17457450.1186/174

H. Burger (✉)
Department of General Practice, University of Groningen, University Medical Centre Groningen, Hanzeplein 1, 9700 RB, Groningen, The Netherlands
e-mail: h.burger@umcg.nl

C. L. H. Bockting
Department of Clinical and Developmental Psychology, University of Groningen, Grote Kruisstraat 2/1, 9712 TS Groningen, The Netherlands

H. Burger · C. Beijers · J. Ormel · P. de Jonge
Department of Psychiatry, University of Groningen, University Medical Centre Groningen, Hanzeplein 1, 9700 RB, Groningen, The Netherlands

T. Verbeek · A. D. Stant · R. P. Stolk · J. Meijer
Department of Epidemiology, University of Groningen, University Medical Centre Groningen, Hanzeplein 1, 9700 RB, Groningen, The Netherlands

M. G. van Pampus
Department of Obstetrics and Gynecology, Onze Lieve Vrouwe Gasthuis, Amsterdam, The Netherlands

M. C. Antonelli (ed.), *Perinatal Programming of Neurodevelopment*, Advances in Neurobiology 10, DOI 10.1007/978-1-4939-1372-5_21,

and neonatal outcomes, as well as the cost-effectiveness of cognitive behavioural therapy relative to usual care.

We will include 300 women with at least moderate levels of anxiety or depression at the end of the first trimester of pregnancy. By including 300 women, we will be able to demonstrate effect sizes of 0.35 or more on the total problems scale of the Child Behaviour Checklist 1.5–5 with alpha 5 % and power (1-beta) 80 %.

Women in the intervention arm are offered 10–14 individual cognitive behavioural therapy sessions, 6–10 sessions during pregnancy and 4–8 sessions after delivery (once a week). Women in the control group receive care as usual.

Primary outcome is behavioural/emotional problems at 1.5 years of age as assessed by the total problems scale of the Child Behaviour Checklist 1.5–5 years.

Secondary outcomes are mental, psychomotor and behavioural development of the child at age 18 months according to the Bayley scales; maternal anxiety and depression during pregnancy and postpartum; and neonatal outcomes such as birth weight, gestational age and Apgar score, health-care consumption and general health status (economic evaluation).

Trial Registration: NTR2242

Abbreviations

ADHD	Attention deficit hyperactivity disorder
ALSPAC	Avon Longitudinal Study of Parents and Children
BSID II	Bayley Scale of Infant Development
CAU	Care as usual
CBCL	Child Behaviour Checklist
CBT	Cognitive behavioural therapy
C-TRF	Caregiver–Teacher Report Form
DALY	Daily adjusted life years
DAS	Dysfunctional Attitudes Scale
DSM	Diagnostic and Statistical Manual of Mental Disorders
ECR	Experiences in Close Relationships Scale
EPDS	Edinburgh Postnatal Depression Scale
EQ-5D	EuroQol
HELLP	Hemolytic anemia, elevated liver enzymes and low platelet count
ITT	Intention to treat
LDS	Language Development Survey
MINI	Mini International Neuropsychiatric Interview
NEO-FFI	NEO Five Factor Inventory
NEO-PI-R	Revised NEO Personality Inventory
NLEQ	Negative Life Events Questionnaire
PROMISES	PRegnancy Outcomes after Maternity Intervention for Stressful EmotionS
QALY	Quality-adjusted life years
RCT	Randomised controlled trial

SCID	Structured Clinical Interview for *DSM-VI* Disorders
SSQ	Social Support Questionnaire
STAI	State and Trait Anxiety Inventory
TiC-P	Trimbos/iMTA Questionnaire for Costs Associated with Psychiatric Illness
UCL	Utrechtse Coping Lijst

21.1 Background

The burden of mental disorders is huge and at least comparable to the burden caused by many severe physical diseases. In the WHO Global Burden of Disease project it was estimated that 50% of all daily adjusted life years (DALYs) in the 15–44 years old are due to nine psychiatry-related conditions (Murray et al. 1996). Depressive disorders are projected to rank second on a list of 15 major diseases in terms of burden of disease in 2030 (Mathers and Loncar 2006). In addition, a substantial part of the costs are caused by new cases, which accounts for 39.2% of the costs at population level (Smit et al. 2006). Therefore, prevention of mental disorders is essential.

Maternal anxiety or depression during pregnancy is an important and potentially modifiable risk factor for cognitive, behavioural and emotional problems among the offspring children (O'Connor et al. 2002; Mäki et al. 2003; Van den Bergh et al. 2005; Davis et al. 2007; Talge et al. 2007; Glover 2011). Around 10–20% of all women are suffering from depression or anxiety during pregnancy (Evans et al. 2001; Marcus et al. 2003; Bennett et al. 2004; Gaynes et al. 2005). The magnitude of the effects of maternal anxiety or depression on the child's psychosocial problems is considerable: It is estimated that up to 22% of the variance in behavioural problems is linked with prenatal anxiety, stress or depression (Van den Bergh et al. 2005). The adverse effects seem to be lasting. For example, antenatal anxiety of the mother was related to behavioural or emotional problems of 4-year-old children, independent of the mother's postnatal depression or anxiety (O'Connor et al. 2002), and higher anxiety levels of the mothers early in pregnancy were related to an increase in attention deficit hyperactivity disorder (ADHD) and other externalizing problems in their 8–9-year-old children (Van den Bergh and Marcoen 2004).

There are several mechanisms through which depression or anxiety during pregnancy could have an adverse effect on the offspring. These mechanisms can be divided into direct and indirect. A direct mechanism that has been researched for decades is one in which depression or anxiety activates the maternal stress system, leading to elevated glucocorticoid levels, which subsequently influence the development and long-term physiology of the foetus' brain by passing the placenta. This direct mechanism falls under the rubric of 'early life programming' and has been a popular hypothesis for the explanation of not only brain disorders but has been suggested to play a role in cardiovascular disease as well (Reynolds 2013). Further, epigenetic variation has been proposed as a mediating mechanism in linking early life exposures to long-term psychological and behavioural outcomes (Monk et al. 2012).

The effect of maternal stress on the developing foetus might also be indirect. Women who suffer from antenatal depression have the tendency to take less good care of themselves (e.g. neglecting personal hygiene, the occurrence of sleeping problems, disturbed drinking and smoking habits, denying prenatal care). These consequences might all influence the development of the foetus (Stewart 2006; O'Keane and Marsh 2007; Andres and Day 2000; Huizink and Mulder 2006). Another indirect way in which depression might influence the mental development of the offspring is when the antenatal depression remains after delivery and turns into a postnatal depression. In this way, mother–child attachment might be endangered, because the mother has a reduced ability to respond to the child. Children from depressed mothers have a higher risk of insecure attachment, which in turn is associated with cognitive, behavioural and emotional problems (Murray et al. 1996; Radke-Yarrow et al. 1985; Cummings and Davies 1994; Martins and Gaffan 2000). In addition, the association between antenatal depression and adverse outcomes in the offspring might be indirect because it could be explained by a shared genetic predisposition between mother and child.

The effectiveness of psychological therapy in the treatment of both depression and anxiety has been shown during the past 50 years, especially for cognitive behavioural therapy (CBT) (Cuijpers et al. 2009; Wampold et al. 2002; Dobson 1989; Gloaguen et al. 1998; Beck 2005). Although guidelines state that medication is an alternative effective treatment, the safety of antidepressants during pregnancy remains insecure (Ross et al. 2013).

Still, it is too early to implement CBT for depressed or anxious women to prevent psychosocial problems in the offspring. This is because in the development of such a preventive strategy, demonstration of the causality and size of the effect of the reduction of symptoms of depression and anxiety on child outcomes is a crucial step, a step that has not been taken to date. This knowledge gap will be filled by the results of the present experimental study.

Whatever the actual mechanisms involved are, there is presently convincing evidence that children whose mothers suffered from anxiety or depression during pregnancy constitute a high-risk group for behavioural and emotional problems. On population level, substantial total mental health gains may be accomplished when depressed or anxious women are adequately treated during their pregnancy, even if the effect size of the treatment is relatively small.

We are currently performing a RCT among pregnant women with symptoms of depression or anxiety to study the effect of CBT as compared to CAU on the offspring's behavioural and emotional problems.

In the CBT arm, we expect more beneficial neonatal outcomes, in particular higher birth weight and less prematurity, which are risk factors for adverse cognitive and behavioural outcomes themselves (Talge et al. 2007). We also anticipate reduced smoking and less drinking, with many physical and mental health benefits for the child as a result (Huizink and Mulder 2006). Since prenatal depression has shown to be related to postnatal depression, we hypothesize that our intervention will also counter postnatal depression, which in turn will benefit the mother—child attachment (Dennis 2005).

Finally, but not unimportantly, the reduction of symptoms of anxiety or depression during pregnancy and the early postnatal period is valuable in itself. CBT may further provide for a safer approach to reducing symptoms in pregnancy than antidepressant medication (Ross et al. 2013). To date, no such study has been performed as far as we are aware of.

21.2 Methods/Design

21.2.1 *Objective*

The aim of the present study is to examine the effect of CBT in women with at least moderate symptoms of anxiety and/or depression at the end of the first trimester of pregnancy, on the extent of total behavioural and emotional problems in their children at 1.5 years of age, as compared with CAU.

21.2.2 *Setting and Design*

The source population consists of all pregnant women in the Netherlands in the first trimester of their pregnancy. Women are recruited in primary, secondary and tertiary obstetric care.

Women are screened for anxiety and depression symptoms at the end of the first trimester of pregnancy. Women with at least moderate symptoms of anxiety and/or depression are either randomised to the intervention group in which they receive 10–14 sessions of CBT, or to the control group in which they receive CAU.

Figure 21.1 shows the detailed design of the study.

21.2.3 *Study Outcome Measures*

The primary outcome in this project is the total emotional and behavioural problems score of the child according to the Child Behaviour Checklist 1.5–5 (CBCL 1.5–5) at 18 months of age.

Secondary outcomes are the child's mental and psychomotor development at 18 months of age; the change in depressive and anxious symptoms in the mother; obstetric variables such as birth weight, gestational age and Apgar score; the sociodemographic and lifestyle factors, such as alcohol use, smoking and education; and cost-effectiveness of the therapy.

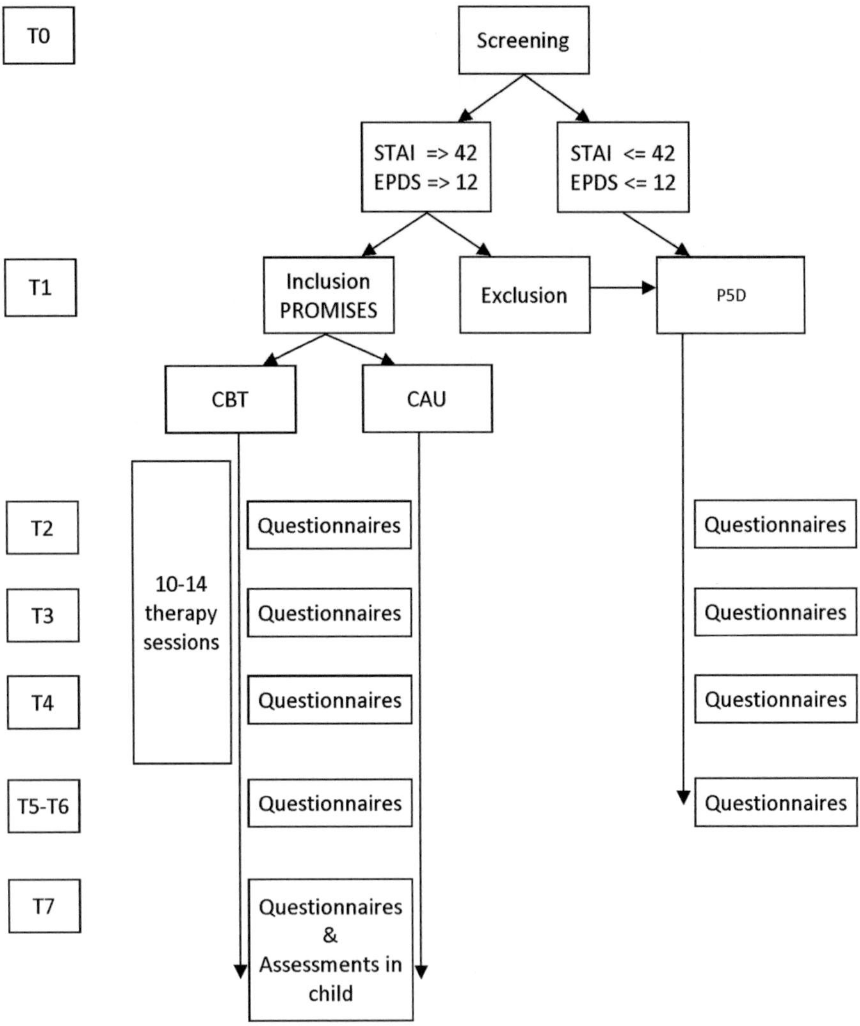

Fig. 21.1 Detailed design of the study examining the effect of cognitive behavioral therapy (CBT) in women. *T0* screening, *T1* baseline assessments, *T2* follow-up assessments at 24 weeks of gestation, *T3* follow-up assessments at 36 weeks of gestation, *T4* follow-up assessments at 6 weeks postnatal, *T5* follow-up assessments at 6 months postnatal, *T6* follow-up assessments at 12 months postnatal, *T7* follow-up assessments at 18 months postnatal, *CBT* cognitive behavioural therapy, *CAU* care as usual, *STAI* 6-item State Trait Anxiety Inventory, *EPDS* Edinburgh Postnatal Depression Scale, *P5D* pregnancy, anxiety and depression study

21.2.4 Sample Size

Studies on the prevention of mental disorders tend to suffer from problems of insufficient statistical power (Cuijpers 2003). In the current study, we aimed to get around this problem by using a continuous primary outcome measure and by including a high-risk group, i.e. selective prevention.

We decided that effect sizes of 0.35 (midpoint of small–medium effect size) or more on the total problems scale of CBCL 1.5–5 are to be detected. With alpha 5 % and power (1-beta) 80 %, we have to include 260 participants in our analyses. To account for some dropout, we aim at 300 women entering the trial. If 50 % eventually meets all criteria and gives informed consent, 600 screen-positives must be identified. The 50 % rate is based on studies with psychological interventions during pregnancy aimed at reducing the occurrence of postnatal depression (Dennis 2005). Given the figures in the literature (Bennett et al. 2004; Marteau and Bekker 1992), we can expect amply 10 % screen-positives on either the anxiety or depression screener. With an estimated 50 % comorbidity between anxiety and depression, this means that approximately 15 % are eligible for the randomisation. Therefore, 4000 women needed to be screened. Assuming a response rate of 75 % (Bowen and Muhajarine 2006), this implicates that 5333 women must be offered screening. To be on the safe side, we aimed at screening 6000 women. During the trial, it appeared that only 25 % rather than 50 % of all screen-positive women meets all criteria and gives informed consent. Therefore, we adjusted the number needed to screen for including 300 women to approximately 12,000.

21.2.5 Inclusion

Women in obstetric care in the Netherlands with a significant level of anxiety (6 item STAI ≥ 42) or at least moderate depressive symptoms (EPDS ≥ 12) in their first trimester are invited to participate in the trial.

21.2.6 Exclusion

Women fulfilling one or more of the following criteria are excluded from participation:

1. Multiple pregnancy. We decided to exclude women with multiple pregnancy as they have a markedly increased obstetric risk; their inclusion would threaten the homogeneity of the study population and thereby decrease the sensitivity to detect effects.
2. High suicidal risk according to the suicidality subscale score on the Mini International Neuropsychiatric Interview (MINI, defined as a positive response on the question on concrete suicide plans).
3. Presently receiving psychotherapy.
4. Substantial physical disease or illegal substance abuse.
5. No mastery of the Dutch language.
6. Having a psychiatric history on bipolar disorder, psychoses and manic disorder.
7. History of in vitro fertilization.

21.2.7 Assessments

Participating women are asked to fill out questionnaires until their child is 1.5 years. This is done at eight time points: The screener at baseline (T0), the additional baseline information (T1), and follow-up questionnaires at 24 and 36 weeks of gestation (T2 and T3), at 6 weeks postpartum (T4), 6 months postpartum (T5), 12 months postpartum (T6) and 18 months postpartum (T7).

At each time point, the levels of anxiety and depression are monitored by the State Trait Anxiety Inventory (STAI) and the Edinburgh Postnatal Depression Scale (EPDS). As depicted in Table 21.1, all other questionnaires are filled out once or at several time points.

For anxiety, we use the Dutch version of the 6-item STAI. This self-report questionnaire is as valid as the full 20-item version and has frequently been used to measure antenatal anxiety (Marteau and Bekker 1992). For the screening on depression, we use the EPDS, which has 10 items (Pop et al. 1992). This is the most frequently used self-report depression screener in the postnatal period, as well as during pregnancy and has been found particularly valid during pregnancy, because this scale omits somatic symptoms (Bowen and Muhajarine 2006).

The following information is obtained from participants. The exact time of administration of the corresponding instrument can be found in Fig. 21.1.

1. Life events before pregnancy are assessed at baseline, using the Negative Life Events Questionnaire (NLEQ, Garnefski et al. 2001).
2. Perceived social support is measured according to the 9-item Social Support Questionnaire (SSQ)-short form (Sarason et al. 1983).
3. General health, socio-economic position, ethnicity, smoking behaviour, alcohol use and psychiatric history (whether the participant has had depression and/or anxiety symptoms before, whether she was treated for this and whether she is presently in treatment for these symptoms) is assessed. Socio-economic position is measured using five indicators: family income, educational level (father and mother) and occupational level (father and mother). This questionnaire is based on a questionnaire used in the Utrecht Health Project (Dutch acronym LRGP: Leidsche Rijn Gezondheids Project, www.lrgp.nl). General health status will also be taken into account according to the EQ-5D (EuroQol Group 1990).
4. Personality is assessed using the NEO Five Factor Inventory (NEO-FFI). The NEO-FFI is a shortened version of the NEO-PR-I Costa and McCrae (1992) and covers the Big Five of personality (neuroticism, extraversion, openness, altruism and conscientiousness). These aspects each contain six subscales. The NEO-FFI contains 60 questions, 2 on each subscale. The present study will add four full subscales to the short version; two subscales of neuroticism, one of extraversion and one of conscientiousness. This is because we expect them to have the strongest association with persistence of depression and/or anxiety. The NEO-FFI is translated and validated in Dutch (Hoekstra et al. 1996).

Table 21.1 Assessments per measurement

	T0	T1	T2	T3	T4	T5	T6	T7
Depression (EPDS)	×	×	×	×	×	×	×	×
Anxiety (STAI)	×	×	×	×	×	×	×	×
Personality NEO-FFI		×		×		×		
Life events before pregnancy (NLEQ)		×						
Life events during pregnancy		×		×		×		
Perceived social support (SSQ)		×						
Coping styles (UCL)		×	×	×	×			
Attitudes (DAS)		×	×	×	×			
Maternal attachment (ECR)		×					×	
Quality of life (EQ-5D)		×	×	×	×	×	×	
Socio-demographic and economic factors		×						
Lifestyle		×		×		×		
Breastfeeding						×		
General health		×						
Health-care consumption		×			×	×	×	
Previous pregnancies		×						
Suicidality (MINI)		×						
Clinical Diagnostic Interview (SCID-II)		×						
Child Behaviour Checklist (CBCL)								×
Child development (BSID-II)								×

5. Information on previous pregnancies, family size and composition, pregnancy-related life events and on reactions on becoming a parent is gathered using questionnaires from the ALSPAC study (www.bristol.ac.uk/alspac).
6. Suicide risk is measured using six screening questions from the MINI (Sheehan et al. 1998).
7. Maternal attachment style is measured according to the ECR (Brennan et al. 1998), which has been translated and validated for the Netherlands by Conradi et al. (2006).
8. Health-care consumption is assessed based on the TIC-P (Hakkaart-van Roijen 2002). This instrument allows reliable recall over the past 6 months (Van den Brink et al. 2005).
9. Coping style is assessed using the Utrechtse Coping Lijst, the UCL (Schreurs and Van de Willige 1988).
10. A Dutch version of the Dysfunctional Attitude Scale (DAS) is used to measure cognitions and attitudes (Douma et al. 1991).
11. Obstetric variables such as gestational age, birth weight and Apgar score; complications such as (pre)eclampsia or HELLP, which is obtained from midwives. Women are asked to give consent for this.
12. Finally, we use the SCID-I to screen for a possible clinical depressive or anxiety disorder (First et al. 2002). The SCID-I is the only questionnaire used that has to be taken in a personal interview.

Besides questionnaires for the mother during her pregnancy and the first 1.5 years postpartum, there are assessments of the child at 1.5 years of age.

One of the assessments concerns the Bayley Scale of Infant Development (BSID-II) (Bayley 1993). This is a formal neuropsychological tool to assess the developmental level of a child between 1 and 42 months. It is individually administered by one of the researchers and consists of three subscales: Cognitive development (mental development index), gross and fine motor development and the behavioural rating scale. This tool is widely used in both research and clinical settings and is considered the best and most applied method for the assessment of the child's development to date (Skovgaard et al. 2004). Importantly, the instrument has shown to be sensitive. In the context of our proposal, maternal anxiety in pregnancy explained as much as 11 % of the variance in the Bayley scores in a study among 2-year-old toddlers by LaPlante et al. (2004).

The second assessment is the CBCL 1.5–5 including the Caregiver–Teacher Report Form (C-TRF) and the Language Development Survey (LDS) (Rescorla 2005). This well-established, reliable and valid scale designed for parents and caregivers comprises seven syndrome scales: Emotionally reactive, anxious depressed, somatic complaints, withdrawn, sleep problems, attention problems and aggressive problems. In addition, it contains scales for internalizing, externalizing and total problems. Symptom scores may further be related to formal DSM-diagnostic criteria. The LDS provides a screen for delays in vocabulary and word combinations.

For the assessment of psychopathology in preschool children, it is essential to obtain information from different sources (Carmen-Wiggins and Carter 2001).

Therefore, we decided to include the C-TRF for the caregivers of the children other than their parents. Parents are asked to hand these lists to the actual caregivers of their children, e.g. grandparents, babysitters, kindergarten-coaches, etc. Relevant in this respect, a review by Skovgaard et al. (2004) underlined the significance of both the developmental aspects (e.g. as measured with the BSID II) and the infant–caregiver relation in the assessment of children 0–3 years of age.

The CBCL has been used successfully in several studies, among others on externalizing problems (LaPlante et al. 2004). It has been translated and standardized for use in around 60 countries, including the Netherlands. The CBCL 1.5–5 is considered a sensitive instrument also deployed in current intervention studies (Robinson et al. 2010; Boers et al. 2007).

Also, mother–child interactions are measured by taping them for 15 min on video and scoring them afterwards on interaction points.

21.2.8 Additional Baseline Data

Women agreeing to participate are asked to provide additional baseline data at T1, as shown in Fig. 21.1.

About half of these questionnaires are sent to the participants in print, the other half can be answered online. All follow-up questionnaires are available online.

After providing baseline data, both in print and online, women are telephoned for the Structured Clinical Interview for *DSM-II* Disorders (SCID-VI). The SCID-II will allow us to study treatment effects additionally according to diagnostic categories rather than symptom levels.

21.2.9 Randomisation

Right after the SCID-II interview, women are randomised 1:1 to either CBT or CAU.

We will create randomisation lists, stratified for parity and socio-economic position, with randomly permuted blocks of random size.

Women randomised to the CAU arm are informed about being at risk of depression or anxiety disorder by the researchers and are advised to contact their general practitioner (GP). A close record is kept of all care provided in the CAU arm.

21.2.10 CBT Intervention

The intervention consists of 10–14 individual sessions: 6–10 sessions during pregnancy and 4–8 sessions after delivery (once a week). The CBT is conducted by registered psychologists, specialized in conducting CBT.

CBT posits that an individual's biased information processing leads to maladaptive feelings and behaviours which can culminate in psychological distress and eventually in psychiatric disorders. The main focus of the proposed intervention is targeted on identifying and changing dysfunctional cognitions and schemata (attitudes) specifically for pregnant, depressed and anxious patients. In CBT, the Socratic dialogue is used aiming to change these dysfunctional cognitions and attitudes permanently. CBT may therefore result in long-term protection against psychosocial problems. It is therefore not surprising that cognitive therapy during the acute phase of depression also appears to be effective in reducing subsequent recurrence rates (Beck 2005).

The first session will focus on the rationale CBT, i.e. the influence of (irrational or dysfunctional) cognitions and attitudes on feelings and behaviours. Additionally, goal setting is initiated. These therapy goals are unique for each patient. The subsequent sessions are targeted at identifying and amending irrational cognitions and attitudes related to pregnancy, delivery, concerns about the (unborn) child and the future family situation. Each session will address specific pregnancy-related cognitions. Additionally, patients are taught how dysfunctional cognitions and attitudes affect adversely feelings and behaviours.

These dysfunctional cognitions and attitudes are challenged and replaced by functional cognitions and attitudes. After each session, the patients are given home work. For example, the patients are asked to register negative experiences, and accompanying cognitions, feelings and behaviours. Finally, in the last two to four sessions, the newly learned cognitions and attitudes are consolidated.

21.2.11 Data Analysis

If necessary, skewed continuous variables will be transformed to normality prior to the analyses. The primary outcome, i.e. the CBCL scores at month 18, will be compared between the treatment arms using the unpaired t test. This test will also be used for detecting differences in the Bayley scores by month 18 and the obstetric variables measured at birth. The latter group of variables will be tested using the Chi-square test if categorical. Differences in attachment style at month 12 will be analysed using analysis of covariance with the baseline variable as a covariate. Continuous outcomes that were measured more than twice (e.g. EPDS and STAI) will be analysed as dependent variables using linear mixed models for fixed and random effects. These models are superior for the analysis of longitudinally correlated data and can optimally deal with missing values (Gibbons et al. 1993). A mixed model ascribes a unique intercept and slope estimate to each individual, while making use of information across individuals for predicting these quantities. In these analyses, a treatment*time variable indicating the effect of the intervention will be included as an independent variable. If despite randomisation important baseline differences exist in prognostically important variables such as the extent of social support or history of life events, they will be adjusted for. Additional analyses will be conducted to demonstrate mediation of the effect of CBT on the child's outcomes by maternal symptom level, alcohol or nicotine consumption in pregnancy, medication use or neonatal outcomes.

The analyses will primarily be carried out according to the intention-to-treat (ITT) principle, i.e. the participants will be analysed according to their randomised allocation, regardless of the actual CBT undergone, or time in study after baseline. Aside from the optimal validity of ITT analyses, they quantify the effects on the outcome measures that would be obtained in practice. The magnitude of the effect measured in an ITT analysis incorporates the effects caused by non-adherence to CBT, behavioural changes, etc. Secondary analyses will be of the 'per protocol' type meaning that they will be restricted to those women that had all of the CBT sessions.

Considering specific target populations, there is evidence that the socio-economically deprived may have more benefit from treatment of depression during pregnancy (O'Keane and Marsh 2007). Therefore, subgroup analyses will be undertaken according to socio-economic position. Subgroup analyses will also be undertaken according to parity.

Differences in effect of CBT between subgroups will be statistically evaluated by testing the treatment by subgroup interaction terms. Effect parameters will be supplied with a 95 % confidence interval.

21.2.12 Economic Evaluation

An economic evaluation will be conducted alongside the trial to assess the cost-effectiveness of CBT compared to CAU in the current study population. Information on costs and health outcomes will be prospectively collected during 24 months (starting at baseline until 18 months after birth) for both mother and child. Two complementary economic analyses will be conducted. The primary outcome measure of the planned cost-effectiveness analysis is the total emotional and behavioural problems score of the child according to the CBCL at 18 months of age.

In the additionally planned cost-utility analysis, QALYs (quality adjusted life years) will be used as the primary outcome measure. The study will be performed from a societal perspective. Medical costs that will be assessed include costs related to CBT, contacts with health-care professionals and medication use. Outside the health-care sector, costs of informal care and productivity losses will be taken into account. Unit prices will largely be based on Dutch standard prices in order to facilitate comparisons with other economic evaluations. Cost-effectiveness acceptability curves will be used to inform decision-makers on the probability that the studied intervention is cost-effective.

This study protocol was approved by the medical ethical committee of the University Medical Centre Groningen.

21.3 Competing Interests

The author(s) declare that they have no competing interests.

21.4 Authors' Contributions

CLHB, JO, RPS, PJ, MGP and HB were responsible for the development of the study design and the funding. CLHB was responsible for the development of the intervention and the training of the psychologists. HB is the trial coordinator responsible for the ongoing management of the trial. ADS was responsible for writing the economic evaluation. JLM wrote the draft manuscript. JLM, CB and TV are responsible for the recruitment of participants. All authors have read and approved of the final manuscript.

Acknowledgments This work is supported by a grant from the Netherlands Organization for Health Research and Development (120520013).

Conflicts of Interest The authors declare no conflicts of interest.

References

Andres RL, Day MC (2000) Perinatal complications associated with maternal tobacco use. Semin Neonatol 5:231–241

Bayley N (1993) Bayley scales of infant development, 2nd edn. Psychological Corporation, San Antonio

Beck AT (2005) The current state of cognitive therapy: a 40-year retrospective. Arch Gen Psychiatry 62:953–959

Bennett HA, Einarson A, Taddio A, Koren G, Einarson TR (2004) Prevalence of depression during pregnancy: systematic review. Obstet Gynecol 103:698–709

Boers KE, Bijlenga D, Mol BW, LeCessie S, Birnie E, Van Pampus MG, Stigter RH, Bloemenkamp KW, Van Meir CA, Van der Post JA, Bekedam DJ, Ribbert LS, Drogtrop AP, Van der Salm PC, Huisjes AJ, Willekes C, Roumen FJ, Scheepers HC, de Boer K, Duvekot JJ, Thornton JG, Scherjon SA (2007) Disproportionate intrauterine growth intervention trial at term: DIGITAT. BMC Pregnancy Childbirth 7:12

Bowen A, Muhajarine N (2006) Prevalence of antenatal depression in women enrolled in an outreach program in Canada. J Obstet Gynecol Neonatal Nurs 35:491–498

Brennan KA, Clark CL, Shaver PR (1998) Self-report measurement of adult attachment: an integrative overview. In: Simpson JA, Rholes WS (eds) Attachment theory and close relationships. The Guilford Press, New York, pp 46–76

Carmen-Wiggins R, Carter AS (2001) Introduction–Assessment of infant and toddler mental health: advances and challenges. J Am Acad Child Adolesc Psychiatry 40:8–10

Conradi HJ, Gerlsma C, Van Duijn M, De Jonge P (2006) Internal and external validity of the experiences in close relationships questionnaire in an American and two Dutch samples. Eur J Psychiatry 20:258–269

Costa PT Jr., McCrae RR (1992) Revised NEO Personality Inventory (NEO-PI-R) and NEO Five-Factor Inventory (NEO-FFI) professional manual. Psychological Assessment Resources, FL

Cuijpers P (2003) Examining the effects of prevention programs on the incidence of new cases of mental disorders: the lack of statistical power. Am J Psychiatry 160:1385–1391

Cuijpers P, Straten A van, Warmerdam L, Andersson G (2009) Psychotherapy versus the combination of psychotherapy and pharmacotherapy in the treatment of depression: a meta-analysis. Depress Anxiety 26:279–288

Cummings EM, Davies PT (1994) Maternal depression and child development. J Child Psychol Psychiatry 35:73–112

Davis EP, Glynn LM, Schetter CD, Hobel C, Chicz-Demet A, Sandman CA (2007) Prenatal exposure to maternal depression and cortisol influences infant temperament. J Am Acad Child Adolesc Psychiatry 46:737–746

Dennis CL (2005) Psychosocial and psychological interventions for prevention of postnatal depression: systematic review. BMJ 331:15

Dobson KS (1989) A meta-analysis of the efficacy of cognitive therapy for depression. J Consult Clin Psych 57:414–419

Douma M, van de Bosch J, Van Dongen PHM, Jansen AE (1991) Het meten van een trekdepressie. Constructie van een Nederlandstalige bewerking van de Dysfunctional Attitude Scale van Arlene Weissman. (Handleiding). St. Louis Marie Jamin, Meerssen

EuroQol Group (1990) EuroQol—a new facility for the measurement of health-related quality of life. Health Policy 16:199–208

Evans J, Heron J, Francomb H, Oke S, Golding J (2001) Cohort study of depressed mood during pregnancy and after childbirth. BMJ 323:257–260

First MB, Spitzer RL, Gibbon M, Williams JBW (2002) Structured clinical interview for DSM-IV-TR Axis I disorders, research version, patient edition. (SCID-I/P). Biometrics Research, New York State Psychiatric Institute, New York

Garnefski N, Kraaij V, Spinhoven P (2001) Negative life events questionnaire, cognitive emotion regulation and emotional problems. Pers Individ Dif 30:1311–1327

Gaynes BN, Gavin N, Meltzer-Brody S, Lohr KN, Swinson T, Gartlehner G, Brody S, Miller WC (2005) Perinatal depression: prevalence, screening accuracy, and screening outcomes. Evid Rep Technol Assess (Summ) 119:1–8

Gibbons RD, Hedeker D, Elkin I, Waternaux C, Kraemer HC, Greenhouse JB, Shea MT, Imber SD, Sotsky SM, Watkins JT (1993) Some conceptual and statistical issues in analysis of longitudinal psychiatric data. Arch Gen Psychiatry 50:739–750

Gloaguen V, Cottraux J, Cucherat M, Blackburn IM (1998) A meta-analysis of the effects of cognitive therapy in depressed patients. J Affect Disord 49:59–72

Glover V (2011) Annual research review: Prenatal stress and the origins of psychopathology: an evolutionary perspective. J Child Psychol Psychiatry 52:356–367

Hakkaart-van Roijen L (2002) Manual Trimbos/iMTA questionnaire for costs associated with psychaitric illness (in Dutch). Institute for Medical Technology Assessment, Rotterdam

Hoekstra HA, Ormel J, De Fruyt F (1996) NEO persoonlijkheidsvragenlijsten NEO-PI-R en NEO-FFI Handleiding. Swets & Zeitlinger, Lisse

Huizink AC, Mulder EJ (2006) Maternal smoking, drinking or cannabis use during pregnancy and neurobehavioral and cognitive functioning in human offspring. Neurosci Biobehav Rev 30:24–41. Epub 2005 Aug 10

Laplante DP, Barr RG, Brunet A, Galbaud du Fort G, Meaney ML, Saucier JF, Zelazo PR, King S (2004) Stress during pregnancy affects general intellectual and language functioning in human toddlers. Pediatr Res 56:400–410

Mäki P, Veijola J, Räsänen P, Joukamaa M, Valonen P, Jokelainen J, Isohanni M (2003) Criminality in the offspring of antenatally depressed mothers: a 33-year follow-up of the Northern Finland 1966 Birth Cohort. J Affect Disord 74:273–278

Marcus SM, Flynn HA, Blow FC, Barry KL (2003) Depressive symptoms among pregnant women screened in obstetrics settings. J Womens Health (Larchmt) 12:373–380

Marteau TM, Bekker H (1992) The development of a six-item short-form of the state scale of the Spielberger State-Trait Anxiety Inventory (STAI). Br J Clin Psychol 31:301–306

Martins C, Gaffan EA (2000) Effects of early maternal depression on patterns of infant-mother attachment: a meta-analytic investigation. J Child Psychol Psychiatry 41:737–746

Mathers CD, Loncar D (2006) Projections of global mortality and burden of disease from 2002 to 2030. PLoS Med 3:e442

Monk C, Spicer J, Champagne FA (2012) Linking prenatal maternal adversity to developmental outcomes in infants: the role of epigenetic pathways. Dev Psychopathol 24:1361–1376

Murray L, Fiori-Cowley A, Hooper R, Cooper P (1996) The impact of postnatal depression and associated adversity on early mother-infant interactions and later infant outcome. Child Dev 67:2512–2526

O'Connor TG, Heron J, Glover V (2002) Antenatal anxiety predicts child behavioral/emotional problems independently of postnatal depression. J Am Acad Child Adolesc Psychiatry 41:1470–1477

O'Keane V, Marsh MS (2007) Depression during pregnancy. BMJ 334:1003–1005

Pop VJ, Komproe IH, van Son MJ (1992) Characteristics of the Edinburgh post natal depression scale in The Netherlands. J Affect Disord 26:105–110

Radke-Yarrow M, Cummings EM, Kuczynski L, Chapman M (1985) Patterns of attachment in two- and three-year-olds in normal families and families with parental depression. Child Dev 56:884–893

Rescorla LA (2005) Assessment of young children using the Achenbach System of Empirically Based Assessment (ASEBA). Ment Retard Dev Disabil Res Rev 11:226–237

Reynolds RM (2013) Glucocorticoid excess and the developmental origins of disease: two decades of testing the hypothesis–2012 Curt Richter Award Winner. Psychoneuroendocrinology 38:1–11

Robinson M, Oddy WH, McLean NJ, Jacoby P, Pennell CE, de Klerk NH, Zubrick SR, Stanley FJ, Newnham JP (2010) Low-moderate prenatal alcohol exposure and risk to child behavioural development: a prospective cohort study. BJOG 117:1139–1150. Epub 2010 May 28

Ross LE, Grigoriadis S, Mamisashvili L, Vonderporten EH, Roerecke M, Rehm J, Dennis CL, Koren G, Steiner M, Mousmanis P, Cheung A (2013) Selected pregnancy and delivery outcomes after exposure to antidepressant medication: a systematic review and meta-analysis. JAMA Psychiatry 70:436–443

Sarason IG, Levine HM, Basham RB, Sarason BR (1983) Assessing social support: the social support questionnaire. J Pers Soc Psychol 44:127–139

Schreurs PGJ, Van de Willige G (1988) Omgaan met problemen en gebeurtenissen: De Utrechtse Coping Lijst (UCL). Swets & Zeitlinger, Lisse

Sheehan DV, Lecrubier Y, Sheehan KH, Amorim P, Janavs J, Weiller E, Hergueta T, Baker R, Dunbar GC (1998) The Mini-International Neuropsychiatric Interview (M.I.N.I.): the development and validation of a structured diagnostic psychiatric interview for DSM-IV and ICD-10. J Clin Psychiatry 59(Suppl 20):22–33

Skovgaard AM, Houmann T, Landorph SL, Christiansen E (2004) Assessment and classification of psychopathology in epidemiological research of children 0–3 years of age: a review of the literature. Eur Child Adolesc Psychiatry 13:337–346

Smit F, Cuijpers P, Oostenbrink J, Batelaan N, de Graaf R, Beekman A (2006) Costs of nine common mental disorders: implications for curative and preventive psychiatry. J Ment Health Policy Econ 9:193–200

Stewart DE (2006) Perinatal depression. Gen Hosp Psychiatry 28:1–2

Talge NM, Neal C, Glover V (2007) Antenatal maternal stress and long-term effects on child neurodevelopment: how and why? J Child Psychol Psychiatry 48:245–261

Van den Bergh BR, Marcoen A (2004) High antenatal maternal anxiety is related to ADHD symptoms, externalizing problems, and anxiety in 8- and 9-year-olds. Child Dev 75:1085–1097

Van den Bergh BR, Mulder EJ, Mennes M, Glover V (2005) Antenatal maternal anxiety and stress and the neurobehavioural development of the fetus and child: links and possible mechanisms. A review. Neurosci Biobehav Rev 29:237–258

Van den Brink M, Van den Hout WB, Stiggelbout AM, Putter H, Van de Velde CJH, Kievit J (2005) Self-reports of health-care utilization: diary or questionnaire? Int J Technol Assess Health Care 21:298–304

Van Zeijl J, Mesman J, Stolk MN, Alink LR, van Ijzendoorn MH, Bakermans-Kranenburg MJ, Juffer F, Koot HM (2006) Terrible ones? Assessment of externalizing behaviors in infancy with the child behavior checklist. J Child Psychol Psychiatry 47:801–810

Wampold BE, Minami T, Baskin TW, Tierney SC (2002) A meta-(re)analysis of the effect of cognitive therapy versus 'other therapies' for depression. J Affect Disorders 68:159–165

Index

M. C. Antonelli (ed.), *Perinatal Programming of Neurodevelopment*, Advances in Neurobiology 10, DOI 10.1007/978-1-4939-1372-5,

Q

R

MIX
Papier aus verantwortungsvollen Quellen
Paper from responsible sources
FSC® C105338

If you have any concerns about our products,
you can contact us on
ProductSafety@springernature.com

In case Publisher is established outside the EU,
the EU authorized representative is:
Springer Nature Customer Service Center GmbH
Europaplatz 3, 69115 Heidelberg, Germany

Printed by Libri Plureos GmbH
in Hamburg, Germany